Essential Laboratory Exercises for General Biology

for Starr's Biology texts

James W. Perry | David Morton | Joy B. Perry

CENGAGE Learning™

Australia • Brazil • Japan • Korea • Mexico • Singapore • Spain • United Kingdom • United States

Essential Laboratory Exercises for General Biology for Starr's Biology texts

James W. Perry | David Morton | Joy B. Perry

Executive Editor:
Maureen Staudt
Michael Stranz

Senior Project Development Manager:
Linda de Stefano

Marketing Specialist:
Sara Mercurio
Lindsay Shapiro

Production/Manufacturing Manager:
Donna M. Brown

PreMedia Supervisor:
Joel Brennecke

Rights & Permissions Specialist:
Kalina Hintz
Todd Osborne

Cover Image:
Getty Images*

ISBN-13: 978-0-495-31005-1

ISBN-10: 0-495-31005-0

Cengage Learning
5191 Natorp Boulevard
Mason, Ohio 45040
USA

Cengage Learning is a leading provider of customized learning solutions with office locations around the globe, including Singapore, the United Kingdom, Australia, Mexico, Brazil, and Japan. Locate your local office at:
international.cengage.com/region

Cengage Learning products are represented in Canada by Nelson Education, Ltd.

For your lifelong learning solutions, visit **www.cengage.com/custom**

Visit our corporate website at **www.cengage.com**

Printed in the United States of America

Contents

Preface

Greetings from the authors! We're happy that you are examining the results of our efforts to assist you and your students. We believe you'll find the information below a valuable introduction to this laboratory manual.

Audience

This manual is designed for students at the college entry level and assumes that the student has no previous college biology or chemistry. The *Lab Manual's* terminology and classification conform to that of Starr and Taggart's *Biology: The Unity and Diversity of Life* and to Starr's *Biology: Concepts and Applications*, although you'll find that the exercises support virtually any biology text used in an introductory course.

Features of This Edition

Inquiry Experiments and the Methods of Science

We realize that the best way to learn science is by doing science. Thus, in this edition, you will find numerous **inquiry-oriented experiments.** Some are extensions of a preceding activity, while others stand alone. Each follows the cognitive techniques whose foundations are laid in the first exercise, "Scientific Method."

Throughout the manual, we place more emphasis on testing predictions generated from hypotheses. We strongly believe that all biology students—both majors and nonmajors—benefit by repeating the logical thought processes of science.

New and Updated Exercises

As always, we strive to provide students with exciting, relevant activities and experiments that allow them to explore some of the rapidly developing areas of biological knowledge. We have added an additional animal diversity exercise and re-ordered and re-grouped some of the microbe, plant and invertebrate phyla.

Updated Taxonomy

An attempt to provide accurate systematic information is like trying to hit a moving target. As new information floods in, our best understanding of taxonomy and systematic relationships sometimes seems to change daily. The taxonomy in our exercises is completely updated and reflects the most widely accepted information at the time of publication.

What's Important to Us

In preparing this lab manual, we paid particular attention to pedagogy, clarity of procedures and terminology, illustrations, and practicality of materials used.

Pedagogy

The exercises are written so the conscientious student can accomplish the objectives of each exercise with minimal input from an instructor. As suggested by the publisher, the procedure sections of the exercises are more detailed and step-by-step than in other manuals. Instructions follow a natural progression of thought so the instructor need not conduct every movement.

We attempted to make each portion of the exercise part of a continuous flow of thought. Thus, we do not wait until the post-lab questions to ask students to record conclusions when it is more appropriate to do so within the body of the procedure. Answers to in-lab questions are to be found in the *Instructor's Manual* as well as online.

Terms required to accomplish objectives are **boldface.** Scientific names and precautionary statements, or those needing emphasis, are *italic*.

The use of scientific names is deemphasized when it is not relevant to understanding the subject. However, these names generally do appear in parentheses because the labels on many prepared microscope slides bear only the scientific name.

Format

Each exercise includes:

1. Objectives: a list of desired outcomes.
2. Introduction: to stimulate student interest, indicate relevance, and provide background.
3. Materials: a list for each portion of the exercise so a student can quickly gather the necessary supplies. Materials are listed "Per student," "Per pair," "Per group," and "Per lab room."
4. Procedures: including safety notes, illustrations of apparatus, figures to be labeled, drawings to be made, tables to record data, graphs to draw, and questions that lead to conclusions. The procedures are listed in easy-to-follow numbered steps.
5. Pre-lab Questions: ten multiple-choice questions that the student should be able to answer after reading the exercise, but prior to entering the laboratory.
6. Post-lab Questions: questions that draw on knowledge gained from doing the exercise and that the student should be able to answer after finishing the exercise. These post-lab questions assess **recall** (preparing students for lab practical assessments), **understanding,** and **application.**

Practical Post-lab Questions

Virtually all courses use laboratory practical examinations. We explain to our students the difference between lecture-type questions, which they need to read and provide an answer based on the written word, and practical-type questions, for which a response depends on observation.

We believe the post-lab questions should draw on the knowledge gained by observation. Consequently, we've incorporated as many illustrations as possible into the post-lab questions. These illustrations typically are similar, but not identical, to those in the procedures. Thus, they assess the student's ability to use knowledge gained during the exercise in a new situation.

Post-lab questions are identified by exercise section. This allows the instructor to easily assign or use only those questions relevant to portions of exercises performed by students. The questions also have been revised and are directly tied to the learning outcomes expected from each exercise.

Flexible Quiz Options

Each exercise has a set of pre-lab questions. We have found through nearly 80 combined years of experience that students left to their own initiative typically come to laboratory unprepared to do the exercise. Few read the exercise beforehand. One solution to this problem is to incorporate some sort of graded pre-exercise activity. At the same time, we recognize that grading a large number of lab papers each week can put an unreasonable burden on instructors. Consequently, we decided on a multiple-choice format, which is easy to grade but still accomplishes the pedagogical goal.

In our own courses, we have students take a pre-lab quiz consisting of the questions in their lab manual in scrambled order to discourage memorization. These scrambled quizzes are reproduced in our *Instructor's Manual.* Our quiz takes about five minutes of lab time and counts as a portion of the lab grade, thus rewarding students for preparation.

Other instructors have told us they use the pre-lab questions to assess learning after the exercise has been completed. We encourage you to be creative with the manual; do what you like best.

Lab Length and Exercise Options

We realize there is wide variation in the amount of time each instructor devotes to laboratory activities. To provide maximum flexibility for the instructor, the procedure portions of the exercises are divided by major headings, and the approximate time it takes to perform each portion indicated. Once the introduction has been studied, portions of the procedures can be deleted or conducted as demonstrations without sacrificing the pedagogy of the exercise as a whole.

It's our experience that if the lecture section covers the topic prior to the lab, students find the exercise much more relevant and understandable. We strive to create this situation in our courses, and thus no time is spent on a lecture-style introduction in the lab itself before the exercise begins. Therefore, we have two to three full hours for real scientific investigation and need to delete very little material to complete most exercises in the time allotted.

Illustrations

Perhaps our illustrations are more noticeable than anything else as you thumb through the manual. We continued our incorporation of a generous number of high-quality color illustrations, including everything a student needs visually to accomplish the objectives of each exercise. While there is no need for students to purchase supplemental publications, students may find the *Photoatlas for Biology*, ISBN 0-534-23556-5, to be a useful reference for this course and other biology courses.

Most illustrations of microscopic specimens are labeled to provide orientation and clarity. A few are unlabeled but are provided with leaders for students to attach labels. In other cases, more can be gained by requiring the student to do simple drawings. Space is included in the manual for these, with boxes for drawings of macroscopic specimens and circles for microscopic specimens.

We include a number of relevant figures from Starr and Taggart's *Biology: The Unity and Diversity of Life* and from Starr's *Biology: Concepts and Applications.* These illustrations further illuminate connections between lab exercise material and biological concepts.

Materials

Most of the equipment and supplies used in the exercises are readily available from biological and laboratory supply houses. Many others can be collected from nature or purchased in local supermarket, discount or office supply stores. (Our department budgets are not large!) We've attempted to keep instrumentation as simple and inexpensive as possible.

Anyone who lives in a temperate climate knows it may be necessary to adjust the sequence of the laboratory exercises to accommodate seasonal availability of certain materials. However, we provided alternatives, including the use of preserved specimens wherever possible, to avoid this problem.

Instructor's Manual

There is no need to worry, "Where can I get that?" or "How do I prepare this?" Our *Instructor's Manual* includes:

- Material and equipment lists for each exercise
- Procedures to prepare reagents, materials, and equipment
- Scheduling information for materials needing advance preparation
- Approximate quantities of materials needed
- Answers to in-text questions
- Answers for pre-lab questions
- Answers for post-lab questions
- Tear-out sheets of pre-lab questions, in scrambled order from those in the lab manual, for those who wish to duplicate them for quizzes
- Vendors and ordering information for supplies

And in the End . . .

We would like to express our special thanks to, Kristina Razmara, Andy Marinkovich, Jack Carey, and other individuals at Cengage Learning, Brooks/Cole and the Wadsworth Group involved in this project, and to Winifred Sanchez and Liah Rose of Interactive Composition Corporation. We would especially like to thank the following reviewers for their helpful insights and suggestions. Their comments allowed us to correct some errors and improve clarity, and so have improved this version: Evelyn Bruce, University of North Alabama; Richard D. Gardner, Southern Virginia University; Sherry Krayesky, University of Louisiana, Lafayette; Kamau W. Mbuthia, Bowling Green State University; Snehlata Pandey, Hampton University; Marie Panec, Moorpark College; Tom Patterson, South Texas College; Jean Rothe, Quinnipiac University; Jennifer Siemantel, Cedar Valley College; Steve Threlkeld, Calhoun State Community College; Liza Vela, University of Texas-Pan American; Mary E. White, Southeastern Louisiana University; Martin Zahn, Thomas Nelson Community College.

There are very few things in life that are perfect. We don't suppose that this lab manual is one of them. We hope your students will enjoy the exercises. We **know** they will learn from them. Perhaps you and they will find places where rephrasing will make the activity better. Please contact us with your opinions and any ideas you wish to share; encourage your students to do likewise.

James W. Perry
Department of Biological Sciences
University of Wisconsin
Fox Valley
1478 Midway Road
Menasha, WI 54952-1297
(920) 832-2610
james.perry@uwc.edu

David Morton
Department of Biology
Frostburg State University
Frostburg, MD 21532-1099
(301) 689-4355
dmorton@frostburg.edu

Joy B. Perry
Department of Biological Sciences
University of Wisconsin
Fox Valley
1478 Midway Road
Menasha, WI 54952-1297
(920) 832-2653
joy.perry@uwc.edu

To the Student

Welcome! You are about to embark on a journey through the cosmos of life. You will learn things about yourself and your surroundings that will broaden and enrich your life. You will have the opportunity to marvel at the microscopic world, to be fascinated by the cellular events occurring in your body at this very moment, and to gain an appreciation for the environment, including the marvelous diversity of the plant and animal world.

We offer a number of suggestions to make your college experience in biology a pleasant one. We have taken the first step toward that goal; we have written a laboratory guide that is user friendly. You will be able to hear the authors speaking as though we were there to share your experience. The authors share a personal belief that the more comfortable we make you feel, the more likely you will share our enthusiasm for biology. It would be naive for us to suppose that each of you will be biology majors at graduation. But one thing we all must realize is that we are citizens of "spaceship Earth." The fate of our spaceship is largely in your hands because you are the decision makers of the future. As has been so aptly stated, "We inherited the earth from our parents and grandparents, but we are only the caretakers for our children and grandchildren."

As caretakers, we need to be informed about the world around us. That's why we enroll in colleges and universities with the hope of gaining a liberal education. In doing so, we establish a basis on which to make educated decisions about the future of the planet. Each exercise in this manual contains a lesson in life that is of a more global nature than the surroundings of your biology laboratory.

To enhance your biology education, take the initiative to give yourself the best possible advantage. Don't miss class. Read your text assignment routinely. And, read the laboratory exercise before you come to the lab.

Each exercise in the manual is organized in the same way:

1. Objectives tell exactly what you should learn from the exercise. If you wish to know what will be on the exam, consult the objectives for each exercise.
2. The Introduction provides background information for the exercise and is intended to stimulate your interest.
3. The Materials list for each portion of the exercise allows you to determine at a glance whether you have all the necessary supplies needed to do the activity.
4. The Procedure for each section, in easy-to-follow step-by-step fashion, describes the activity. Within the procedure, spaces are provided to make required drawings. Questions are posed with space for answers, asking you to draw conclusions about an activity you are engaged in. You'll find a lot of illustrations, most of which are labeled and others which are not but have leaders for you to attach labels. The terms to be used as labels are found in the procedure and in a list accompanying the illustration. We believe it best for you to sometimes make a simple drawing, and have inserted boxes or circles for your sketches. Where appropriate, tables and graphs are present to record your data.
5. Pre-lab questions can be answered easily by simply reading the exercise. They're meant to "set the stage" for the lab period by emphasizing some of the more salient points.
6. Post-lab questions are intended to be answered after the laboratory is completed. Some are straightforward interpretations of what you have done, while others require additional thought and perhaps some research in your textbook. In fact, some have no "right" or 'wrong" answer at all!

It is our experience that students are much too reluctant to ask questions for fear of appearing stupid. Remember, there is no such thing as a stupid question. Speak up! Think of yourselves as "basic learners" and your instructors as "advanced learners." Interact and ask questions so that you and your instructors can further your/their respective educations.

Laboratory Supplies and Procedures

Materials and Supplies Kept in the Lab at All Times

The following materials will always be available in the lab room. Familiarize yourself with their location prior to beginning the exercises.

- Compound light microscopes
- Dissection microscopes
- Glass microscope slides
- Coverslips

- Lens paper
- Tissue wipes
- Plastic 15-cm rulers
- Dissecting needles
- Razor blades
- Assorted glassware-cleaning brushes
- Detergent for washing glassware
- Distilled water
- Hand soap
- Paper towels
- Safety equipment (see separate list)

Laboratory Safety

None of the exercises in this manual are inherently dangerous. Some of the chemicals are corrosive (causing burns to the skin) and others are poisonous if ingested or inhaled in large amounts. Contact with your eyes by otherwise innocuous substances may result in permanent eye injury. **Remember, once your sight is lost, it's probably lost forever.** Locate the following safety items and then study the list of basic safety rules.

1. Eyewash bottle or eye bath

   Should any substance be splashed in your eyes, wash them thoroughly.

2. Fire extinguisher

   Read the directions for use of the fire extinguisher.

3. Fire blanket

   Should someone's clothing catch fire, wrap the blanket around the individual and roll the person on the floor to smother the flames.

4. First-aid kit

   Minor injuries such as small cuts can be treated effectively in the lab. Open the first-aid kit to determine its contents.

5. Safety goggles

   Eye protection should be worn during the more experimental exercises.

Safety Rules

1. Do not eat, drink, or smoke in the laboratory.
2. Wash your hands with soap and warm water before leaving the laboratory.
3. When heating a test tube, point the mouth of the tube away from yourself and other people.
4. Always wear shoes in the laboratory.
5. Keep extra books and clothing in designated places so your work area is as uncluttered as possible.
6. If you have long hair, tie it back when in the laboratory.
7. Read labels carefully before removing substances from a container. Never return a substance to a container.
8. Discard used chemicals and materials into appropriately labeled containers. Certain chemicals should not be washed down the sink; these will be indicated by your instructor.

Caution: Report all accidents and spills to your instructor immediately!

Instructions for Washing Laboratory Glassware

1. Place contents to be discarded in proper waste container as described in exercise.
2. Rinse glassware with tap water.
3. Add a small amount of glassware cleaning detergent.
4. Scrub using an appropriately sized brush.
5. Rinse with tap water until detergent disappears.
6. Rinse three times with distilled water (dH_2O).
7. Allow to dry in inverted position on drying rack (if available).

When glassware is clean, dH_2O sheets off rather than remaining on the surface in droplets.

EXERCISE 1

The Scientific Method

OBJECTIVES

After completing this exercise, you will be able to

1. define *scientific method, mechanist, vitalist, cause and effect, induction, deduction, experimental group, control group, independent variable, dependent variable, controlled variables, correlation, theory, principle, bioassay;*
2. explain the nature of scientific knowledge;
3. describe the basic steps of the scientific method;
4. state the purpose of an experiment;
5. explain the difference between cause and effect and correlation;
6. describe the design of a typical research article in biology;
7. describe the use of bioassays.

INTRODUCTION

To appreciate biology or, for that matter, any body of scientific knowledge, you need to understand how the **scientific method** is used to gather that knowledge. We use the scientific method to test the predictions of possible answers to questions about nature in ways that we can duplicate or verify. Answers supported by test results are added to the body of scientific knowledge and contribute to the concepts presented in your textbook and other science books. Although these concepts are as up-to-date as possible, they are always open to further questions and modifications.

One of the roots of the scientific method can be found in ancient Greek philosophy. The natural philosophy of Aristotle and his colleagues was mechanistic rather than vitalistic. A **mechanist** believes that only natural forces govern living things, along with the rest of the universe, while a **vitalist** believes that the universe is at least partially governed by supernatural powers. Mechanists look for interrelationships between the structures and functions of living things, and the processes that shape them. Their explanations of nature deal in **cause and effect**—the idea that one thing is the result of another thing (for example, Fertilization of an egg initiates the developmental process that forms an adult.). In contrast, vitalists often use purposeful explanations of natural events (The fertilized egg strives to develop into an adult.). Although statements that ascribe purpose to things often feel comfortable to the writer, try to avoid them when writing lab reports and scientific papers.

Aristotle and his colleagues developed three rules to examine the laws of nature: Carefully observe some aspect of nature; examine these observations as to their similarities and differences; and produce a principle or generalization about the aspect of nature being studied (for example, All mammals nourish their young with milk.).

The major defect of natural philosophy was that it accepted the idea of *absolute truth.* This belief suppressed the testing of principles once they had been formulated. Thus, Aristotle's belief in spontaneous generation, the principle that some life can arise from nonliving things (say maggots from spoiled meat), survived over 2000 years of controversy before being discredited by Louis Pasteur in 1860. Rejection of the idea of absolute truth coupled with the testing of principles either by experimentation or by further pertinent observation is the essence of the modern scientific method.

1.1 Modern Scientific Method *(About 70 min.)*

Although there is not one universal scientific method, Figure 1-1 illustrates the general process.

MATERIALS

Per lab room:

- blindfold
- plastic beakers with an inside diameter of about 8 cm stuffed with cotton wool
- four or five liquid crystal thermometers

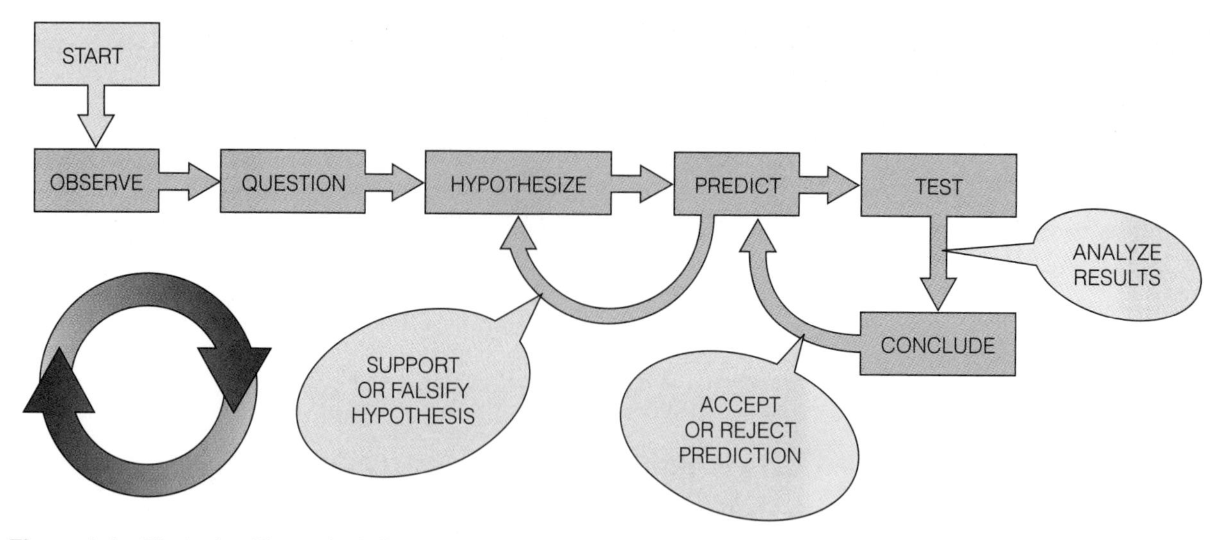

Figure 1-1 The scientific method. Support or falsification of the hypothesis usually necessitates further observations, adjustment to the question, and modification of the hypothesis. Once started, the scientific method cycles over and over again, each turn further refining the hypothesis.

PROCEDURE

Step 1. Observation. As with natural philosophy, the scientific method starts with careful observation. An investigator may make observations from nature or from the written words of other investigators, which are published in books or research articles in scientific journals and are available in the storehouse of human knowledge, the library. One subject we all have some knowledge of is the human body. The first four rows of Table 1-1 list some observations about the human body. The fifth row is blank so that you can fill in the steps of the scientific method for either the bioassay described at the end of the exercise, another observation about the human body, or anything else you and your instructor wish to investigate.

Step 2. Question. In the second step of the scientific method, *we ask a question* about these observations. The quality of this question will depend on how carefully the observations were made and analyzed. Table 1-1 includes questions raised by the listed observations.

Step 3. Hypothesis. Now we *construct a hypothesis*—that is, we derive by inductive reasoning a possible answer to the question. **Induction** is a logical process by which all known observations are combined and considered before producing a possible answer. Table 1-1 includes examples of hypotheses.

Step 4. Prediction. In this step we *formulate a prediction*—we assume the hypothesis is correct and predict the result of a test that reveals some aspect of it. This is deductive or "if-then" reasoning. **Deduction** is a logical process by which a prediction is produced from a possible answer to the question asked. Table 1-1 lists a prediction for each hypothesis.

Step 5. Experiment or Pertinent Observations. In this step we *perform an experiment or make pertinent observations* to test the prediction.

(a) Choose, along with the other members of your lab group, one prediction from Table 1-1. Coordinate your choice with the other lab groups so that each group tests a different prediction.

(b) In an experiment of classical design, the individuals or items under study are divided into two groups: an **experimental group** that is treated with (or possesses) the independent variable and a **control group** that is not (or does not). Sometimes there is more than one experimental group. Sometimes subjects participate in both groups, experimental and control, and are tested both with and without the treatment.

In any test there are three kinds of variables. The **independent variable** is the treatment or condition under study. The **dependent variable** is the event or condition that is measured or observed when the results are gathered. The **controlled variables** are all other factors, which the investigator attempts to keep the same for all groups under study.

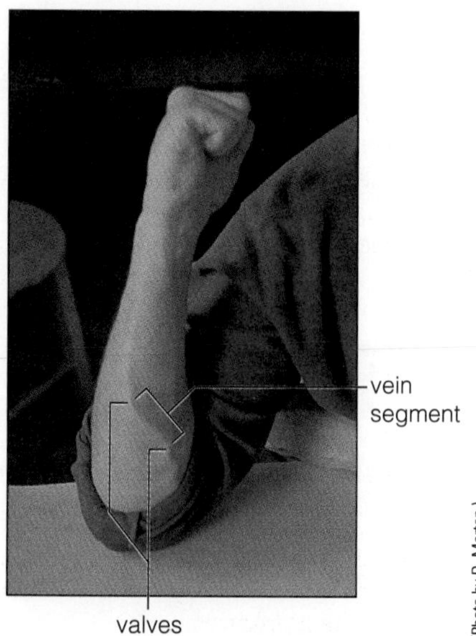

Figure 1-2 Veins under the skin.

TABLE 1-1 Some Observations About the Human Body

Observation	Question	Hypothesis	Prediction
I. Veins containing blood are seen under the skin. Swellings present along the vein are often located where veins join together.[a]	What is the function of the swellings?	Swellings contain one-way valves that allow blood to flow only toward the heart.	If these valves are present, then blood flows only from vein segments farther from the heart to the next segments nearer the heart and never in the opposite direction.
II. People have two ears.	What is the advantage of having two ears?	Two ears allow us to locate the sources of sounds.	If the hypothesis is correct, then blocking hearing in one ear will impair our ability to determine a sound's source.[b]
III. People can hold their breath for only a short period of time.	What factor forces a person to take a breath?	The buildup of carbon dioxide derived from the body's metabolic activity stimulates us to take a breath.	If the hypothesis is correct, then people will hold their breath a shorter time just after exercise compared to when they are at rest.
IV. Normal body temperature is 98.6°F.	Is all of the body at the same 98.6°F temperature?	The skin, or at least some portion of it, is not 98.6°F.	If the hypothesis is correct, then a liquid crystal thermometer will record different temperatures on the forehead, back of neck, and forearm.
V. ________________	________________	________________	________________

[a] The portion of the vein between swellings is called a segment and is illustrated in Figure 1-2.
[b] This is especially the case for high-pitched sounds because higher frequencies travel less easily directly through tissues and bones.

Note: **The italicized statements show how the scientific method is applied to the predictions in Table 1-1 or give examples of the scientific method in practice.**

To test the prediction that blocking hearing in one ear impairs our ability to point out a sound's source (row II in Table 1-1), a group of subjects is tested first with no ears blocked (control group) and then with one ear blocked (experimental group). The independent variable is the blocking of one ear; the dependent variable is the ability to point out a sound's source; and the controlled variables are the standard conditions used for each trial—same test sound, same background noise, same procedure for each subject (same blocked ear, same instructions, same sequence of trials, same time between trials), and recruitment of appropriate subjects.

Sometimes the best tests of the predictions of a hypothesis are not actual experiments but further pertinent observations. One of the most important principles in biology, Darwin's theory of natural selection, was developed by this nonexperimental approach. Although they are a little more difficult to form, the nonexperimental approach also has variables.

To test the prediction that a liquid crystal thermometer will record different temperatures on the forehead, back of neck, and forearm (row IV of Table 1-1), the independent variable is location; the dependent variable is temperature; and the controlled variables are using the same thermometer to measure skin temperature at the three locations and measuring all of the subjects at rest.

INSTRUCTIONS. Members of groups testing predictions in rows I, II, and IV of Table 1-1 read items 1, 2, or 3, respectively. Then all groups describe in item 4 an experiment or a pertinent observation to test their prediction.

1. To test the prediction in row I of Table 1-1, determine the direction of blood flow between vein segments as follows:
   - **(a)** With arms dangling at their sides, identify in as many group members as possible a vein segment either on the back of the hands or on the forearms.
   - **(b)** Demonstrate that blood flows in vein segments: Note their collapse due to gravity speeding up blood flow when each subject's arm is raised above the level of the heart.
   - **(c)** Lower the arm to a position below the heart and stop blood flow through the vein segment by permanently pressing a finger on the swelling farthest from the heart.
   - **(d)** Use a second finger to squeeze the blood out of the vein segment past the next swelling toward the heart.
   - **(e)** Remove the second finger and note whether blood flows back into the vein segment.
   - **(f)** Remove the first finger and note whether blood flows back into the vein segment.
   - **(g)** Repeat c–e, only reverse the order of the swellings—first stop blood flow through the swelling nearest the heart, then squeeze the blood out of the vein segment in a direction away from the heart past the next swelling.
2. To test the prediction in row II of Table 1-1, you will need a blindfold and a beaker stuffed with cotton wool to block hearing in one ear. Quantify how well a blindfolded group member can point out the source of a high-pitched sound by estimating the angle (0–180°) between the line from source to subject and the line along which the subject points to identify the source.
3. To test the prediction in row IV, you will need liquid crystal thermometers to measure skin temperature.
4. Describe your experiment or pertinent observation:

   ____

   ____

   ____

   ____

   ____

   ____

   ____

   ____

5. Describe the variables involved with testing the prediction your group chose from Table 1-1.
   - Independent variable—

     ____

     ____

   - Dependent variable—

     ____

     ____

   - Controlled variables—

     ____

     ____

6. After your instructor's approval, perform the procedure you outlined in item 4 and record your results in Table 1-2. You may not need all of the rows and columns to accommodate your data.
7. Describe or present your results in one of the bar charts shown on the next page.

Step 6. Conclusion. *To make a conclusion*—the last step in one cycle of the scientific method—you use the results of the experiment or pertinent observations to evaluate your hypothesis. If your prediction does not occur, it is rejected and your hypothesis or some aspect of it is falsified. If your prediction does occur, you may conditionally accept your prediction and your hypothesis is supported. However, you can never completely accept or reject any hypothesis; all you can do is state a probability that one is correct or incorrect. To quantify this probability, scientists use a branch of mathematics called *statistical analysis.*

Even if the prediction is rejected, this does not necessarily mean that the treatment caused the result. A coincidence or the effect of some unforeseen and thus uncontrolled variable could be causing the result. For this reason, the results of experiments and observations must be *repeatable* by the original investigator and others.

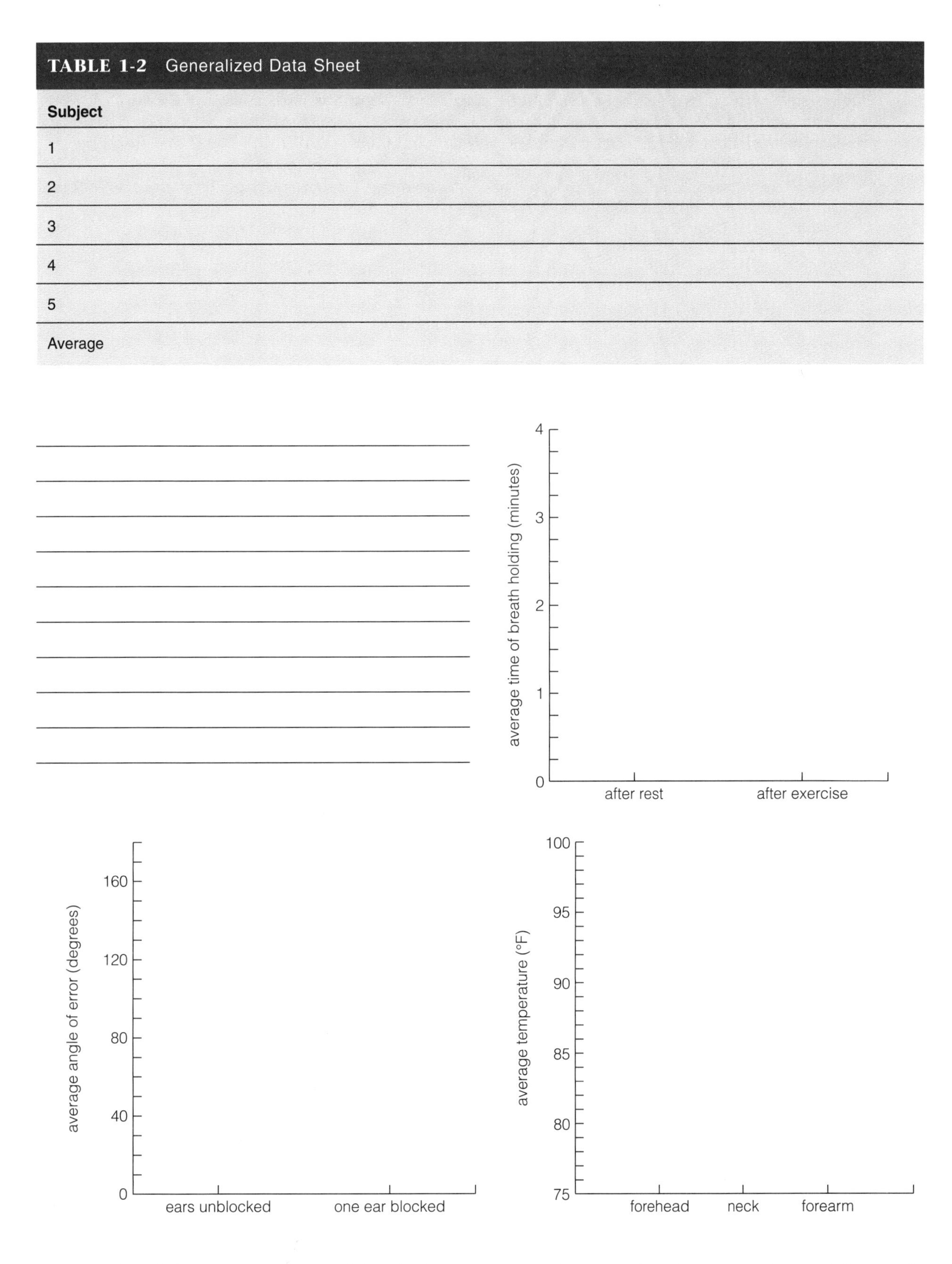

TABLE 1-2 Generalized Data Sheet

Subject
1
2
3
4
5
Average

Even if the results are repeatable, this does not necessarily mean that the treatment caused the result. *Cause and effect,* especially in biology, is rarely proven in experiments. We can, however, say that the treatment and result are correlated. A **correlation** is a relationship between the independent and the dependent variables.

Severe narrowing of a coronary artery branch reduces blood flow to the heart muscle downstream. This region of heart muscle gets insufficient oxygen and cannot contract and may die, resulting in a heart attack. The initial cause is the narrowing of the artery and the final effect is the heart attack. Perhaps the heart attack victim smoked cigarettes. Smoking cigarettes is one of several factors that make a person more likely to have a heart attack. This is based on a correlation between smoking and heart attacks in the general population but we cannot say for sure that the smoking caused the heart attack.

Write a likely conclusion for your experiment or pertinent observation. Statistics are not required, but if you know how, apply the correct statistical test before writing your conclusion.

____

____

____

THEORIES and PRINCIPLES: When exhaustive experiments and observations consistently support an important hypothesis, it is accepted as a **theory**. A theory that stands the test of time may be elevated to the status of a **principle**. Theories and principles are always considered when new hypotheses are formulated. However, like hypotheses, theories and principles can be modified or even discarded in the light of new knowledge. Biology, like life itself, is not static but is constantly changing.

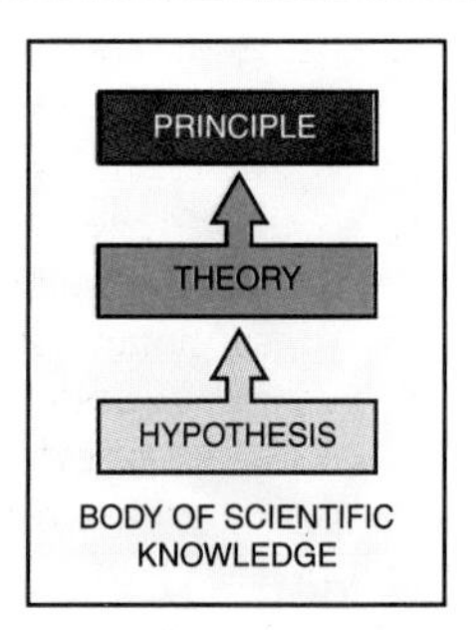

1.2 Research Article *(About 20 min.)*

The account of one or several related cycles of the scientific method is usually initially reported in depth in a research article published in a scientific journal. The goal of the scientific community is to be cooperative as well as competitive. Writing research articles allows scientists to share knowledge. They provide enough information so that other scientists can repeat the experiments or pertinent observations they describe. The journal *Science* along with several others presents its research articles in narrative form, and many of the details of the scientific method are understood and not stated. However, adherence to the modern scientific method is expected, and the scientific community understands that it is as important to expose mistakes as it is to praise new knowledge.

MATERIALS

Per student:

- a typical research article in biology
- a research article from *Science*

PROCEDURE

1. Check the design of a typical research article in biology and list the titles of its various sections.
   - **(a)** Example: Abstract-summary of paper ____________________
   - **(b)** ____________________
   - **(c)** ____________________
   - **(d)** ____________________
   - **(e)** ____________________
   - **(f)** ____________________
2. Read the article, then fill in the blanks in the following statements or answer the question.
   - **(a)** Any changes in the dependent variable are described in the ____________________ section.
   - **(b)** Which section contains the details necessary to repeat this experiment or observation?

     ____

(c) Which section contains the questions being asked, the predictions, or the hypotheses?

____

(d) The ______________________ section contains the conclusions.

3. Look at an article from the journal *Science.* What steps of the scientific method are included in the narrative?

____

____

____

1.3 Bioassay *(Portions of two days)*

You no doubt know through various media sources of the myriad chemicals that pervade modern life. These substances can be beneficial or toxic to plants or animals. It's obvious that drinking gasoline is bad for you, but this is an extreme example. The broad questions are, What effects do new chemicals have on living organisms, and at what exposure level do they exert these effects? To answer these questions, we perform experiments known as bioassays. Bioassays are used by the pharmaceutical industry to test new drugs. Agricultural firms determine the effectiveness of new fertilizers and herbicides with bioassays. Some industries measure the effects their waste discharges on aquatic organisms with bioassays. A **bioassay** establishes the quantity of a substance that results in a defined *effective dose* (ED)—that is, the dose that produces a particular effect. In the case of toxic substances, the standard measurement of toxicity is known as an LD_{50} (the lethal dose causing the death of 50% of the organisms exposed to the substance).

In this experiment you will test the hypothesis that many everyday substances affect germination of the seeds of Wisconsin Fast Plants™, members of the mustard family. Predictions are that if this is so then the timing of seed germination will change, and if germination occurs, the extent of germination as measured by the length of roots and stems will change. If possible, you will use your results to estimate ED or LD_{50} for the substance tested.

MATERIALS

Per student:

- fine-pointed, water-resistant marker ("Sharpie" or similar)
- 35-mm film canister with lid prepunched with four holes
- 4 microcentrifuge tubes with caps
- 4 paper-towel wicks
- disposable pipet and bulb
- forceps
- 8 RCBr seeds (Wisconsin Fast Plants ™)
- 15-cm plastic ruler

Per student group (6):

- solution of test substance (paint thinner, household cleaner, perfume, coffee, vinegar, and so on)
- distilled water (dH_2O) in dropping bottle

PROCEDURE

1. Using a fine-pointed, water-resistant marker, label the caps and sides of the four microcentrifuge tubes 1.0, 0.1, 0.01, and C (for "Control").
2. Place 10 drops of the test solution into the tube labeled 1.0. This tube now contains the full concentration (100%) of the test solution. A test solution can be anything you and your instructor decide to test. Write the name of your test substance. ______________________________
3. Make *serial dilutions* of the solution, producing concentrations of 10% and 1% of the test substance (Figure 1-3). Start by adding nine drops of distilled water (dH_2O) to the tubes labeled 0.1 and 0.01.
4. Remove a small quantity of the solution from tube 1.0 with the disposable pipet and place *one* drop into tube 0.1. Mix the contents thoroughly by "thipping" the side of the microcentrifuge tube (flicking the sides of the bottom of the tube with the index finger of your dominant hand while holding the sides of the top of the tube between the index finger and thumb of your other hand). Return any solution remaining in the pipet to tube 1.0. Tube 0.1 now contains a concentration 10% of that in tube 1.0.

5. Now, from tube 0.1, remove a small quantity of the solution with the disposable pipet and place *one* drop in tube 0.01. Mix the contents thoroughly and return any solution left in the pipet to tube 0.1. Tube 0.01 now contains a 1% concentration of the original.
6. From the dropping bottle, place 10 drops of dH_2O into tube C. This is the control for the experiment; it contains none of the test solution.
7. Insert a paper towel wick into each microcentrifuge tube (Figure 1-4). With your forceps, place two RCBr seeds near the top of each wick. Close the caps of the tubes. Do not allow any of the wick to protrude outside the cap.
8. Carefully insert the microcentrifuge assay tubes through the holes in the film canister lid (Figure 1-5). Set the experiment aside in the location indicated by your instructor.

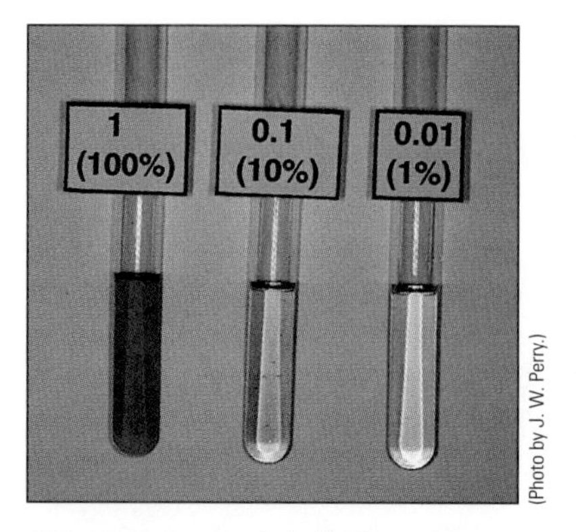

Figure 1-3 Serial dilution of test substance.

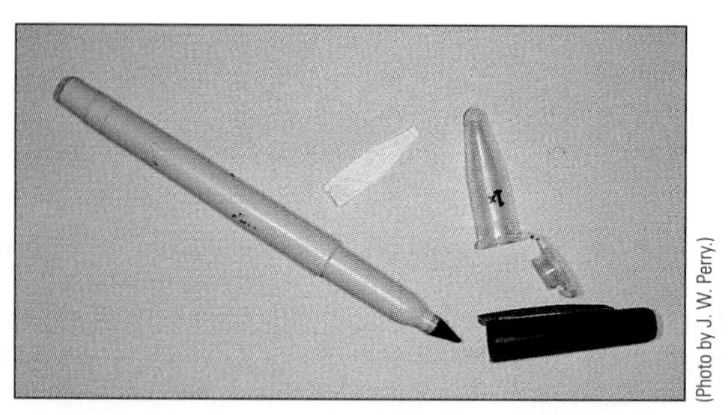

Figure 1-4 One of the four assay tubes.

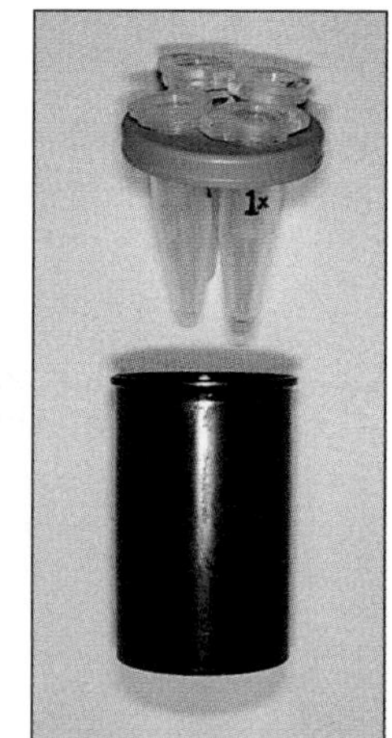

Figure 1-5 Experimental apparatus.

9. At this point, write a prediction for your experiment in Table 1-2.
10. State your experimental variables.
    - Independent variable—

      ____

      ____

    - Dependent variable—

      ____

      ____

    - Controlled variables—

      ____

      ____

11. After 24 hours or more, examine your experiment to see how many and which seeds germinated. Record your results in Table 1-3. Record seeds as "germinated" (G) or "did not germinate" (DNG), and for germinated seeds indicate the extent of germination by measuring and recording the length in millimeters (mm) of the roots and shoots of seed 1 (S1) and seed 2 (S2) for each tube.
12. You have studied the effect of your substance on only a small number of seeds. To make your analysis more reliable, you should analyze a larger number of seeds. In an effort to increase reliability, pool your data with those students who tested the same substance. Calculate the averages and record them in Table 1-3.
13. If germination occurred in more than one group, graph the results in Figure 1-6, plotting the growth of roots and shoots at each concentration.

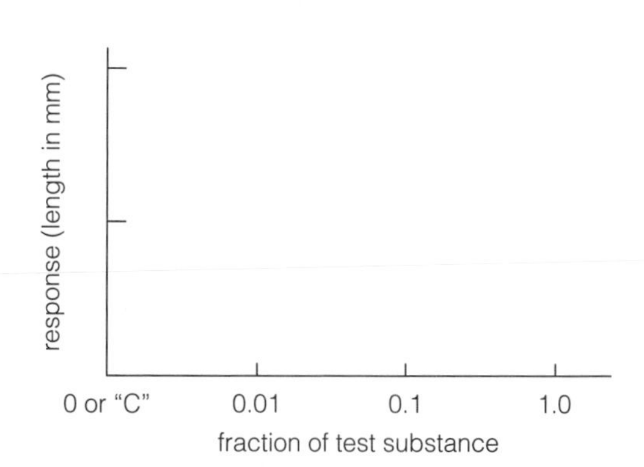

Figure 1-6 Effect of ______________ on germination of RCBr seeds (○= roots and ● = shoots).

TABLE 1-3 Effect of ____________ on Germination of RCBr Seeds

Prediction:							
		My Experiment				Class Averages	
		Roots		Shoots (mm)			
Tube	Germination	S1	S2	S1	S2	Roots	Shoots
C (dH_2O)							
1.0							
0.1							
0.01							

14. If germination occurred in the lower concentrations but not the higher ones, what effective dose prevents germination? Alternatively, could you have defined any other effective doses?

____

____

____

15. Considering your germination results, estimate or make a statement about the LD_{50} for your test substance. If negative effects on germination were observed, describe how you could modify the procedure to better determine the LD_{50}.

____

____

16. Make a conclusion regarding the effect of the test substance on germination of RCBr seeds.

____

____

17. Does your conclusion allow you to accept or reject your prediction? Is the hypothesis supported or falsified? Explain your answer.

____

____

PRE-LAB QUESTIONS

_____ **1.** The natural philosophy of Aristotle and his colleagues was
(a) mechanistic
(b) vitalistic
(c) a belief in absolute truth
(d) a and c

_____ **2.** A person who believes that the universe is at least partially controlled by supernatural powers can best be described as a(n)
(a) teleologist
(b) vitalist
(c) empiricist
(d) mechanist

_____ **3.** The first step of the scientific method is to
(a) ask a question
(b) construct a hypothesis
(c) observe carefully
(d) formulate a prediction

_____ **4.** Which series of letters lists the first four steps of the scientific method (see question 3) in the correct order?
(a) a, b, c, d
(b) a, b, d, c
(c) c, a, b, d
(d) d, c, a, b

_____ **5.** In an experiment the subjects or items being investigated are divided into the experimental group and
(a) the nonexperimental group
(b) the control group
(c) the statistics group
(d) none of these choices

_____ **6.** The variables that investigators try to keep the same for both the experimental and the control groups are
(a) independent
(b) controlled
(c) dependent
(d) a and c

_____ **7.** Variables that are *always* different between the experimental and the control groups are
(a) independent
(b) controlled
(c) dependent
(d) a and c

_____ **8.** The results of an experiment
(a) don't have to be repeatable
(b) should be repeatable by the investigator
(c) should be repeatable by other investigators
(d) must be both b and c

_____ **9.** The detailed report of an experiment is usually published in a
(a) newspaper
(b) book
(c) scientific journal
(d) magazine

_____ **10.** Bioassays can be used to
(a) test new drugs
(b) determine the effectiveness of new fertilizers and herbicides
(c) measure the effects their waste discharges have on aquatic organisms
(d) do all of these choices

Name ______________________ Section Number ______________________

EXERCISE 1

The Scientific Method

POST-LAB QUESTIONS

Introduction

1. How does the modern scientific method differ from the natural philosophy of the ancient Greeks?

1.1 Modern Scientific Method

2. List the six steps of one full cycle of the scientific method.
   a. ______________________
   b. ______________________
   c. ______________________
   d. ______________________
   e. ______________________
   f. ______________________
3. What is tested by an experiment?

4. Within the framework of an experiment, describe the
   a. independent variable— ______________________
   b. dependent variable— ______________________
   c. controlled variables— ______________________
5. Is the statement, "In most biology experiments, the relationship between the independent and the dependent variable can best be described as cause and effect," true or false? Explain your answer.

6. Is a scientific principle taken as absolutely true? Explain your answer.

1.2 Research Article

7. What is the function of research articles in scientific journals?

1.3 Bioassay

8. Define the design and structure of a bioassay.

Food for Thought

9. Describe how you have applied or could apply the scientific method to an everyday problem.

10. Do you think the differences between religious and scientific knowledge make it difficult to debate points of perceived conflict between them? Explain your answer.

Measurement

OBJECTIVES

After completing this exercise, you will be able to

1. define *length, volume, meniscus, mass, density, temperature, thermometer;*
2. recognize graduated cylinders, beakers, Erlenmeyer flasks, different types of pipets, and a triple beam balance;
3. measure and estimate length, volume, and mass in metric units;
4. explain the concept of temperature;
5. measure and estimate temperature in degrees Celsius;
6. explain the advantages of the metric system of measurement.

INTRODUCTION

One requirement of the scientific method is that results be repeatable. As numerical results are more precise than purely written descriptions, scientific observations are usually made as measurements. Of course, sometimes a written description without numbers is the most appropriate way to describe a result.

2.1 Metric System *(About 60 min.)*

Logically, units in the ideal system of measurement should be easy to convert from one to another (for example, inches to feet or centimeters to meters) and from one related measurement to another (length to area, and area to volume). The metric system meets these requirements and is used by the majority of citizens and countries in the world. Universally, science educators and researchers prefer it. In most nonmetric countries, governments have launched programs to hasten the conversion to metrics. Any country that fails to do so could be at a serious economic and scientific disadvantage. In fact, the U.S. Department of Defense adopted the metric system in 1957, and all cars made in the United States have metric components.

The metric reference units are the **meter** for length, the **liter** for volume, the **gram** for mass, and the **degree Celsius** for temperature. Regardless of the type of measurement, the same prefixes are used to designate the relationship of a unit to the reference unit. Table 2-1 lists the prefixes we will use in this and subsequent exercises.

As you can see, the metric system is a decimal system of measurement. Metric units are 10, 100, 1000, and sometimes 1,000,000 or more times larger or smaller than the reference unit. Thus, it's relatively easy to convert from one measurement to another either by multiplying or dividing by 10 or a multiple of 10:

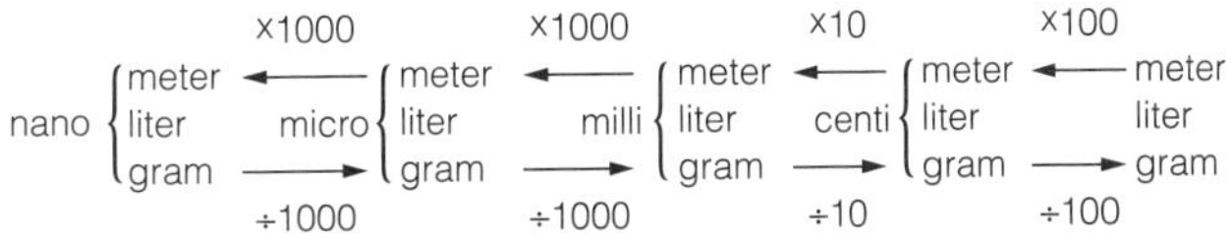

nano {meter, liter, gram} ← ×1000 / ÷1000 → micro {meter, liter, gram} ← ×1000 / ÷1000 → milli {meter, liter, gram} ← ×10 / ÷10 → centi {meter, liter, gram} ← ×100 / ÷100 → meter, liter, gram

TABLE 2-1 Prefixes for Metric System Units

Prefix of Unit (Symbol)	Part of Reference Unit
nano (n)	1/1,000,000,000 = 0.000000001 = 10^{-9}
micro (μ)	1/1,000,000 = 0.000001 = 10^{-6}
milli (m)	1/1000 = 0.001 = 10^{-3}
centi (c)	1/100 = 0.01 = 10^{-2}
kilo (k)	1000 = 10^{3}

In this section, we examine the metric system and compare it to the American Standard system of measurement (feet, quarts, pounds, and so on).

MATERIALS

Per student pair:

- 30-cm ruler with metric and American (English) Standard units on opposite edges
- 250-mL beaker made of heat-proof glass
- 250-mL Erlenmeyer flask
- 3 graduated cylinders: 10-mL, 25-mL, 100-mL
- 1-quart jar or bottle marked with a fill line
- one-piece plastic dropping pipet (not graduated) or Pasteur pipet and bulb
- graduated pipet and safety bulb or filling device
- 1-pound brick of coffee
- ceramic coffee mug
- 1-gallon milk bottle
- metric tape measure
- 1-L measuring cup
- nonmercury thermometer(s) with Celsius (°C) and Fahrenheit (°F) scales (about 220–110°C)
- hot plate
- 3 boiling chips
- thermometer holder

Per student group:

- a triple beam balance

Per lab room:

- source of distilled water (dH_2O)
- metric bathroom scale
- source of ice

PROCEDURE

A. Length *(15 min.)*

Length is the measurement of a real or imaginary line extending from one point to another. The standard unit is the meter, and the most commonly used related units of length are

1000 millimeters (mm) = 1 meter (m)
100 centimeters (cm) = 1 m
1000 m = 1 kilometer (km)

For orientation purposes, the yolk of a chicken egg is about 3 cm in diameter. Since the differences between these metric units are based on 10 or multiples of 10, it's fairly easy to convert a measurement in one unit to another.

1. For example, if you want to convert 1.7 km to centimeters, your first step is to determine how many centimeters there are in 1 km. Remember, like units may be cancelled.

$$\frac{100\text{ cm}}{1\text{ m}} \times \frac{1000\text{ m}}{1\text{ km}} = \frac{100{,}000\text{ cm}}{1\text{ km}}$$

The second step is to multiply the number by this fraction.

$$\frac{1.7\text{ km}}{1} \times \frac{100{,}000\text{ cm}}{1\text{ km}} = 170{,}000\text{ cm}$$

The last calculation can also be done quickly by shifting the decimal point five places to the right.

1.700000 km = 170,000.0 cm

Alternatively, to convert 1.7 km to centimeters, we add exponents.

$$\frac{1.7\text{ km}}{1} \times \frac{10^2\text{ cm}}{\text{m}} \times \frac{10^3\text{ m}}{\text{km}} = 1.7 \times 10^5\text{ cm} = 170{,}000\text{ cm}$$

Using the method most comfortable for you, calculate how many millimeters there are in 4.8 m.

__________ mm

Now let's convert 17 mm to meters.

step 1 $$\frac{1\text{ m}}{100\text{ cm}} \times \frac{1\text{ cm}}{10\text{ mm}} = \frac{1\text{ m}}{1000\text{ mm}}$$

step 2 $$\frac{17\text{ mm}}{1} \times \frac{1\text{ m}}{1000\text{ mm}} = 0.017\text{ m}$$

Alternatively, you can shift the decimal point three places to the left,

0017.0 mm = 0.017 m

or add exponents:

$$\frac{1.7 \text{ mm}}{1} \times \frac{10^{-2} \text{ m}}{\text{cm}} \times \frac{10^{-1} \text{ cm}}{\text{mm}} = 17 \times 10^{-3} \text{ m} = 0.017 \text{ m}$$

Calculate how many kilometers there are in 16 cm. ________ km

2. Precisely measure the length of this page in centimeters to the nearest tenth of a centimeter with the metric edge of a ruler. Note that nine lines divide the space between each centimeter into 10 millimeters.

   The page is ______ cm long.

   Calculate the length of this page in millimeters, meters, and kilometers.

   ______ mm ______ m ______ km

   Now repeat the above measurement using the American Standard edge of the ruler. Measure the length of this page in inches to the nearest eighth of an inch.

   ______ in

   Convert your answer to feet and yards.

   ______ ft ______ yd

   Explain why it is much easier to convert units of length in the metric system than in the American Standard system.

   ____

   ____

   ____

B. Volume *(20 min.)*

Volume is the space an object occupies. The standard unit of volume is the liter (L), and the most commonly used subunit, the milliliter (mL). There are 1000 mL in 1 liter. A chicken egg has a volume of about 60 mL.

The volume of a box is the height multiplied by the width multiplied by the depth. The amount of water contained in a cube with sides 1 cm long is 1 cubic centimeter (cc), which for all practical purposes equals 1 mL (Figure 2-1).

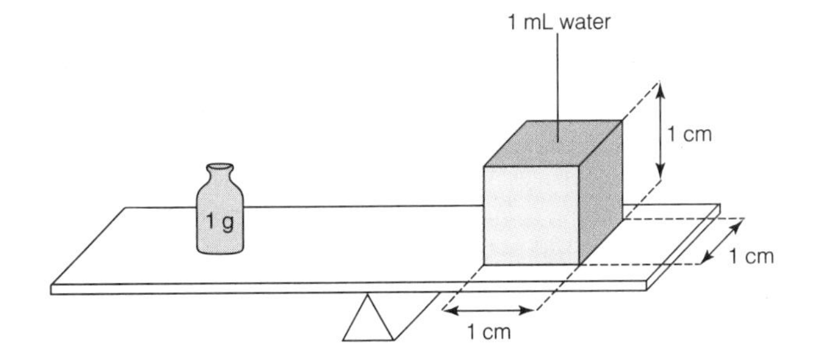

Figure 2-1 Relationship among the units of length, volume, and mass in the metric system.

1. How many milliliters are there in 1.7 L?

   ______ mL

   How many liters are there in 1.7 mL?

   ______ L

2. Use Figure 2-2 to recognize **graduated cylinders, beakers, Erlenmeyer flasks,** and the different types of **pipets.** Some of these objects are made of glass; some are plastic. Some will be calibrated in milliliters and liters; others will not be.

3. Pour some water into a 100-mL graduated cylinder and observe the boundary between fluid and air, the **meniscus.** Surface tension makes the meniscus curved, not flat. The high surface tension of water is due to its cohesive and adhesive or "sticky" properties.

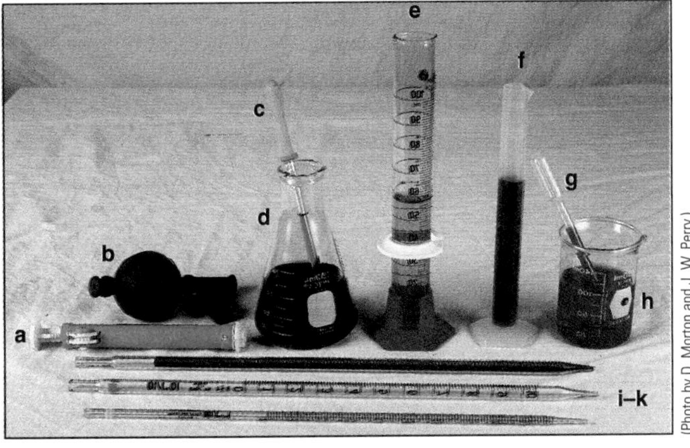

Figure 2-2 Apparatuses commonly used to measure volume: (**a**) pipet filling device, (**b**) pipet safety bulb, (**c**) Pasteur pipet and bulb, (**d**) Erlenmeyer flask, (**e**) glass graduated cylinder, (**f**) plastic graduated cylinder, (**g**) plastic dropping pipet, (**h**) beaker, (**i–k**) graduated pipets.

Draw the meniscus in the plain cylinder outlined in Figure 2-3. The correct reading of the volume is at the *lowest* point of the meniscus.

4. Using the 100-mL graduated cylinder, pour water into a 1-quart jar or bottle. About how many milliliters of water are needed to fill the vessel up to the line?

   ______ mL

5. Pipets are used to transfer small volumes from one vessel to another. Some pipets are not graduated (for example, Pasteur pipets and most one-piece plastic dropping pipets); others are graduated.
   (a) Fill a 250-mL Erlenmeyer flask with distilled water.
   (b) Use a plastic dropping pipet or Pasteur pipet with a bulb to withdraw some water.
   (c) Count the number of drops needed to fill a 10-mL graduated cylinder to the 1-mL mark. Record this number in Table 2-2.
   (d) Repeat steps b and c two more times and calculate the average for your results in Table 2-2.
   (e) Explain why the average of three trials is more accurate than if you only do the procedure once.

   ____

   ____

   ____

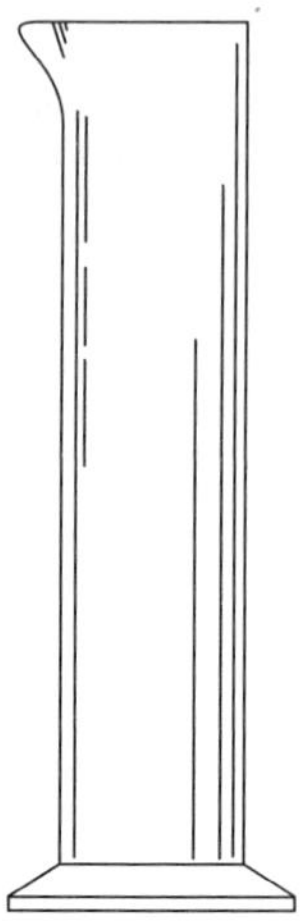

Figure 2-3 Draw a meniscus in this plain cylinder.

TABLE 2-2 Estimate of the Number of Drops in 1 mL

Trial	Drops/mL
1	
2	
3	
Total	
Average	

C. *Mass* (*25 min.*)

Mass is the quantity of matter in a given object. The standard unit is the **kilogram** (kg), and other commonly used units are the milligram (mg) and gram (g). There are 1,000,000 mg in 1 kg and 1000 g in 1 kg. A chicken egg has a mass of about 60 g. Note that the following discussion avoids the term *weight.* This is because weight depends on the gravitational field in which the matter is located. For example, you weigh less on the moon but your mass remains the same as it is on earth. Although it is technically incorrect, mass and weight are often used interchangeably.

1. How many milligrams are there in 1 g?

   ______ mg

   Convert 1.7 g to milligrams and kilograms.

   ______ mg ______ kg

2. A 1-cc cube, if filled with 1 mL of water, has a mass of 1 g (Figure 2-1). The mass of other materials depends on their **density** (water is defined as having a density of 1). The density of any substance is its mass divided by its volume.

   Approximately how many liters are present in 1 cubic meter (m^3) of water? As each of the sides of 1 m^3 are 100 cm in length, it's easy to calculate the number of cubic centimeters (that is, 100 cm $\times$ 100 cm $\times$ 100 cm = 1,000,000 cc). Now just change cubic centimeters to milliliters and convert 1,000,000 mL to liters.

   ______ L

   What is its mass in kilograms?

   ______ kg

In your "mind's eye," contemplate calculating how many pounds there are in a cubic yard of water. It's easier to convert between different units of the metric system than between those of the American Standard system.

3. Determine the mass of an unknown volume of water.* Mass may be measured with a **triple beam balance,** which gets its name from its three beams (Figure 2-4). A movable mass hangs from each beam.
   (a) Slide all of the movable masses to zero. Note that the middle and back masses each click into the leftmost notch.
   (b) Clear the pan of all objects and make sure it is clean.
   (c) The balance marks should line up, indicating that the beam is level and that the pan is empty. If the balance marks don't line up, rotate the zero adjust knob until they do.
   (d) Place a 250-mL beaker on the pan. The right side of the beam should rise. Slide the mass on the middle beam until it clicks into the notch at the 100-g mark. If the right end of the beam tilts down below the stationary balance mark, you have added too much mass. Move the mass back a notch. If the right end remains tilted up, additional mass is needed. Keep adding 100-g increments until the beam tilts down; then move the mass back one notch. Repeat this procedure on the back beam, adding 10 g at a time until the beam tilts down, and then backing up one notch. Next, slide the front movable mass until the balance marks line up.
   (e) The sum of the masses indicated on the three beams gives the mass of the beaker. Nine unnumbered lines divide the space between the numbered gram markings on the front beam into 10 sections, each representing 0.1 g. Record the mass of the beaker to the nearest tenth of a gram in Table 2-3.
   (f) Add an unknown amount of water and repeat the above procedure. Record the mass of the beaker and water in Table 2-3.
   (g) Calculate the mass of the water alone by subtracting the mass of the beaker from that of the combined beaker and water. Record it in Table 2-3. Do not dispose of the water yet.
   (h) Now measure the volume of the water in milliliters with a graduated cylinder.
   What is the volume? _______ mL

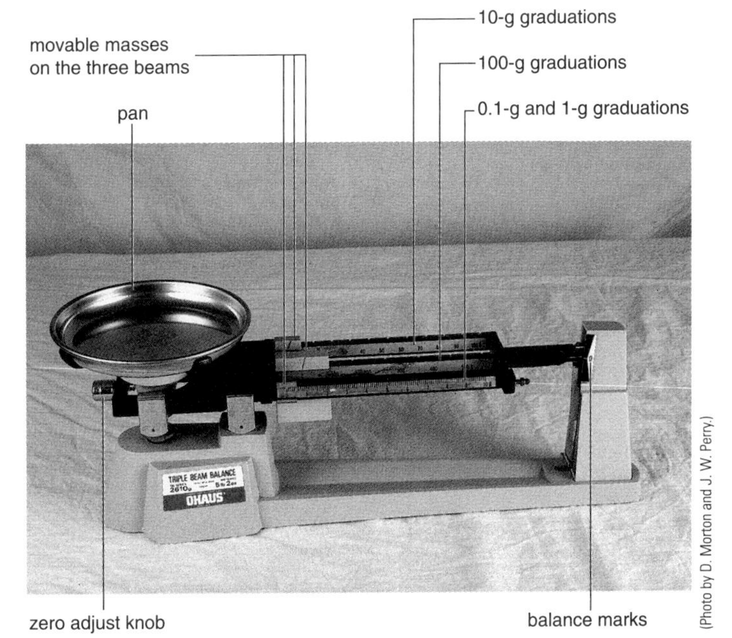

Figure 2-4 Triple beam balance.

TABLE 2-3 Weighing an Unknown Quantity of Water with a Triple Beam Balance

Objects	Masses (g)
Beaker and water	
Beaker	
Water	

4. Using the triple beam balance, determine the mass of (that is, weigh) a pound of coffee in grams.

   ______ g

*Modified from C. M. Wynn and G. A. Joppich, *Laboratory Experiments for Chemistry: A Basic Introduction*, 3rd ed. Wadsworth, 1984.

D. Estimating Length, Volume, and Mass (*10 min.*)

Now that you have experience using metric units, let's try estimating the measurements of some everyday items. Also, you may consult the metric/American Standard conversion table in Appendix 1 at the end of the lab manual.

1. Estimate the length of your index finger in centimeters. ______ cm
2. Estimate your lab partner's height in meters. ______ m
3. How many milliliters will it take to fill a ceramic coffee mug? ______ mL
4. How many liters will it take to fill a 1-gallon milk bottle? ______ L
5. Estimate the weight of some small personal item (for example, loose change) in grams. ______ g
6. Estimate your or your lab partner's weight in kilograms. ______ kg
7. Transfer your estimates to Table 2-4.
8. Then check your results using either a ruler, metric tape measure, 100-mL graduated cylinder, 1-L measuring cup, triple beam balance, or metric bathroom scale, recording your measurements in Table 2-4. Complete Table 2-4 by calculating the difference between each estimate and measurement.

TABLE 2-4 Differences Between Estimates and Measurements

Number	Estimate	Measurement	Estimate − Measurement
1	cm	cm	cm
2	m	m	m
3	mL	mL	mL
4	L	L	L
5	g	g	g
6	kg	kg	kg

E. Temperature (*About 20 min.*)

The degree of hot or cold of an object is termed **temperature.** More specifically, it is the average kinetic energy of molecules. Heat always flows from high to low temperatures. This is why hot objects left at room temperature always cool to the surrounding or ambient temperature, while cold objects warm up. Consequently, to keep a heater hot and the inside of a refrigerator cold requires energy. **Thermometers** are instruments used to measure temperature.

1. Using a thermometer with both *Celsius* (°C) and *Fahrenheit* (°F) scales, measure room temperature and the temperature of cold and hot running tap water. Record these temperatures in Table 2-5.
2. Fill a 250-mL beaker with ice about three-fourths full and add cold tap water to just below the ice. Wait for 3 minutes, measure the temperature, and record it in Table 2-5. Remove the thermometer and discard the ice water into the sink.
3. Fill the beaker with warm tap water to about three-fourths full and add three boiling chips. Use a thermometer holder to clip the thermometer onto the rim of the beaker so that the bulb of the thermometer is halfway into the water. Boil the water in the beaker by placing it on a hot plate. After the water boils, record its temperature in Table 2-5. Turn off the hot plate and let the water and beaker cool to below 50°C before pouring the water into the sink.
4. To convert Celsius degrees to Fahrenheit degrees, multiply by 9/5 and add 32. Is 4°C the temperature of a hot or cool day? ______
   What temperature is this in degrees Fahrenheit? ______ °F
5. To convert Fahrenheit degrees to Celsius degrees, subtract 32 and multiply by 5/9. What is body temperature, 98.6°F, in degrees Celsius? ______ °C

Caution

Your instructor will give you specific instructions on how to set up the equipment in your lab for boiling water.

TABLE 2-5 Comparison of Celsius and Fahrenheit Temperatures

Location	°C	°F
Room		
Cold running tap water		
Hot running tap water		
Ice water		
Boiling water		

6. In summary, the formulas for these temperature conversions are:

   °F = $\frac{9}{5}$ °C + 32

   °C = $\frac{5}{9}$ (°F – 32)

2.2 Micromeasurement *(About 30 min.)*

More and more procedures used in molecular biology and related areas require the measurement of very small masses and volumes. The following procedure introduces you to use of the electronic balance and micropipets or micropipetters. We'll do this as an experiment in part by answering the question, In the hands of students, how exact are instruments that deliver small volumes in doing what they say they do? Our hypothesis is that under these conditions these instruments perform within the manufacturer's stated specifications for the volumes they deliver.

MATERIALS

Per student group (table):

- standardized electronic balance capable of weighing 0.001 g and with at least a readability of 0.001 g, a repeatability (standard deviation) of ± 0.001 g, and a linearity of 0.002 g
- micropipets or micropipetters with matching tips (capable of delivering 10-μL, 50-μL, and 100-μL volumes)
- nonabsorbent weighing dish
- small test tube in a beaker or rack
- scientific calculator with instructions

Per lab room:

- source of distilled water (dH_2O)

PROCEDURE

1. Remember that in Section 2.1 we established a relationship between the volume and mass of water. If the mass of 1 μL of water is 1 g (Figure 2-1), how much does 1 μL weigh?

   ______ g or ______ mg

   The electronic balances have been standardized prior to lab. If the micropipets or micropipetters in the hands of students perform within manufacturer's specifications, then we can predict what 100 μL, 50 μL, and 10 μL delivered from one of these instruments will weigh. In Table 2-6, write a prediction for this experiment.
2. Fill your test tube with distilled water.
3. Turn on the electronic balance and wait a moment (usually about 5 seconds) until it is stable.
4. Place a weighing dish on the pan.
5. Press the zero or tare key of the balance. Note that the digital display reads all zeros.

Cautions on the Use of Micropipetters

- ***Never adjust the micropipetter above or below the stated range or volume limit of measurement.***
- ***Never use the micropipetter without a tip in place, as this will damage the internal piston mechanism.***
- ***Never invert the micropipetter with a filled tip, as fluid could run up into its chamber.***
- ***Never allow the plunger to snap back from the depressed position, as this could damage the piston. After drawing up or dispensing the fluid, release the plunger slowly.***
- ***Never immerse the barrel of the micropipetter in fluid.***

6. Use a 100-μL micropipet or a micropipetter adjusted to 100 μL and deliver this volume into the weighing dish. Wait a second for the balance to stabilize, then read and record the mass in grams or milligrams in the correct column of Table 2-6. If you are using a toploading balance, the last digit may continue to change. In this case, use the middle number of the range of numbers displayed.
7. Tare or zero your balance and repeat step 6 for a total of 10 trials.
8. Tare or zero your balance and repeat steps 6 and 7 for the 50- and 10-μL volumes.
9. Use a scientific calculator to determine the average or mean and standard deviation for each column.
10. Plot your points on the graph and the means ± standard deviations on the bar chart:

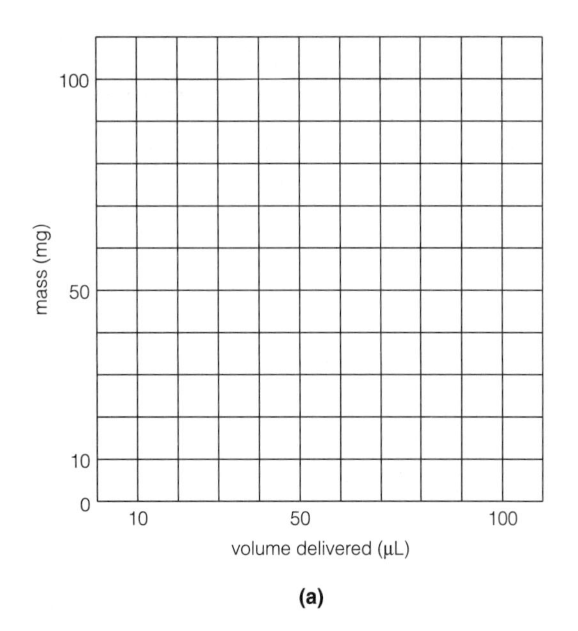

(a)

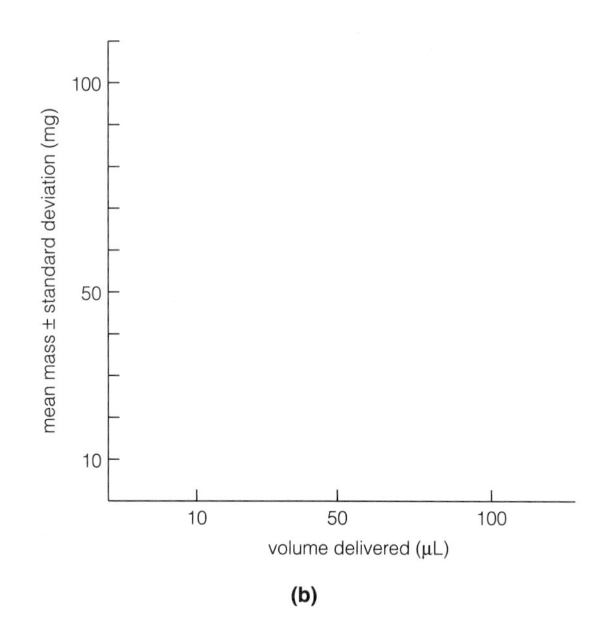

(b)

11. Write your conclusion in Table 2-6.

TABLE 2-6 Comparison of Volumes of Water with Their Masses

Prediction:

Trial	100 µL	50 µL	10 µL
1			
2			
3			
4			
5			
6			
7			
8			
9			
10			
Mean			
Standard deviation			

Conclusion:

PRE-LAB QUESTIONS

_____ **1.** The metric system is the measurement system of choice for
(a) science educators
(b) science researchers
(c) the citizens of most countries in the world
(d) all of the above

_____ **2.** A kilowatt, a unit of electrical power, is
(a) ten watts
(b) a hundred watts
(c) a thousand watts
(d) a million watts

_____ **3.** A millicurie, a unit of radioactivity, is
(a) a tenth of a curie
(b) a hundredth of a curie
(c) a thousandth of a curie
(d) a millionth of a curie

_____ **4.** Length is the measurement of
(a) a line, extending from one point to another
(b) the space an object occupies
(c) the quantity of matter present in an object
(d) the degree of hot or cold of an object

_____ **5.** Volume is the measurement of
(a) a line, extending from one point to another
(b) the space an object occupies
(c) the quantity of matter present in an object
(d) the degree of hot or cold of an object

_____ **6.** Mass is the measurement of
(a) a line, extending from one point to another
(b) the space an object occupies
(c) the quantity of matter present in an object
(d) the degree of hot or cold of an object

_____ **7.** If 1 cc of a substance has a mass of 1.5 g, its density is
(a) 0.67
(b) 1.0
(c) 1.5
(d) 2.5

_____ **8.** If your mass is 70 kg on the earth, how much is your mass on the moon?
(a) less than 70 kg
(b) more than 70 kg
(c) 70 kg
(d) none of the above

_____ **9.** Above zero degrees, the actual number of degrees Celsius for any given temperature is _____________ the degrees Fahrenheit.
(a) higher than
(b) lower than
(c) the same as
(d) a or b

_____ **10.** A thermometer measures
(a) the degree of hot or cold
(b) temperature
(c) a and b
(d) none of the above

Name ______________________________ Section Number ______________________

EXERCISE 2

Measurement

POST-LAB QUESTIONS

Introduction

1. What is the importance of measurement to science?

2.1 Metric System

2. Convert 1.24 m to millimeters, centimeters, and kilometers.

   ______ mm

   ______ cm

   ______ km

3. Observe the following carefully and read the volume.

   ______ mL

(Photo by D. Morton.)

4. Construct a conversion table for mass. Construct it so that if you wish to convert a measurement from one unit to another, you multiply it by the number at the intersection of the original unit and the new unit.

	New Unit		
Original Unit	**mg**	**g**	**kg**
mg	1		
g		1	
kg			1

5. How is it possible for objects of the same volume to have different masses?

6. Today, many packaged items have the volume or weight listed in both American Standard and metric units. Before your next lab period, find and list 10 such items.

Item	American Standard	Metric

7. Each °F is ______ (larger, smaller) than each °C.

2.2 Micromeasurement

8. Describe what it means to tare or zero an electronic balance. What purpose does this function serve?

Food for Thought

9. How are length, area, and volume related in terms of the three dimensions of space?

10. If you were to choose between the metric and American Standard systems of measurement for future generations, which one would you choose? Set aside your familiarity with the American Standard system and consider their ease of use and degree of standardization with the rest of the world.

Per student group (4):

- dropper bottle of distilled water (dH_2O)

PROCEDURE

1. Carefully use a razor blade to cut a number of *very thin shavings* from a cork stopper. Place them on a glass microscope slide.
2. Gently add a drop of distilled water.
3. Place one end of a glass coverslip to the right or left of the specimen so that the rest of the slip is held at a 45° angle over the specimen (Figure 3-9a).

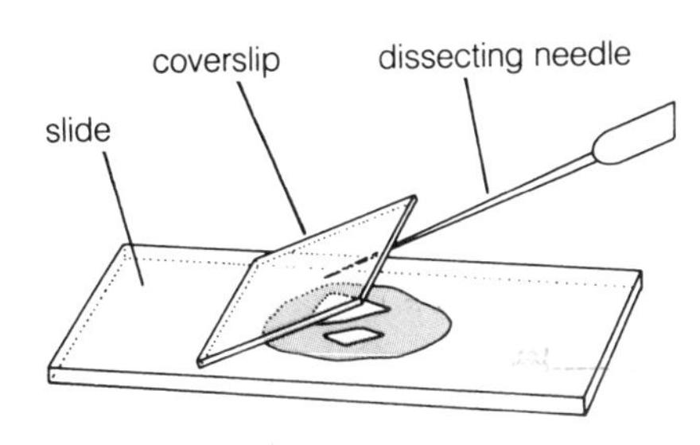

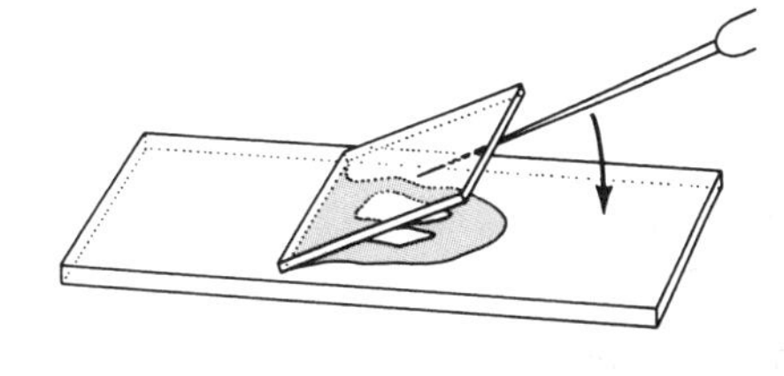

Figure 3-9 How to make a wet mount.

4. Slowly lower the coverslip with a dissecting needle so as not to trap air bubbles (Figure 3-9b).
5. Observe the wet mount, first at low magnification and then with higher power. Air may be trapped either in the cork or as free bubbles (Figure 3-10). Trapped air will appear dark and refractive around its edges. This effect is due to sharply bending rays of light. Draw what you see in Figure 3-11. Note the total magnification used to make the drawing.
6. Clean and replace the slide and coverslip as indicated by your instructor.

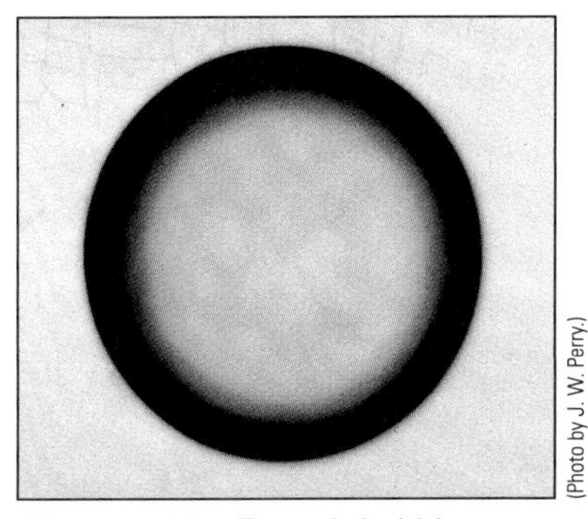

(Photo by J. W. Perry.)

Figure 3-10 Free air bubble (250×).

3.3 Microscopic Observations *(About 20 min.)*

Examining the microscopic world is both challenging and fun. Most of the macroscopic world has been explored, but the microscopic world is barely touched. Yet the microbes in it are essential to our very existence. So be an explorer and see what you can discover!

MATERIALS

Per student:

- compound microscope, lens paper, a bottle of lens-cleaning solution (optional), a lint-free cloth (optional)
- glass microscope slide
- glass coverslip

Per student group (4):

- pond water or some other mixed culture in a dropper bottle
- dropper bottle of Protoslo®

Per lab room:

- reference books for the identification of microorganisms

PROCEDURE

1. Obtain a drop of pond water or other mixed culture from the bottom of the bottle.
2. Add a drop of Protoslo®. This methyl cellulose solution slows down any swimming microorganisms.
3. Make a wet mount (Figure 3-9).

4. Observe the wet mount with your compound microscope. Start at the upper-left corner of the coverslip and scan the wet mount with the low-power objective. When you find something interesting, focus on it and switch to the medium-power objective and then, if necessary, the high-dry objective.
5. Draw what you find on Figure 3-12 and note the total magnification.
6. Attempt to identify it using the resource books provided you by your instructor. If successful, write its name under your drawing.
7. Clean and replace the slide and coverslip as indicated by your instructor.
8. Put away your compound microscope as described on page 27.

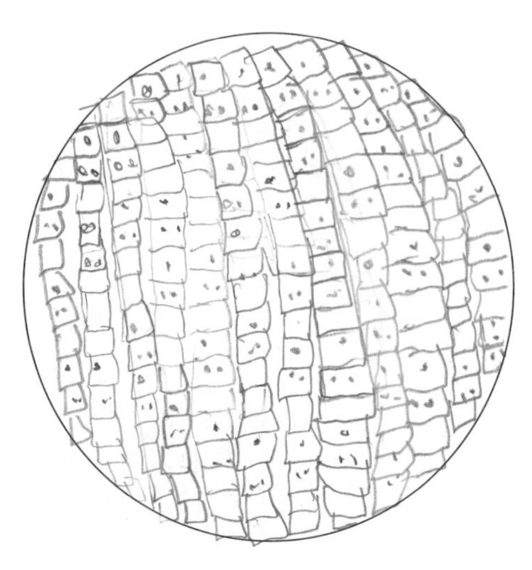

Figure 3-11 Drawing of the microscopic structure in a cork shaving (_______×).

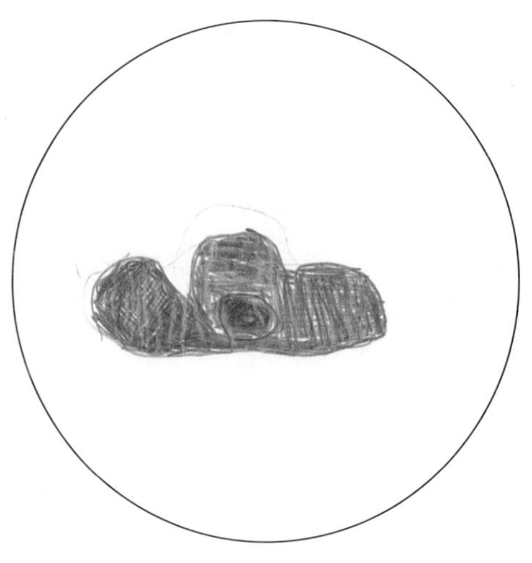

Figure 3-12 Drawing of microorganisms (_______×).

3.4 Dissecting Microscope *(About 10 min.)*

Dissecting microscopes (Figure 3-13) have a large working distance between the specimen and the objective lens. They are especially useful in viewing larger specimens (including thicker slide-mounted specimens) and in manipulating the specimen (when dissection of a small structure or organism is required, for example).

The large working distance also allows for illumination of the specimen from above (reflected light) as well as from below (transmitted light). Reflected light shows up surface features on the specimen better than transmitted light does.

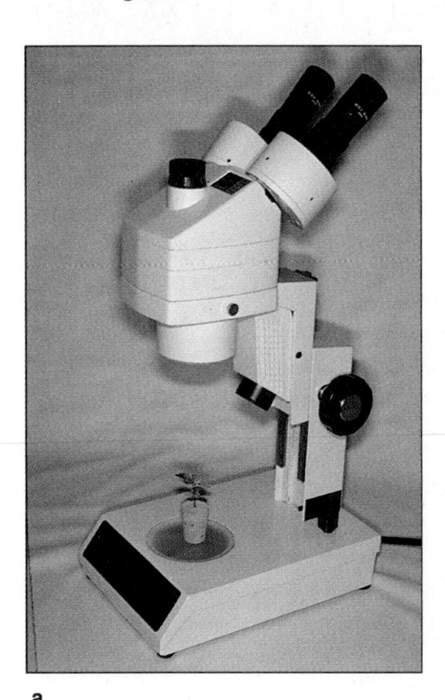

a

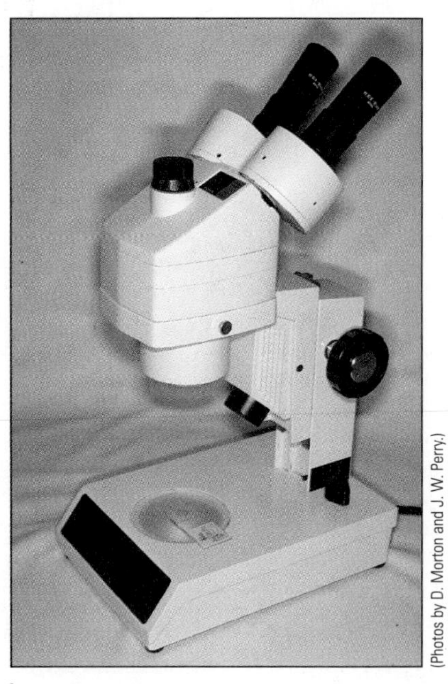

b

(Photos by D. Morton and J. W. Perry.)

Figure 3-13 Dissecting microscope (**a**) reflected light, (**b**) transmitted light.

MATERIALS

Per student group:

- dissecting microscope
- specimens appropriate for viewing with the dissecting microscope (for example, a prepared slide with a whole mount of a small organism, bread mold, an insect mounted on a pin stuck in a cork, a small flower)

PROCEDURE

1. Under a dissecting microscope, view one or more of the specimens provided by your instructor. What is the magnification range of this microscope?

   _________× to _________×

2. Is the image of the specimen inverted as in the compound light microscope? (yes or no) _____________
3. Describe the type of illumination used by your dissecting microscope. Is there a choice?

   ____

   ____

3.5 Other Microscopes *(About 10 min.)*

In future exercises you will examine pictures taken with other types of microscopes (Figure 3-14). Some will be of living cells taken with a phase-contrast microscope or similar instrument, including those using the Nomarski process. Others will be of very thin-sectioned, heavy metal–stained specimens taken with a transmission electron microscope or TEM (Figure 3-15c and d). Still others will be of precious metal–coated surfaces produced by signals from the scanning electron microscope or SEM (Figure 3-16). Table 3-6 summarizes the technology and use of these microscopes.

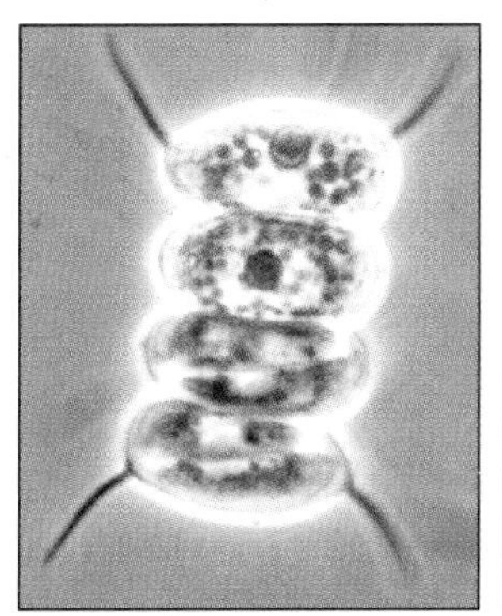

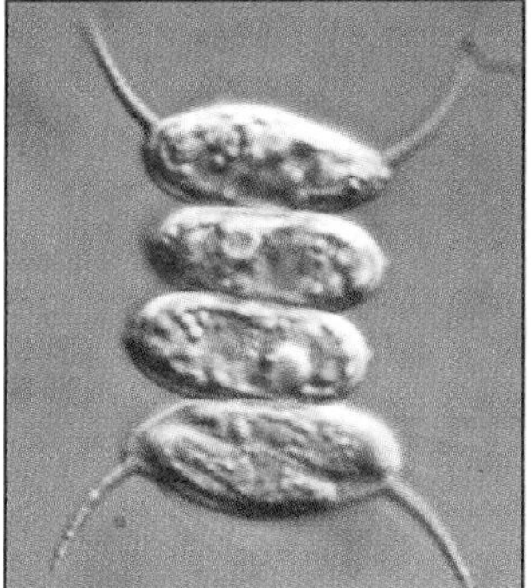

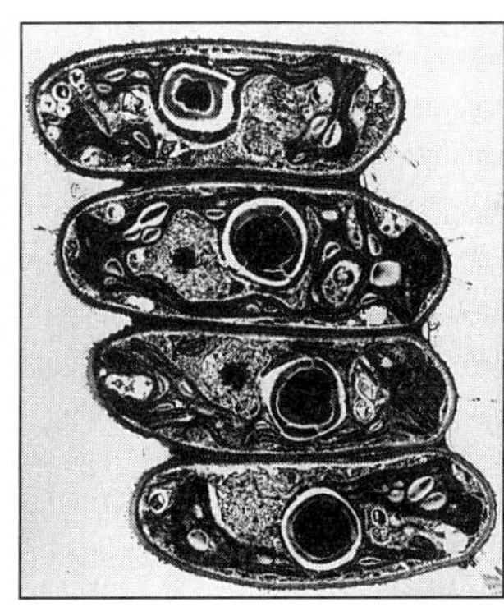

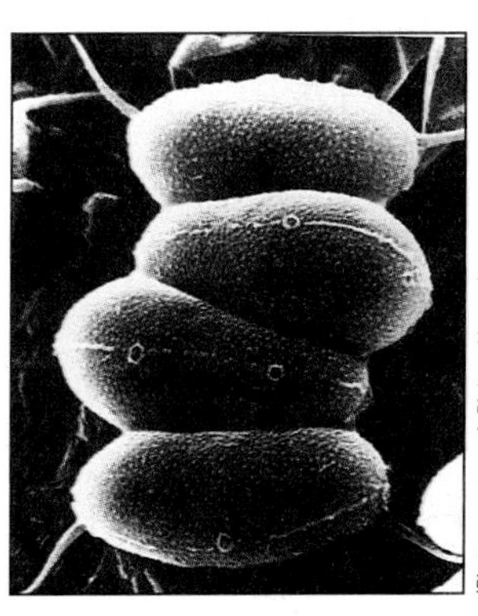

(Photos courtesy J. Pickett-Heaps.)

a phase contrast **b** Nomarski process **c** transmission electron **d** scanning electron

Figure 3-14 How different types of microscopes reveal detail in cells of the green alga *Scenedesmus.*

TABLE 3-6 Other Microscopes

Microscope	Technology	Use
Phase-contrast and Nomarski process	Converts phase differences in light to differences in contrast.	Observation of low contrast specimens (often living).
Transmission electron	Increases resolving power by using electrons in a vacuum and magnetic lenses instead of light and glass lenses, respectively.	Preservation of greater specimen detail allows for magnifications up to 1,000,000× or more (usually dead materials).
Scanning electron	Forms TV-like picture from a secondary electron signal, which is emitted from surface points excited by a thin beam of electrons drawn across the surface in a raster pattern.	Investigation of the fine structure of surfaces (usually dead specimens).

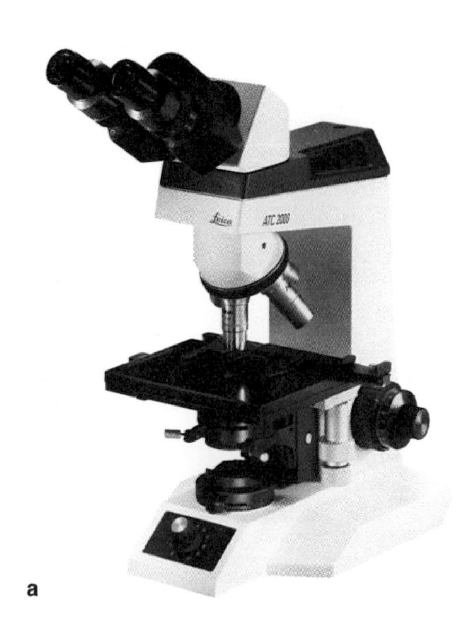

a

c

path of light rays (bottom to top) to eye

Ocular lens enlarges primary image formed by objective lenses.

prism that directs rays to ocular lens

Objective lenses (those closest to specimen) form the primary image. Most compound light microscopes have several.

stage (holds microscope slide in position)

Condenser lenses focus light rays onto specimen.

illuminator

microscope base housing source of illumination

b

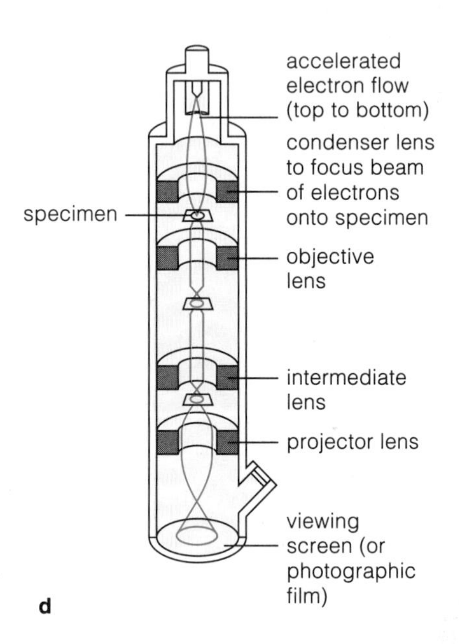

d

Figure 3-15 External view (**c**) and diagram (**d**) of a transmission electron microscope. Similar illustrations of a compound light microscope (**a, b**) are provided for comparing the similarities and differences of these microscopes. (After Starr and Taggart, 2001; (**a**) Leica Microsystems, Inc., Deerfield, IL; (**b**) George Musil/Visuals Unlimited.)

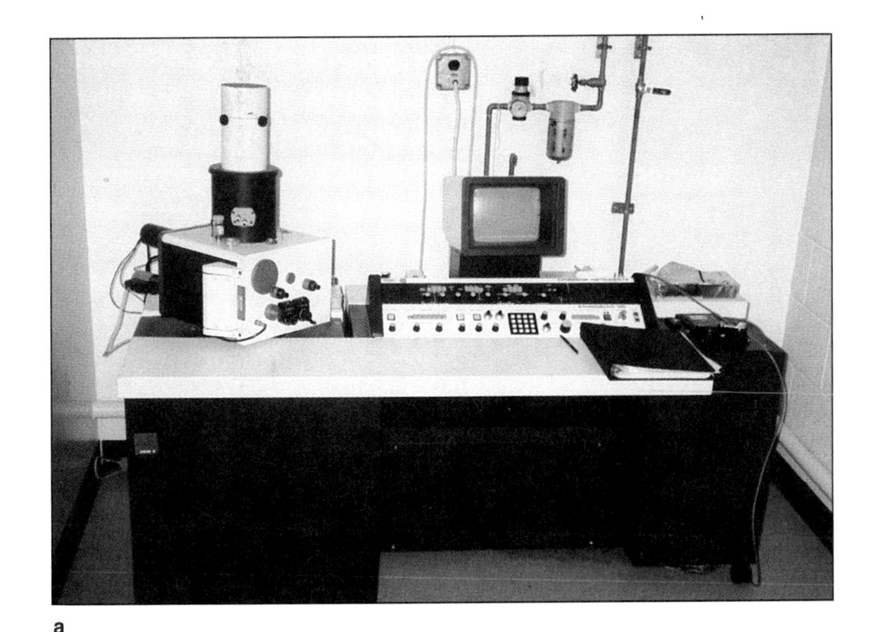

a

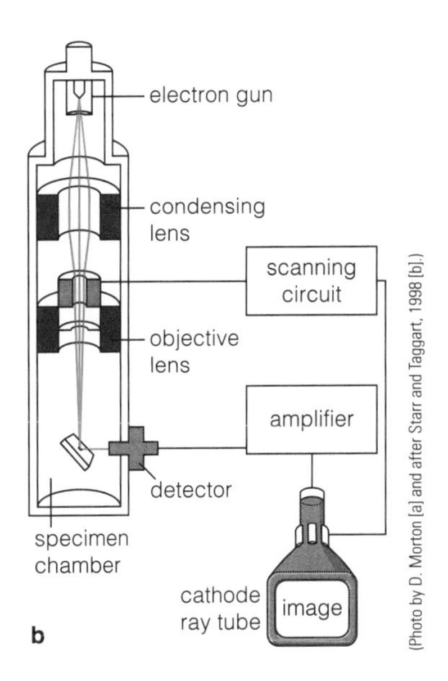

Figure 3-16 External view (**a**) and diagram (**b**) of a scanning electron microscope.

MATERIALS

Per student group:

- photographs of TEM micrographs (negatives) or digital images
- photographs of SEM negatives or digital images

PROCEDURE

1. Examine some photographs of TEM micrographs (negatives) or digital images. The darker areas are more electron-dense in the specimen than the lighter areas.
2. Now look at some photographs of SEM negatives or digital images. The lighter areas correspond to the emission of greater numbers of secondary electrons from that part of the specimen; the darker areas emit less.
3. What type of microscope (compound light, dissecting, phase-contrast, TEM, or SEM) would you use to examine the specimens listed in Table 3-7?

TABLE 3-7 Microscope Use

Specimen	Microscope
Living surface of the finger	
Dye-stained slide of a section of the finger	
Gold-coated bacteria on a single cell of the finger	
Unstained section of a biopsy from the finger	
Heavy metal–stained, very thin section of the finger	

PRE-LAB QUESTIONS

____ **1.** Magnification
(a) is the amount that an object's image is enlarged
(b) is the extent to which detail in an image is preserved during the magnifying process
(c) is the degree to which image details stand out against their background
(d) focuses radiation emanating from an object to produce an image

____ **2.** Resolving power
(a) is the amount that an object's image is enlarged
(b) is the extent to which detail in an image is preserved during the magnifying process
(c) is the degree to which image details stand out against their background
(d) focuses radiation emanating from an object to produce an image

____ **3.** A lens
(a) is the amount that an object's image is enlarged
(b) is the extent to which detail in an image is preserved during the magnifying process
(c) is the degree to which image details stand out against their background
(d) focuses radiation emanating from an object to produce an image

____ **4.** Contrast
(a) is the amount that an object's image of a object is enlarged
(b) is the extent to which detail in an image is preserved during the magnifying process
(c) is the degree to which image details stand out against their background
(d) focuses radiation emanating from an object to produce an image

____ **5.** The maximum useful magnification for a light microscope is about
(a) 100×
(b) 1000×
(c) 10,000×
(d) 100,000×

____ **6.** The two image-forming lenses of a compound light microscope are
(a) the condenser and objective
(b) the condenser and ocular
(c) the objective and ocular
(d) none of these choices

____ **7.** Dyes are usually added to sections of biological specimens to increase
(a) resolving power
(b) magnification
(c) contrast
(d) all of the above

____ **8.** If the magnification of the two image-forming lenses are both 10×, the total magnification of the image will be
(a) 1×
(b) 10×
(c) 100×
(d) 1000×

____ **9.** The distance through which a microscopic specimen can be moved and still have it remain in focus is called the
(a) field of view
(b) working distance
(c) depth of field
(d) magnification

____ **10.** Electron microscopes differ from light microscopes in that
(a) electrons are used instead of light
(b) magnetic lenses replace glass lenses
(c) the electron path has to be maintained in a high vacuum
(d) a, b, and c are all true

Name ______________________________ Section Number ______________________

EXERCISE 3

Microscopy

POST-LAB QUESTIONS

3.1 Compound Light Microscope

1. What is the function of the following parts of a compound light microscope?
   a. condenser lens

   b. iris diaphragm

   c. objective

   d. ocular

2. In order, list the lenses in the light path between a specimen viewed with the compound light microscope and its image on the retina of the eye.

3. What happens to contrast and resolving power when the aperture of the condenser (that is, the size of the hole through which light passes before it reaches the specimen) of a compound light microscope is decreased?

4. What happens to the field of view in a compound light microscope when the total magnification is increased?

5. Describe the importance of the following concepts to microscopy.

   a. magnification

   b. resolving power

   c. contrast

6. Which photomicrograph of unstained cotton fibers was taken with the iris diaphragm closed? ________

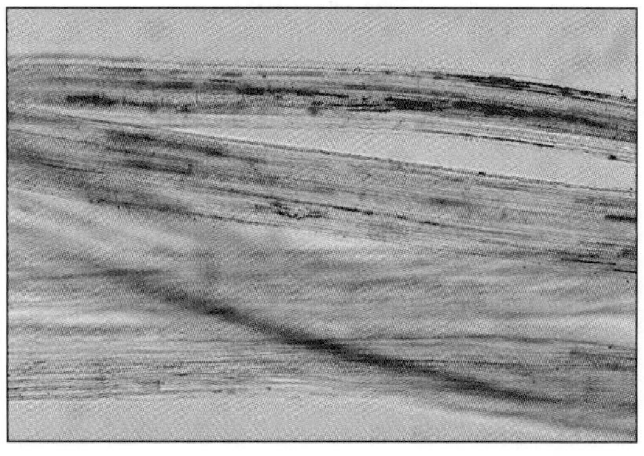

(Photo by J. W. Perry.)

a

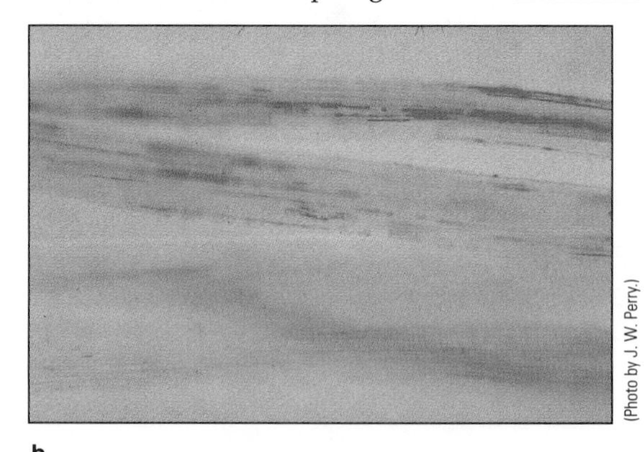

(Photo by J. W. Perry.)

b

7. Describe how you would care for and put away your compound light microscope at the end of lab.

3.2 How to Make a Wet Mount

8. Describe how to make a wet mount.

3.4 Dissecting Microscope, 3.5 Other Microscopes

9. A camera mounted on a ________________________ microscope took this photo of a cut piece of cork.

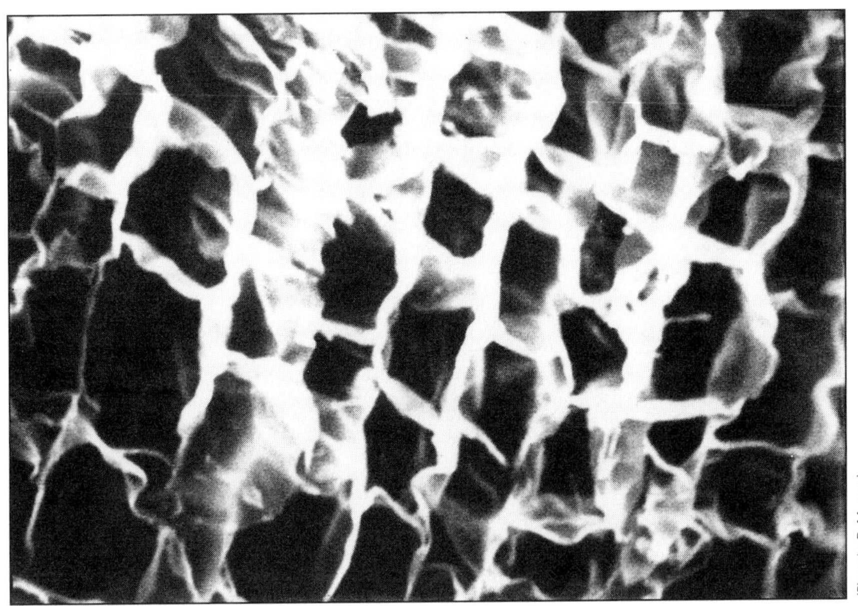

(Photo by D. Morton.)

(516×).

Food for Thought

10. Why were humans unaware of microorganisms for most of their history?

EXERCISE 4

Macromolecules and You: Food and Diet Analysis

OBJECTIVES

After completing this exercise, you will be able to

1. define *macromolecule, vitamin, mineral, carbohydrate, monosaccharide, disaccharide, polysaccharide, lipid, protein, amino acid, calorie;*
2. describe the basic structures of carbohydrates, lipids, glycerides, and proteins;
3. recognize positive and negative tests for carbohydrates, lipids, and proteins;
4. identify the roles that carbohydrates, lipids, proteins, minerals, and vitamins play in the body's construction and metabolism;
5. test food substances to determine the presence of biologically important macromolecules;
6. identify common dietary sources of nutrients;
7. identify the components and relative proportions of a healthy diet with respect to nutrients and calories.

INTRODUCTION

Humans are omnivores, animals who consume a wide variety of food items from several trophic levels. As with all consumers, our food items provide us with the energy stored in the chemical bonds of the food molecules, plus the raw carbon-based materials from which cellular components are built.

Food nutrients include minerals and vitamins, plus the larger molecules known as carbohydrates (which include sugars, starches, and cellulose), lipids (fats and oils), and proteins. As it happens, these last nutrients are three of the four major groups of biological **macromolecules,** large organic molecules of which all cells are made. (The fourth group is the nucleic acids that store and control the genetic instructions within a cell. These crucial macromolecules are not usually utilized by the body as nutrients, and so will not be considered in this exercise.)

Carbohydrates, lipids, and proteins supply the materials from which our cells and tissues are constructed and the energy (measured in kilocalories) to run our metabolic processes. **Vitamins** are necessary organic molecules that our bodies do not construct from other molecules; we require vitamins in our diets in only small amounts. **Minerals** are required inorganic (noncarbon-containing) nutrients such as calcium and potassium. Good health, then, depends upon ingesting the proper balance and quantities of these major nutrients.

In this exercise, you will test for the presence of carbohydrates, lipids, and proteins, and then use your knowledge to determine the macromolecular composition of various common food products. You will also analyze your own diet and compare it to a diet recommended for maintaining good health.

4.1 Identification of Macromolecules

You will use some simple tests for carbohydrates, lipids, and proteins in a variety of substances, including food products. Most of the reagents used are not harmful; however, observe all precautions listed and perform the experiments only in the proper location as identified by your instructor.

A. Carbohydrates

A **carbohydrate** is a simple sugar or a larger molecule composed of multiple sugar units. Carbohydrates are composed of carbon, hydrogen, and oxygen. Those composed of a single sugar molecule are called **monosaccharides.** Examples include ribose and deoxyribose (components of our genetic material), fructose (sometimes called fruit sugar), and glucose (a sugar that commonly serves as the most immediate source of cellular energy needs; see Figure 4-1). Monosaccharides are easily used within cells as energy sources.

Monosaccharides can be bonded together. If two monosaccharides are joined, they form a **disaccharide.** Examples of disaccharides are sucrose (common table sugar), maltose (found in many seeds), and lactose (milk sugar). If more than two monosaccharides are bonded together, the resulting large carbohydrate molecule is called a **polysaccharide.**

Animals, including humans, store glucose in the form of glycogen (Figure 4-2), a highly branched polysaccharide chain of glucose molecules. Starch (Figure 4-2), a key glucose-storing polysaccharide in plants, is an important food molecule for humans. Plant cells are surrounded by a tough cell wall made of another chain formed of glucose molecules, cellulose (Figure 4-2). Our digestive system does not break cellulose apart very well. Fibrous plant materials thus provide bulky cellulose fiber, which is necessary for good intestinal health.

Carbohydrate digestion occurs in humans as digestive enzymes produced in the salivary glands, pancreas, and lining of the small intestine break polysaccharides and disaccharides into monosaccharides such as glucose. The simple sugars are small enough to be absorbed into intestinal cells and then carried throughout the bloodstream.

Many small carbohydrate molecules react upon heating with a copper-containing compound called Benedict's reagent, changing the reagent color from blue to orange or red. A disaccharide may or may not react with Benedict's solution, depending upon how the bonding of the component monosaccharides took place.

Other tests are available for some polysaccharides, the most common of which is Lugol's test for starch. In this test, a dilute solution of potassium iodide reacts with the starch molecule to form a deep blue (nearly black) color product.

(After Starr, 2000.)

Figure 4-1 Structural diagrams of glucose.

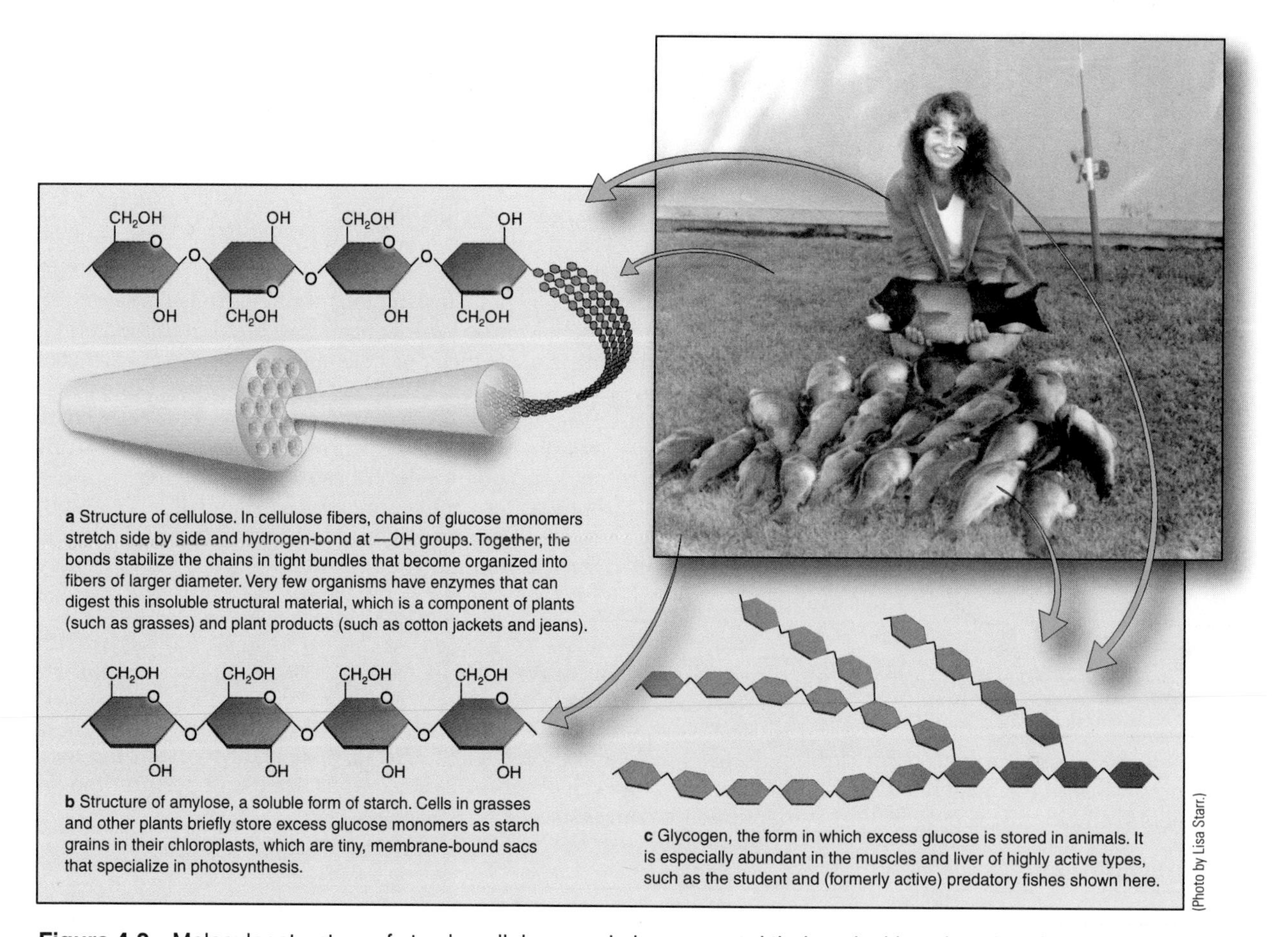

a Structure of cellulose. In cellulose fibers, chains of glucose monomers stretch side by side and hydrogen-bond at —OH groups. Together, the bonds stabilize the chains in tight bundles that become organized into fibers of larger diameter. Very few organisms have enzymes that can digest this insoluble structural material, which is a component of plants (such as grasses) and plant products (such as cotton jackets and jeans).

b Structure of amylose, a soluble form of starch. Cells in grasses and other plants briefly store excess glucose monomers as starch grains in their chloroplasts, which are tiny, membrane-bound sacs that specialize in photosynthesis.

c Glycogen, the form in which excess glucose is stored in animals. It is especially abundant in the muscles and liver of highly active types, such as the student and (formerly active) predatory fishes shown here.

(Photo by Lisa Starr.)

Figure 4-2 Molecular structure of starch, cellulose, and glycogen, and their typical locations in a few organisms.

A.1. Test for Sugars Using Benedict's Solution (*About 20 min.*)

MATERIALS

Per student group (4):

- china marker
- 11 test tubes in test tube rack
- test tube clamp
- hot plate *or* ring stand with wire gauze support and Bunsen burner
- 250- or 400-mL beaker with boiling beads or stones
- stock bottles* of
  - Benedict's solution
  - distilled water (dH_2O)
  - glucose solution
  - maltose solution
  - lactose solution
  - sucrose solution
  - fructose solution
  - starch solution
  - lemon juice
  - unsweetened orange juice
  - colorless nondiet soda
  - colorless diet soda
- vortex mixer (optional)

PROCEDURE

1. Half-fill the beaker with tap water and apply heat with a hot plate or burner to bring the water to a gentle boil.
2. Using the china marker, number the test tubes 1–11.
3. Pipet 2 mL of the correct stock solutions into the test tubes as described in Table 4-1. (If you don't know how to use the Pi-pump, check with your instructor before proceeding.)
4. Add 2 mL of Benedict's solution to each test tube, agitate the mixture by shaking the tubes from side to side or with a vortex mixer, if available, and record the color of the mixture in Table 4-1 in the column "Initial Color."
5. Heat the tubes in the boiling water bath for 3 minutes. Remove the tubes with the test tube clamp, and record any color changes that have taken place in the column "Color After Heating." Also record your conclusions regarding the presence or absence of simple carbohydrates.

TABLE 4-1 Benedict's Test for Sugars

Tube	Contents	Initial Color	Color After Heating	Conclusion
1	Distilled water			
2	Glucose			
3	Maltose			
4	Lactose			
5	Sucrose			
6	Starch			
7	Fructose			
8	Lemon juice			
9	Orange juice			
10	Nondiet soda			
11	Diet soda			

*Each stock bottle should have its own 2- or 5-mL pipet fitted with a Pi-pump.

A.2. Test for Starch Using Lugol's Solution *(About 15 min.)*

MATERIALS

Per student group (4):

- china marker
- 7 test tubes in test tube rack
- dropper bottle containing Lugol's solution
- stock bottles* of
  - distilled water (dH_20)
  - glucose solution
  - maltose solution
  - lactose solution
  - sucrose solution
  - starch solution
  - cream
- vortex mixer (optional)

PROCEDURE

1. Using the china marker, number the test tubes 1–7.
2. Pipet 3 mL of the correct stock solutions into the test tubes as described in Table 4-2.
3. Record the color of each solution in the column "Initial Color" in Table 4-2.
4. Add 9 drops of Lugol's solution to each test tube, agitate the mixture by shaking the tubes from side to side or with a vortex mixer, if available, and record the color of the mixture in the column "Color After Adding Lugol's Solution." Also record your conclusions regarding the presence or absence of starch in each test solution.

TABLE 4-2 Lugol's Test for Starch

Tube	Contents	Initial Color	Color After Adding Lugol's Solution	Conclusion
1	Distilled water			
2	Glucose			
3	Maltose			
4	Lactose			
5	Sucrose			
6	Starch			
7	Cream			

5. Hypothesize about why you got the result you did with the Lugol's test for each substance:

   dH_2O ______

   glucose ______

   maltose ______

   lactose ______

   sucrose ______

   starch ______

   Does cream contain starch? ______

B. Lipids

Lipids are oily or greasy compounds insoluble in water, but dissolvable in nonpolar solvents such as ether or chloroform. Those lipids having fatty acids—hydrocarbon chains with a carboxyl (—COOH) group at one end—combine with glycerol to form glycerides, a rich source of stored energy. Lipids with three fatty acid chains attached to a glycerol backbone are called triglycerides (Figure 4-3).

*Each stock bottle should have its own 5-mL pipet fitted with a Pi-pump.

Lipids provide long-term energy storage in cells and are very diverse. Lipid digestion occurs primarily in the small intestine where bile produced by the liver breaks lipid globules into smaller droplets, and then pancreatic enzymes break large lipid molecules into smaller components for absorption. The lipid components are then transported throughout the body in lymph, the fluid that bathes the tissues.

Substances we think of as fats and oils are examples of lipids. One of the most simple tests for lipids is to determine whether they leave a grease spot on a piece of uncoated paper, such as a grocery sack. A test commonly used to identify fats and oils in microscopic preparations—Sudan IV—can also be used to indicate their presence in a test tube.

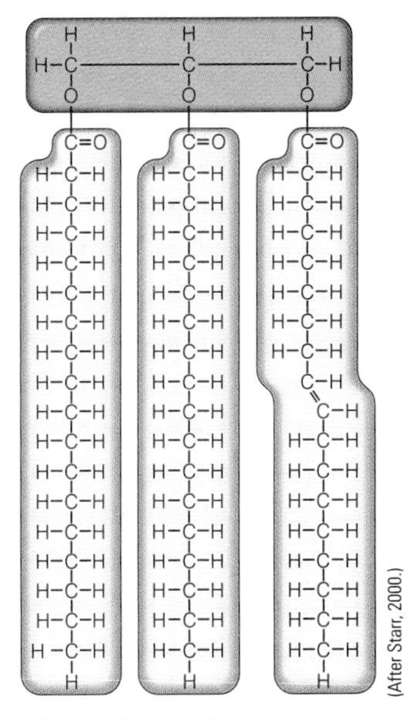

Figure 4-3 A triglyceride.

B.1. Uncoated Paper Test *(About 20 min.)*

MATERIALS

Per student group (4):

- china marker
- dropper bottles of
  - distilled water (dH_2O)
  - vegetable oil
- stock bottles* of
  - hamburger juice
  - onion juice
  - colorless nondiet soft drink
  - cream
- dropper bottle of Sudan IV
- vortex mixer (optional)
- piece of uncoated paper (grocery bag)

PROCEDURE

1. Place a drop of distilled water and a drop of vegetable oil on a piece of grocery bag. Set aside to dry for 10 minutes.
2. After 10 minutes, describe the appearance of each spot on the paper.

   distilled water ______________________________

   vegetable oil ______________________________

3. Place a drop of each test substance on a labeled spot on the grocery bag piece. Set aside to dry for 10 minutes.
4. After 10 minutes, describe the appearance of each spot on the paper.

   hamburger juice ______________________________

   onion juice ______________________________

   colorless nondiet soft drink ______________________________

   cream ______________________________

5. What can you infer about the lipid content of each substance?

   ____

   ____

   ____

B.2. Sudan IV Test *(About 20 min.)*

MATERIALS

Per student group (4):

- china marker
- dropper bottles of
  - distilled water (dH_2O)
  - vegetable oil
- 6 test tubes in test tube rack
- stock bottles* of
  - hamburger juice
  - onion juice
  - colorless nondiet soft drink
  - cream
- dropper bottle of Sudan IV
- vortex mixer (optional)

*Each stock bottle should have its own 5-mL pipet fitted with a Pi-pump.

PROCEDURE

1. With a china marker, number the test tubes 1–6.
2. Add 3 mL of dH_2O to each test tube and then 3 mL of each substance indicated in Table 4-3 to the appropriate test tube.
3. Add 9 drops of Sudan IV to each tube, agitate by shaking the tube side, to side and then add 2 mL of dH_2O to each tube.
4. Record your results in Table 4-3.

TABLE 4-3 Test for Lipids with Sudan IV

Tube	Contents	Observations After Addition of Sudan IV	Conclusion
1	Distilled water		
2	Vegetable oil		
3	Hamburger juice		
4	Onion juice		
5	Soft drink		
6	Cream		

Which test, the uncoated paper test or the Sudan IV test, do you think is most sensitive to small quantities of lipids?

____

What are the limitations of these tests?

____

____

C. Proteins

Proteins are a diverse group of macromolecules with a wide range of functions in the human body. Many are structural components of muscle, bone, hair, fingernails, and toenails, among other tissues. Others are enzymes that speed up cellular reactions that would otherwise take years to occur. Movement of structures within cells (such as occurs during cell division) and of sperm cells is also associated with proteins.

All proteins are chains of **amino acids,** whose general structure is illustrated in Figure 4-4.

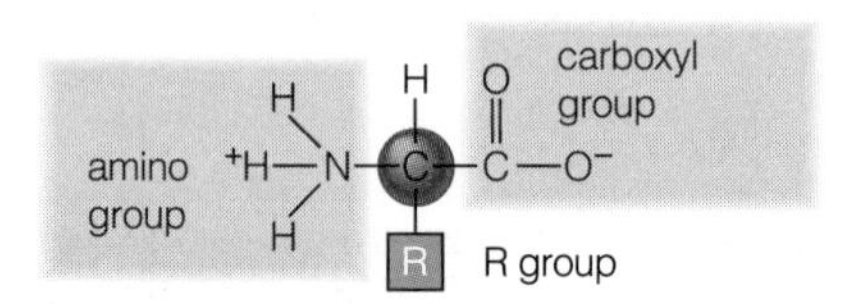

Figure 4-4 General structure of an amino acid. The R group consists of one or more atoms that make each kind of amino acid unique. (After Starr, 2000.)

Though there are only about 20 amino acids, innumerable kinds of proteins result from different amino acid sequences. A protein begins to form when two or more amino acids are linked together by peptide bonds (bonding between the amino group of one amino acid and the acid group of another), forming a polypeptide chain (Figure 4-5). Subsequently, the chain folds and twists, with links formed between adjacent parts of the chain. Humans cannot synthesize about half of the amino acids; these essential amino acids must be obtained through our foods.

Protein digestion begins in the stomach lining and continues in the small intestine, where various enzymes break protein molecules first into protein fragments, and then into amino acids, which are absorbed across the small intestine wall. The amino acids are then transported throughout the body in the blood.

Several tests are used for proteins, including Biuret reagent, which indicates the presence of peptide bonds. The greater the number of peptide bonds, the more intense the color reaction with Biuret.

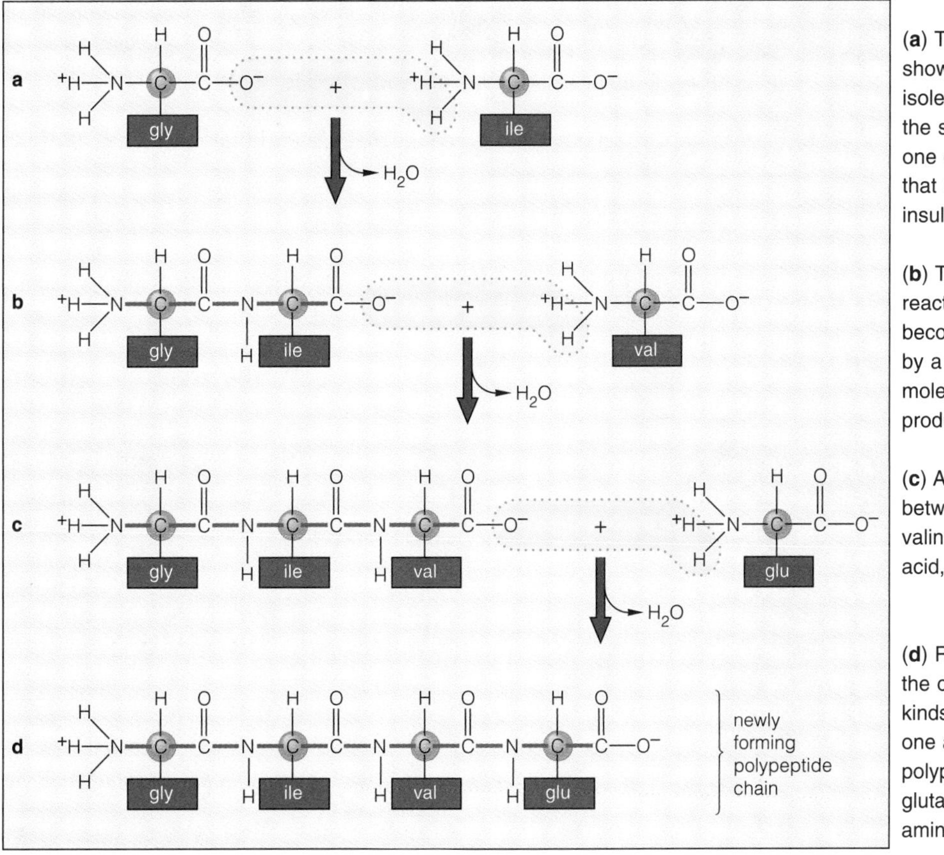

(**a**) The first two amino acids shown are glycine (gly) and isoleucine (ile). They are at the start of the sequence for one of two polypeptide chains that make up the protein insulin in cattle.

(**b**) Through a condensation reaction, the isoleucine becomes joined to the glycine by a peptide bond. A water molecule forms as a by-product of the reaction.

(**c**) A peptide bond forms between the isoleucine and valine (val), another amino acid, and water again forms.

(**d**) Remember, DNA specifies the order in which the different kinds of amino acids follow one another in a growing polypeptide chain. In this case, glutamate (glu) is the fourth amino acid specified.

(After Starr, 2000.)

Figure 4-5 Peptide bond formation during protein synthesis.

C.1. Biuret Test for Proteins with Biuret Reagent (*About 15 min.*)

MATERIALS

Per student group (4):

- china marker
- 5 test tubes in test tube rack
- stock bottles* of
  - Biuret reagent
  - distilled water (dH_2O)
  - starch solution
  - egg albumin
  - glucose solution
  - colorless soft drink
- vortex mixer (optional)

PROCEDURE

1. Using the china marker, number the test tubes 1 through 5.
2. Pipet 1 mL of the correct stock solutions into the test tubes as described in Table 4-4.
3. Add 10 drops of Biuret reagent to each test tube and agitate the mixture by shaking the tubes from side to side or with a vortex mixer, if available.
4. Wait 2 minutes and then record the color of the mixture in Table 4-4, in the column "Color Reaction." Also enter your conclusion about the presence of protein in the substance.

D. Testing Unknown Food Substances (*About 30 min.*)

Your instructor will provide you with several samples of unidentified food substances.

PROCEDURE

1. Following the previous procedures, perform tests for carbohydrates (Benedict's test and Lugol's solution test), lipids (uncoated paper and Sudan IV tests), and proteins (Biuret reagent) on each item.
2. Record your test results (+ or – for each test) in Table 4-5.

*Each stock bottle should have its own 2- or 5-mL pipet fitted with a Pi-pump.

TABLE 4-4 Test for Protein with Biuret Reagent

Tube	Contents	Color Reaction	Conclusion
1	Distilled water		
2	Starch		
3	Egg albumin		
4	Glucose		
5	Soft drink		

TABLE 4-5 Tests for Unknown Food Substances

Unknown Sample Designation	Tests for Carbohydrates		Tests for Lipids		Test for Proteins
	Benedict's Test	Lugol's Test	Uncoated Paper Test	Sudan IV	Biuret Test

Note: **After completing all experiments, take your dirty glassware to the sink and wash it following the directions given in "*Instructions for Washing Laboratory Glassware,*" page x. Invert the test tubes in the test tube rack so they drain. Tidy up your work area, making certain all equipment used in this exercise is there for the next class.**

4.2 The Food Pyramid and Diet Analysis

The foods we eat are usually complex mixtures of various macromolecules, vitamins, and minerals that supply energy and raw materials to the body. However, food must be digested before most nutrients are available within cells. Within the digestive system, macromolecules are hydrolyzed to smaller components (proteins broken down to amino acids, for example). These smaller molecules are absorbed across the intestinal wall, transported throughout the body, and used to build and power cells. We truly *are* what we eat.

The U.S. Department of Agriculture issued its "MyPyramid" food guide in 2005 to provide daily dietary guidelines for average people of good health. Several nutrition recommendations underlie MyPyramid: moderation in portion sizes (many American portions are 2 to 3 times larger than recommended); balance among food groups to ensure that all key nutrients are included; and emphasis on daily exercise for a multitude of health benefits.

The major food groups are the grains, vegetables, fruits, dairy and soy, meat and beans, and vegetable oils. There is also an allotment of "discretionary calories" for each person. Why is it important to include each of these in our diets? What do we gain from each type of food?

Grains are high in the complex carbohydrates (starches and cellulose) that provide ready energy sources. The USDA recommends that you "make half your grains whole," meaning that *at least* half your grain intake should be whole grains, which supply important vitamins and minerals, plus the fiber needed for a healthy digestive system. Whole grains include oatmeal, brown rice, and foods whose first ingredients include the words *whole grain* or *whole wheat.*

Vegetables and **fruits** are especially important sources of the vitamins and minerals needed for a healthy diet, as well as complex carbohydrates and **phytochemicals**, a catchall term for a diverse group of molecules whose benefits we are just beginning to understand. Diets that include plenty of vegetables and fruits help prevent a variety of illnesses, including heart disease, stroke, and certain types of cancer. To meet your requirements for these important foods, mix it up. Eat more vegetables and fruits of every color and variety, especially the vegetables that provide high volume and high nutritional value with low calorie impact (unless they're deep-fried).

Low-fat **dairy** and **soy** products are important sources of calcium and provide significant amounts of proteins. Protein-rich, vitamin-D-fortified dairy and soy foods, along with weight-bearing exercise, build and maintain strong bones.

Smaller quantities are recommended of the "meat and bean" group, which includes poultry, fish, and nuts. These protein-rich foods supply essential amino acids, vitamins, and minerals, and also help maintain the "full" feeling between meals. Choose beans, fish, and lean meats and poultry, along with lower quantities of calorie-rich nuts, rather than fatty or deep-fried choices with unhealthy lipid levels.

Vegetable oils have been promoted to their own food group in MyPyramid, and include monounsaturated vegetable oils (olive oil and canola oil, for example) and polyunsaturated oils (corn oil, soybean oil). These oils are considered healthy in moderation, unlike the solid saturated fats (sour cream, butter, palm oil) and especially the harmful trans fats (partially hydrogenated vegetable oils) that are found in many processed and restaurant foods. Aim for zero trans fats in your diet.

Discretionary calories can be used to eat more foods from any food group or to eat higher-calorie forms of a food or beverage with added sugars or fats. Such foods are high in calorie-dense lipids and the simple sugars that are the "lighter fluid" (quickly used energy source) of metabolism. While some lipids are essential nutrients, other food groups on the pyramid provide those lipids in abundance. Foods that provide abundant fats and simple sugars usually contain few other nutrients and so are not recommended dietary components.

The dietary guidelines of MyPyramid vary depending upon your gender and activity level, with more calories of food energy needed to power your cells as your activity level increases. Use Table 4-6 to determine your recommended average daily calorie allowance.

TABLE 4-6 Approximate Daily Calorie Requirements

FEMALES	Sedentary (less than 30 min. moderate activity daily)	Moderate (30 to 60 min. moderate activity daily)	Active (more than 60 min. moderate activity daily)
Ages 16 to 18	1800	2000	2400
Ages 19 to 30	2000	2200	2400
Ages 31 to 50	1800	2000	2200
MALES			
Ages 16 to 18	2400	2800	3200
Ages 19 to 30	2400	2800	3000
Ages 31 to 50	2200	2600	3000

Use Table 4-7 to find out recommendations for consumption of each food group for your calorie level.

Follow the MyPyramid guidelines and eat a variety of foods to enjoy a healthy diet that provides balanced nutrients without excess calories.

A. Food Diary

In this activity you will analyze *your* diet, so that you can see in which areas you maintain good nutrition, and which areas need improvement.

PROCEDURE (to be completed before class, about 15 minutes per day)
For 2 days during the week preceding this activity, keep a complete diary of *every* food and drink item you consume other than water, unsweetened coffee or tea, or other zero-calorie beverages. In Table 4-8, record what you eat and drink, and how much (the portion size: ounces, cups, teaspoons, and so on). Enter this data in the first two columns. Be as honest, specific, and descriptive as you can. Also keep track separately of which items of processed foods contain trans fats (this information will be found on the food label). Do this for *each* meal and snack.

TABLE 4-7 Recommended Daily Food Group Intakes								
Calorie level	1800	2000	2200	2400	2600	2800	3000	3200
Grains	6 oz	6 oz	7 oz	8 oz	9 oz	10 oz	10 oz	10 oz
Vegetables	2 cups (c)	2½ c	3 c	3 c	3½ c	3½ c	4 c	4 c
Fruits	1½ c	2 c	2 c	2 c	2 c	2½ c	2½ c	2½ c
Dairy & Soy	3 c	3 c	3 c	3 c	3 c	3 c	3 c	3 c
Meat & Beans	5 oz	5½ oz	6 oz	6½ oz	6½ oz	7 oz	7 oz	7 oz
Vegetable Oils	5 tsp	6 tsp	6 tsp	7 tsp	8 tsp	8 tsp	10 tsp	11 tsp
Discretionary calories*	195	267	290	362	410	426	512	650

*Discretionary calories are those "extra" calories, usually from fats and sugars, that remain when all the other food group portions and nutrients are consumed.

Eat your usual diet. Don't change your eating habits for this exercise.

It's often difficult to determine portion sizes. To do this accurately, record the information on food product labels regarding serving size and calorie content of each item whenever possible. You also may want to use a measuring cup for this activity. For nonlabeled foods, use the following guidelines:

Portion	Approximate Size of Item
1 ounce	one slice processed cheese; two dice
3 ounces meat	one deck of playing cards or computer mouse
½ cup	one tennis ball or one ice cream scoop
1 serving fruit	one baseball (*not* the larger softball)
1 cup	volume of a woman's fist
1 teaspoon	top half of thumb
1 ounce grains	½ cup oatmeal or rice or 1 cup cereal flakes or ½ English muffin or 1 slice commercial bread

TABLE 4-8 Food Diary

Food or Beverage Item	Total Portion Size	Food Group Servings in Each Food Item						
		Grains (ounces)	Vegetables (cups)	Fruits (cups)	Dairy & Soy (cups)	Meat & Beans (ounces)	Oils (teaspoons)	Calories
2-day total:								
2-day average:								

B. In-Class Food Diary Analysis *(About 1.5 hours)*

MATERIALS

Per student group (4):

- food diary data
- diet analysis books and/or computers with diet analysis software
- colored paper squares
- glue or tape

PROCEDURE

1. Complete the food diary table (Table 4-8) in class if necessary, listing everything you've ingested, including both the food group and the number of servings. Then determine how each food item is allocated as servings of the various MyPyramid food groups: grains, vegetables, fruits, dairy and soy, meat and bean, and vegetable oils. You can find portion information for common foods in online sources and/or the books provided in your classroom.
2. Determine and record the calorie content of each item using the resources provided in class.
3. Average the food group servings and calorie content over the 2 days of data collection, and use the averaged data for the rest of this activity.
4. Construct a food pyramid to visualize the MyPyramid recommendations for your gender and activity level. Use the colored paper blocks, with each block corresponding to one serving unit (ounce, cup, or teaspoon, depending on the food group involved). Glue or tape the pyramid to the bottom of page 67 in the following order, from bottom up: orange = grain group servings (ounces); blue = vegetable group servings (cups); green = fruit group servings (cups); yellow = dairy and soy group servings (cups); brown = meat and bean group servings (ounces); red = vegetable oil servings (teaspoons); and purple = discretionary calories (100 calories per block).
5. Now construct your **personal food pyramid** in the same way on the next page, using 2-day average data from your food diary.
6. Answer the following questions AFTER you have completed your 2-day food diary and pyramids:
   - **(a)** What is your typical level of physical activity (sedentary, moderate, active)? ______________
   - **(b)** What is your approximate daily calorie requirement? ______________
   - **(c)** How does your caloric intake compare with the recommendations for your gender and activity level?

     ____

     ____

     ____

     ____
   - **(d)** What food groups are you eating too much of? What specific kinds of nutrient molecules are you thus overconsuming? What are the potential consequences of this excess for your health?

     ____

     ____

     ____

     ____
   - **(e)** What food groups are you not eating enough of? What kinds of specific nutrient molecules are thus underrepresented in your diet? What are the potential consequences of this deficiency for your health?

     ____

     ____

     ____

     ____
   - **(f)** What proportion of your grain servings included whole grains? ______________
     How many different *kinds* of vegetables and fruits did you consume? ______________
     How many food items contained saturated or trans fats? ______________
   - **(g)** Given all the above information, describe two *reasonable* changes you could make to improve your diet and your health.

     ____

     ____

     ____

     ____

Personal Food Pyramid

My Recommended Food Pyramid

PRE-LAB QUESTIONS

_____ **1.** A carbohydrate consists of
(a) amino acid units
(b) one or more sugar units
(c) lipid droplets
(d) glycerol

_____ **2.** A protein is made up of
(a) amino acid units
(b) one or more sugar units
(c) lipid droplets
(d) Biuret solution

_____ **3.** Benedict's solution is commonly used to test for
(a) proteins
(b) certain carbohydrates
(c) nucleic acids
(d) lipids

_____ **4.** Glycogen is
(a) a polysaccharide
(b) a storage carbohydrate
(c) found in human tissues
(d) all of the above

_____ **5.** To test for starch, one would use
(a) Benedict's solution
(b) uncoated paper
(c) Sudan IV
(d) Lugol's solution

_____ **6.** Rich sources of stored energy that are dissolvable in organic solvents are
(a) carbohydrates
(b) proteins
(c) glucose
(d) lipids

_____ **7.** Rubbing a substance on uncoated paper should reveal if it is a
(a) lipid
(b) carbohydrate
(c) protein
(d) sugar

_____ **8.** Proteins consist of
(a) monosaccharides linked in chains
(b) amino acid units
(c) polysaccharide units
(d) condensed fatty acids

_____ **9.** Biuret reagent will indicate the presence of
(a) peptide bonds
(b) proteins
(c) amino acids units linked together
(d) all of the above

_____ **10.** The largest number of food servings in your daily diet should be from
(a) meats and beans
(b) dairy products
(c) vegetables and fruits
(d) vegetable oils

Name ______________________ Section Number ______________________

EXERCISE 4

Macromolecules and You: Food and Diet Analysis

POST-LAB QUESTIONS

A.1. Test for Sugars Using Benedict's Solution

1. Let's suppose you are teaching science in a part of the world without ready access to a doctor and you're worried that you may have developed diabetes. (Diabetics are unable to regulate blood glucose levels, and glucose accumulates in blood and urine.) What could you do to gain an indication of whether or not you have diabetes?

2. The test tubes in the photograph contain Benedict's solution and two unknown substances that have been heated. What do the results indicate?

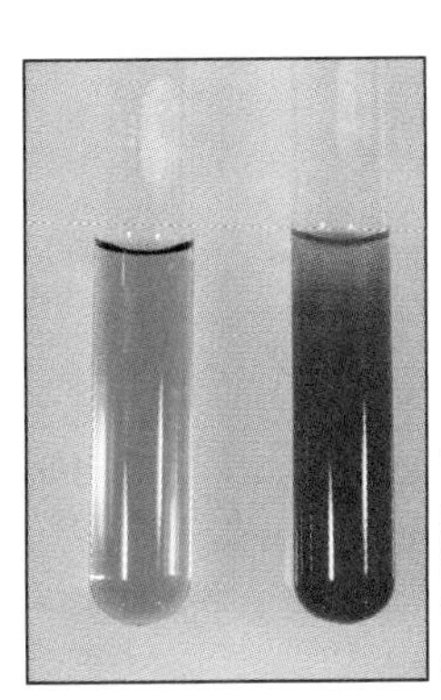

(Photo by J. W. Perry.)

3. How could you verify that a soft-drink can contains diet soda rather than soda sweetened with fructose?

A.2. Test for Starch Using Lugol's Solution

4. Observe the photomicrograph accompanying this question. This section of a potato tuber has been stained with Lugol's iodine solution. When you eat french fries, the potato material is broken down in your small intestine into what small subunits?

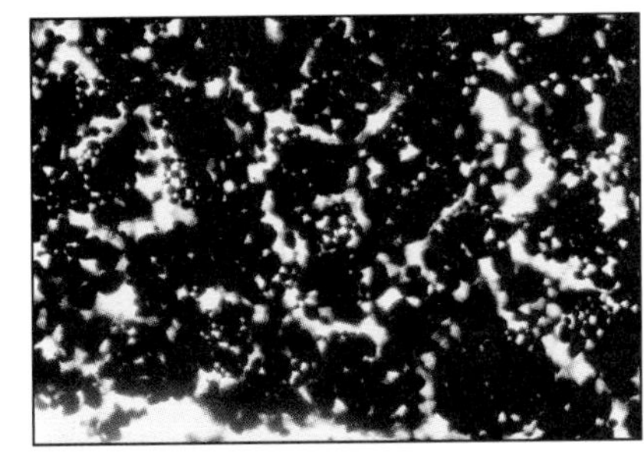

(Photo by J. W. Perry.)

B.2. Sudan IV Test

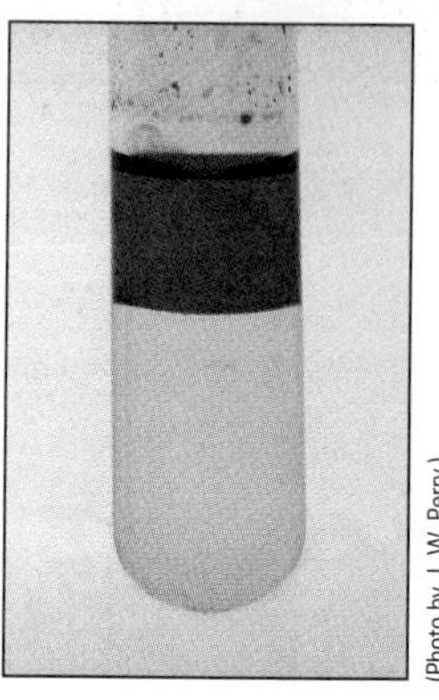
(Photo by J. W. Perry.)

5. The test tube in the photograph contains water at the bottom and another substance that has been stained with Sudan IV at the top. What is the macromolecular composition of this stained substance?

6. You are given a sample of an unknown food. Describe how you would test it for the presence of lipids.

7. You wish to test the same unknown food for the presence of sugars. Describe how you would do so.

Food for Thought

8. What is the purpose of the distilled water sample in each of the chemical tests in this exercise?

9. Many health food stores carry enzyme preparations that are intended to be ingested orally (by mouth) to supplement existing enzymes in various organs like the liver, heart, and muscle. Explain why these preparations are unlikely to be effective as advertised.

10. A young child grows rapidly, with high levels of cell division and high energy requirements. If you were planning the child's diet, which food groups would you emphasize, and why? Which food groups would you deemphasize, and why?

EXERCISE 5

Structure and Function of Living Cells

OBJECTIVES

After completing this exercise, you will be able to

1. define *cell, cell theory, prokaryotic, eukaryotic, nucleus, cytomembrane system, organelle, multinucleate, cytoplasmic streaming, sol, gel, envelope;*
2. list the structural features shared by all cells;
3. describe the similarities and differences between prokaryotic and eukaryotic cells;
4. identify the cell parts described in this exercise;
5. state the function for each cell part;
6. distinguish between plant and animal cells;
7. recognize the structures presented in **boldface** in the procedure sections.

INTRODUCTION

Structurally and functionally, all life has one common feature: All living organisms are composed of **cells.** The development of this concept began with Robert Hooke's seventeenth-century observation that slices of cork were made up of small units. He called these units "cells" because their structure reminded him of the small cubicles that monks lived in. Over the next 100 years, the **cell theory** emerged. This theory has three principles: (1) All organisms are composed of one or more cells; (2) the cell is the basic *living* unit of organization; and (3) all cells arise from preexisting cells.

Although cells vary in organization, size, and function, all share three structural features: (1) All possess a **plasma membrane** defining the boundary of the living material; (2) all contain a region of **DNA** (deoxyribonucleic acid), which stores genetic information; and (3) all contain **cytoplasm,** everything inside the plasma membrane that is not part of the DNA region.

With respect to internal organization, there are two basic types of cells, **prokaryotic** and **eukaryotic.** Study Table 5-1, comparing the more important differences between prokaryotic and eukaryotic cells. The Greek word *karyon* means "kernel," referring to the nucleus. Thus, *prokaryotic* means "before a nucleus," while *eukaryotic* indicates the presence of a "true nucleus." Prokaryotic cells typical of bacteria, cyanobacteria, and archaea are believed to be similar to the first cells, which arose on Earth 3.5 billion years ago. Eukaryotic cells, such as those that comprise the bodies of protists, fungi, plants, and animals, probably evolved from prokaryotes.

This exercise will familiarize you with the basics of cell structure and the function of prokaryotes (prokaryotic cells) and eukaryotes (eukaryotic cells).

5.1 Prokaryotic Cells *(About 20 min.)*

MATERIALS

Per student:

- dissecting needle
- compound microscope
- microscope slide
- coverslip

Per student pair:

- distilled water (dH_2O) in dropping bottle

Per student group (table):

- culture of a cyanobacterium (either *Anabaena* or *Oscillatoria*)

Per lab room:

- 3 bacterium-containing nutrient agar plates (demonstration)
- 3 demonstration slides of bacteria (coccus, bacillus, spirillum)

Jessica Williams

TABLE 5-1 Comparison of Prokaryotic and Eukaryotic Cells

	Cell Type	
Characteristic	**Prokaryotic**	**Eukaryotic**
Genetic material	Located within cytoplasm, not bounded by a special membrane Consists of a single molecule of DNA	Located in **nucleus,** a double membrane-bounded compartment within the cytoplasm Numerous molecules of DNA combined with protein Organized into chromosomes
Cytoplasmic structures	Small ribosomes Photosynthetic membranes arising from the plasma membrane (in some representatives only)	Large ribosomes **Cytomembrane system,** a system of connected membrane structures **Organelles,** membrane-bounded compartments specialized to perform specific functions
Kingdoms represented	Bacteria Archaea	Protista Fungi Plantae Animalia

PROCEDURE

1. Observe the culture plate with bacteria growing on the surface of a nutrient medium. Can you see the individual cells with your naked eye?

   no

2. Observe the microscopic preparations of bacteria on *demonstration* next to the culture plate. The three slides represent the three different shapes of bacteria. Which objective lenses are being used to view the bacteria?

   100x

Can you discern any detail within the cytoplasm?

no

In the space provided in Figure 5-1, sketch what you see through the microscope. Record the magnification you are using in the blank provided in the figure caption. Then record the approximate size of the bacterial cells. (Return to page 34 of Exercise 3 if you've forgotten how to estimate the size of an object being viewed through a microscope.)

3. Study Figure 5-2, a three-dimensional representation of a bacterial cell. Now examine the electron micrograph of the bacterium *Escherichia coli* (Figure 5-3). Locate the **cell wall,** a structure chemically distinct from the wall of plant cells but serving the same primary function to contain and protect the cell's contents.
4. Find the **plasma membrane,** which is lying flat against the internal surface of the cell wall and is difficult to distinguish.
5. Look for two components of the **cytoplasm: ribosomes,** electron-dense particles (they appear black) that give the cytoplasm its granular appearance, and a relatively electron-transparent region (appears light) containing fine threads of DNA called the **nucleoid.**

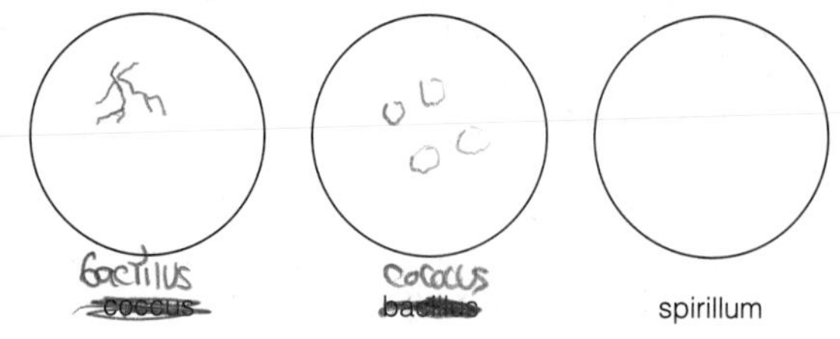

Figure 5-1 Drawing of several bacterial cells (100 ×). Approximate size = ______ μm.

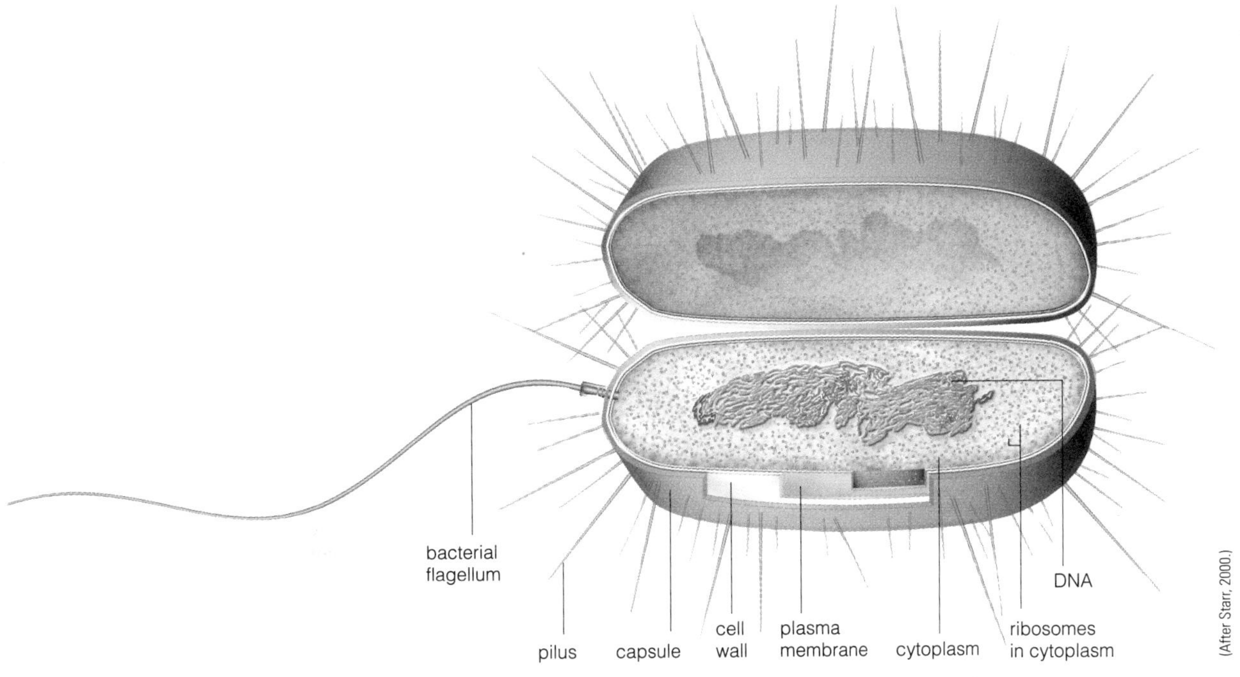

Figure 5-2 Three-dimensional representation of a bacterial cell as seen with the electron microscope.

Another type of prokaryotic cell is exemplified by cyanobacteria, such as *Oscillatoria* and *Anabaena.* Cyanobacteria (sometimes called blue-green algae) are commonly found in water and damp soils. They obtain their nutrition by converting the sun's energy through photosynthesis.

cell wall
cytoplasm with ribosomes
DNA region
plasma membrane
(Photo courtesy of G. Cohen-Bazire.)

Figure 5-3 Electron micrograph of the bacterium *Escherichia coli* (28,300×).

6. With a dissecting needle, remove a few filaments from the cyanobacterial culture, placing them in a drop of water on a clean microscope slide.
7. Place a coverslip over the material and examine it first with the low-power objective and then using the high-dry objective (or oil-immersion objective, if your microscope is so equipped).
8. In the space provided in Figure 5-4, sketch the cells you see at high power. Estimate the size of a *single* cyanobacterial cell and record the magnification you used to make your drawing.
9. Now examine the electron micrograph of *Anabaena* (Figure 5-5), which identifies the **cell wall, cytoplasm,** and **ribosomes.** The cyanobacteria also possess membranes that function in photosynthesis. Identify these **photosynthetic membranes,** which look like tiny threads within the cytoplasm.
10. Look at the captions for Figures 5-3 and 5-5. Judging by the magnification of each electron micrograph, which cell is larger, the bacterium *E. coli* or the cyanobacterium *Anabaena*?

____

____

____

Figure 5-4 Drawing of several prokaryotic cells of a cyanobacterium (__________×). Approximate size = _______ μm.

Because the electron micrograph of *Anabaena* is of relatively low magnification, the plasma membrane is not obvious, but if you could see it, it would be found just under the cell wall.

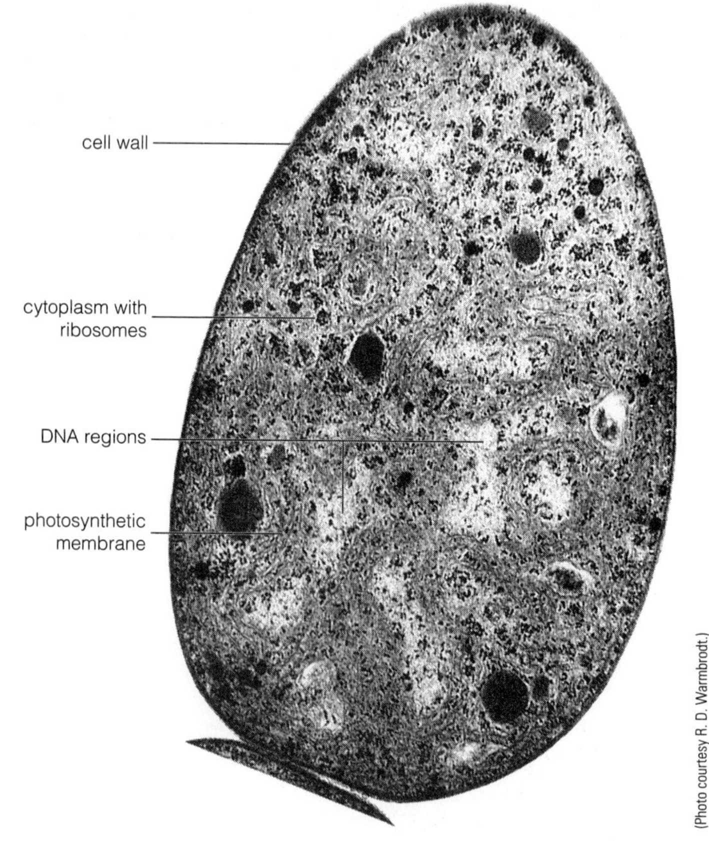

Figure 5-5 Electron micrograph of *Anabaena* (11,600×).

5.2 Eukaryotic Cells
(About 75 min.)

MATERIALS

Per student:

- textbook
- toothpick
- microscope slide
- coverslip
- culture of *Physarum polycephalum*
- compound microscope
- forceps
- dissecting needle

Per student pair:

- methylene blue in dropping bottle
- distilled water (dH_2O) in dropping bottle

Per student group (table):

- *Elodea* in water-containing culture dish
- onion bulb
- tissue paper

Per lab room:

- model of animal cell
- model of plant cell

A. Protist Cells as Observed with the Light Microscope

The slime mold *Physarum polycephalum* is in the Kingdom Protista. *Physarum* is a unicellular organism, so it contains all the metabolic machinery for independent existence.

PROCEDURE

1. Place a plain microscope slide on the stage of your compound microscope. This will serve as a platform on which you can place a culture dish.
2. Now obtain a petri dish culture of *Physarum,* remove the lid, and place it on the platform. Observe initially with the low-power objective and then with the medium-power objective. Place a coverslip over part of the organism before rotating the high-dry objective into place. (This prevents the agar from getting on the lens.)

Physarum is **multinucleate,** meaning that more than one nucleus occurs within the cytoplasm. Unfortunately, the nuclei are tiny; you won't be able to distinguish them from other granules in the cytoplasm.

3. Locate the **plasma membrane,** which is the outer boundary of the cytoplasm. Once again, the resolving power of your microscope is not sufficient to allow you to actually view the membrane.
4. Watch the cytoplasm of the organism move. This intracellular motion is known as **cytoplasmic streaming.** Although not visible with the light microscope without using special techniques, contractile proteins called **microfilaments** are believed responsible for cytoplasmic streaming.
5. Note that the outer portion of the cytoplasm appears solid; this is the **gel** state of the cytoplasm. Notice that the granules closer to the interior are in motion within a fluid; this portion of the cytoplasm is in the **sol** state. Movement of the organism occurs as the sol-state cytoplasm at the advancing tip pushes against the plasma membrane, causing the region to swell outward. The sol-state cytoplasm flows into the region, converting to the gel state along the margins.
6. In Figure 5-6, sketch the portion of *Physarum* that you have been observing and label it.

B. Experiment: Cytoplasmic Streaming *(About 25 min.)*

As you might predict, temperature affects many cellular processes. You may have observed that snakes and insects, being ectotherms (animals that gain heat from the environment and unlike humans, not primarily from metabolic activities), in nature are relatively sluggish during cold weather. Is the same true for other organisms, like the slime mold?

This simple experiment addresses the hypothesis that *cold slows cytoplasmic streaming in Physarum polycephalum.* Before starting this experiment, you may wish to review the discussion in Exercise 1, "The Scientific Method."

Figure 5-6 Drawing of a portion of *Physarum* (________×).

MATERIALS

Per experimental group:

- culture of *Physarum polycephalum*
- compound microscope
- container with ice *or* refrigerator
- timer *or* watch with second hand
- Celsius thermometer

PROCEDURE

1. Place the *Physarum* culture on the stage of your compound microscope as described in Section A.
2. Time the duration of cytoplasmic streaming in one direction and then in the other direction. Do this for five cycles of back-and-forth motion. Calculate the average duration of flow in either direction. Record the temperature and your observations in Table 5-2.
3. Remove your culture from the microscope's stage, replace the cover, and place it and the thermometer in a refrigerator or atop ice for 15 minutes.
4. While you are waiting, in Table 5-2 write a prediction for the effect on the duration of cytoplasmic streaming by reducing the temperature of the culture of *Physarum polycephalum.*
5. After 15 minutes have elapsed, remove the culture from the cold treatment, record the temperature of the experimental treatment, and repeat the observations in step 2.
6. Record your observations and make a conclusion in Table 5-2, accepting or rejecting the hypothesis.

TABLE 5-2 Effect of Temperature on Cytoplasmic Streaming

Prediction:		
Temperature (°C)	**Time**	**Observations and Duration of Directional Flow (sec.)**

A logical question to ask at this time is *why* temperature has the effect you observed. If you perform Exercise 8, "Enzymes: Catalysts of Life," you may be able to make an educated guess (another hypothesis).

C. Animal Cells Observed with the Light Microscope

PROCEDURE

1. *Human cheek cells.* Using the broad end of a clean toothpick, gently scrape the inside of your cheek. Stir the scrapings into a drop of distilled water on a clean microscope slide and add a coverslip. Dispose of used toothpicks in the jar containing alcohol.
2. Because the cells are almost transparent, decrease the amount of light entering the objective lens to increase the contrast. (See Exercise 3, page 33.) Find the cells using the low-power objective of your microscope; then switch to the high-dry objective for detailed study.

3. Find the **nucleus,** a centrally located spherical body within the **cytoplasm** of each cell.
4. Now stain your cheek cells with a dilute solution of methylene blue, a dye that stains the nucleus darker than the surrounding cytoplasm. To stain your slide, follow the directions illustrated in Figure 5-7.

Without removing the coverslip, add a drop of the stain to one edge of the coverslip. Then draw the stain under the coverslip by touching a piece of tissue paper to the *opposite* side of the coverslip.

5. In Figure 5-8, sketch the cheek cells, labeling the **cytoplasm, nucleus,** and the location of the **plasma membrane.** (A light microscope cannot resolve the plasma membrane, but the boundary between the cytoplasm and the external medium indicates its location.) Many of the cells will be folded or wrinkled due to their thin, flexible nature. Estimate and record in your sketch the size of the cells. (The method for estimating the size is found in Part I page 34.)

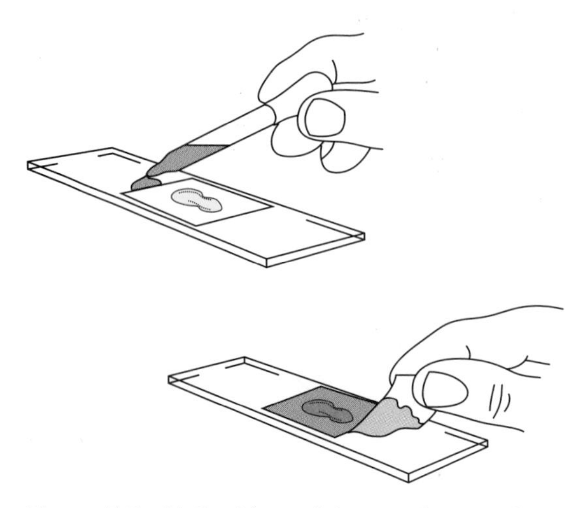

Figure 5-7 Method for staining specimen under coverslips of microscope slide.

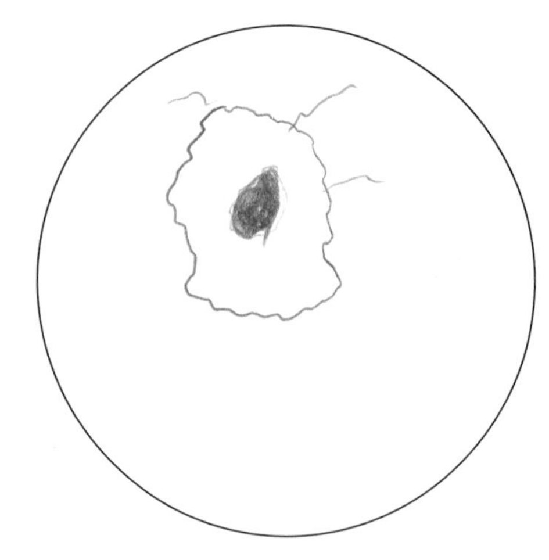

Figure 5-8 Drawing of human cheek cells (_100_ ×). Approximate size = ________ µm.
Labels: cytoplasm, nucleus, plasma membrane

D. Animal Cells as Observed with the Electron Microscope

Studies with the electron microscope have yielded a wealth of information on the structure of eukaryotic cells. Structures too small to be seen with the light microscope have been identified. These include many **organelles,** structures in the cytoplasm that have been separated ("compartmentalized") by enclosure in membranes. Examples of organelles are the nucleus, mitochondria, endoplasmic reticulum, and Golgi bodies. Although the cells in each of the six kingdoms have some peculiarities unique to that kingdom, electron microscopy has revealed that all cells are fundamentally similar.

PROCEDURE

1. Study Figure 5-9, a three-dimensional representation of an animal cell.
2. With the aid of Figure 5-9, identify the parts on the model of the animal cell that is on *demonstration.*
3. Figure 5-10 is an electron micrograph (EM) of an animal cell (kingdom Animalia). Study the electron micrograph and, with the aid of Figure 5-9 and any electron micrographs in your textbook, label each structure listed.
4. Pay particular attention to the membranes surrounding the nucleus and mitochondria. Note that these two are each bounded by *two* membranes, which are commonly referred to collectively as an **envelope.**
5. Using your textbook as a reference, list the function for the following cellular components:

   (a) plasma membrane ________________________________

   ____

   (b) cytoplasm ________________________________

   ____

   (c) nucleus (the plural is *nuclei*) ________________________________

   ____

   (d) nuclear envelope ________________________________

   ____

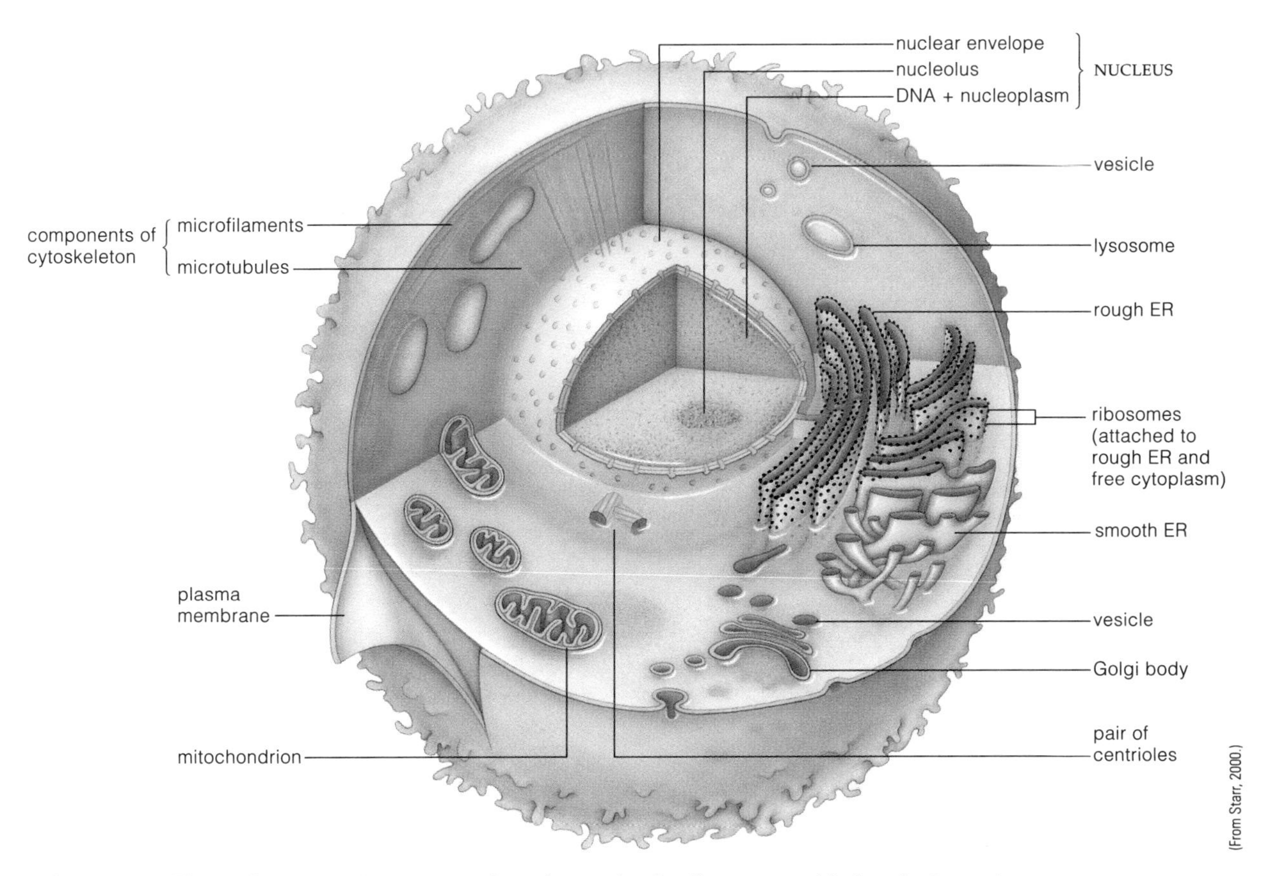

Figure 5-9 Three-dimensional representation of an animal cell as seen with the electron microscope.

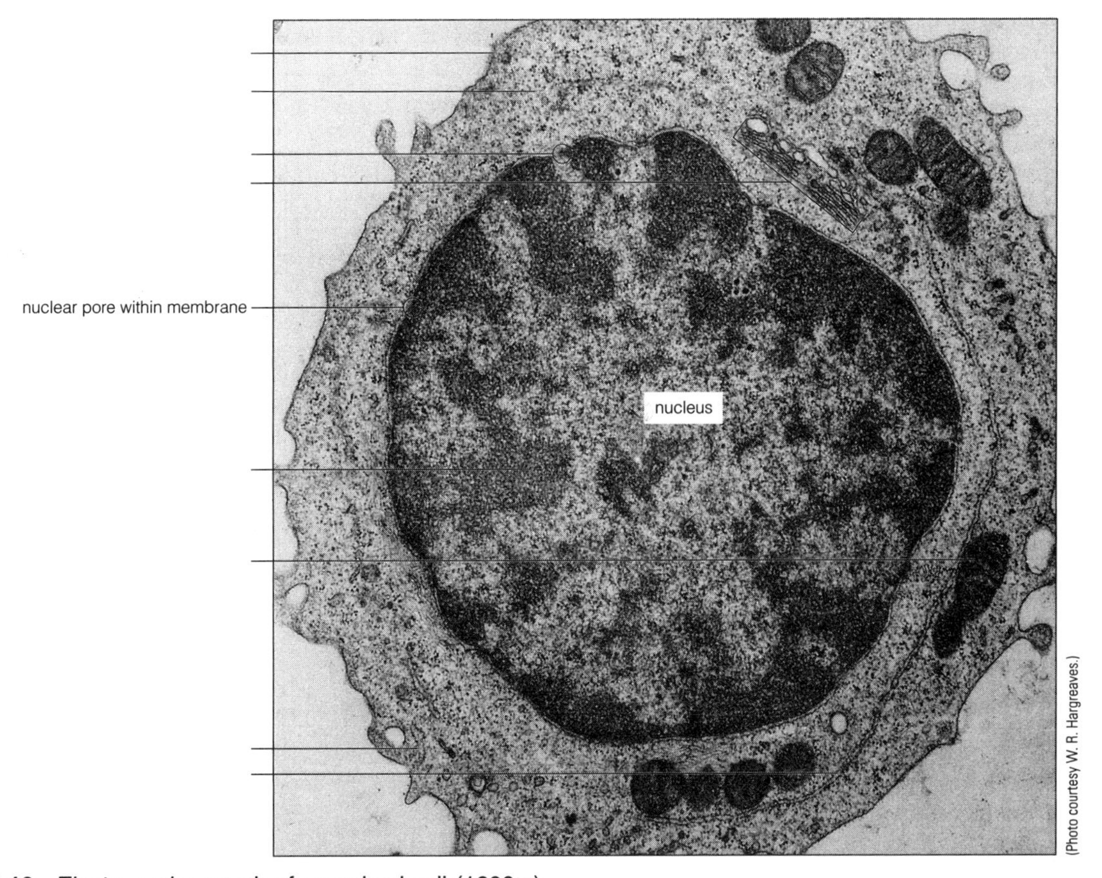

Figure 5-10 Electron micrograph of an animal cell (1600×).
Labels: plasma membrane, cytoplasm, nuclear envelope, nuclear pore, chromatin, rough ER, smooth ER, Golgi body, mitochondrion

(e) nuclear pores ____

____

(f) chromatin ____

____

(g) nucleolus (the plural is *nucleoli*) ____

____

(h) rough endoplasmic reticulum (RER) ____

____

(i) smooth endoplasmic reticulum (SER) ____

____

(j) Golgi body ____

____

(k) mitochondrion (the plural is *mitochondria*)

____

E. Plant Cells Seen with the Light Microscope

E.1. Elodea leaf cells

Young leaves at the growing tip of *Elodea* are particularly well suited for studying cell structure because these leaves are only a few cell layers thick.

PROCEDURE

1. With a forceps, remove a single young leaf, mount it on a slide in a drop of distilled water, and cover with a coverslip.
2. Examine the leaf first with the low-power objective. Then concentrate your study on several cells using the high-dry objective. Refer to Figure 5-11.
3. Observe the abundance of green bodies in the cytoplasm. These are the **chloroplasts,** organelles that function in photosynthesis and that are typical of green plants.
4. Locate the numerous dark lines running parallel to the long axis of the leaf. These are the air-containing **intercellular spaces.**
5. Find the **cell wall,** a structure distinguishing plant from animal cells, visible as a clear area surrounding the cytoplasm.
6. After the cells have warmed a bit, notice the **cytoplasmic streaming** taking place. Movement of the chloroplasts along the cell wall is the most obvious visual evidence of cytoplasmic streaming. Microfilaments (much too small to be seen with your light microscope) are responsible for this intracellular motion.
7. Remember that you are looking at a three-dimensional object. In the middle portion of the cell is the large, clear **central vacuole,** which can take up from 50% to 90% of the cell interior. Because the vacuole in *Elodea* is transparent, it cannot be seen with the light microscope.
8. The chloroplasts occur in the cytoplasm surrounding the vacuole, so they will appear to be in different locations, depending on where you focus in the cell. Focus in the upper or lower surface and observe that the chloroplasts appear to be scattered throughout the cell.

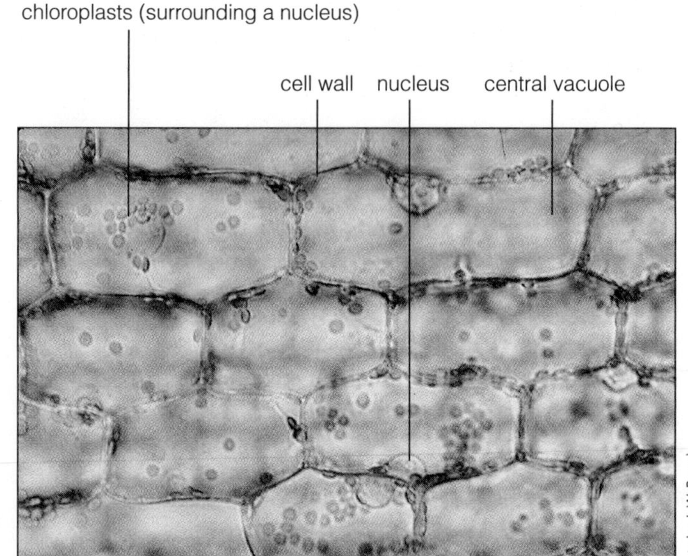

Figure 5-11 *Elodea* cells (400×).

9. Now focus in the center of the cell (by raising or lowering the objective with the fine focus knob), and note that the chloroplasts lie in a thin layer of cytoplasm along the wall.
10. Locate the **nucleus** within the cytoplasm. It will appear as a clear or slightly amber body that is slightly larger than the chloroplasts. (You may need to examine several cells to find a clearly defined nucleus.)
11. Describe the three-dimensional shape of the *Elodea* leaf cell.

____

____

12. What are the shapes of the chloroplasts and nucleus? ____

____

13. Now add a drop of methylene blue stain to make the cell wall more obvious. Add the stain as shown in Figure 5-7.
14. Look for the very, very tiny **mitochondria.** (If you have an oil-immersion lens on your microscope, you should use that lens.)
15. Compare the size of the mitochondria to chloroplasts:

____

____

E.2. Onion scale cells

1. Make a wet mount of a colorless scale of an onion bulb, using the technique described in Figure 5-12. The *inner* face of the scale is easiest to remove, as shown in Figure 5-12d.
2. Observe your preparation with your microscope, focusing first with the low-power objective. Continue your study, switching to the medium-power and finally the high-dry objective. Refer to Figure 5-13.
3. Identify the **cell wall** and **cytoplasm.**
4. Find the **nucleus,** a prominent sphere within the cytoplasm.
5. Examine the nucleus more carefully at high magnification. Within it, find one or more nucleoli (the singular is *nucleolus*). Nucleoli are rich in a nucleic acid known as RNA (ribonucleic acid), while the nucleus as a whole is largely DNA (deoxyribonucleic acid), the genetic material.
6. You may see numerous **oil droplets** within the cytoplasm, visible in the form of granulelike bodies. These oil droplets are a form of stored food material. You may be surprised to learn that onion scales are actually leaves! Which cellular components present in *Elodea* leaf cells are absent in onion leaf cells?

____

7. If you are using the pigmented tissue from a red onion, you should see a purple pigment located in the vacuole. In this case, the cell wall appears as a bright line.
8. In Figure 5-14, sketch and label several cells from onion scale leaves.

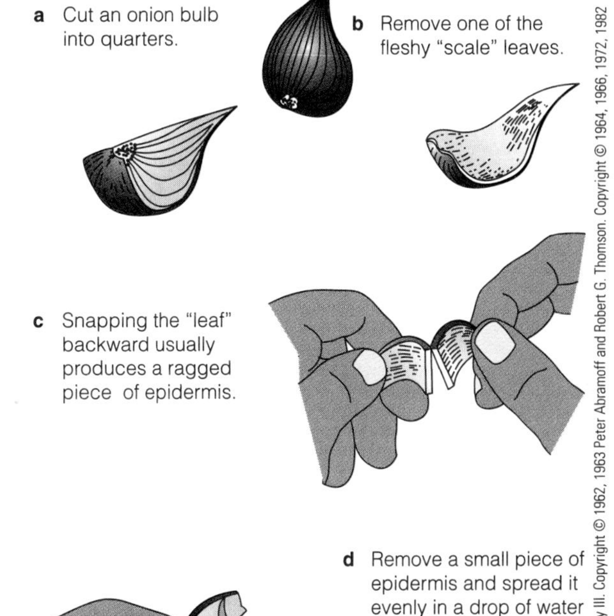

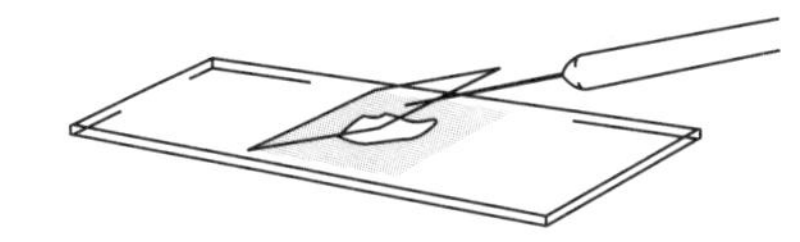

Figure 5-12 Method for obtaining onion scale cells.

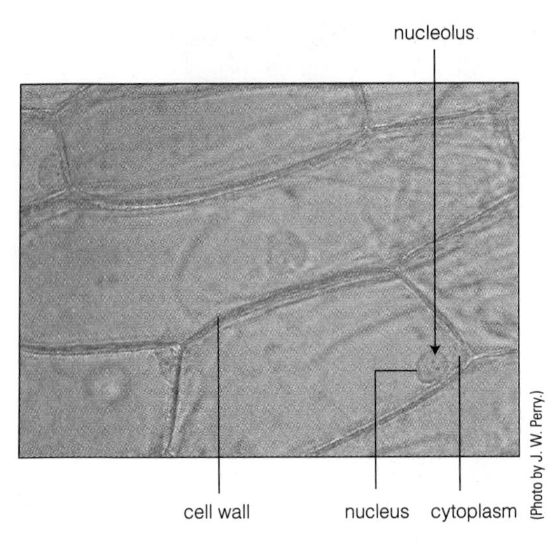

Figure 5-13 Onion bulb leaf cells (67×).
Labels: cell wall, cytoplasm, nucleus, nucleolus, oil droplets

Figure 5-14 Drawing of onion scale cells (______×).
Labels: cell wall, cytoplasm, nucleus

F. Plant Cells as Seen with the Electron Microscope

The electron microscope has made obvious some of the unique features of plant cells.

PROCEDURE

1. Study Figure 5-15, a three-dimensional representation of a typical plant cell.
2. With the aid of Figure 5-15, identify the structures present on the model of a plant cell that is on *demonstration.*
3. Now examine Figure 5-16, a transmission electron micrograph from a corn leaf. Label all of the structures listed. *Caution: Many plant cells do not have a large central vacuole. This is one of them.* Notice that the chloroplast has an envelope, just as do the nucleus and mitochondria.
4. With the help of Figure 5-15 and any transmission electron micrographs and text in your textbook, list the function of the following structures.

   (a) cell wall ____

   ____

   (b) chloroplast ____

   ____

   (c) vacuole ____

   ____

   (d) vacuolar membrane ____

   ____

   (e) plasma membrane ____

   ____

   (f) cytoplasm ____

   ____

   (g) nucleus ____

   ____

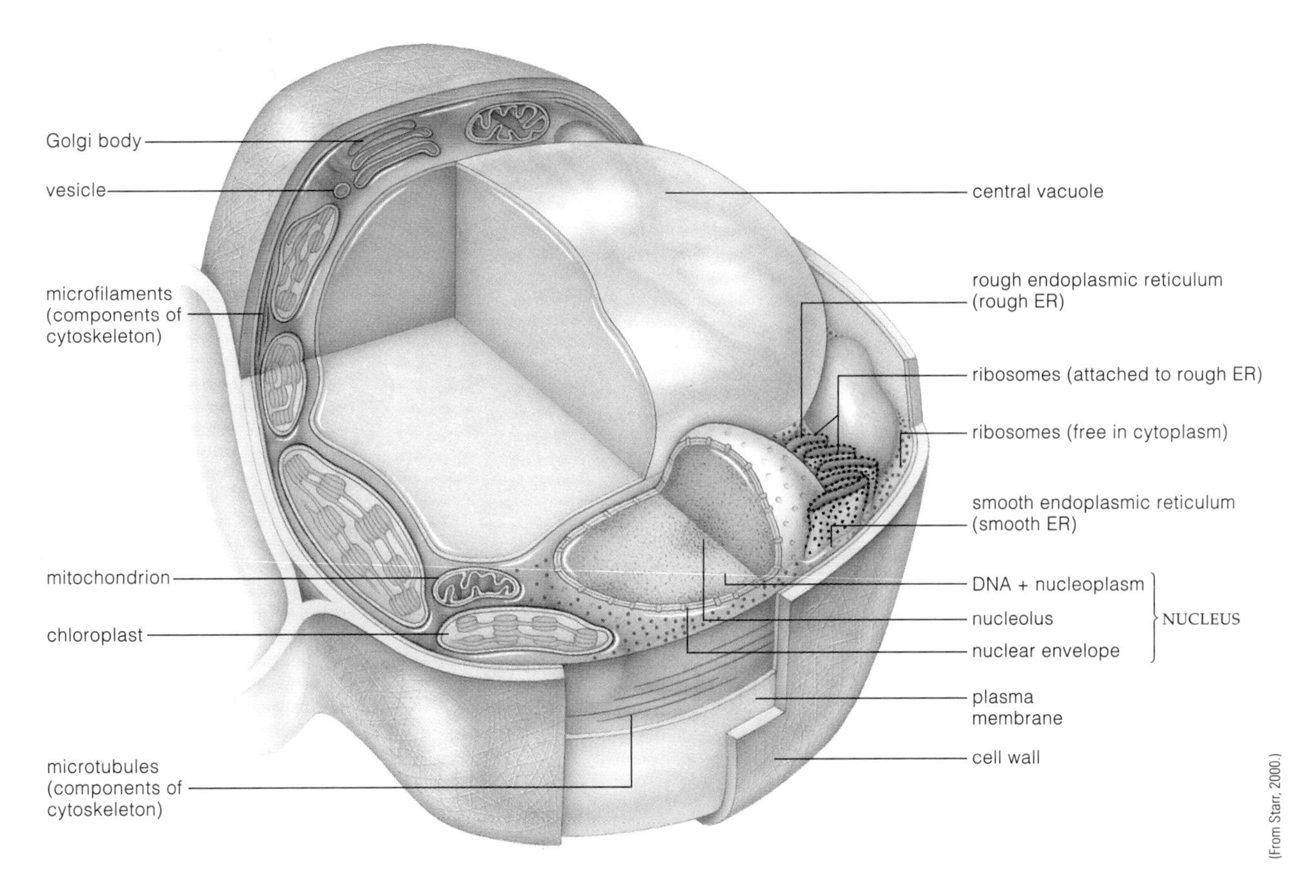

Figure 5-15 Three-dimensional representation of a plant cell as seen with the electron microscope.

(h) nuclear envelope ______________________________

____

(i) nuclear pore ______________________________

____

(j) chromatin ______________________________

____

(k) nucleolus ______________________________

____

(l) rough endoplasmic reticulum (RER) ______________________________

____

(m) smooth endoplasmic reticulum (SER) ______________________________

____

(n) Golgi body ______________________________

____

(o) mitochondrion ______________________________

____

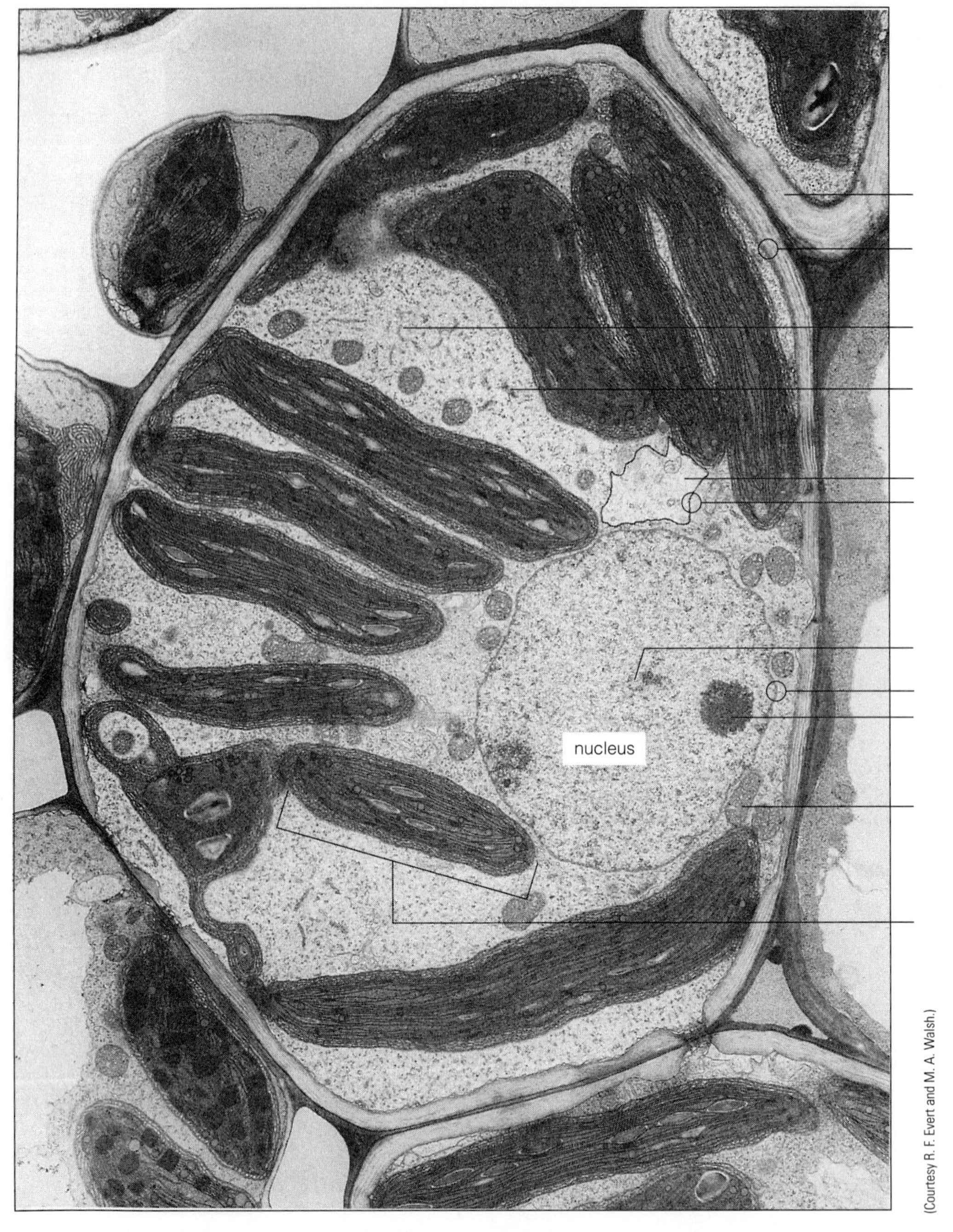

(Courtesy R. F. Evert and M. A. Walsh.)

Figure 5-16 Electron micrograph of a corn leaf cell (2700×).
Labels: cell wall, chloroplast, vacuole, vacuolar membrane, plasma membrane, nuclear envelope, chromatin, nucleolus, endoplasmic reticulum (ER), Golgi body, mitochondrion

PRE-LAB QUESTIONS

c. 1. The person who first used the term *cell* was
   (a) Darwin
   (b) Leeuwenhoek
   (c) Hooke
   (d) Watson

c. 2. All cells contain
   (a) a nucleus, plasma membrane, and cytoplasm
   (b) a cell wall, nucleus, and cytoplasm
   (c) DNA, plasma membrane, and cytoplasm
   (d) mitochondria, plasma membrane, and cytoplasm

b. 3. Prokaryotic cells *lack*
   (a) DNA
   (b) a true nucleus
   (c) a cell wall
   (d) none of the above

b. 4. The word *eukaryotic* refers specifically to a cell containing
   (a) photosynthetic membranes
   (b) a true nucleus
   (c) a cell wall
   (d) none of the above

a. 5. A bacterium is an example of a
   (a) prokaryotic cell
   (b) eukaryotic cell
   (c) plant cell
   (d) all of the above

d. 6. Methylene blue
   (a) is used to kill cells that are moving too quickly to observe
   (b) renders cells nontoxic
   (c) is a portion of the electromagnetic spectrum used by green plant cells
   (d) is a biological stain used to increase contrast of cellular constituents

b. 7. Components typical of plant cells but not of animal cells are
   (a) nuclei
   (b) cell walls
   (c) mitochondria
   (d) ribosomes

c. 8. A central vacuole
   (a) is found only in plant cells
   (b) may take up between 50% and 90% of the cell's interior
   (c) is both of the above
   (d) is none of the above

a. 9. The intercellular spaces between plant cells
   (a) contain air
   (b) are responsible for cytoplasmic streaming
   (c) are nonexistent
   (d) contain chloroplasts

d. 10. An envelope
   (a) surrounds the nucleus
   (b) surrounds mitochondria
   (c) consists of two membranes
   (d) does all of the above

Name ______________________ Section Number ______________________

EXERCISE 5

Structure and Function of Living Cells

POST-LAB QUESTIONS

5.1 Prokaryotic Cells

1. Did all living cells that you saw in lab contain mitochondria?

2. Below is a high-magnification photomicrograph of an organism you observed in this exercise. Each rectangular box is a single cell. What organelle is absent from each cell that makes it "prokaryotic?"

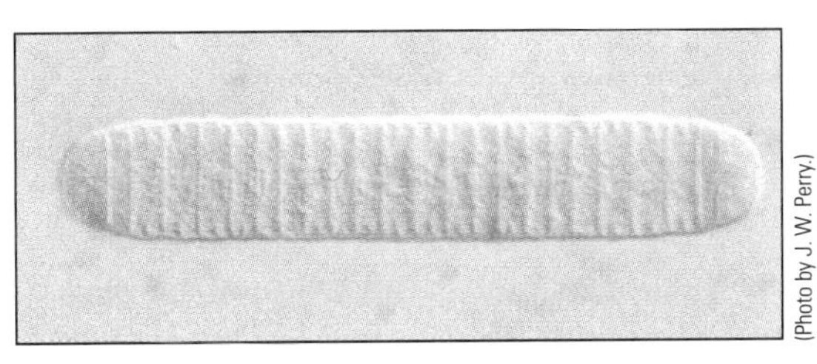

(750×).

5.2 Eukaryotic Cells

3. Is it possible for a cell to contain more than one nucleus? Explain.

4. When students are asked to distinguish between an animal cell and a plant cell, they typically answer that plant cells contain chloroplasts and animal cells do not. If you were the professor reading that answer, what sort of credit would you give and why?

5. Describe a major distinction between most plant cells and animal cells.

6. Observe the electron micrograph to the right. Is the cell prokaryotic or eukaryotic?

   Identify the labeled structures.

   A. ______________________

   B. ______________________

   C. ______________________

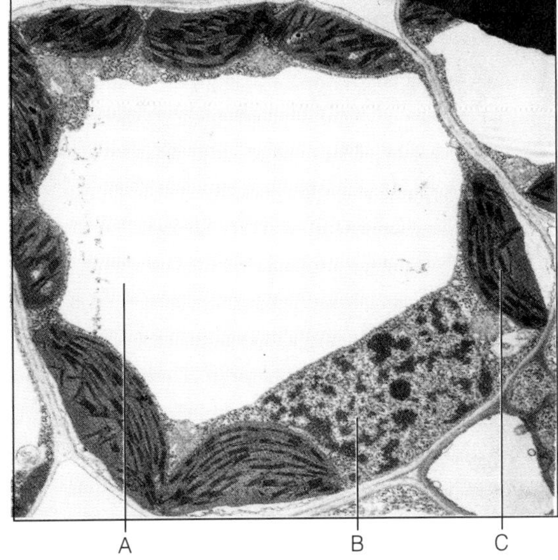

(4800×).

7. Look at the photomicrograph to the right, which was taken with a technique that gives a three-dimensional impression. Identify the structures labeled A, B, and C.

A. ______________________________

B. ______________________________

C. ______________________________

C B A

(750×).

(Photo by J. W. Perry.)

8. Is the electron micrograph below a plant or an animal cell?

____

Identify structures labeled A and B.

A. ______________________________

B. ______________________________

9. What are the numerous "wavy lines" within the cell (labeled C)? ______________________________

____

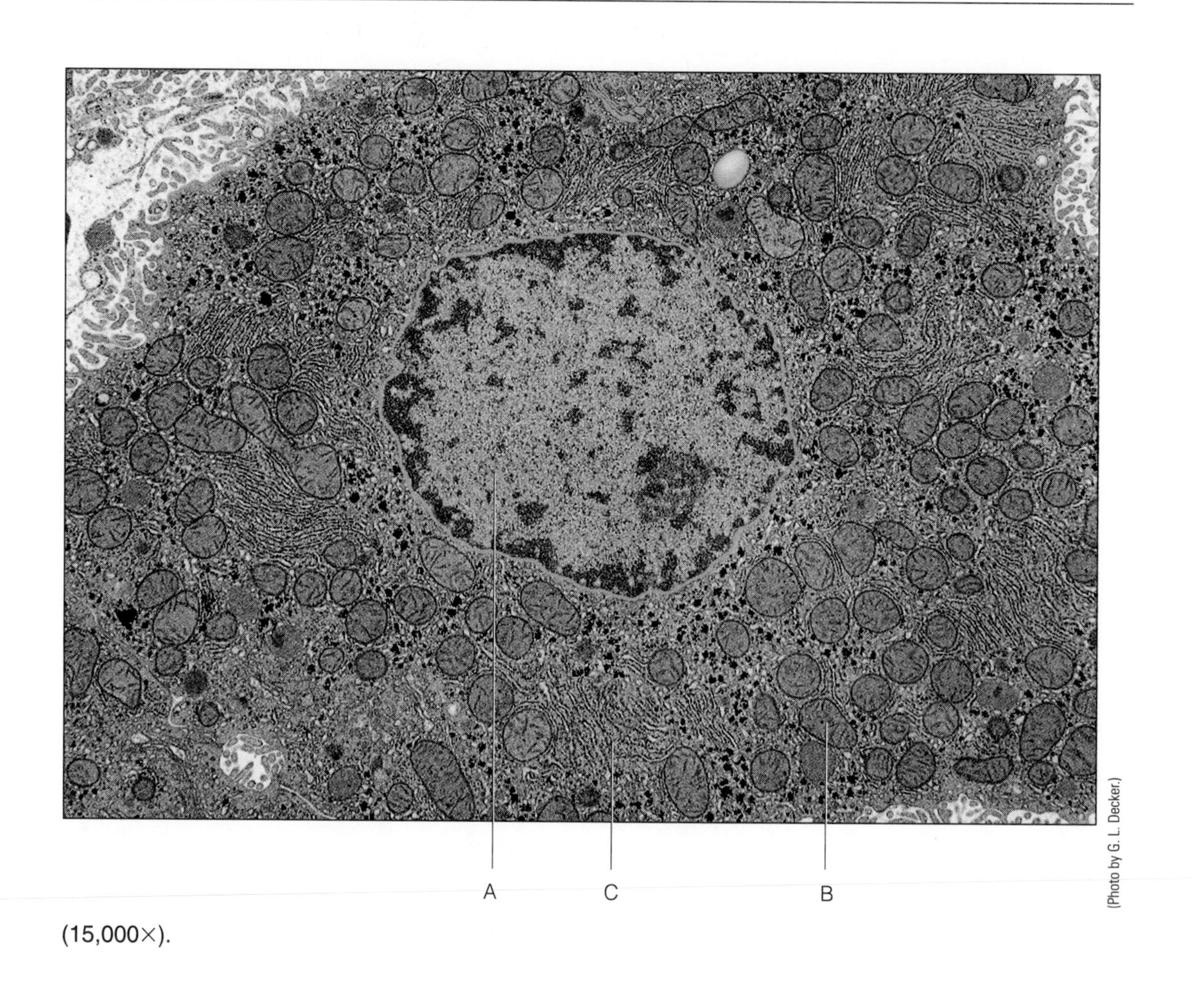

(15,000×).

(Photo by G. L. Decker.)

10. What structure(s) found in plant cells is (are) primarily responsible for cellular support?

Food for Thought

11. What structural differences did you observe between prokaryotic and eukaryotic cells?

12. Are the cells in the electron micrograph below prokaryotic or eukaryotic? How do you know?

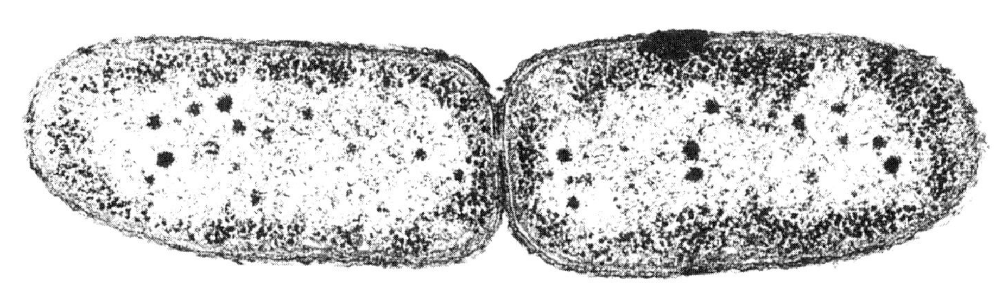

(Photo by J. J. Cardamone, Jr., University of Pittsburgh/BPS.)

EXERCISE 6

Diffusion, Osmosis, and the Functional Significance of Biological Membranes

OBJECTIVES

After completing this exercise, you will be able to

1. define *solvent, solute, solution, selectively permeable, diffusion, osmosis, concentration gradient, equilibrium, turgid, plasmolyzed, plasmolysis, turgor pressure, tonicity, hypertonic, isotonic, hypotonic;*
2. describe the structure of cellular membranes;
3. distinguish between diffusion and osmosis;
4. determine the effects of concentration and temperature on diffusion;
5. describe the effects of hypertonic, isotonic, and hypotonic solutions on red blood cells and *Elodea* leaf cells.

INTRODUCTION

Water is a great environment. Earthly life is believed to have originated in the water. Without it, life as we know it would cease to exist. Recently, the discovery of water in meteorites originating within our solar system has fueled speculation that life may not be unique to earth.

Living cells are made up of 75–85% water. Virtually all substances entering and leaving cells are dissolved in water, making it the **solvent** most important for life processes. The substances dissolved in water are called **solutes** and include such substances as salts and sugars. The combination of a solvent and dissolved solute is a **solution.** The cytoplasm of living cells contains numerous solutes, like sugars and salts, in solution.

All cells possess membranes composed of a phospholipid bilayer that contains different kinds of embedded and surface proteins. Look at Figure 6-1 to get an idea of the complexity of a cellular membrane.

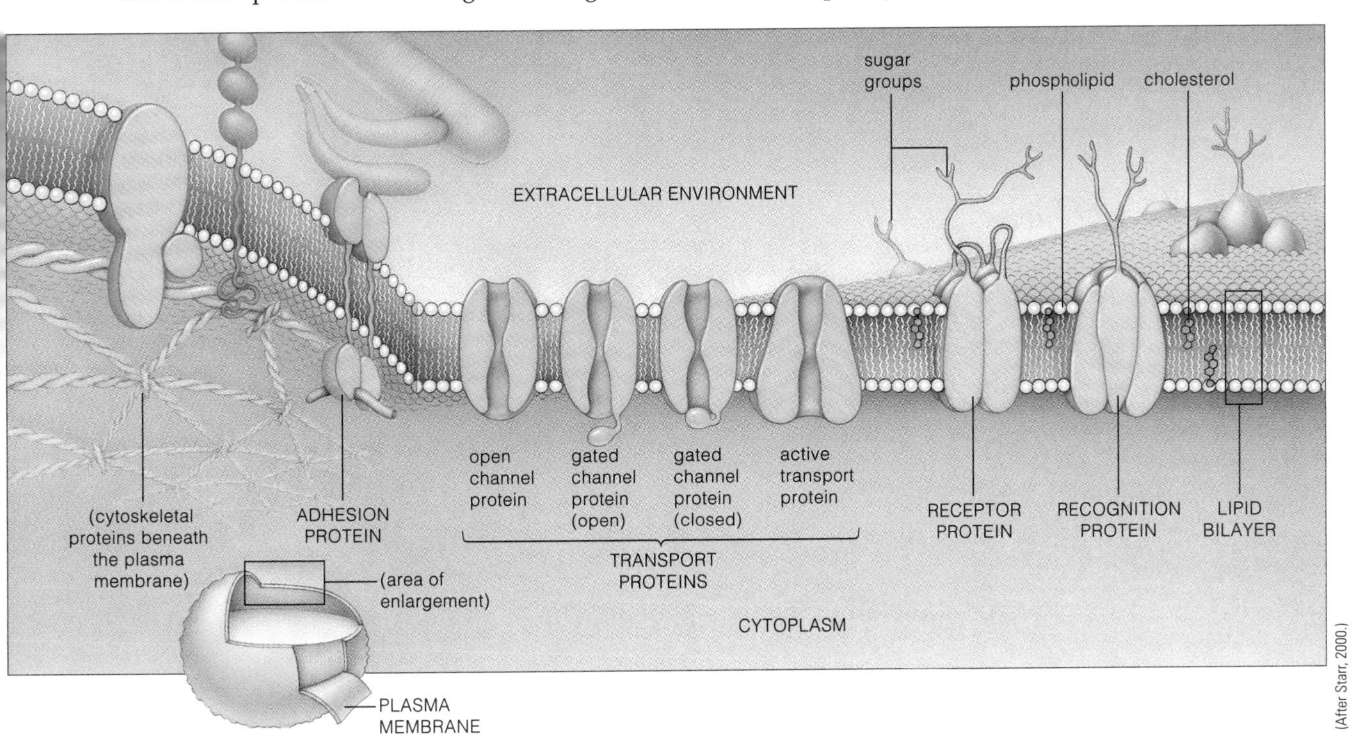

Figure 6-1 Artistic rendering of cutaway view of part of the plasma membrane.

Membranes are boundaries that solutes must cross to reach the cellular site where they will be utilized in the processes of life. These membranes regulate the passage of substances into and out of the cell. They are **selectively permeable,** allowing some substances to move easily while completely excluding others.

The most simple means by which solutes enter the cell is **diffusion,** the movement of solute molecules from a region of high concentration to one of lower concentration. Diffusion occurs without the expenditure of cellular energy. Once inside the cell, solutes move through the cytoplasm by diffusion, sometimes assisted by cytoplasmic streaming.

Water (the solvent) also moves across the membrane. **Osmosis** is the movement of *water* across selectively permeable membranes. Think of osmosis as a special form of diffusion, one occurring from a region of higher *water* concentration to one of lower *water* concentration.

The difference in concentration of like molecules in two regions is called a **concentration gradient.** Diffusion and osmosis take place *down* concentration gradients. Over time, the concentration of solvent and solute molecules becomes equally distributed, the gradient ceasing to exist. At this point, the system is said to be at **equilibrium.**

Molecules are always in motion, even at equilibrium. Thus, solvent and solute molecules continue to move because of randomly colliding molecules. However, at equilibrium there is no *net change* in their concentrations.

This exercise introduces you to the principles of diffusion and osmosis.

***Note:* If Sections 6.2 and 6.3 are to be done during this lab period, start them before doing any other activity in this exercise.**

6.1 Experiment: Rate of Diffusion of Solutes *(About 10 min.)*

Solutes move within a cell's cytoplasm largely because of diffusion. However, the rate of diffusion (the distance diffused in a given amount of time) is affected by such factors as temperature and the size of the solute molecules. In this experiment, you will discover the effects of these two factors in gelatin (the substance of Jell-O®), a substance much like cytoplasm and used to simulate it in this experiment.

MATERIALS

Per student:

- metric ruler

Per student group (table):

- 1 set of 3 screw-cap test tubes, in rack, each half-filled with 5% gelatin, to which the following dyes have been added: potassium dichromate, aniline blue, Janus green; labeled with each dye and marked "5°C"
- 1 set of 3 screw-cap test tubes, in rack, as above but marked "Room Temperature"

Per lab room:

- 5°C refrigerator

PROCEDURE

Two sets of three screw-cap test tubes have been half-filled with 5% gelatin; and 1 mL of a dye has been added to each test tube. Set 1 is in a 5°C refrigerator; set 2 is at room temperature. Record the time at which your instructor tells you the experiment was started: ____________________

1. Remove set 1 from the refrigerator and compare the distance the dye has diffused in corresponding tubes of each set.

Caution

Be certain the cap to each tube is tight!

2. Invert and hold each tube vertically in front of a white sheet of paper. Use a metric ruler to measure how far each dye has diffused from the gelatin's surface. Record this distance in Table 6-1.
3. Determine the *rate* of diffusion for each dye by using the following formula:

rate of diffusion = distance ÷ elapsed time (hours)

Time experiment ended: ____________

Time experiment started: ____________

Elapsed time: ____________ hours

TABLE 6-1 Effect of Temperature on Diffusion Rates of Various Solutes

	Set 1 (5°C)		Set 2 (Room Temp.)	
Solute (dye)	**Distance (mm)**	**Rate**	**Distance (mm)**	**Rate**
Potassium dichromate (MW = 294)[a]				
Janus green (MW = 511)				
Aniline blue (MW = 738)				

[a]MW = molecular weight, a reflection of the mass of a substance. To determine MW, add the atomic weights of all elements in a compound.

Which of the solutes diffused the slowest (regardless of temperature)? ______

Which diffused the fastest? ______

What effect did temperature have on the rate of diffusion? ______

Make a conclusion about the diffusion of a solute in a gel, relating the rate of diffusion to the molecular weight of the solute and to temperature.

______

______

Note: **Return set 1 to the refrigerator.**

6.2 Experiment: Osmosis *(About 20 min. for setup)*

Osmosis occurs when different concentrations of water are separated by a selectively permeable membrane. One example of a selectively permeable membrane within a living cell is the plasma membrane. In this experiment, you will learn about osmosis using dialysis membrane, a selectively permeable cellulose sheet that permits the passage of water but obstructs passage of larger molecules. If you examined the membrane with a scanning electron microscope, you would see that it is porous; it thus prevents molecules larger than the pores from passing through the membrane.

MATERIALS

Per student group (4):

- four 15-cm lengths of dialysis tubing, soaking in dH_2O
- eight 10-cm pieces of string or waxed dental floss
- ring stand and funnel apparatus (Figure 6-2)
- 25-mL graduated cylinder
- 4 small string tags
- china marker
- four 400-mL beakers

Per student group (table):

- dishpan half-filled with dH_2O
- paper toweling
- balance

Per lab room:

- source of dH_2O (at each sink)
- 15% and 30% sucrose solutions
- scissors (at each sink)

PROCEDURE

Work in groups of four for this experiment.

1. Obtain four sections of dialysis tubing, each 15 cm long, that have been presoaked in dH_2O. Recall that the dialysis tubing is permeable to water molecules but not to sucrose.
2. Fold over one end of each tube and tie it tightly with string or dental floss.
3. Attach a string tag to the tied end of each bag and number them 1–4.

4. Slip the open end of the bag over the stem of a funnel (Figure 6-2). Using a graduated cylinder* to measure volume, fill the bags as follows:
   Bag 1. 10 mL of dH_2O Bag 3. 10 mL of 30% sucrose
   Bag 2. 10 mL of 15% sucrose Bag 4. 10 mL of dH_2O
5. As each bag is filled, force out excess air by squeezing the bottom end of the tube.
6. Fold the end of the bag and tie it securely with another piece of string or dental floss.
7. Rinse each filled bag in the dishpan containing dH_2O; gently blot off the excess water with paper toweling.
8. Weigh each bag to the nearest 0.5 g.
9. Record the weights in the column marked "0 min." in Table 6-2.
10. Number four 400-mL beakers with a china marker.
11. Add 200 mL of dH_2O to beakers 1–3.
12. Add 200 mL of 30% sucrose solution to beaker 4.
13. Place bags 1–3 in the correspondingly numbered beakers.
14. Place bag 4 in the beaker containing 30% sucrose.
15. After 15 minutes, remove each bag from its beaker, blot off the excess fluid, and weigh each bag.
16. Record the weight of each bag in Table 6-2.
17. Return the bags to their respective beakers immediately after weighing.
18. Repeat steps 15–17 at 30, 45 and 60 minutes from time zero.

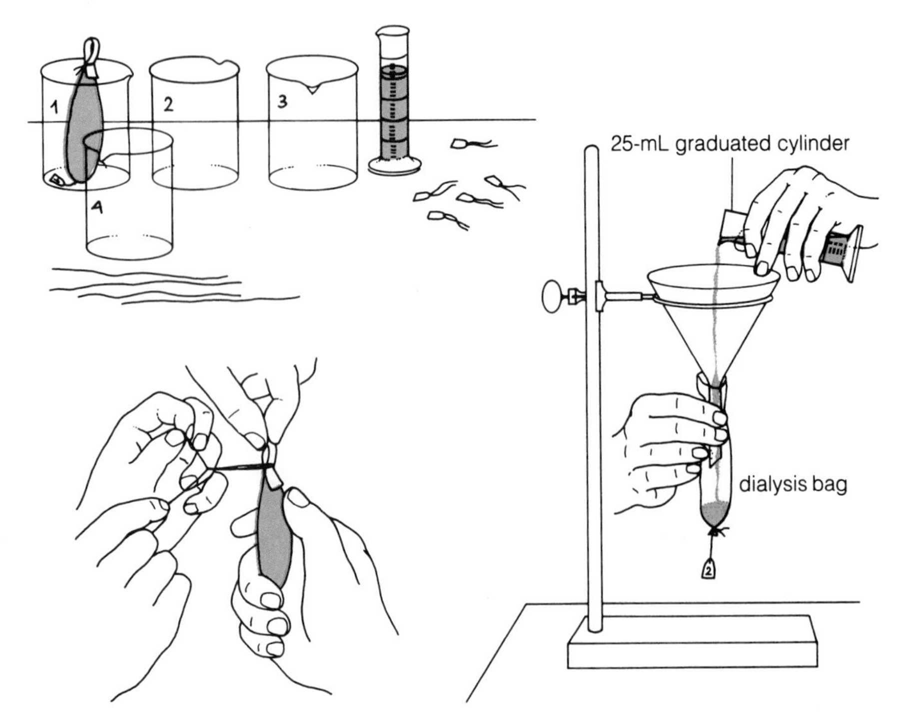

Figure 6-2 Method for filling dialysis bags.

At the end of the experiment, take the bags to the sink, cut them open, pour the contents down the drain, and discard the bags in the wastebasket. Pour the contents of the beakers down the drain and wash them according to the instructions given on page x.

Make a *qualitative* statement about what you have observed. ______________________________

____

____

____

Was the direction of *net* movement of water in bags 2–4 into or out of the bags? ________________

____

____

Which bag gained the most weight? Why? ______________________________________

____

____

*Be sure to rinse the cylinder if it has been used to measure sucrose.

TABLE 6-2 Change in Weight as a Consequence of Osmosis

Bag	Bag Contents	Beaker Contents	Bag Weight (g)					Weight Change (g)
			0 min.	15 min.	30 min.	45 min.	60 min.	
1	dH_2O	dH_2O						
2	15% sucrose	dH_2O						
3	30% sucrose	dH_2O						
4	dH_2O	30% sucrose						

6.3 Experiment: Selective Permeability of Membranes *(About 15 min. for setup)*

Dialysis tubing is a selectively permeable material that provides a means to demonstrate the movement of substances through cellular membranes.

MATERIALS

Per student group (4):

- 1 25-cm length of dialysis tubing, soaking in dH_2O
- two 10-cm pieces of string or waxed dental floss
- bottle of 1% soluble starch in 1% sodium sulfate (Na_2SO_4)
- dishpan half-filled with dH_2O
- 400-mL graduated beaker
- ring stand and funnel apparatus (Figure 6-2)
- bottle of 1% albumin in 1% sodium chloride (NaCl)
- 8 test tubes
- test tube rack
- china marker
- 25-mL graduated cylinder
- iodine (I_2KI) solution in dropping bottle
- 2% barium chloride ($BaCl_2$) in dropping bottle
- 2% silver nitrate ($AgNO_3$) in dropping bottle
- Biuret reagent in dropping bottle
- albustix reagent strips (optional)
- scissors

Per lab room:

- series of 4 test tubes in test tube rack demonstrating positive tests for starch, sulfate ion, chloride ion, protein

PROCEDURE

Work in groups of four.

1. Obtain a 25-cm section of dialysis tubing that has been soaked in dH_2O.
2. Fold over one end of the tubing and tie it securely with string or dental floss to form a leakproof bag (Figure 6-2).
3. Slip the open end of the bag over the stem of a funnel and fill the bag approximately half full with 25 mL of a solution of 1% soluble starch in 1% sodium sulfate (Na_2SO_4).
4. Remove the bag from the funnel; fold and tie the open end of the bag.
5. Rinse the tied bag in a dishpan partially filled with dH_2O.
6. Pour 200 mL of a solution of 1% albumin (a protein) in 1% sodium chloride (NaCl) into a 400-mL beaker.
7. Place the bag into the fluid in the beaker.
8. Record the time: ____________________
9. With a china marker, label eight test tubes, numbering them 1–8.
10. Seventy-five minutes after the start of the experiment, pour 20 mL of the *beaker contents* into a *clean* 25-mL graduated cylinder.
11. Decant (pour out) 5 mL from the graduated cylinder into each of the first four test tubes.

12. Perform the following tests, recording your results in Table 6-3. Your instructor will have a series of test tubes showing positive tests for starch, sulfate and chloride ions, and proteins. You should compare your results with the known positives.
    (a) *Test for starch.* Add several drops of iodine solution (I_2KI) from the dropper bottle to test tube 1. If starch is present, the solution will turn blue-black.
    (b) *Test for sulfate ion.* Add several drops of 2% barium chloride ($BaCl_2$) from the dropper bottle to test tube 2. If sulfate ions (SO^{-4}) are present, a white precipitate of barium sulfate ($BaSO_4$) will form.
    (c) *Test for chloride ion.* Add several drops of 2% silver nitrate ($AgNO_3$) from the dropper bottle to test tube 3. A milky-white precipitate of silver chloride (AgCl) indicates the presence of chloride ions (Cl^2).
    (d) *Test for protein.* Add several drops of Biuret reagent from the dropper bottle to test tube 4. If protein is present, the solution will change from blue to pinkish-violet. The more intense the violet hue, the greater the quantity of the protein.
    An alternative method for determining the presence of protein is the use of albustix reagent strips. Presence of protein is indicated by green or blue-green coloration of the paper.
13. Wash the graduated cylinder, using the technique described on page x.
14. Thoroughly rinse the bag in the dishpan of dH_2O.
15. Using scissors, cut the bag open and empty the contents into the 25-mL graduated cylinder.
16. Decant 5-mL samples into each of the four remaining test tubes.
17. Perform the tests for starch, sulfate ions, chloride ions, and protein on tubes 5–8, respectively.
18. Record the results of this series of tests in Table 6-4.

To which substances was the dialysis tubing permeable?

___

What physical property of the dialysis tubing might explain its differential permeability?

___

19. Discard contents of test tubes and beaker down sink drain. Wash glassware by using the technique described on page x.
20. Discard dialysis tubing in wastebasket.

TABLE 6-3 Results of Tests for Substances in Beaker[a]

	At Start of Experiment	After 75 min.
Starch	–	
Sulfate ion	–	
Chloride ion	+	
Albumin	+	

[a]Contents of beaker: (+) = presence, (−) = absence.

TABLE 6-4 Results of Tests for Substances in Dialysis Bag[a]

	At Start of Experiment	After 75 min.
Starch	+	
Sulfate ion	+	
Chloride ion	–	
Albumin	–	

[a]Contents of dialysis bag: (+) = presence, (–) = absence.

6.4 Experiment: Plasmolysis in Plant Cells *(About 15 min.)*

Plant cells are surrounded by a rigid cell wall, composed primarily of the glucose polymer, cellulose. Recall from Exercise 6 that many plant cells have a large central vacuole surrounded by the vacuolar membrane. The vacuolar membrane is selectively permeable. Normally, the solute concentration within the cell's central vacuole is greater than that of the external environment. Consequently, water moves into the cell, creating **turgor pressure,** which presses the cytoplasm against the cell wall. Such cells are said to be **turgid.** Many nonwoody plants (like beans and peas) rely on turgor pressure to maintain their rigidity and erect stance.

In this experiment, you will discover the effect of external solute concentration on the structure of plant cells.

MATERIALS

Per student:

- forceps
- 2 microscope slides
- 2 coverslips
- compound microscope

Per student group (table):

- *Elodea* in tap water
- 2 dropping bottles of dH_2O
- 2 dropping bottles of 20% sodium chloride (NaCl)

PROCEDURE

1. With a forceps, remove two young leaves from the tip of an *Elodea* plant.
2. Mount one leaf in a drop of distilled water on a microscope slide and the other in 20% NaCl solution on a second microscope slide.
3. Place coverslips over both leaves.
4. Observe the leaf in distilled water with the compound microscope. Focus first with the medium-power objective and then switch to the high-dry objective.
5. Label the photomicrograph of turgid cells (Figure 6-3).
6. Now observe the leaf mounted in 20% NaCl solution. After several minutes, the cell will have lost water, causing it to become **plasmolyzed.** (This process is called **plasmolysis.**) Label the plasmolyzed cells shown in Figure 6-4.

Figure 6-3 Turgid *Elodea* cells (400×). (Photo by J. W. Perry.)
Labels: cell wall, chloroplasts in cytoplasm, central vacuole

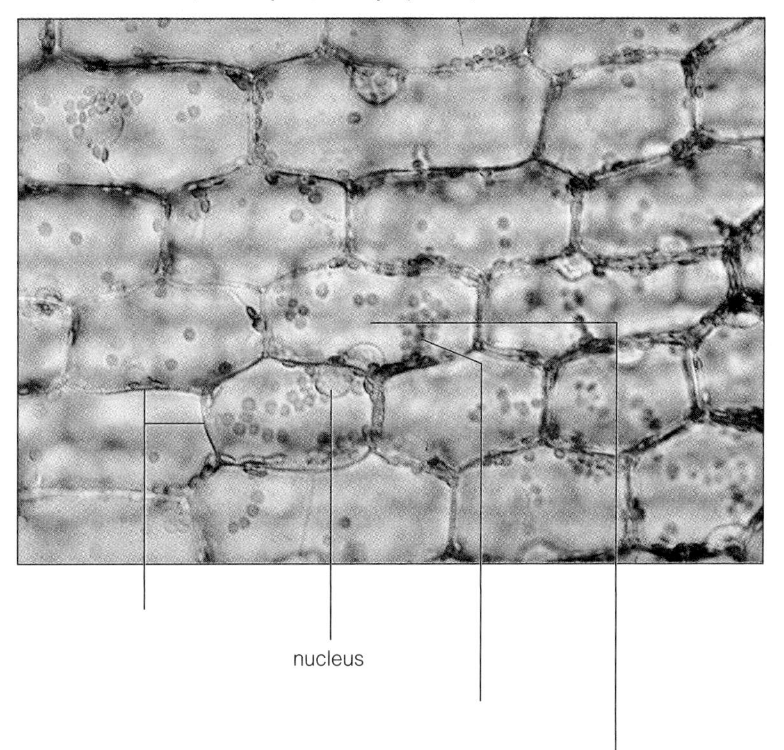

Tonicity describes one solution's solute concentration compared to that of another solution. The solution containing the lower concentration of solute molecules than another is **hypotonic** *relative to the second solution.* Solutions containing equal concentrations of solute are **isotonic** to each other, while one containing a greater concentration of solute relative to a second one is **hypertonic.**

Were the contents of the vacuole in the *Elodea* leaf in distilled water hypotonic, isotonic, or hypertonic compared to the dH_2O? ____________________

Was the 20% NaCl solution hypertonic, isotonic, or hypotonic relative to the cytoplasm? ____________________

____

If a hypotonic and a hypertonic solution are separated by a selectively permeable membrane, in which direction will the water move? ____________________

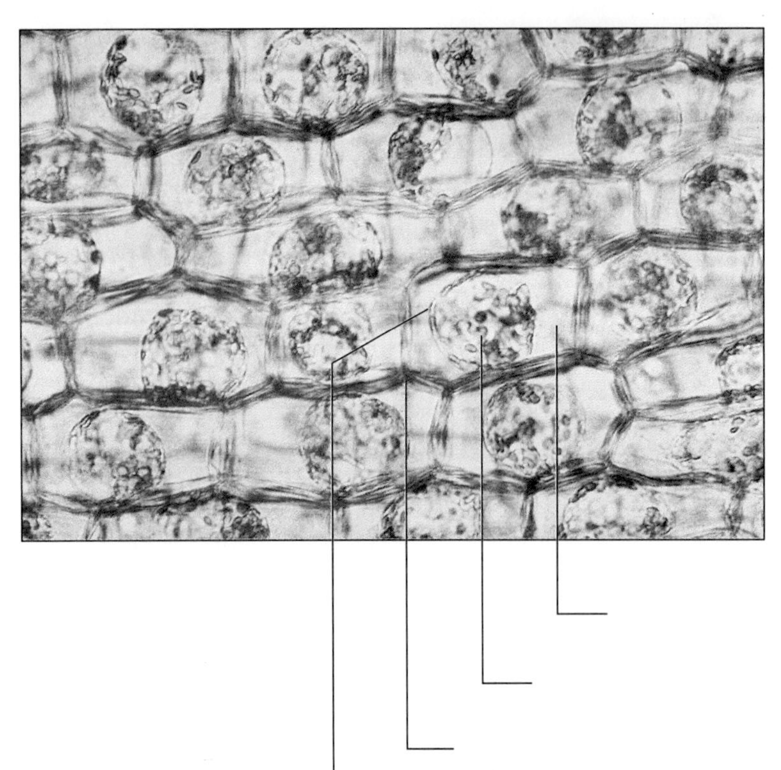

Figure 6-4 Plasmolyzed *Elodea* cells (400×). (Photo by J. W. Perry.)
Labels: cell wall, chloroplasts in cytoplasm, plasma membrane, space (between cell wall and plasma membrane)

Name two selectively permeable membranes that are present within the *Elodea* cells and that were involved in the plasmolysis process.

1. ______________________________

2. ______________________________

6.5 Experiment: Osmotic Changes in Red Blood Cells *(About 15 min.)*

Animal cells lack the rigid cell wall of a plant. The external boundary of an animal cell is the selectively permeable plasma membrane. Consequently, an animal cell increases in size as water enters the cell. However, since the plasma membrane is relatively fragile, it ruptures when too much water enters the cell. This is because of excessive pressure pushing out against the membrane. Conversely, if water moves out of the cell, it becomes plasmolyzed and looks spiny.

In this experiment, you will use red blood cells to discover the effects of osmosis in animal cells.

MATERIALS

Per student:

- compound microscope

Per student group (4):

- 3 clean screw-cap test tubes
- test tube rack
- metric ruler
- china marker
- bottle of 0.9% sodium chloride (NaCl)
- bottle of 10% NaCl
- bottle of dH_2O
- 3 disposable plastic pipets
- 3 clean microscope slides
- 3 coverslips

Per student group (table):

- bottle of sheep blood (in ice bath)

Per lab room:

- source of dH_2O

PROCEDURE

Work in groups of four for this experiment, but do the microscopic observations individually.

1. Observe the scanning electron micrographs in Figure 6-5.

Figure 6-5a illustrates the normal appearance of red blood cells. They are biconcave disks; that is, they are circular in outline with a depression in the center of both surfaces. Cells in an isotonic solution will appear like these blood cells.

Figure 6-5b shows cells that have been plasmolyzed. (In the case of red blood cells, plasmolysis is given a special term, *crenation;* the blood cell is said to be *crenate.*)

Figure 6-5c represents cells that have taken in water but have not yet burst. (Burst red blood cells are said to be *hemolyzed,* and of course they can't be seen.) Note their swollen, spherical appearance.

2. Obtain three clean screw-cap test tubes.
3. Lay test tubes 1 and 2 against a metric ruler and mark lines indicating 5 cm *from the bottom of each tube.*
4. Fill each tube as follows:
   Tube 1: 5 cm of 0.9% sodium chloride (NaCl)
   5 drops of sheep blood
   Tube 2: 5 cm of 10% NaCl
   5 drops of sheep blood
5. Lay test tube 3 against a metric ruler and mark lines indicating 0.5 cm and 5 cm *from the bottom of the tube.*
6. Fill tube 3 to the 0.5-cm mark with 0.9% NaCl, and to the 5-cm mark with dH_2O. Then add 5 drops of sheep blood. Enter the contents of each tube in the appropriate column of Table 6-5.

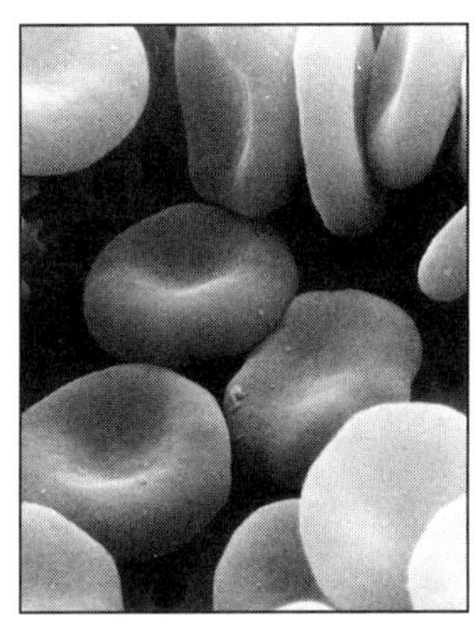

a Red blood cells in an isotonic solution ("normal")

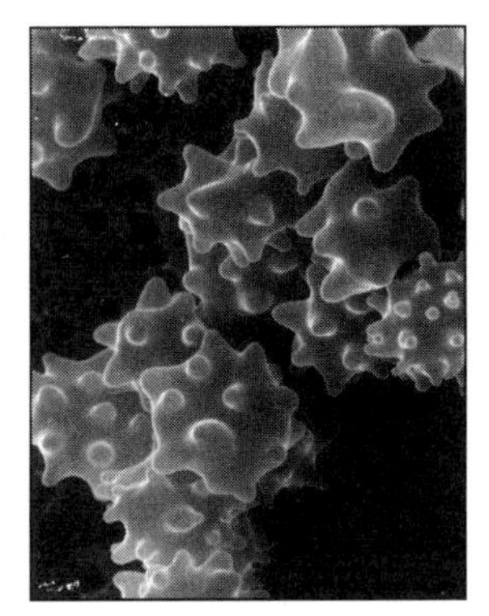

b Red blood cells in a hypertonic solution ("crenate")

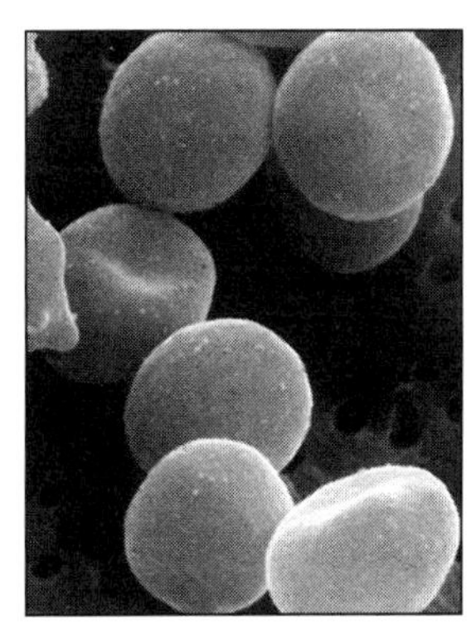

c Red blood cells in a hypotonic solution

Figure 6-5 Scanning electron micrographs of red blood cells. (Photos from M. Sheetz, R. Painter, and S. Singer. Reproduced from *The Journal of Cell Biology,* 1976, 70:193, by copyright permission of the Rockefeller University Press and M. Sheetz.)

7. Replace the caps and mix the contents of each tube by inverting several times (Figure 6-6a).
8. Hold each tube flat against the printed page of your lab manual (Figure 6-6b). *Only if the blood cells are hemolyzed should you be able to read the print.*
9. In Table 6-5, record your observations in the column "Print Visible?"
10. Number three clean microscope slides.
11. With three *separate* disposable pipets, remove a small amount of blood from each of the three tubes. Place 1 drop of blood from tube 1 on slide 1, 1 drop from tube 2 on slide 2, and 1 drop from tube 3 on slide 3.
12. Cover each drop of blood with a coverslip.
13. Observe the three slides with your compound microscope, focusing first with the medium-power objective and finally with the high-dry objective. (Hemolyzed cells are virtually unrecognizable; all that remains are membranous "ghosts," which are difficult to see with the microscope.)
14. In Figure 6-7, sketch the cells from each tube. Label the sketches, indicating whether the cells are normal, plasmolyzed (crenate), or hemolyzed.

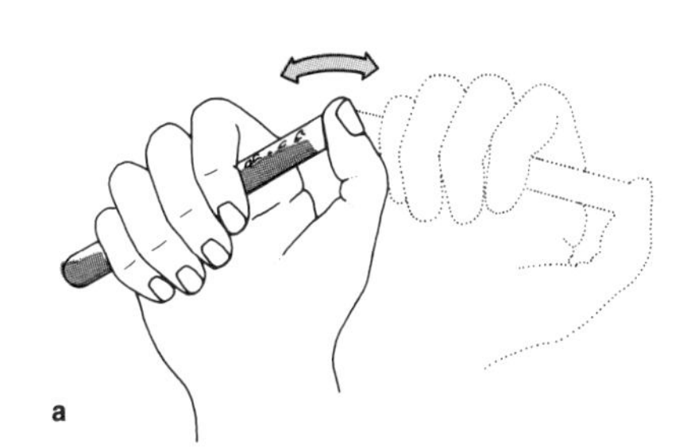

(After Abramoff and Thomson, 1982.)

Figure 6-6 Method for studying effects of different solute concentrations on red blood cells.

15. Record the microscopic appearance in Table 6-5.
16. Record the relative tonicity of the sodium chloride solutions you added to the test tubes in Table 6-5.

Why do red blood cells burst when put in a hypotonic solution whereas *Elodea* leaf cells do not?

____

____

____

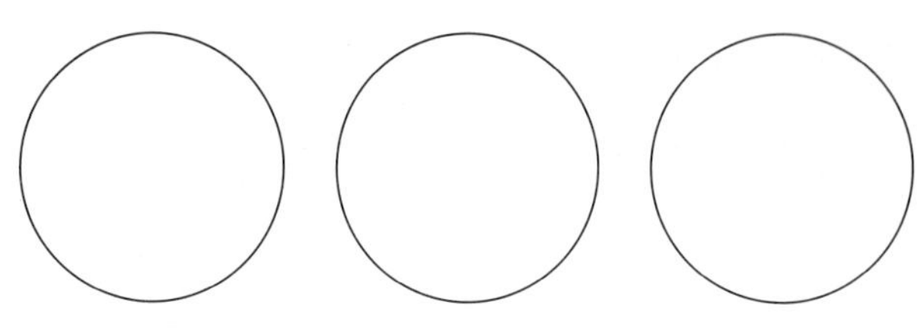

Figure 6-7 Microscopic appearance of red blood cells in different solute concentrations (_____×).
Labels: normal, plasmolyzed (crenate), hemolyzed

After completing all experiments, take your dirty glassware to the sink and wash it as directed on page x. Invert the test tubes in the test tube rack so they drain. Reorganize your work area, making certain all materials used in this exercise are present for the next class.

TABLE 6-5 Effect of Salt Solutions on Red Blood Cells

Tube	Contents	Print Visible?	Microscopic Appearance of Cells	Tonicity of External Solution[a]
1				
2				
3				

[a]With respect to that inside the red blood cell at the start of the experiment.

6.6 Experiment: Determining the Concentration of Solutes in Cells

(About 20 min. for setup)

If you've done the previous experiments of this exercise, you now know that water flows into or out of cells in response to the concentration of solutes within the cells. But you might logically ask at this point how much solute is present in a typical cell. While the answer varies from cell to cell, a simple experiment enables you to determine the osmotic concentration in the cells of a potato tuber.

MATERIALS

Per student group (4):

- five 250-mL beakers
- large potato tuber
- china marker
- single-edge razor blades *or* paring knife
- metric ruler
- potato peeler

Per table:

- balance
- paper toweling
- bottles containing solutions of 0.15, 0.20, 0.25, 0.30, and 0.35 M sucrose

PROCEDURE

1. With the china marker, label the five 250-mL beakers with the concentrations of sucrose solution.
2. Pour about 100 mL of each solution into its respective beaker.
3. Peel the potato and then cut it into five 3-cm cubes (3 cm on each side).
4. Without delay, weigh each cube to the nearest 0.01 g. Record the weights in Table 6-6.
5. Place one cube in each beaker and allow it to remain there for a minimum of 30 minutes, longer if time is available.

6. After the experimental period has elapsed, remove each cube, one at a time, and blot it lightly but thoroughly with the paper toweling.
7. Weigh each cube and record its final weight in Table 6-6. Then calculate and record the weight loss or gain.
8. Calculate the percent change in weight by dividing the *initial weight* by the *final weight*.

TABLE 6-6 Determining the Solute Concentration in Potato Tuber Cells

	Weight			
Solution	**Initial**	**Final**	**Change**	**Percent Change**
0.15 M				
0.20 M				
0.25 M				
0.30 M				
0.35 M				

The cube with the lowest percentage of weight loss or gain is in a solution that most closely approximates the solute concentration of the cells within the potato tuber. Of course, most of the solute within the tuber is in the form of starch, and our experimental solution is sucrose. The results of this experiment indicate that the *concentration* of the solute, but not the *type* of solute, is important for osmosis to occur.

What was the approximate concentration of solute in the potato tuber? ______________________________

____

Which concentration resulted in the *greatest* percentage change? ______________________________

____

Make a statement that relates the amount of water loss or gain to the concentration of the solute. __________

____

____

____

PRE-LAB QUESTIONS

_____ **1.** If one were to identify the most important compound for sustenance of life, it would probably be
(a) salt
(b) $BaCl_2$
(c) water
(d) I_2KI

_____ **2.** A solvent is
(a) the substance in which solutes are dissolved
(b) a salt or sugar
(c) one component of a biological membrane
(d) selectively permeable

_____ **3.** Diffusion
(a) is a process requiring cellular energy
(b) is the movement of molecules from a region of higher concentration to one of lower concentration
(c) occurs only across selectively permeable membranes
(d) is none of the above

_____ **4.** Cellular membranes
(a) consist of a phospholipid bilayer containing embedded proteins
(b) control the movement of substances into and out of cells
(c) are selectively permeable
(d) are all of the above

_____ **5.** An example of a solute would be
(a) Janus green B
(b) water
(c) sucrose
(d) both a and c

_____ **6.** Dialysis membrane is
(a) selectively permeable
(b) used in these experiments to simulate cellular membranes
(c) permeable to water but not to sucrose
(d) all of the above

_____ **7.** Specifically, osmosis
(a) requires the expenditure of cellular energy
(b) is diffusion of water from one region to another
(c) is diffusion of water across a selectively permeable membrane
(d) is none of the above

_____ **8.** Which of the following reagents does *not* fit with the substance being tested for?
(a) Biuret reagent protein
(b) $BaCl_2$ starch
(c) $AgNO_3$ chloride ion
(d) albustix protein

_____ **9.** When the cytoplasm of a plant cell is pressed against the cell wall, the cell is said to be
(a) turgid
(b) plasmolyzed
(c) hemolyzed
(d) crenate

_____ **10.** If one solution contains 10% NaCl and another contains 30% NaCl, the 30% solution is
(a) isotonic
(b) hypotonic
(c) hypertonic
(d) plasmolyzed, with respect to the 10% solution

Name ____________________ Section Number ____________________

EXERCISE 6

Diffusion, Osmosis, and the Functional Significance of Biological Membranes

POST-LAB QUESTIONS

6.1 Experiment: Rate of Diffusion of Solutes

1. You want to dissolve a solute in water. Without shaking or swirling the solution, what might you do to increase the rate at which the solute would go into solution? Relate your answer to your method's effect on the motion of the molecules.

6.2 Experiment: Osmosis

2. If a 10% sugar solution is separated from a 20% sugar solution by a selectively permeable membrane, in which direction will there be a net movement of water?

3. Based on your observations in this exercise, would you expect dialysis membrane to be permeable to sucrose? Why?

6.4 Experiment: Plasmolysis in Plant Cells

4. You are having a party and you plan to serve celery, but your celery has gone limp, and the stores are closed. What might you do to make the celery crisp (turgid) again?

5. Why don't plant cells undergo osmotic lysis?

6. This drawing represents a plant cell that has been placed in a solution.
   a. What *process* is taking place in the direction of the arrows? What is happening at the cellular level when a wilted plant is watered and begins to recover from the wilt?

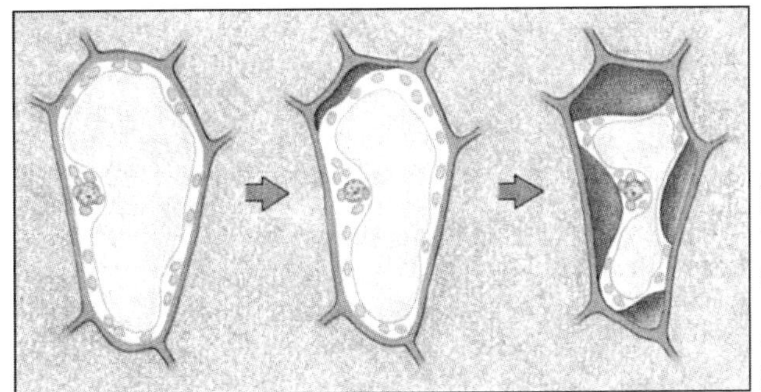

(After Starr and Taggart, 1989.)

   b. Is the solution in which the cells have been placed hypotonic, isotonic, or hypertonic relative to the cytoplasm?

6.5 Experiment: Osmotic Changes in Red Blood Cells

7. A human lost at sea without fresh drinking water is effectively lost in an osmotic desert. Why would drinking salt water be harmful?

Food for Thought

8. How does diffusion differ from osmosis?

9. Plant fertilizer consists of numerous different solutes. A small dose of fertilizer can enhance plant growth, but overfertilization can kill the plant. Why might overfertilization have this effect?

10. What does the word *lysis* mean? (*Now* does the name of the disinfectant Lysol® make sense?)

EXERCISE 7

Enzymes: Catalysts of Life

OBJECTIVES

After completing this exercise, you will be able to

1. define *catalyst, enzyme, activation energy, enzyme–substrate complex, substrate, product, active site, denaturation, cofactor;*
2. explain how an enzyme operates;
3. recognize benzoquinone as a brown substance formed in damaged plant tissue;
4. indicate the substrates for the enzyme catechol oxidase;
5. describe the effect of temperature on the rate of chemical reactions in general and on enzymatically controlled reactions in particular;
6. describe the effect that an atypical pH may have on enzyme action;
7. indicate how a cofactor might operate and identify a cofactor for catechol oxidase.

INTRODUCTION

Life as we know it is impossible without enzymes. The energy required by your muscles simply to open your lab manual would take years to accumulate without enzymes. Due to the presence of enzymes, the myriad chemical reactions occurring in your cells at this very moment are being completed in a fraction of a second rather than the years or even decades that would be otherwise required.

Enzymes are proteins that function as biological catalysts. A **catalyst** is a substance that lowers the amount of energy necessary for a chemical reaction to proceed. You might think of this so-called **activation energy** as a mountain to be climbed. Enzymes decrease the size of the mountain, in effect turning it into a molehill (Figure 7-1).

By lowering the activation energy, an enzyme affects the *rate* at which reaction occurs. Enzyme-boosted reactions may proceed from 100,000 to 10 million times faster than they would without the enzyme.

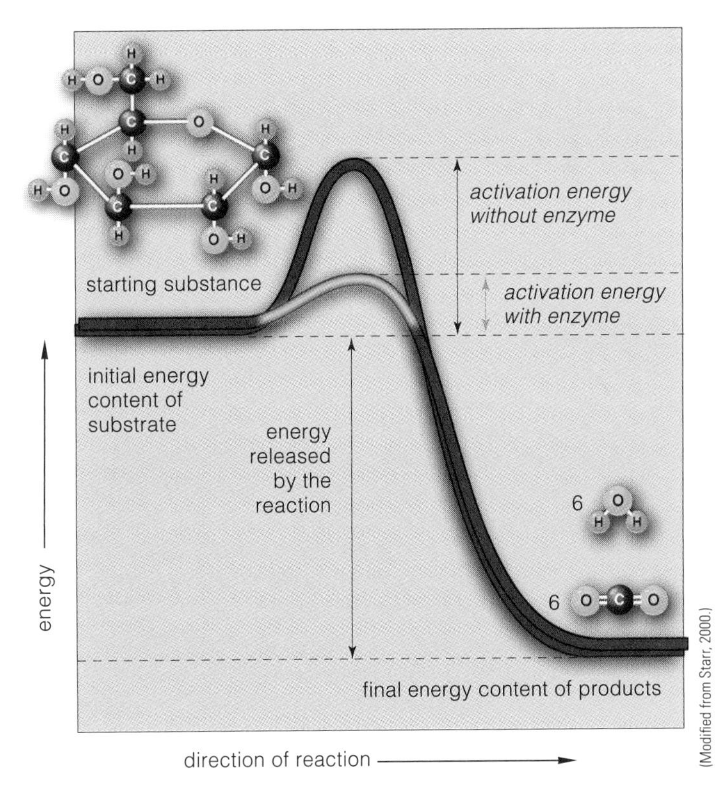

Figure 7-1 Enzymes and activation energy.

In an enzyme-catalyzed reaction, the reactant (the substance being acted upon) is called the **substrate.** Substrate molecules combine with enzyme molecules to form a temporary **enzyme–substrate complex**. **Products** are formed, and the enzyme molecule is released unchanged. Thus, the enzyme is not used up in the process and is capable of catalyzing the same reaction again and again. This can be summarized as follows:

$$\textbf{substrate} \xrightarrow{\text{enzyme}} \textbf{enzyme–substrate complex} \longrightarrow \textbf{products + enzyme}$$

Before we proceed, let's visualize an enzyme. Look at Figure 7-2. Using common, everyday items, think of an enzyme as a key that unlocks a lock. Imagine that you have a number of locks, and all use the same key. You would only need one key to unlock them.

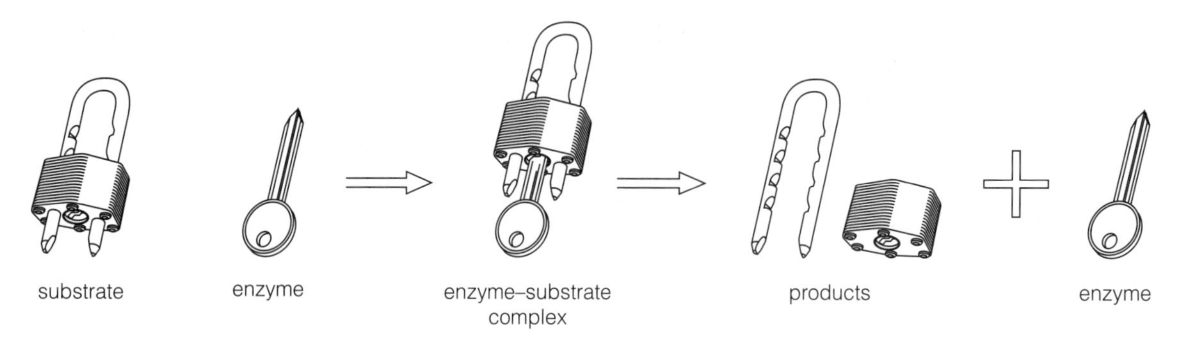

Figure 7-2 Action of an enzyme.

Changing the shape of the key just a tiny bit may still allow the key to function in the lock, but you may have to fumble with the key a bit to get the lock open. Changing the key more results in an inability to open the lock. Similarly, a change in the shape of an enzyme alters its function. We will examine a number of factors in this exercise to determine their effects on enzyme action, including

1. temperature
2. pH (hydrogen-ion concentration of the environment)
3. specificity (how discriminating the enzyme is in catalyzing different potential substrates)
4. cofactor necessity (the need for a metallic ion for enzyme activity)

Although thousands of enzymes are present within cells, we'll examine only one: catechol oxidase (also known as tyrosinase).

Work in groups of four for all sections in this exercise.

7.1 Using a Spectronic 20 (Spec 20) to Determine Color Changes

(About 15 min.)

If you are not using a spectrophotometer, an instrument that measures color change, skip the following 11-step procedure. (See Appendix 2 for an explanation of how the Spec 20 works.)

MATERIALS

Per student:

- disposable plastic gloves

Per student group (4):

- ice bath with wash bottle of potato extract containing catechol oxidase
- bottle of dH_2O
- china marker
- Spec 20 spectrophotometer
- 1-mL pipet and bulb
- 5-mL pipet and bulb

PROCEDURE: ZEROING THE SPEC 20

1. Obtain a clean test tube designed to fit in the Spec 20.
2. With a wax pencil, label the top of it with a C for "control."
3. If the tube does not already have a vertical line on it, place one on it with the wax pencil. This mark must face the front of the sample holder.
4. Using the 1-mL pipet and bulb, measure 1 mL of potato extract into test tube C.
5. Using the 5-mL pipet and bulb, add 5 mL of dH_2O to the test tube.
6. Place your gloved thumb over the mouth of the test tube and invert the tube to mix the contents.
7. Adjust the wavelength knob (top, right) of the Spec 20 to 540 nm.
8. Rotate the lower *left* absorbance adjustment so the needle reads infinity (∞) on the bottom scale.
9. Clean the surface of the tube by wiping it with a tissue and insert the tube into the Spec 20 sample holder with the vertical mark facing front (toward you).
10. Rotate the lower *right* absorbance adjustment so the needle reads zero.
11. You have now zeroed the Spec 20 for this exercise. Remove tube C and set it aside.

The previous procedure is done to account for the fact that the potato extract has color and absorbs light. In your experiments, you will be measuring the change in the amount of light absorbed by potato extract before and after various experimental treatments.

7.2 Formation and Detection of Benzoquinone *(About 20 min.)*

Catechol oxidase is an enzyme that catalyzes the production of benzoquinone and water from catechol:

$$\text{catechol} + \tfrac{1}{2} O_2 \xrightarrow{\text{catechol oxidase}} \text{benzoquinone} + H_2O$$

(substrates) (enzyme in potato extract) (product)

This is an oxidation reaction, with catechol and oxygen as the substrates. Hence, the enzyme gets its preferred name, *catechol oxidase.* (This suffix *-ase* is a tipoff that the substance is an enzyme.)

Catechol and catechol oxidase are present in the cells of many plants, although in undamaged tissue they are separated in different compartments of the cells. Injury causes mixing of the substrate and enzyme, producing benzoquinone, a brown substance.

You've probably noticed the brown coloration of a damaged apple or the blackening of an injured potato tuber (Figure 7-3). Benzoquinone inhibits the growth of certain microorganisms that cause rot.

In this section, you will form the product, benzoquinone, and establish a color intensity scale or absorbance standard that you will use in subsequent experiments.

(Photo by J. W. Perry.)

Figure 7-3 Potato showing browning due to benzoquinone production. The potato on the left was sliced immediately before this photo was taken. The one on the right had been cut and exposed to the oxygen in the air for several minutes before being photographed.

MATERIALS

Per student:

- disposable plastic gloves

Per student group (4):

- 3 test tubes
- test tube rack
- metric ruler
- ice bath with wash bottle of potato extract containing catechol oxidase
- wash bottle containing 1% catechol solution
- bottle of dH_2O
- china marker
- warmed up and zeroed Spec 20, optional; see Appendix 2

Per lab room:

- 40°C waterbath
- vortex mixer (optional)

PROCEDURE

Caution

Some of the chemicals (catechol, hydroquinone) used in these experiments can be hazardous to your health if they are ingested or taken in through your skin. Wear disposable plastic gloves for all experiments.

1. With a china marker, label three test tubes 2_A, 2_B, and 2_C. Place your initials on each test tube for later identification.
2. Lay the test tubes against a metric ruler and mark lines on the tubes corresponding to 1 cm and 2 cm *from the bottom* of each tube.
3. Fill each tube as follows:

   Tube 2_A: 1 cm of potato extract containing catechol oxidase
   1 cm of 1% catechol solution

   Tube 2_B: 1 cm of potato extract containing catechol oxidase
   1 cm of dH_2O

   Tube 2_C: 1 cm of 1% catechol solution
   1 cm of dH_2O

Note: **Be certain to return potato extract to ice bath IMMEDIATELY after use in this and subsequent experiments.**

4. Shake all tubes (using a vortex mixer if available).
5. Record the color of the solution in each tube in the "time 0" spaces of Table 7-1.
6. Place the tubes in a 40°C waterbath.
7. After 10 minutes, the catechol should be completely oxidized. Remove the tubes from the waterbath. Record the *color* of the substance in each test tube in Table 7-1.
   What you do next depends on whether you're using a spectrophotometer. Proceed with *either* step 8 *or* 9.
8. *Color-intensity method:* Consider the color of the product in tube 2_A to be a 5 on a color intensity scale of 0–5, while the color of the substance in tubes 2_B and 2_C is a 0. In Table 7-2, record the color intensity for tubes 2_B and 2_C as 0, and that of tube 2_A as 5.
   You will use this scale to make comparisons in Sections 7.3–7.6. Keep the contents in tubes 2_A, 2_B, and 2_C, and refer to them as you make comparisons in subsequent experiments.
9. *Spec 20 method:*
   (a) Pipet 6 mL of the contents of each tube into six separate Spec 20 tubes.
   (b) Clean off each tube.
   (c) Zero the Spec 20 using the contents of tube 2_C.
   (d) Insert tubes containing the contents of tube 2_A and then tube 2_B, determine the absorbance of each, and record them in Table 7-2.

TABLE 7-1 Formation and Detection of Benzoquinone: Record Color

Time	Tube 2_A: Potato Extract and Catechol	Tube 2_B: Potato Extract and Water	Tube 2_C: Catechol and Water
0 min.			
10 min.			

What is the brown-colored substance that appeared in tube 2_A? ______________________

What is the substrate for the reaction that occurred in tube 2_A? ______________________

What is the product of the reaction in tube 2_A? ______________________

What substances do tubes 2_B and 2_C lack that account for the absence of the brown-colored substance?

2_B ______________ 2_C ______________

What is the purpose of tubes 2_B and 2_C? ______________________

TABLE 7-2 Color-Intensity Scale or Absorbance

Intensity/Absorbance	Tube	Color of Product
	2_C	
	2_B	
	2_A	

7.3 Experiment: Enzyme Specificity *(About 20 min.)*

Generally, enzymes are substrate-specific, acting on one particular substrate or a small number of structurally similar substrates. This specificity is due to the three-dimensional structure of the enzyme. For the enzyme–substrate complex to form, the structure of the substrate must very closely complement that of the enzyme's

active site. The active site is a special region of the enzyme where the substrate binds. The active site has a small amount of moldability, so that the active site and substrate become fully complementary to each other, as shown in Figure 7-4.

Think again about the lock and key analogy (Figure 7-2). If the lock and key are not complementary, the lock won't open. But how exact a fit is necessary?

In this experiment, you will determine the ability of the enzyme catechol oxidase to catalyze the oxidation of two different but structurally similar substrates: catechol and hydroquinone. First, examine the chemical structure of each compound:

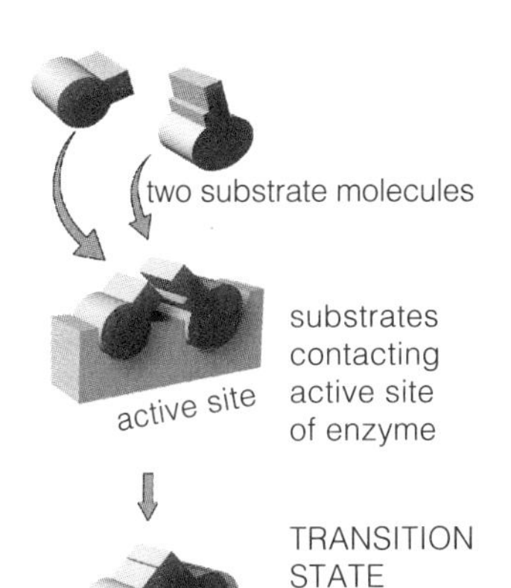

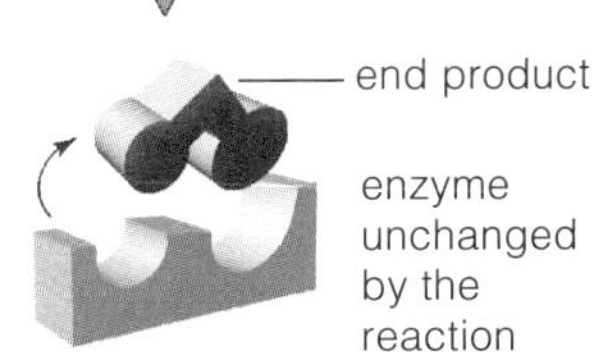

Figure 7-4 Induced-fit model of enzyme–substrate interactions.

OH OH
OH
OH
catechol hydroquinone

You need not memorize these structural formulas, but do notice that both are ring structures with two hydroxyl (—OH) groups attached.

Keep this in mind as you do the next experiment, in which you will determine how specific (discriminating) catechol oxidase is for particular substrates.

MATERIALS

Per student group (4):

- 3 test tubes
- test tube rack
- metric ruler
- china marker
- wash bottle containing 1% catechol
- wash bottle containing 1% hydroquinone
- ice bath with wash bottle of potato extract containing catechol oxidase
- warmed up and zeroed Spec 20, optional; see Appendix 2

Per lab room:

- 40°C waterbath
- vortex mixer (optional)

PROCEDURE

1. With a china marker, label three clean test tubes 3_A, 3_B, and 3_C. Include your initials for identification.
2. Lay the test tubes against a metric ruler and mark lines indicating 1 cm and 2 cm *from the bottom* of each test tube.
3. Fill each tube as follows:
   Tube 3_A: 1 cm of potato extract containing catechol oxidase
   1 cm of 1% catechol
   Tube 3_B: 1 cm of potato extract containing catechol oxidase
   1 cm of 1% hydroquinone
   Tube 3_C: 1 cm of potato extract containing catechol oxidase
   1 cm of dH_2O
4. Gently shake the test tubes to mix the contents.
5. Compare the color intensity of the solution in each test tube *with the standards produced in Section 7.2* and record them at time 0 in Table 7-3.
6. Place the test tubes in a 40°C waterbath.

This experiment addresses the hypothesis that *the structure of a substrate determines how well an enzyme acts upon the substrate.*

7. While you wait for the experiment to run its course, write your prediction of the outcome in Table 7-3.
8. After 10 minutes, remove the test tubes from the water bath and examine them. Choose step 9 *or* 10 and proceed.
9. *Color-intensity method:* Record the color intensity (scale 0–5) of each tube's contents in Table 7-3.

TABLE 7-3 Specificity of Catechol Oxidase for Different Substrates			
Prediction:			
	Relative Color Intensity or Absorbance		
Time	Tube 3_A: Catechol	Tube 3_B: Hydroquinone	Tube 3_C: dH_2O
0 min.			
10 min.			
Conclusion:			

10. *Spec 20 method:*
   (a) Pipet 6 mL of the contents of each tube into three labeled Spec 20 tubes.
   (b) Clean off each tube.
   (c) Zero the Spec 20 using the contents of tube 3_C.
   (d) Insert tubes containing the contents of tube 3_A and then tube 3_B, determine the absorbance of each, and record them in Table 7-3.

11. Upon which substrate does catechol oxidase work best, forming the most benzoquinone in the shortest amount of time?

____

12. Based on your knowledge of the structures of the two substrates, what apparently determines the specificity of catechol oxidase?

____

____

13. Why was tube 3_C included in this experiment?

____

____

14. Record your conclusion in Table 7-3, either accepting or rejecting the hypothesis.

7.4 Experiment: Effect of Temperature on Enzyme Activity *(25 min.)*

The rate at which chemical reactions take place is largely determined by the temperature of the environment. *Generally, for every 10°C rise in temperature, the reaction rate doubles.* Within a rather narrow range, this is true for enzymatic reactions also. However, because enzymes are proteins, excessive temperature alters their structure, destroying their ability to function. When an enzyme's structure is changed sufficiently to destroy its function, the enzyme is said to be **denatured.** Most enzymatically controlled reactions have an *optimum* temperature and pH—that is, one temperature and pH where activity is maximized.

In this experiment, you will determine the temperature range over which the enzyme catechol oxidase is able to catalyze its substrate. You will also determine the best (optimum) temperature for the reaction.

MATERIALS

Per student group (4):

- 6 test tubes
- test tube rack
- metric ruler
- china marker
- wash bottle containing 1% catechol
- ice bath with wash bottle of potato extract containing catechol oxidase
- three 400-mL graduated beakers
- heat-resistant glove
- Celsius thermometer
- warmed up and zeroed Spec 20, optional; see Appendix 2

Per student group (table):

- hot plate *or* burner, tripod support, wire gauze, and matches or striker
- boiling chips

Per lab room:

- source of room-temperature water
- three waterbaths: 40°C, 60°C, 80°C
- vortex mixer (optional)

PROCEDURE

1. Half fill one 400-mL beaker with tap water. Add a few boiling chips and turn on the hotplate to the highest temperature setting, *or,* if your lab is equipped with burners, light the burner. Bring the water to a boil, then turn the heat down so that the water just continues to boil.
2. Put 150 mL of tap water into a second beaker and add ice to the water.
3. Half fill a third beaker with water from the source at room temperature.
4. With a china marker, label six test tubes 4_A–4_F. Include your initials for identification.
5. Lay the test tubes against a metric ruler and mark off lines indicating 1 cm and 2 cm *from the bottom* of each tube.
6. Fill each tube to the 1-cm mark with potato extract containing catechol oxidase.
7. **(a)** Place tube 4_A in the 400-mL beaker of ice water.
   Measure and record the water temperature: ____________°C
   (b) Place tube 4_B in the 400-mL beaker containing room-temperature water.
   Room temperature: ________________°C
   (c) Place tube 4_C in the 40°C waterbath.
   (d) Place tube 4_D in the 60°C waterbath.
   (e) Place tube 4_E in the 80°C waterbath.
   (f) Place tube 4_F in the 400-mL beaker containing boiling water.
   Temperature of boiling water: _______________°C
8. Allow the test tubes to remain at the various temperatures for 5 minutes.
9. Remove the tubes and add catechol to the 2-cm line on each. Agitate the tubes (with a vortex mixer if available) to mix the contents.
10. *Color-intensity method:* In Table 7-4, record the relative color intensity (scale 0–5) of the solution in each tube, using the standard established in Section 7.2. Return each tube to its respective temperature bath immediately after recording.

This experiment addresses the hypothesis that *the temperature of a substrate and an enzyme determines the amount of product that is formed.*

Caution

Wear a heat-resistant glove when handling heated glassware.

11. While you wait for the experiment to run its course, write your prediction of the outcome of the experiment in Table 7-4.
12. Shake periodically (by hand) all tubes over the next 10 minutes.
13. After 10 minutes, remove the test tubes from the water baths. Choose step 14 *or* 15 and proceed.
14. *Color-intensity method:* Record the color intensity (scale 0–5) of each tube's contents in Table 7-4.

TABLE 7-4 Effect of Temperature on Enzyme Activity

Prediction:						
	Relative Color Intensity or Absorbance					
Time	Tube 4_A	Tube 4_B	Tube 4_C	Tube 4_D	Tube 4_E	Tube 4_F
0 min.						
10 min.						
Conclusion:						

15. *Spec 20 method:*
    (a) Pipet 6 mL of the contents of each tube into six labeled Spec 20 tubes.
    (b) Clean off each tube.
    (c) Zero the Spec 20 using the contents of tube 2_C.
    (d) Insert tubes containing the contents of each tube, recording them in Table 7-4.
16. Plot the data from Table 7-4 for the 10-minute reading in Figure 7-5.
17. Over what temperature *range* is catechol oxidase active?

____

18. What is the *optimum* temperature for activity of this enzyme?

____

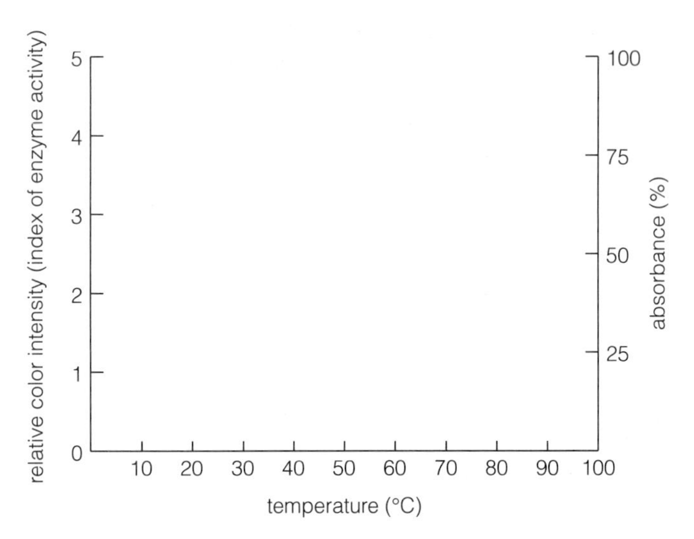

Figure 7-5 Effect of temperature on catechol oxidase activity.

19. What happens to enzyme activity at very high temperatures?

____

____

20. Record your conclusion in Table 7-4, either accepting or rejecting the hypothesis.

7.5 Experiment: Effect of pH on Enzyme Activity *(25 min.)*

Another factor influencing the rate of enzyme catalysis is the hydrogen-ion concentration (pH) of the solution. Like temperature, pH affects the three-dimensional shape of enzymes, thus regulating their function. Most enzymes operate best when the pH of the solution is near neutrality (pH 7). Others, however, have pH optima in the acidic or basic range, corresponding to the environment in which they normally function.

In this experiment, you will determine the pH range over which the enzyme catechol oxidase is able to catalyze its substrate. You will also determine the optimum pH for the reaction.

MATERIALS

Per student group (4):

- 7 test tubes
- test tube rack
- metric ruler
- china marker
- wash bottle containing 1% catechol
- ice bath with wash bottle of potato extract containing catechol oxidase
- warmed up and zeroed Spec 20, optional; see Appendix 2

Per lab room:

- 40°C waterbath
- phosphate buffer series, pH 2–12 (2, 4, 6, 7, 8, 10, 12)
- vortex mixer (optional)

PROCEDURE

1. With a china marker, label seven test tubes 5_A–5_G. Include your initials for identification.
2. Lay the test tubes against a metric ruler and mark lines indicating 4 cm, 5 cm, and 6 cm *from the bottom* of each tube.
3. Take your test tubes to the location of the phosphate buffer series and fill each tube according to the following directions:

4. Return to your work area and add 1 cm of potato extract containing catechol oxidase to each of the seven tubes (thus bringing the total volume of each to the 5-cm mark). Agitate the tubes by hand.
5. Add 1% catechol to each of the seven tubes, bringing the total volume to the 6-cm mark. Agitate the contents of the tubes, using a vortex mixer if available.
6. In Table 7-5 at time 0, record the relative color intensity of each tube *immediately after adding the 1% catechol.*
7. Place the tubes in the 40°C waterbath.
8. Agitate the tubes periodically over the next 10 minutes.

Tube	Fill to the 4-cm Mark with Buffer of
5_A	pH 2
5_B	pH 4
5_C	pH 6
5_D	pH 7
5_E	pH 8
5_F	pH 10
5_G	pH 12

This experiment addresses the hypothesis that *the pH of a substrate and an enzyme determines the amount of product that is formed.*

9. While you wait for the experiment to run its course, write your prediction of the outcome in Table 7-5.
10. After 10 minutes, remove the test tubes from the water baths. Choose step 11 *or* 12 and proceed.
11. *Color-intensity method:* Record the color intensity (scale 0–5) of each tube's contents in Table 7-5.

TABLE 7-5 Effect of pH on Enzyme Activity

Prediction:							
	Relative Color Intensity or Absorbance						
Time	**Tube 5_A (pH 2)**	**Tube 5_B (pH 4)**	**Tube 5_C (pH 6)**	**Tube 5_D (pH 7)**	**Tube 5_E (pH 8)**	**Tube 5_F (pH 10)**	**Tube 5_G (pH 12)**
0 min.							
10 min.							
Conclusion:							

12. *Spec 20 method:*
    (a) Pipet 6 mL of the contents of each tube into seven labeled Spec 20 tubes.
    (b) Clean off each tube.
    (c) Zero the Spec 20 using the contents of tube 2_C from Section 7.2.
    (d) Insert each tube, determine the absorbance, and record it in Table 7-5.
13. Plot the data from Table 7-5 for your 10-minute reading in Figure 7-6.
14. Over what pH *range* does catechol oxidase catalyze catechol to benzoquinone?

____

15. What is the *optimum* pH for catechol oxidase activity?

____

16. Record your conclusion in Table 7-5, either accepting or rejecting the hypothesis.

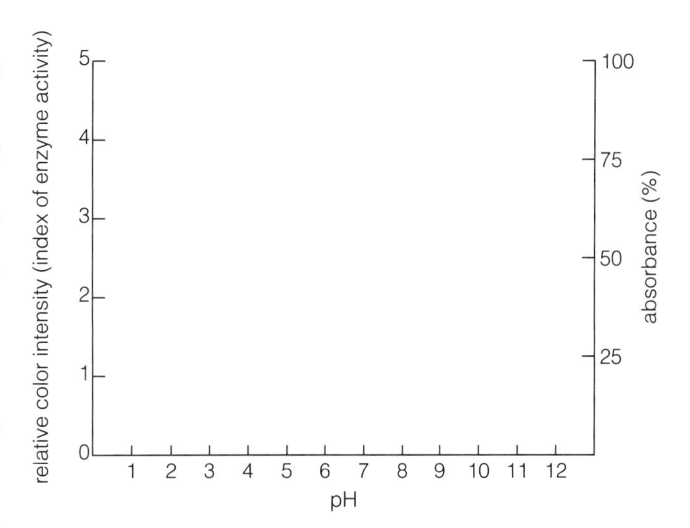

Figure 7-6 Effect of pH on catechol oxidase activity.

7.6 Experiment: Necessity of Cofactors for Enzyme Activity *(15 min.)*

Some enzymatic reactions occur only when the proper *cofactors* are present. **Cofactors** are nonprotein organic molecules and metal ions that are part of the structure of the active site, making the formation of the enzyme–substrate complex possible.

In this experiment, you will use phenylthiourea (PTU), which binds strongly to copper, to remove copper ions. Thus, you'll be able to determine whether copper is a necessary cofactor necessary for producing benzoquinone from catechol.

MATERIALS

Per student group (4):

- 2 test tubes
- test tube rack
- metric ruler
- china marker
- ice bath with wash bottle of potato extract containing catechol oxidase
- wash bottle containing 1% catechol solution
- bottle of dH_2O
- china marker
- scoopula (small spoon)
- phenylthiourea crystals in small screw-cap bottle
- warmed up and zeroed Spec 20, optional; see Appendix 2

Per lab room:

- 40°C waterbath
- vortex mixer (optional)
- bottle of 95% ethanol (at each sink)
- tissues (at each sink)

PROCEDURE

1. With a china marker, label two test tubes 6_A and 6_B. Include your initials for identification.
2. Lay the test tube against a metric ruler and mark lines indicating 1 cm and 2 cm *from the bottom* of each tube.
3. Add potato extract containing catechol oxidase to the 1-cm mark of each test tube.
4. Using a scoopula, add five crystals of phenylthiourea (PTU) to tube 6_A. Do not add anything to tube 6_B.
5. Agitate the contents of both test tubes frequently by hand during the next 5 minutes.
6. Add 1% catechol to the 2-cm mark of both test tubes and agitate the contents of the tubes, using a vortex mixer if available.
7. *Color-intensity method:* In Table 7-5 at time 0, record the relative color intensities (scale of 0–5).

Caution

PTU is poisonous.

TABLE 7-6 Is Copper a Cofactor for Catechol Oxidase?

Prediction:

	Relative Color Intensity or Absorbance	
Time	**Tube 6_A: with PTU**	**Tube 6_B: without PTU**
0 min.		
10 min.		

Conclusion:

8. Place the tubes in a 40°C waterbath. Agitate the tubes several times during the next 10 minutes.

This experiment addresses the hypothesis that *a cofactor is necessary for the action of the enzyme catecol oxidase.*

9. While you wait for the experiment to run its course, write your prediction of the outcome in Table 7-6.
10. After 10 minutes, remove the test tubes from the water baths. Choose step 11 *or* 12 and proceed.
11. *Color-intensity method:* Record the color intensity (scale 0–5) of each tube's contents in Table 7-6.

12. *Spec 20 method:*
    (a) Pipet 6 mL of the contents of each tube into seven separate Spec 20 tubes.
    (b) Clean off each tube.
    (c) Zero the Spec 20 using the contents of tube 2_C from Section 7.2.
    (d) Insert each tube in the Spec 20, determine the absorbance, and record it in Table 7-6.
13. Did benzoquinone form in tube 6_A? In tube 6_B? ______________________________
14. From this experiment, what can you conclude about the necessity for copper for catechol oxidase activity?

    ____
15. What substance used in this experiment contained copper?

    ____
16. Record your conclusion in Table 7-5, either accepting or rejecting the hypothesis.

Note: **After completing all experiments, take your dirty glassware to the sink and wash it following directions on page x. Use 95% ethanol to remove the china marker. Invert the test tubes in the test tube racks so they drain. Tidy up your work area, making certain all materials used in this exercise are there for the next class.**

PRE-LAB QUESTIONS

_____ **1.** Enzymes are
(a) biological catalysts
(b) agents that speed up cellular reactions
(c) proteins
(d) all of the above

_____ **2.** Enzymes function by
(a) being consumed (used up) in the reaction
(b) lowering the activation energy of a reaction
(c) combining with otherwise toxic substances in the cell
(d) adding heat to the cell to speed up the reaction

_____ **3.** The substance that an enzyme combines with is
(a) another enzyme
(b) a cofactor
(c) a coenzyme
(d) the substrate

_____ **4.** Enzyme specificity refers to the
(a) need for cofactors for some enzymes to function
(b) fact that enzymes catalyze one particular substrate or a small number of structurally similar substrates
(c) effect of temperature on enzyme activity
(d) effect of pH on enzyme activity

_____ **5.** For every 10°C rise in temperature, the rate of most chemical reactions will
(a) double
(b) triple
(c) increase by 100 times
(d) stop

_____ **6.** When an enzyme becomes denatured, it
(a) increases in effectiveness
(b) loses its requirement for a cofactor
(c) forms an enzyme–substrate complex
(d) loses its ability to function

_____ **7.** An enzyme may lose its ability to function because of
(a) excessively high temperatures
(b) a change in its three-dimensional structure
(c) a large change in the pH of the environment
(d) all of the above

_____ **8.** pH is a measure of
(a) an enzyme's effectiveness
(b) enzyme concentration
(c) the hydrogen-ion concentration
(d) none of the above

_____ **9.** Catechol oxidase
(a) is an enzyme found in potatoes
(b) catalyzes the production of catechol
(c) has as its substrate benzoquinone
(d) is a substance that encourages the growth of microorganisms

_____ **10.** The relative color intensity used in the experiments of this exercise
(a) is a consequence of production of benzoquinone
(b) is an index of enzyme activity
(c) may differ depending on the pH, temperature, or presence of cofactors, respectively
(d) is all of the above

Name ______________________ Section Number ______________________

EXERCISE 7

Enzymes: Catalysts of Life

POST-LAB QUESTIONS

7.4 Experiment: Effect of Temperature on Enzyme Activity

1. Eggs can contain bacteria such as *Salmonella.* Considering what you've learned in this exercise, explain how cooking eggs makes them safe to eat.

2. As you demonstrated in this experiment, high temperatures inactivate catechol oxidase. How is it that some bacteria live in the hot springs of Yellowstone Park at temperatures as high as 73°C?

3. Why do you think high fevers alter cellular functions?

4. Some surgical procedures involve lowering a patient's body temperature during periods when blood flow must be restricted. What effect might this have on enzyme-controlled cellular metabolism?

5. At one time, it was believed that individuals who had been submerged under water for longer than several minutes could not be resuscitated. Recently this has been shown to be false, especially if the person was in cold water. Explain why cold-water "drowning" victims might survive prolonged periods under water.

7.5 Experiment: Effect of pH on Enzyme Activity

6. Explain what happens to catechol oxidase when the pH is on either side of the optimum.

7. What would you expect the pH optimum to be for an enzyme secreted into your stomach?

Food for Thought

8. Is it necessary for a cell to produce one enzyme molecule for every substrate molecule that needs to be catalyzed? Why or why not?

9. Explain the difference between *substrate* and *active site.*

10. The photo shows slices of two apples. The one on the left sat on the counter for 15 minutes prior to being photographed. The one on the right was sliced immediately prior to the photo being taken.

a. Explain as thoroughly as possible what you see and why the two slices differ.

(Photo by J. W. Perry.)

b. If you don't want a cut apple to brown, what can you do to prevent it?

Photosynthesis: Capture of Light Energy

OBJECTIVES

After completing this exercise, you will be able to

1. define *photosynthesis, autotroph, heterotroph, chlorophyll, chromatogram, absorption spectrum, carotenoid;*
2. describe the role of carbon dioxide in photosynthesis;
3. determine the effect of light and carbon dioxide on photosynthesis;
4. determine the wavelengths absorbed by pigments;
5. identify the pigments in spinach chloroplast extract;
6. identify the carbohydrate produced in geranium leaves during photosynthesis;
7. identify the structures composing the chloroplast and indicate the function of each structure in photosynthesis.

INTRODUCTION

Photosynthesis is the process by which light energy converts inorganic compounds to organic substances with the subsequent release of elemental oxygen. It may very well be the most important biological event sustaining life. Without it, most living things would starve, and atmospheric oxygen would become depleted to a level incapable of supporting animal life. Ultimately, the source of light energy is the sun, although on a small scale we can substitute artificial light.

Nutritionally, two types of organisms exist in our world, autotrophs and heterotrophs. **Autotrophs** (*auto* means self, *troph* means feeding) synthesize organic molecules (carbohydrates) from inorganic carbon dioxide. The vast majority of autotrophs are the photosynthetic organisms that you're familiar with—plants, as well as some protistans and bacteria. These organisms use light energy to produce carbohydrates. (A few bacteria produce their organic carbon compounds chemosynthetically, that is, using chemical energy.)

By contrast, **heterotrophs** must rely directly or indirectly on autotrophs for their nutritional carbon and metabolic energy. Heterotrophs include animals, fungi, many protistans, and most bacteria.

In both autotrophs and heterotrophs, carbohydrates originally produced by photosynthesis are broken down by *cellular respiration* (Exercise 10), releasing the energy captured from the sun for metabolic needs.

The photosynthetic reaction is often conveniently summarized by the equation:

$$\underset{\text{water}}{12H_2O} + \underset{\text{carbon dioxide}}{6CO_2} \xrightarrow{\text{light energy}} \underset{\text{oxygen}}{6O_2} + \underset{\text{glucose}}{C_6H_{12}O_6} + \underset{\text{water}}{6H_2O}$$

114 - 120

Although glucose is often produced during photosynthesis, it is usually converted to another transport or storage compound unless it is to be used immediately for carbohydrate metabolism. In plants and many protistans, the most common storage carbohydrate is *starch*, a compound made up of numerous glucose units linked together. Starch is designated by the chemical formula $(C_6H_{12}O_6)_n$, where n indicates a large number. Most plants transport carbohydrate as sucrose.

The following experiments will acquaint you with the principles of photosynthesis.

8.1 Test for Starch *(About 10 min.)*

As indicated by the overall formula of photosynthesis, one end product is a carbohydrate (CH_2O). But a number of different carbohydrates have the empirical formula CH_2O. In this section, you will perform a simple test to visually distinguish between two different carbohydrates and water.

MATERIALS

Per student group (4):

- 1 dropper bottle each of
  - iodine (I_2KI) solution
  - starch solution
  - glucose solution
  - dH_2O
- depression (spot) plate

PROCEDURE

1. Place a couple drops of starch, glucose, and distilled water in three different depressions of the spot plate. Now add a drop of iodine solution to each.
2. Record your observations. How can you identify the presence of starch?

   Observations: ______________________________

   ____

   ____

8.2 Experiment: Effects of Light and Carbon Dioxide on Starch Production

(About 30 min.)

In this experiment, you will perform a test to determine the environmental conditions necessary for photosynthesis and starch production. This experiment addresses the hypothesis that *photosynthesis proceeds only in the presence of light and carbon dioxide.*

MATERIALS

Per student group (4):

- two 400-mL beakers
- square of aluminum foil
- hot plate in fume hood
- heat-resistant glove
- petri dish halves
- bottle of iodine solution
- bottle of 95% ethanol (EtOH)
- forceps

Per lab room:

- source of dH_2O
- Fast Plants™, 9–10 days old, grown for 4 days in three different environments:
  - I. normal conditions with both light and carbon dioxide
  - II. in dark, with normal carbon dioxide
  - III. in light, but with carbon dioxide removed

PROCEDURE

1. Carefully observe and record any differences in appearance among the three sets of plants in Table 8-1.
2. Write a prediction regarding starch presence, and thus photosynthesis activity, for each growing condition in Table 8-1.
3. Pigments present in the plants must be removed before a test for starch presence can be performed. Kill the plants and extract the pigments:
   - **(a)** With a china marker, label the beakers A (for alcohol) and dH_2O.
   - **(b)** Add about 150 mL distilled water to the dH_2O beaker, set it on the hot plate, and turn on the hot plate to the highest setting. Allow the water to come to a boil.
   - **(c)** Completely remove 1–2 plants of each treatment from its growing container. Wash off all soil from the roots. Keep plants of each treatment separate and labeled.
   - **(d)** Alcohol has a much lower boiling point than water, and so takes very little time to come to a boil. When the water is boiling, put about 150 mL of alcohol in the A beaker, set it also on the hot plate and bring to a boil. Keep the alcohol beaker covered with aluminum foil as much as possible throughout the lab to prevent excess evaporation.
   - **(e)** Place the plants from one treatment in the beaker of boiling water for about 1 minute. This kills the tissue and breaks down internal membranes.
   - **(f)** Use the long forceps to move the wilted plants from the water into the boiling alcohol. This will extract the photosynthetic pigments from the plant tissues. When the pigments have been extracted, the liquid will appear green, and the plant will appear to be mostly bleached.

Caution

Ethanol is highly flammable. Use only electric hot plates, never open flame. Also, never let a beaker boil dry. Add more liquid, or remove the beaker from the burner, and place it on a pad of folded paper towels.

TABLE 8-1 Effect of Light and Carbon Dioxide on Starch Presence

Fast Plants™ Growing Condition	Appearance	Prediction	Starch Presence and Location
I. Normal conditions with both light and carbon dioxide			
II. In dark, with normal carbon dioxide			
III. In light, but with carbon dioxide removed			

(g) Remove the plants from the alcohol with forceps, and dip momentarily in the boiling water to soften.

4. Test the plants for the presence and localization of starches:
   (a) Place killed, depigmented plants in petri dishes filled with iodine solution.
   (b) Let the plants soak in the iodine solution for a couple minutes, rinse, and float in water in another petri dish in order to observe the pattern of staining.
5. Repeat the pigment extraction and staining process for plants of the other two treatments, being careful to keep the treatments separate and identified.
6. Remove the A beaker from the hot plate, and turn the heat off.
7. In Figure 8-1, sketch a plant from each experimental treatment, shading in the portions that stained dark. Be careful to note where any dark staining occurs. Record your written observations in Table 8-1.

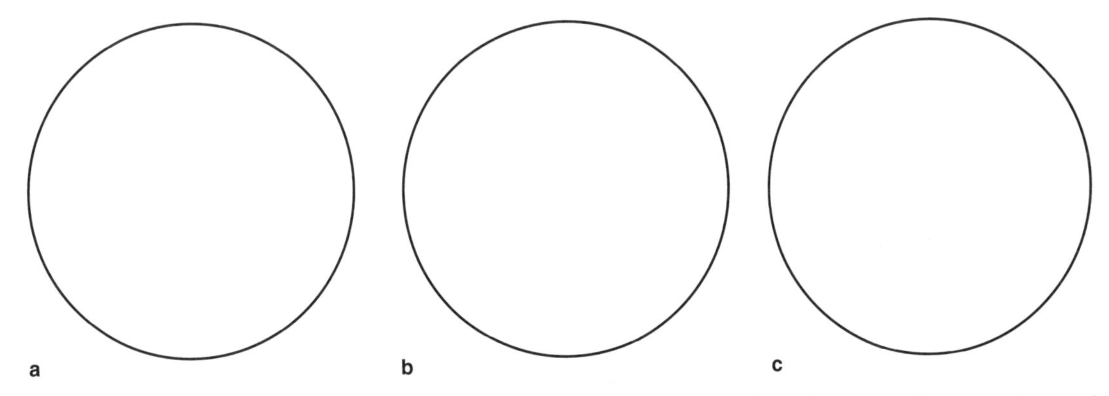

Figure 8-1 Distribution of starch in Fast Plants™. (**a**) Normal light and CO_2; (**b**) dark, normal CO_2; (**c**) light, no CO_2.

8. What does this staining pattern in plants of the three treatments indicate?

____

____

____

9. What conclusion can you draw about the effect of light on the presence of starch?

____

____

10. What conclusion can you draw about the effect of carbon dioxide on the presence of starch?

____

____

11. Write a conclusion either accepting or rejecting the hypothesis.

____

8.3 Experiment: Relationship Between Light and Photosynthetic Products

(About 15 min.)

This experiment addresses the hypothesis that *light is necessary for photosynthesis to proceed.*

MATERIALS

Per student group (4):

- china marker
- two 400-mL graduated beakers
- hot plate in fume hood
- heat-resistant glove
- bottle of 95% ethanol
- forceps
- 2 petri dishes
- I_2KI solution in foil-wrapped stock bottle

Per lab room:

- source of dH_2O
- light-grown geranium plant or leaves of geranium plant with "masks"
- dark-grown geranium plant or leaves of geranium plant with "masks" (kept in dark place)

PROCEDURE

Work in groups of four.

1. Observe the two geranium plants available. One plant has been growing in bright light for several hours; the other has been kept in the dark for a day or more. Both leaves have had an area of the lamina (blade) masked by an opaque design.
2. Write a prediction regarding starch presence, and thus photosynthesis activity, for each growing condition and leaf treatment area in Table 8-2.

TABLE 8-2 Relationship Between Light and Starch Production

Geranium Plant Growing Condition	Prediction		Starch Presence and Location	
	Masked Areas	Unmasked Areas	Masked Areas	Unmasked Areas
Light-grown				
Dark-grown				

3. Select a leaf from one of the two geranium plants. Pigments present in the plants must be removed before a test for starch presence can be performed. Kill the leaf and extract the pigments:
   (a) With a china marker, label the beakers A (for alcohol) and dH_2O.
   (b) Add about 150 mL distilled water to the dH_2O beaker, set it on the hot plate, and turn on the hot plate to the highest setting. Allow the water to come to a boil.
   (c) Remove a treated leaf from each plant. Keep each separate and labeled. Remove the opaque cover from the leaf before proceeding.

(d) Alcohol has a much lower boiling point than water, and so takes very little time to come to a boil. When the water is boiling, put about 150 mL of alcohol in the A beaker, set it also on the hot plate, and bring to a boil. Keep the alcohol beaker covered with aluminum foil as much as possible throughout the lab to prevent excess evaporation.

(e) Place the leaf from one treatment in the beaker of boiling water for about 1 minute. This kills the tissue and breaks down internal membranes.

(f) Use the long forceps to move the wilted leaf from the water into the boiling alcohol. This will extract the photosynthetic pigments from the plant tissues. When the pigments have been extracted, the liquid will appear green, and the leaf will appear to be mostly bleached.

(g) Remove the leaf from the alcohol with forceps, dip momentarily in the boiling water to soften.

Caution

Ethanol is highly flammable. Use only electric hot plates, never open flame. Also, never let a beaker boil dry. Add more liquid, or remove the beaker from the burner, and place it on a pad of folded paper towels.

4. Test the plants for the presence and localization of starches:
   (a) Place killed, depigmented leaf in a petri dish filled with iodine solution.
   (b) Let the leaf soak in the iodine solution for a couple minutes, rinse, and float in water in another petri dish in order to observe the pattern of staining.
5. Show the distribution of the stain in the leaf by shading in and labeling Figure 8-2. In the blank provided in the legend for Figure 8-2, record the *substance* that I_2KI stains.
6. Repeat the pigment extraction and staining process for a leaf of the other treatment. Show the distribution of stain in this leaf by shading in and labeling Figure 8-2.
7. Remove the A beaker from the hot plate, and turn the heat off.

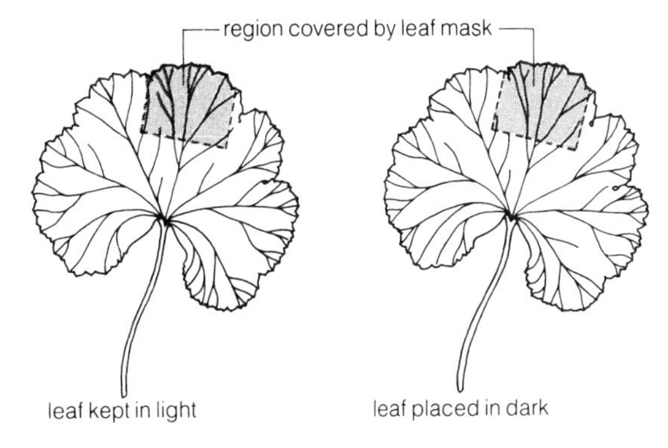

Figure 8-2 Distribution of the photosynthetic product ______________.

What does the blue-black coloration of the leaf indicate? ______________________________

____

Why did the masked area fail to stain? ______________________________

____

Write a conclusion accepting or rejecting the hypothesis. ______________________________

____

8.4 Experiment: Necessity of Photosynthetic Pigments for Photosynthesis

(About 15 min.)

Coleus plants are widely planted ornamentals that are popular for their striking foliage color patterns. Observe the plants available in the lab and note their wide variety and attractiveness. This experiment addresses the hypothesis that *chlorophyll is necessary for photosynthesis to occur.*

MATERIALS

Per student group (4):
- colored pencils or pens
- two 400-mL beakers
- hot plate in fume hood
- heat-resistant glove
- petri dish halves
- bottle of iodine solution
- bottle of 95% ethanol (EtOH)
- forceps

Per lab room:
- source of dH_2O
- variegated *Coleus* plants

PROCEDURE

1. **Obtain a leaf of variegated *Coleus*.** In the left-hand circle, carefully sketch the leaf, indicating the distribution of each color on the leaf with colored pencils or pens. Green coloration is due to chlorophyll, the major photosynthetic pigment. Pink colors are caused by water-soluble anthocyanin pigments (not involved in photosynthesis), and yellows are formed by carotenoid pigments. Be sure that you look at both surfaces of the leaf in case pigment distribution differs.

   Make a prediction regarding starch presence and photosynthetic activity for each pigmentation area:

   ____

   ____

2. Kill and extract the pigments from the leaf:
   (a) With a china marker, label the beakers A (for alcohol) and dH_2O.
   (b) Add about 150 mL distilled water to the dH_2O beaker, set it on the hot plate, and turn on the hot plate to the highest setting. Allow the water to come to a boil.
   (c) Alcohol has a much lower boiling point than water, and so takes very little time to come to a boil. When the water is boiling, put about 150 mL of alcohol in the A beaker, set it also on the hot plate, and bring to a boil. Keep the alcohol beaker covered with aluminum foil as much as possible throughout the lab to prevent excess evaporation.
   (d) Place the leaf from one treatment in the beaker of boiling water for about 1 minute. This kills the tissue and breaks down internal membranes.
   (e) Use the long forceps to move the wilted leaf from the water into the boiling alcohol. This will extract the photosynthetic pigments from the plant tissues. When the pigments have been extracted, the liquid will appear green, and the leaf will appear to be mostly bleached.
   (f) Remove the leaf from the alcohol with forceps, and dip momentarily in the boiling water to soften.

> ***Caution***
>
> ***Ethanol is highly flammable. Use only electric hot plates, never open flame. Also, never let a beaker boil dry. Add more liquid, or remove the beaker from the burner, and place it on a pad of folded paper towels.***

3. Test the leaf for the presence and localization of starch:
   (a) Place the killed, depigmented leaf in a petri dish filled with iodine solution.
   (b) Let the leaf soak in the iodine solution for a couple minutes, rinse, and float in water in another petri dish in order to observe the pattern of staining.
4. On the right-hand side of the space below, resketch the leaf, indicating the pattern of staining with iodine.

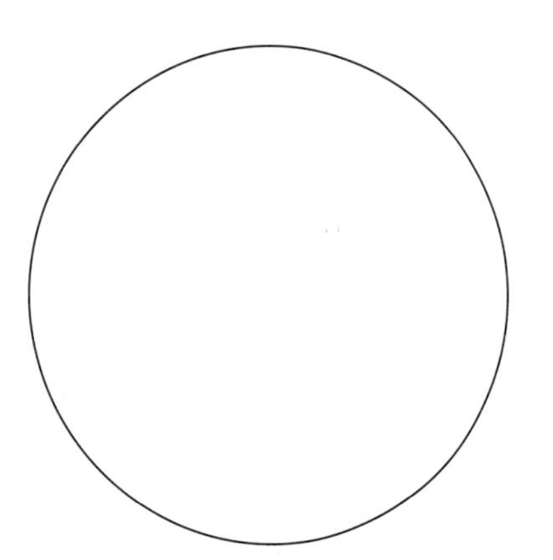

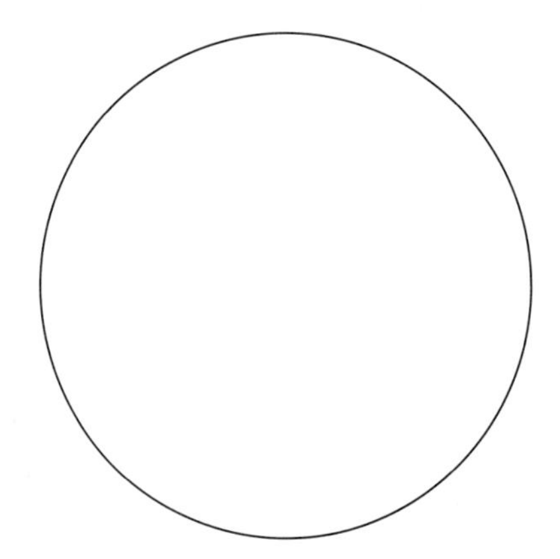

5. How does the pattern of starch storage relate to the distribution of chlorophyll?

   ____

   ____

6. Write a conclusion either accepting or rejecting the hypothesis.

   ____

   ____

8.5 Experiment: Absorption of Light by Chloroplast Extract *(10 min.)*

We tend to think of sunlight as being white. However, as you will see in this experiment, white light consists of a continuum of wavelengths. If we see light of just one wavelength, that light will appear colored.

When light hits a pigmented surface, some of the wavelengths are absorbed and others are reflected or transmitted. In this experiment, you will discover *which* wavelengths are absorbed, transmitted, or reflected by *particular* pigments, among them the photosynthetic pigment **chlorophyll.**

MATERIALS

Per lab room:

- several spectroscope setups (Figure 8-3a)
- sets of colored pencils (violet, blue, green, yellow, orange, red)
- colored filters (blue, green, red)
- small test tube containing pigment extract

Per student:

- hand-held spectroscope (optional; Figure 8-3b)

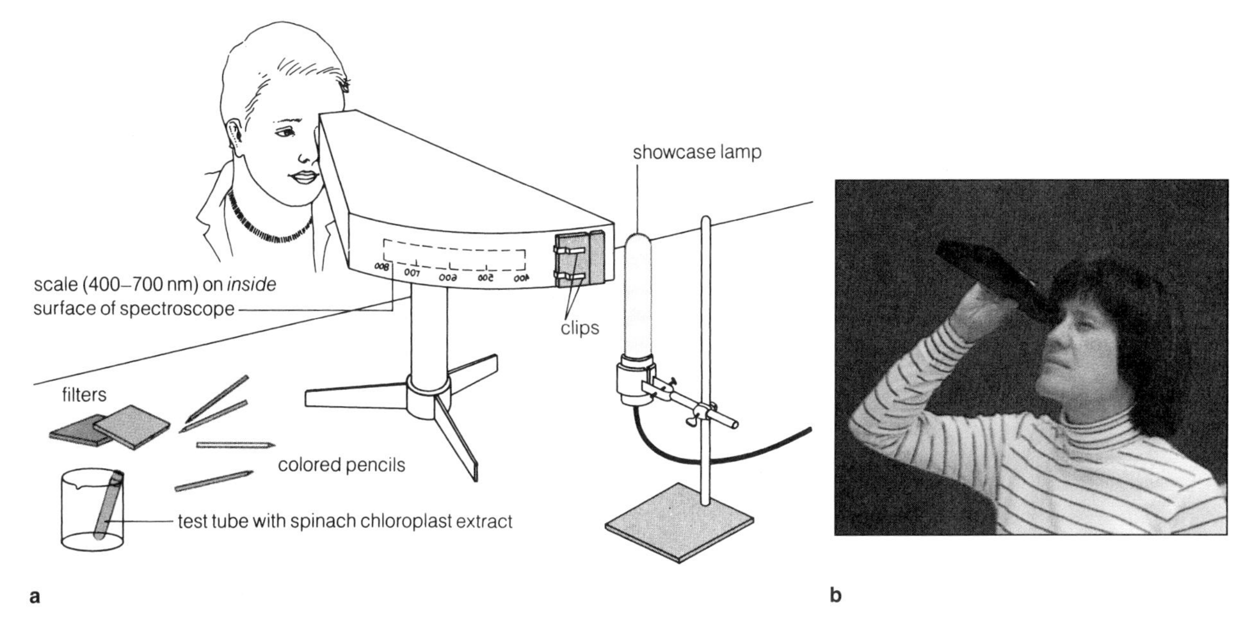

Figure 8-3 Use of a spectroscope. (**a**) Table-mounted. (After Abramoff and Thomson, 1982.) (**b**) Hand-held. (Photo by J. W. Perry.)

PROCEDURE

Work alone.

One way to separate light into its component parts is to view the light through a spectroscope. The spectroscope contains a prism that causes a spectrum of colors to form. A nanometer scale is imposed on the spectrum to indicate the wavelength of each component of white light.

1. Observe the spectrum of white light given off by an incandescent bulb through the spectroscope. With the colored pencils provided, record the positions of the colors violet, blue, green, yellow, orange, and red on the scale in Figure 8-4.
2. Observe the spectrum produced by the three colored filters using the spectroscope.

Which color or colors are absorbed when a red filter is placed between the light and the prism?

____

____

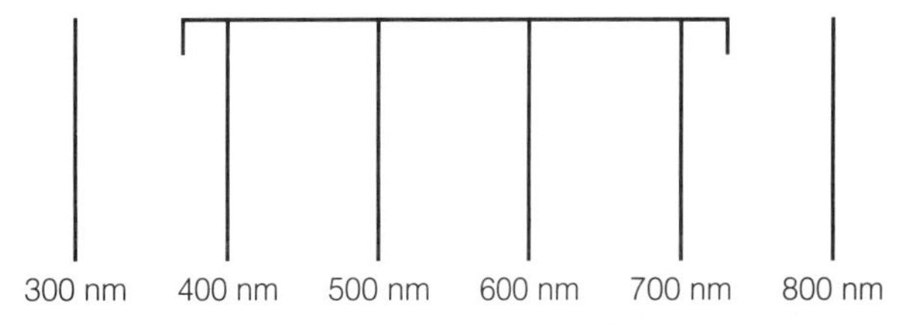

Figure 8-4 Spectrum of white light.

When a blue filter is used? ________________

A green filter? ________________

Make a general statement concerning the color of a pigment (filter) and the absorption of light by that pigment.

____

____

3. Now obtain a small test tube containing spinach chloroplast pigment extract and place it between the light source and the spectroscope. By adjusting the height of the tube so that the upper portion of the light passes through the pigment extract and the lower portion is white light, you can compare the absorption spectrum of the pigment extract with the spectrum of white light.

An **absorption spectrum** is a spectrum of light waves absorbed by a particular pigment. By contrast, the wavelengths that pass through the pigment extract and are visible in the spectroscope make up the *transmission spectrum* of the pigment.

4. Using the colored pencils, record the transmission spectrum of the chloroplast extract on the scale in Figure 8-5.

How does the absorption spectrum of the chloroplast extract compare with the absorption spectrum of the green filter?

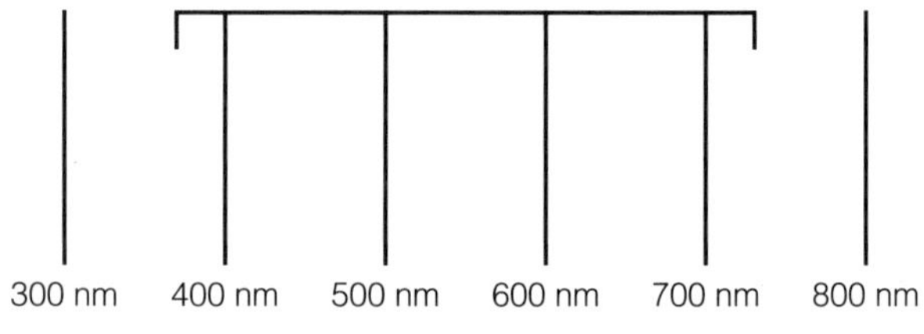

Figure 8-5 Transmission spectrum of chloroplast extract.

____

____

How might you explain the difference in absorption by the green filter and by the chloroplast pigment extract? (You might want to do Section 8.6 before answering this question.)

____

____

8.6 Separation of Photosynthetic Pigments by Paper Chromatography

(15 min.)

Paper chromatography allows substances to be separated from one another on the basis of their physical characteristics. A chloroplast pigment extract has been prepared for you by soaking spinach leaves in cold acetone and ethanol. Although the extract appears green, other pigments present may be masked by the chlorophyll. In this activity, you will use paper chromatography to separate any pigments present. Separation occurs due to the solubility of the pigment in the chromatography solvent and the affinity of the pigments for absorption to the paper surface. The finished product, showing separated pigments, is called a **chromatogram.**

MATERIALS

Per student:

- chromatography paper, 3 cm × 15 cm sheet
- metric ruler

Per student pair:

- chloroplast pigment extract in foil-wrapped dropping bottle
- chromatography chamber containing solvent
- colored pencils (green, blue-green, yellow, orange)

PROCEDURE

1. Obtain a 3 cm × 15 cm sheet of chromatography paper. *Touch only the edges of the paper,* because oil from your fingers can interfere with development of the chromatogram.
2. Using a ruler, make a *pencil* line (do *not* use ink) about 2 cm from the bottom of the paper.
3. Load the paper by applying a droplet of the chloroplast pigment extract near the center of the pencil line. Allow the pigment spot to dry for about 30 seconds. Five to eight applications of extract on the same spot are necessary to get enough pigment for a good chromatogram. Be certain to allow the pigment to air-dry between applications.

Respiration: Energy Conversion

OBJECTIVES

After completing this exercise, you will be able to

1. define *metabolism, reaction, metabolic pathway, respiration, ATP, aerobic respiration, alcoholic fermentation;*
2. give the overall balanced equations for aerobic respiration and alcoholic fermentation;
3. distinguish among the inputs, products, and efficiency of aerobic respiration and those of fermentation;
4. explain the relationship between temperature and goldfish respiration rate;
5. identify the structures and list the functions of each part of a mitochondrion.

INTRODUCTION

The first law of thermodynamics states that energy can neither be created nor destroyed, only converted from one form to another. Because all living organisms have a constant energy requirement, they have mechanisms to gather, store, and use energy. Collectively, these mechanisms are called **metabolism.** A single, specific reaction that starts with one compound and ends up with another compound is a **reaction,** and a sequence of such reactions is a **metabolic pathway.**

In Exercise 9, we investigated the metabolic pathways by which green plants capture light energy and use it to make carbohydrates such as glucose. Carbohydrates are temporary energy stores. The process by which energy stored in carbohydrates is released to the cell is **respiration.**

Both autotrophs and heterotrophs undergo respiration. Photoautrotrophs such as plants utilize the carbohydrates they have produced by photosynthesis to build new cells and maintain cellular machinery. Heterotrophic organisms may obtain materials for respiration in two ways: by digesting plant material or by digesting the tissues of animals that have previously digested plants.

Several different forms of respiration have evolved. The specific respiration pathway used depends on the specific organism and/or environmental conditions. In this exercise, we will consider two alternative pathways: (1) **aerobic respiration,** an oxygen-dependent pathway common in most organisms; and (2) **alcoholic fermentation,** an ethanol-producing process occurring in some yeasts.

Perhaps the most important aspect to remember about these two processes is that aerobic respiration is by far the most energy-efficient. **Efficiency** refers to the amount of energy captured in the form of ATP relative to the amount available within the bonds of the carbohydrate. **ATP, adenosine triphosphate,** is the so-called universal energy currency of the cell. Energy contained within the bonds of carbohydrates is transferred to ATP during respiration. This stored energy can be released later to power a wide variety of cellular reactions.

For aerobic respiration, the general equation is

$$\underset{\text{glucose}}{C_6H_{12}O_6} + \underset{\text{oxygen}}{6O_2} \xrightarrow{\text{enzymes}} \underset{\text{carbon dioxide}}{6CO_2} + \underset{\text{water}}{6H_2O} + \underset{\text{chemical energy}}{36ATP^*}$$

If glucose is completely broken down to CO_2 and H_2O, about 686,000 calories of energy are released. Each ATP molecule produced represents about 7500 calories of usable energy. The 36 ATP represent 270,000 calories of energy (36 × 7500 calories). Thus, aerobic respiration is about 39% efficient [(270,000/686,000) × 100%].

By contrast, fermentation yields only 2 ATP. Thus, these processes are only about 2% efficient [(2 × 7500/686,000) × 100%]. Obviously, breaking down carbohydrates by aerobic respiration gives a bigger payback than the other means.

*Depending on th[illegible]sue, as many as 38 ATP may be produced.

9.1 Aerobic Respiration

During the process of aerobic respiration, relatively high-energy carbohydrates are broken down in stepwise fashion, ultimately producing the low-energy products of carbon dioxide and water and transferring released energy into ATP. But what is the role of oxygen?

During aerobic respiration, the carbohydrate undergoes a series of oxidation–reduction reactions. Whenever one substance is oxidized (loses electrons), another must be reduced (accept, or gain, those electrons). The final electron acceptor in aerobic respiration is oxygen. Tagging along with the electrons as they pass through the electron transport process are protons (H^+). When the electrons and protons are captured by oxygen, water (H_2O) is formed:

$$2H^+ + 2e^- + \frac{1}{2} O_2 \longrightarrow H_2O$$

In the following experiments, we examine aerobic respiration in two sets of seeds.

A. Experiment: Carbon Dioxide Production *(About 25 min. for setup, 1 3/4 hr to complete)*

Seeds contain stored food material, usually in the form of some type of carbohydrate. When a seed germinates, the carbohydrate is broken down by aerobic respiration, liberating the energy (ATP) required for each embryo to grow into a seedling.

Two days ago, one set of dry pea seeds was soaked in water to start the germination process. Another set was not soaked. In this experiment, you will compare carbon dioxide production between germinating pea seeds, germinating pea seeds that have been boiled, and ungerminated (dry) pea seeds.

This experiment investigates the hypothesis that *germinating seeds produce carbon dioxide from aerobic respiration.*

MATERIALS

Per student group (4):

- 600-mL beaker
- hot plate *or* burner, wire gauze, tripod, and matches
- heat-resistant glove
- 3 respiration bottle apparatuses (Figure 9-1)
- china marker
- phenol red solution

Per lab room:

- germinating pea seeds
- ungerminated (dry) pea seeds

PROCEDURE

Work in groups of four.

1. Place about 250 mL of tap water in a 600-mL beaker, put the beaker on a heat source, and bring the water to a boil.
2. Obtain three respiration bottle setups (Figure 9-1). With a china marker, label one "Germ" for germinating pea seeds, the second "Germ-Boil" for those you will boil, and the third "Ungerm" for ungerminated seeds.
3. From the class supply, obtain and put enough germinating pea seeds into the two appropriately labeled respiration bottles to fill them approximately halfway. Fill the third bottle half full with ungerminated (dry) pea seeds.
4. Dump the germinating peas from the "Germ-Boil" bottle into the boiling waterbath; continue to boil for 5 minutes. After 5 minutes, turn off the heat source, put on a heat-resistant glove, and remove the water bath. Pour the water off into the sink and cool the boiled peas by pouring cold water into the beaker. Allow 5 minutes for the peas to cool to room temperature; then pour off the water. Now replace the peas into the "Germ-Boil" respiration bottle.

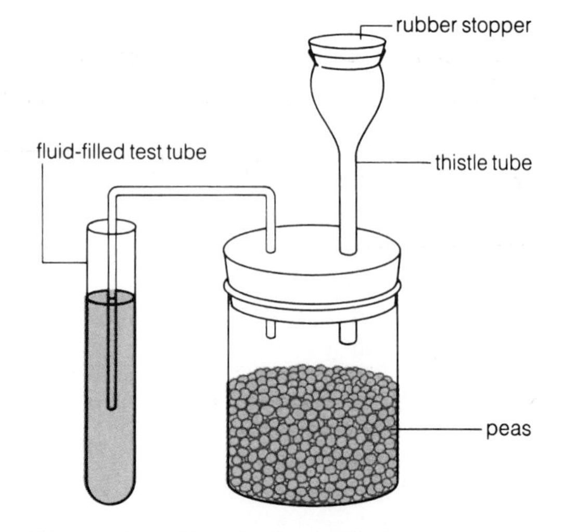

Figure 9-1 Respiration bottle apparatus.

5. Fit the rubber stopper with attached glass tubes into the respiration bottles. Add enough water to the test tube to cover the end of the glass tubing that comes out of the respiration bottle. (This keeps gases from escaping from the respiration bottle.)
6. Insert rubber stoppers into the thistle tubes.
7. Set the three bottles aside for the next 1 1/4 hours and do the other experiments in this exercise.

8. Make a prediction about carbon dioxide production in each of the three bottles:

____

____

____

Now start the next series of experiments while you allow this one to proceed.

9. After 1 1/4 hours, pour the water in each test tube into the sink and replace it with an equal volume of dilute phenol red solution. Phenol red solution, which should appear pinkish in the stock bottle, will be used to test for the presence of carbon dioxide (CO_2) within the respiration bottles. If CO_2 is bubbled through water, carbonic acid (H_2CO_3) forms:

$$CO_2 + H_2O \longrightarrow H_2CO_3$$

Phenol red solution is mostly water. When the phenol red solution is basic (pH > 7), it is pink; when it is acidic (pH < 7), the solution is yellow. The phenol red solution in the stock bottle is

_____________ (color); therefore, the stock solution is

_____________ (acidic/basic).

10. Put several hundred milliliters of tap water in the 600-mL beaker.
11. Remove the stopper plugging the top of the thistle tube and *slowly* pour water from the beaker into each thistle tube. The water will force out gases present in the bottles. If CO_2 is present, the phenol red will become yellow.
12. Record your observations in Table 9-1.

TABLE 9-1 CO_2 Evolution by Pea Seeds

Pea Seeds	Indicator Color (Phenol Red)	Conclusion (CO_2 Present or Absent)
Germinating—unboiled		
Germinating—boiled		
Ungerminated		

Which set(s) of seeds underwent respiration? _______________________

What happened during boiling that caused the results you found? (*Hint:* Think "enzymes.")

____

Write a conclusion, accepting or rejecting the hypothesis.

____

____

B. Experiment: Oxygen Consumption *(About 1 1/2 hr.)*

One set of pea seeds has been soaked in water for the past 48 hours to initiate germination. In this section, you will measure oxygen consumption to answer the question, "Do germinating peas have a higher respiratory rate than ungerminated peas?"

MATERIALS

Per student group (4):

- volumeter (Figure 9-2)
- china marker
- 80 germinating pea seeds
- 80 ungerminated (dry) pea seeds
- glass beads
- nonabsorbent cotton
- metric ruler
- bottle of potassium hydroxide (KOH) pellets
- 1/4 teaspoon measure
- marker fluid in dropping bottle

PROCEDURE

Work in groups of four.

1. Obtain a volumeter set up as in Figure 9-2. Skip to step 6 if your instructor has already assembled the volumeters as described by steps 2–5.

2. Remove the test tubes from the volumeter. With a china marker, number the tubes and then fill as follows:
   Tube 1: 80 germinating (soaked) pea seeds
   Tube 2: 80 ungerminated (dry) pea seeds plus enough glass beads to bring the total volume equal to that in tube 1
   Tube 3: Enough glass beads to equal the volume of tube 1

Both temperature and pressure affect gas pressure within a closed tube. Tube 3 serves as a thermobarometer and is used as a control to correct experimental readings to account for changes in temperature and barometric pressure taking place during the experiment.

3. Pack cotton *loosely* into each tube to a thickness of about 1.5 cm above the peas/beads.
4. Measure out 1 cubic centimeter (cm^3) (about 1/4 teaspoon) of KOH pellets and pour them atop the cotton.

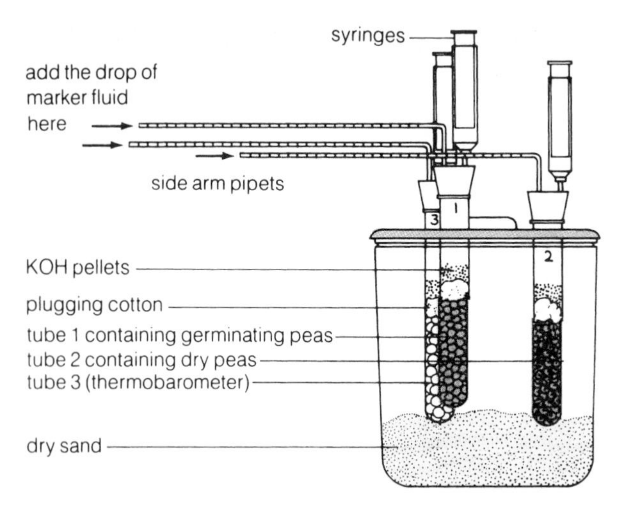

Figure 9-2 Volumeter.

Caution

Potassium hydroxide can cause burns. Do not get any on your skin or clothing. If you do, wash immediately with copious amounts of water.

Potassium hydroxide absorbs CO_2 given off during aerobic respiration. Since the volumeter measures change in gas volume, any gas *given off* during respiration must be removed from the tube so an accurate measure of O_2 consumption can be made.

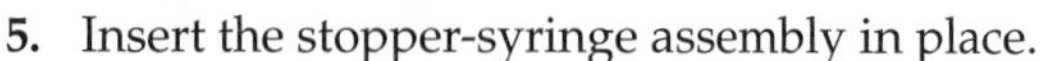

5. Insert the stopper-syringe assembly in place.
6. Add a small drop of marker fluid to each side arm pipet by touching the dropper to the end of each. The drop should be taken into the side arm by capillary action. Gently withdraw the plunger of each syringe and adjust the position of the drop so it is between 0.80 and 0.90 cm^3 on the scale of the graduated pipet.

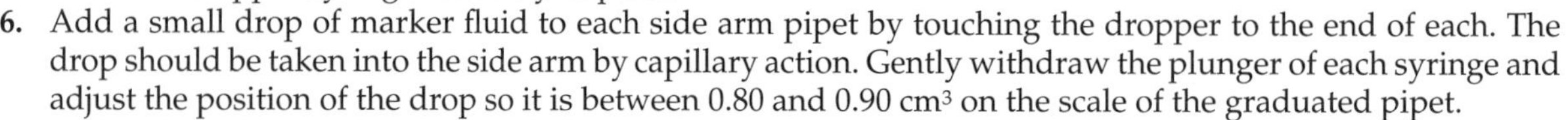

7. Adjust each side arm pipet so it is parallel to the table top. Wait 5 minutes before starting data collection.
8. In Table 9-2 at time 0, record the position of the marker droplet within each pipet. Record readings for each tube at 5-minute intervals for the next 60 minutes, keeping track of whether changes are positive (movement toward the test tube) or negative (movement away from the test tube).

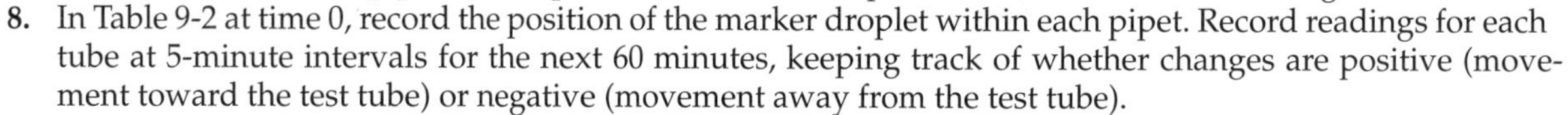

If respiration is rapid and the marker drop moves too near the end of the scale to read, *carefully* use the syringe to readjust its position so it is between 0.80 and 0.90 cm^3 on the scale of the graduated pipet again. Note this in Table 9-2 and continue to record readings every 5 minutes.

9. To determine change in volume of gas within each tube at each sampling, subtract each subsequent reading from the previous reading.
10. Determine cumulative volume change (cumulative oxygen consumption) by adding each volume change to the previous volume-change measurement. The final figure represents the total oxygen consumption (in mL) in that tube.
11. At the end of the experiment, correct for any volume changes caused by changes in temperature or barometric pressure by using the reading obtained from the thermobarometer. If the thermobarometric marker moves *toward* the test tube (decrease in volume), *subtract* the volume change from the total oxygen consumption measurement of tubes 1 and 2. If the marker droplet moves *away* from the test tube (increase in volume), *add* the volume change to the last total oxygen consumption measurement for tubes 1 and 2.
12. In Figure 9-3, graph the consumption of oxygen over time. Use a + for data points of germinating peas, a • for dry peas.

How do the respiratory rates for germinating and nongerminating seeds compare?

____

____

How do you account for this difference?

____

____

It takes 820 cm^3 of oxygen to completely oxidize 1 g of glucose. How much glucose are the 80 germinating peas consuming per hour? (Recall that 1 cm^3 = 1 mL.)

TABLE 9-2 Respiratory Rate as Measured by Oxygen Consumption

	Tube 3: Thermobarometer		Tube 1: Germinating Peas			Tube 2: Dry Peas		
Time (min.)	Reading	Total Change in Volume	Reading	Total Change in Volume	Total Oxygen Consumption	Reading	Total Change in Volume	Total Oxygen Consumption
0		0		0	0		0	0
5								
10								
15								
20								
25								
30								
35								
40								
45								
50								
55								
60								

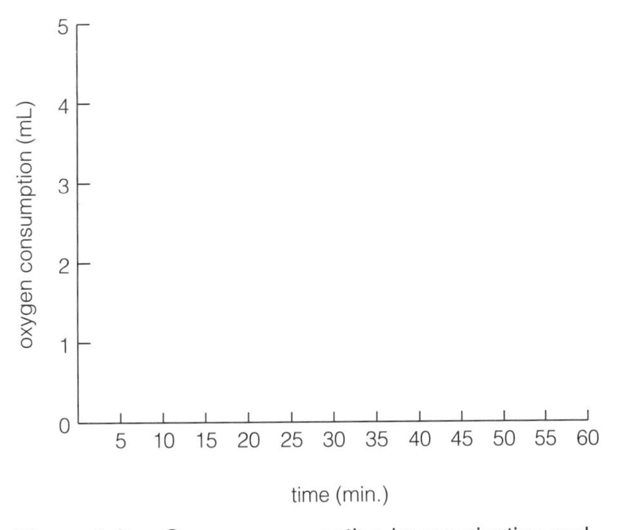

Figure 9-3 Oxygen consumption by germinating and nongerminating pea seeds.

9.2 Fermentation *(About 20 min. to set up, 1½ hr. to complete)*

Despite relatively low energy yield, fermentation provides sufficient energy for certain organisms to survive. Alcoholic fermentation by yeast is the basis for the baking, wine-making, and brewing industries. It's been said that yeast and alcoholic fermentation made Milwaukee famous.

The chemical equation for this process is

$$C_6H_{12}O_6 \longrightarrow 2CH_3CH_2OH + CO_2 + 2ATP$$

glucose — ethanol — carbon dioxide — energy

Starch (amylose), a common storage carbohydrate in plants, is a polymer consisting of a chain of repeating glucose ($C_6H_{12}O_6$) units. The polymer has the chemical formula $(C_6H_{12}O_6)n$,* where *n* represents a large number. Starch is broken down by the enzyme amylase into individual glucose units. To summarize:

$$(C_6H_{12}O_6)n \xrightarrow{\text{amylase}} C_6H_{12}O_6 + C_6H_{12}O_6 + C_6H_{12}O_6 + \ldots$$

starch — glucose

This section demonstrates the action of yeast cells on carbohydrates.

MATERIALS

Per student group (4):

- china marker
- three 50-mL beakers
- 25-mL graduated cylinder
- bottle of 10% glucose
- bottle of 1% starch
- 0.5% amylase in bottle fitted with graduated pipet
- 3 glass stirring rods
- 1/4 teaspoon measure (optional)
- 3 fermentation tubes
- 15-cm metric ruler

Per lab room:

- 0.5-g pieces of fresh yeast cake
- scale and weighing paper (optional)
- 37°C incubator

PROCEDURE

Work in groups of four.

1. Using a china marker, number three 50-mL beakers.
2. With a *clean* 25-mL graduated cylinder, measure out and pour 15 mL† of the following solutions into each beaker:

Note: **Wash the graduated cylinder between solutions.**

Beaker 1: 15 mL of 10% glucose
Beaker 2: 15 mL of 1% starch
Beaker 3: 15 mL of 1% starch; then, using the graduated pipet to measure, add 5 mL of 0.5% amylase.

3. Wait 5 minutes and then to each beaker add a 0.5-g piece of fresh cake yeast. Stir with *separate* glass stirring rods.
4. When each is thoroughly mixed, pour the contents into three correspondingly numbered fermentation tubes (Figure 9-4). Cover the opening of the fermentation tube with your thumb and invert each fermentation tube so that the "tail" portion is filled with the solution.
5. Write a prediction about gas production in each tube.

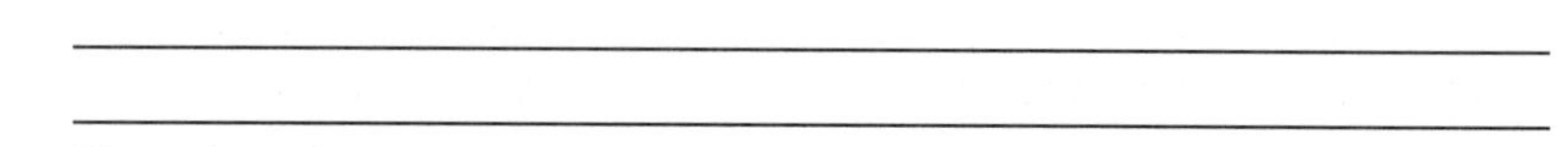

6. Place the tubes in a 37°C incubator.
7. At intervals of 20, 40, and 60 minutes after the start of the experiment, remove the tubes and, using a metric ruler, measure the distance from the tip of the tail to the fluid level. Record your results in Table 9-3. Calculate the volume of gas evolved using the formula below Table 9-3. (If time is short, do your calculations later.)

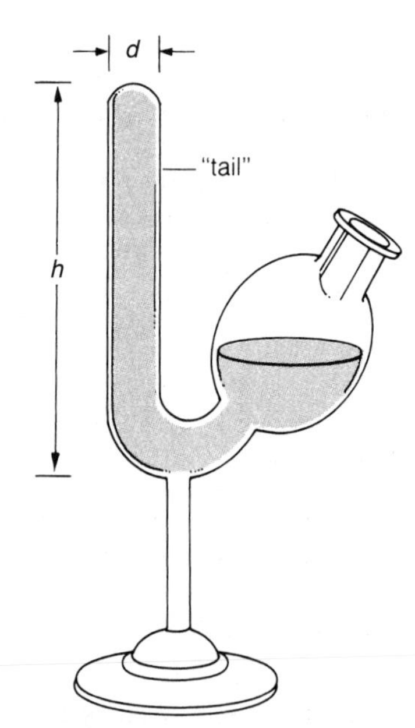

Figure 9-4 Fermentation tube.

*A number of carbohydrates share this same chemical formula but differ slightly in the arrangement of their atoms. These carbohydrates are called *structural isomers*.

†The amount of fluid needed to fill the fermentation tube depends on its size. Your instructor may indicate the required volume.

TABLE 9-3 Evolution of Gas by Yeast Cells

Tube	Solution	Distance from Tip of Tube to Fluid Level (mm) 20 min.	40 min.	60 min.	Volume of Gas Evolved (mm^3)
1	10% glucose + yeast				
2	1% starch + yeast				
3	1% starch + yeast + amylase				

To calculate the volume of gas evolved, use the following equation: $V = \pi r^2 h$, where $\pi = 3.14$, r = radius of tail of fermentation tube [$r = \frac{1}{2} d$ (diameter)], h = distance from top of tail to level of solution.

Did your results conform to your predictions? If not, speculate on reasons why this might be so.

____

____

What gas accumulates in the tail portion of the fermentation tube?

____

9.3 Experiment: Effect of Temperature on Goldfish Respiration *(About 1 hr.)*

All organisms respond to changes in their environment, often by changing their rate of respiration. Aerobic respiration is not completely efficient in converting glucose energy into ATP energy, and some energy is released as heat. Our bodies can respond to a decrease in environmental temperature (within limits) by increasing the rate of aerobic respiration in order to maintain a stable body temperature. We are considered *endothermic* animals, meaning that we control the heat balance from within our bodies.

Body temperature in *ectothermic* animals rises and falls with their surroundings and they are, figuratively speaking, at the mercy of their environment to a greater extent than endotherms. Activity levels may be closely associated with external temperatures. However, ectotherms may make behavioral changes to adjust to changing external temperatures.

Changes in behavior can be correlated with changes in carbohydrate metabolism. The previous experiments allowed you to determine which gases are used and produced during respiration. The respiratory cycle in fish, which accomplishes gas exchange between the environment and the fish, occurs when the water, laden with dissolved oxygen, enters the mouth and is forced out over the gill filaments when the mouth is closed. The oxygen dissolved in the water diffuses into the capillaries in the gill filaments and enters the blood.

Carbon dioxide, one of the products of respiration, diffuses from the blood across the gill filaments and into the water. The gill cover opens to allow the carbon dioxide–laden water to leave the gill chamber, thus completing the respiratory cycle.

By observing the action of the gill cover (determining the gill beat rate) in fishes (which are ectothermic animals) in this experiment, you will determine the effect of temperature on respiratory rate and the correlated metabolism rate. This experiment tests the hypothesis that *fish respiratory rates vary with temperature.*

MATERIALS

Per student group (4):

- aquarium (battery jar or 4-L beaker) filled with dechlorinated water at room temperature
- one goldfish
- Celsius thermometer with clip
- plastic dishpan
- crushed ice
- clock

PROCEDURE

1. Make a prediction regarding the outcome of your experiment. Identify the independent and dependent variables, and enter them in Table 9-4.

2. Place the goldfish in the aquarium with sufficient water to cover the dorsal (top) fin. Clip a thermometer on the edge of the container so you can read it without disturbing the fish. Take care to disturb the fish as little as possible.
3. Place the aquarium in the plastic dishpan. Determine the initial temperature of the water, and record it in Table 9-4.
4. Observe the movement of the operculum (gill cover). Count the movements of the operculum at the starting temperature for 1 minute. Repeat this procedure two more times. Record this gill beat rate data plus the average rate in Table 9-4.
5. Gradually add ice to the plastic container around the aquarium until the water temperature in it has been reduced to 25°C. If the starting temperature of the water is lower than 25°C, lower the temperature to 20°C. Wait 1 minute for the fish to adjust to the new temperature. Repeat step 4 and determine the gill beat rate.
6. Continue to lower the temperature of the water surrounding the fish with additional ice around the aquarium, waiting 1 minute each time a temperature is reached to allow acclimatization by the fish. Count operculum movements as described in step 5 at 15°C, 10°C, and 5°C. Record all data in Table 9-4.
7. At the end of the experiment, remove the aquarium containing the goldfish from the plastic container, and allow it to warm up gradually to room temperature.
8. Graph your average data points on Figure 9-5 to illustrate the manner in which goldfish gill beat rates ("breaths" per minute)—and hence cellular respiration—vary depending on water temperature.

TABLE 9-4 Operculum Movement of Goldfish as a Function of Temperature

Prediction:

Independent Variable:

Dependent Variable:

RESULTS:

Trial	Starting Temperature ______°C	25°C	20°C	15°C	10°C	5°C
1						
2						
3						
Average						

9. Make a conclusion about your results, accepting or rejecting the hypothesis.

____

____

____

____

What controls might you have used for this experiment?

____

____

____

____

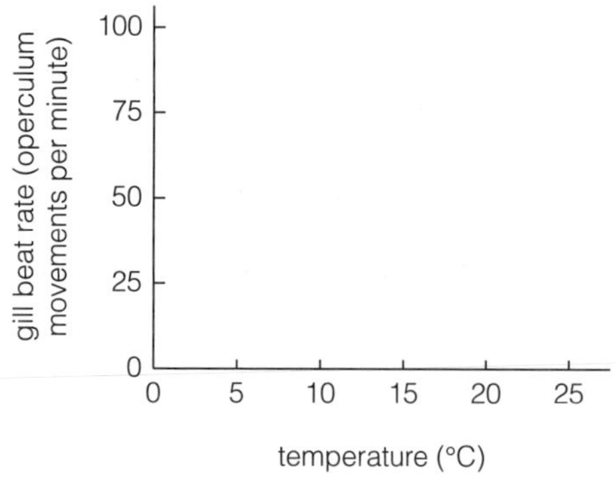

At which temperature was the gill beat rate highest?

____

Figure 9-5 Relationship between gill beat rate and temperature in goldfish.

What is the relationship between the increase in water temperature and gill beat rate in goldfish? Why?

____

____

9.4 Ultrastructure of the Mitochondrion

1. Study Figure 9-6b, which shows the three-dimensional structure of a mitochondrion, the respiratory organelle of all living eukaryotic cells. The mitochondrion has frequently been referred to as the "powerhouse of the cell," because most of the cell's chemical energy (ATP) is produced here.

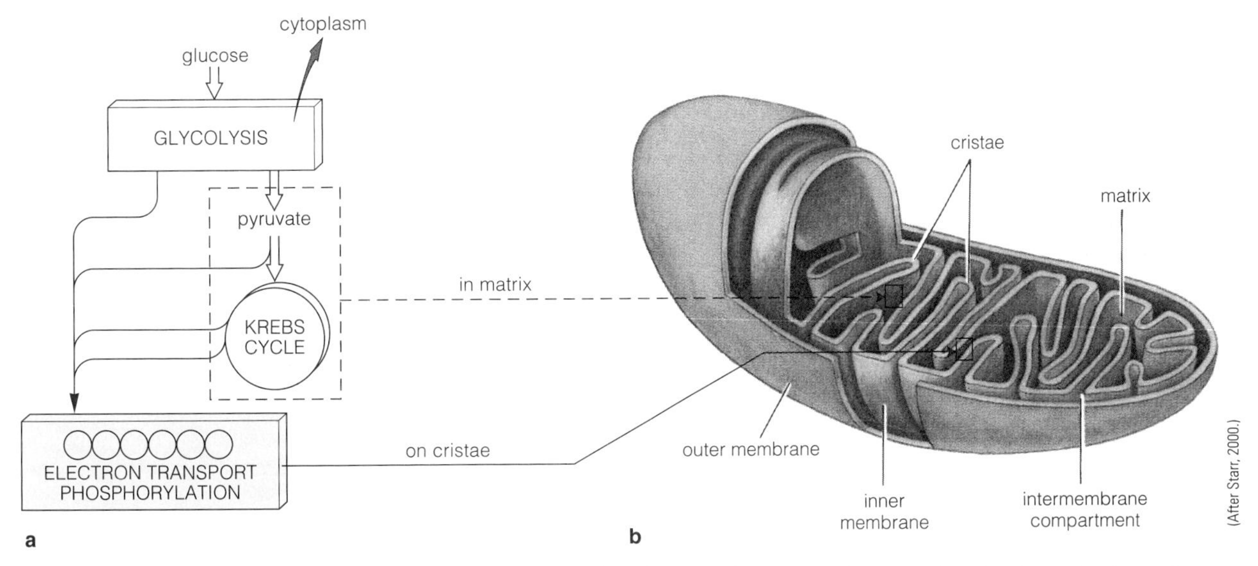

Figure 9-6 (**a**) The pathways in aerobic respiration. (**b**) The membranes and compartments of a mitochondrion.

2. Now observe Figure 9-7, a high-magnification electron micrograph of a mitochondrion. Identify and label the **outer membrane** separating the organelle from the cytoplasm.
3. Note the presence of an inner membrane, folded into fingerlike projections. Each projection is called a **crista** (the plural is *cristae*). The folding of the inner membrane greatly increases the surface area on which many of the chemical reactions of aerobic respiration take place. Label the crista.
4. Identify and label the **outer compartment,** the space between the inner and outer membranes. The outer compartment serves as a reservoir for hydrogen ions.
5. Finally, identify and label the **inner compartment** (filled with the *matrix*), the interior of the mitochondrion.
6. Study Figure 9-6a, a diagram of the pathways in aerobic respiration.

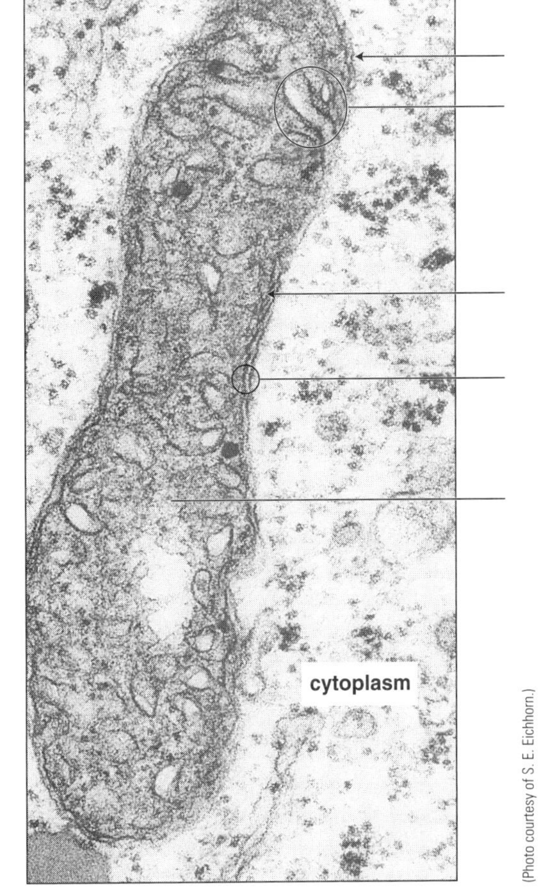

Figure 9-7 Transmission electron micrograph of a mitochondrion (18,600×).
Labels: outer membrane, inner membrane, crista, intermembrane compartment, matrix

Now that you know the structure of the mitochondrion, you can visualize the events that take place to produce the chemical energy needed for life. Glycolysis, the first step in *all* respiration pathways, takes place in the cytoplasm. Pyruvate, a carbohydrate, and energy carriers formed during glycolysis enter the mitochondrion during aerobic respiration, moving through both the outer and inner membranes to the matrix within the inner compartment.

Within the inner compartment, the pyruvate is broken down in the Krebs cycle, forming more energy-carrier molecules as well as CO_2. A small amount of ATP is also produced during these reactions.

Electron transport molecules are embedded on the inner membrane, and ATP production occurs as hydrogen ions cross from the outer compartment to the inner compartment. Water is also formed. This accomplishes the third portion of aerobic respiration, electron transport phosphorylation.

Note: **After completing all labs, take your dirty glassware to the sink and wash it following the directions given in *"Instructions for Washing Laboratory Glassware,"* page x. Invert the test tubes in the test tube racks so that they drain. Tidy up your work area, making certain all equipment used in this exercise is there for the next class.**

PRE-LAB QUESTIONS

_____ **1.** A metabolic pathway is
(a) a single, specific reaction that starts with one compound and ends up with another
(b) a sequence of chemical reactions that are part of the metabolic process
(c) a series of events that occur only in autotrophs
(d) all of the above

_____ **2.** The "universal energy currency" of the cell is
(a) O_2
(b) $C_6H_{12}O_6$
(c) ATP
(d) H_2O

_____ **3.** Products of aerobic respiration include
(a) glucose
(b) oxygen
(c) carbon dioxide
(d) starch

_____ **4.** "Efficiency" of a respiration pathway refers to the
(a) number of steps in the pathway
(b) amount of CO_2 produced relative to the amount of carbohydrate entering the pathway
(c) amount of H_2O produced relative to the amount of carbohydrate entering the pathway
(d) amount of ATP energy produced relative to the energy content of the carbohydrate entering the pathway

_____ **5.** The purpose of the thermobarometer in a volumeter is to
(a) judge the amount of O_2 evolved during respiration
(b) determine the volume changes as a result of respiration
(c) indicate oxygen consumption by germinating pea seeds
(d) indicate volume changes resulting from changes in temperature or barometric pressure

_____ **6.** Phenol red is used in the experiments as
(a) an O_2 indicator
(b) a CO_2 indicator
(c) a sugar indicator
(d) an enzyme

_____ **7.** As temperatures rise, the body temperatures of _____ organisms rise.
(a) ectothermic
(b) endothermic
(c) autotrophic
(d) heterotrophic

_____ **8.** Which of the following enzymes breaks down starch into glucose?
(a) kinase
(b) maltase
(c) fructase
(d) amylase

_____ **9.** Oxygen is necessary for life because
(a) photosynthesis depends on it
(b) it serves as the final electron acceptor during aerobic respiration
(c) it is necessary for glycolysis
(d) of all of the above

_____ **10.** Yeast cells undergoing alcoholic fermentation produce
(a) ATP
(b) ethanol
(c) CO_2
(d) all of the above

Name ______________________________ Section Number ______________________

EXERCISE 9

Respiration: Energy Conversion

POST-LAB QUESTIONS

9.1 Aerobic Respiration

1. Explain the role of the following components in the experiment on carbon dioxide production (page 132).

   germinating pea seeds ______________________

   ungerminated (dry) pea seeds ______________________

   germinating, boiled pea seeds ______________________

   phenol red solution ______________________

2. If you performed the experiment on oxygen consumption (page 133) without adding KOH pellets to the test tubes, what results would you predict? Why?

9.2 Fermentation

3. Sucrose (table sugar) is a disaccharide composed of glucose and fructose. Glycogen is a polysaccharide composed of many glucose subunits. Which of the following fermentation tubes would you expect to produce the greatest gas volume over a 1-hour period? Why?

   Tube 1: glucose plus yeast

   Tube 2: sucrose plus yeast

   Tube 3: glycogen plus yeast

4. Bread is made by mixing flour, water, sugar, and yeast to form a dense dough. Why does the dough rise? What gas is responsible for the holes in bread?

9.3 Experiment: Effect of Temperature on Goldfish Respiration

5. Warmer water holds less dissolved oxygen than cooler water. Use this information plus that which you gleaned from this experiment to more fully explain why the gill beat rate of a fish increases as water temperature increases.

9.4 Ultrastructure of the Mitochondrion

6. Examine the electron micrograph of the mitochondrion on the right.

a. What portions of aerobic respiration occur in region b?

____

b. What substance is produced as hydrogen ions cross from the space between the inner and outer membranes into region b?

____

c. What portion of cellular respiration takes place in the cytoplasm *outside* of this organelle?

____

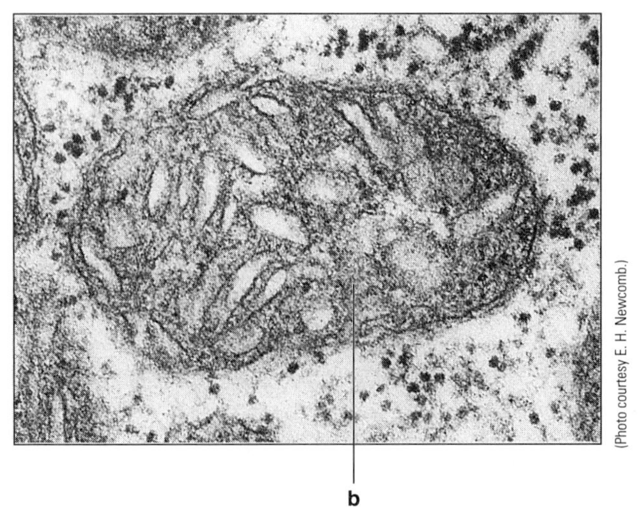

(20,000×).

Food for Thought

7. Oxygen is used during aerobic respiration. What biological process is the source of the oxygen?

8. Compare aerobic respiration and fermentation in terms of

a. efficiency of obtaining energy from glucose

b. end products

9. How would you explain this statement: "The ultimate source of our energy is the sun"?

10. The first law of thermodynamics seems to conflict with what we know about ourselves. For example, after strenuous exercise we run out of "energy." We must eat to replenish our energy stores. Where has that energy gone? What form has it taken?

Mitosis and Cytokinesis: Nuclear and Cytoplasmic Division

OBJECTIVES

After completing this exercise, you will be able to

1. define *fertilization, zygote, DNA, chromosome, mitosis, cytokinesis, nucleoprotein, sister chromatid, centromere, meristem;*
2. identify the stages of the cell cycle;
3. distinguish between mitosis and cytokinesis as they take place in animal and plant cells;
4. identify the structures involved in nuclear and cell division (those in **boldface**) and describe the role each plays.

INTRODUCTION

"All cells arise from preexisting cells." This is one tenet of the cell theory. It's easy to understand this concept if you think of a single-celled *Amoeba* or bacterium. Each cell divides to give rise to two entirely new individuals, and it is fascinating that each of us began life as *one* single cell and developed into this astonishingly complex animal, the human. Our first cell has *all* the hereditary information we'll ever get.

In higher plants and animals, **fertilization,** the fusion of egg and sperm nuclei, produces a single-celled **zygote**. The zygote divides into two cells, these two into four, and so on to produce a multicellular organism. During cell division, each new cell receives a complete set of hereditary information and an assortment of cytoplasmic components.

Recall from Exercise 6 that there are two basic cell types, prokaryotic and eukaryotic. The genetic material of both consists of **DNA (deoxyribonucleic acid).** In prokaryotes, the DNA molecule is organized into a single circular **chromosome.** Prior to cell division, the chromosome duplicates. Then the cell undergoes **fission,** the splitting of a preexisting cell into two, with each new cell receiving a full complement of the genetic material.

In eukaryotes, the process of cell division is more complex, primarily because of the much more complex nature of the hereditary material. Here the chromosomes consist of DNA and proteins complexed together within the nucleus. Cell division is preceded by duplication of the chromosomes and usually involves two processes: **mitosis** (nuclear division) and **cytokinesis** (cytoplasmic division). Whereas mitosis results in the production of two nuclei, both containing identical chromosomes, cytokinesis ensures that each new cell contains all the metabolic machinery necessary for sustenance of life.

In this exercise, we consider only what occurs in eukaryotic cells.

Dividing cells pass through a regular sequence of events called the cell cycle (Figure 10-1). Notice that the majority of the time is spent in interphase and that actual nuclear division—mitosis—is but a brief portion of the cycle.

Interphase is comprised of three parts (Figure 10-1): the G1 period, during which cytoplasmic growth takes place; the S period, when the DNA is duplicated; and the G2 period, when structures directly involved in mitosis are synthesized.

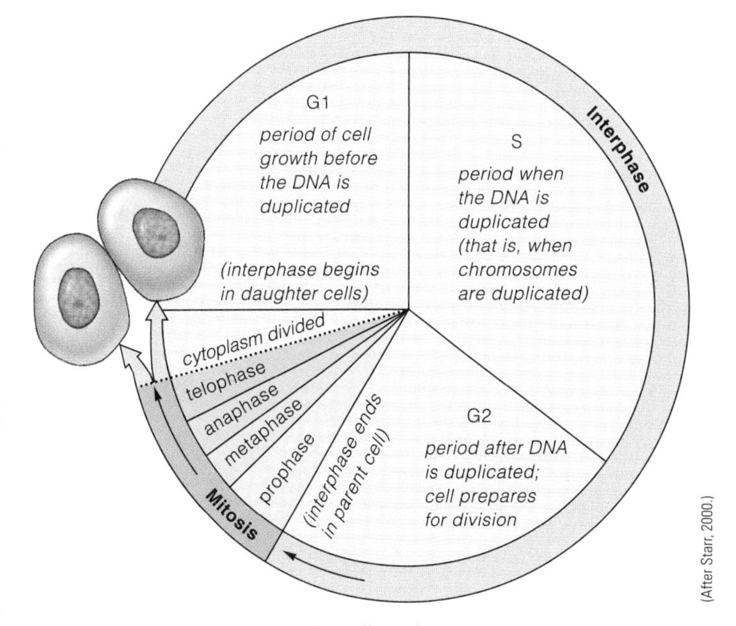

Figure 10-1 The eukaryotic cell cycle.

Unfortunately, because of the apparent relative inactivity that early microscopists observed, interphase was given the misnomer "resting stage." In fact, we know now that interphase is anything but a resting period. The cell is producing new DNA, assembling proteins from amino acids, and synthesizing or breaking down carbohydrates. In short, interphase is a very busy time in the life of a cell.

10.1 Chromosomal Structure *(About 25 min. without drying time)*

During much of a cell's life, each DNA–protein complex, the **nucleoprotein,** is extended as a thin strand within the nucleus. In this form, it is called **chromatin.** Prior to the onset of nuclear division, the genetic material duplicates itself. As nuclear division begins, the chromatin condenses. The two identical condensed nucleoproteins are called **sister chromatids** and are attached at the **centromere.** The centromere gives the appearance of dividing each chromatid into two "arms." Collectively, the two attached sister chromatids are referred to as a *duplicated chromosome*. Before looking at actual chromosomes, let's examine the background of the materials we'll use.

In 1951, Henrietta Lacks died from cervical cancer. Prior to her death, some of the cancerous (tumor) cells were removed from her body and grown in culture. The cells were allowed to divide repeatedly in this artificial culture environment and resulted in the formation of what is known as a *cell line*. These HeLa cells (the abbreviation coming from *He*nrietta *La*cks) live on today and are used for research on cell and tumor growth. They're also used to show chromosomal structure in the biology laboratory.

Normal human body cells contain 23 pairs of chromosomes (46 total chromosomes). Human cells of tumor origin (such as the HeLa cells) often produce greater chromosome numbers when grown in culture. Cells containing up to 200 chromosomes have been observed, although 50–70 per cell are most frequently found. In this section, you will observe the chromosomes of the descendant tumor cells of Henrietta Lacks.

Note: **There is no danger from this activity. The tumor-producing properties are nontransmissible, and the cells have been killed and preserved.**

MATERIALS

Per student:

- clean microscope slide prechilled in cold 40% methanol
- coverslip
- disposable pipet
- paper toweling
- compound microscope (with oil-immersion objective preferable)

Per student pair:

- metric ruler
- tube of HeLa cells
- dropper bottle containing Permount mounting medium or dH_2O

Per lab room:

- Coplin staining jars containing stains 1 and 2
- 1-L beaker containing dH_2O

PROCEDURE

Refer to Figure 10-2.

1. Place a paper towel on your work surface.
2. Remove a clean glass microscope slide from the cold methanol and lean it against a surface at a 45° angle, with one short edge resting on the paper towel.
3. Most of the cells are at the bottom of the culture tube. Gently resuspend the cells by inserting your pipet and squeezing the bulb. This expels air from the pipet, which disperses the cells. Remove a small cell sample with the pipet. Holding the pipet about 18–36 cm above the slide, allow 8–10 drops of the cell suspension to "splat" onto the upper edge of the wet slide and to tumble down the slide.

   Note: **The slide must be wet when "splatting" takes place.**

4. Allow the cells to *air-dry completely*.
5. You will now stain the slide three times with both stains, for *1 second each time*. The time in the stain is critical. Count "one thousand one"; this will be 1 second.

Go to the location of the staining solutions and dip the slide in stain 1 for 1 second. Withdraw the slide and then dip twice more. Drain the slide of the excess stain, blotting the bottom edge of the slide on the paper toweling before proceeding.

6. Immediately dip the slide in stain 2 for 1 second. (Repeat twice more.) Drain and blot the excess stain as before.

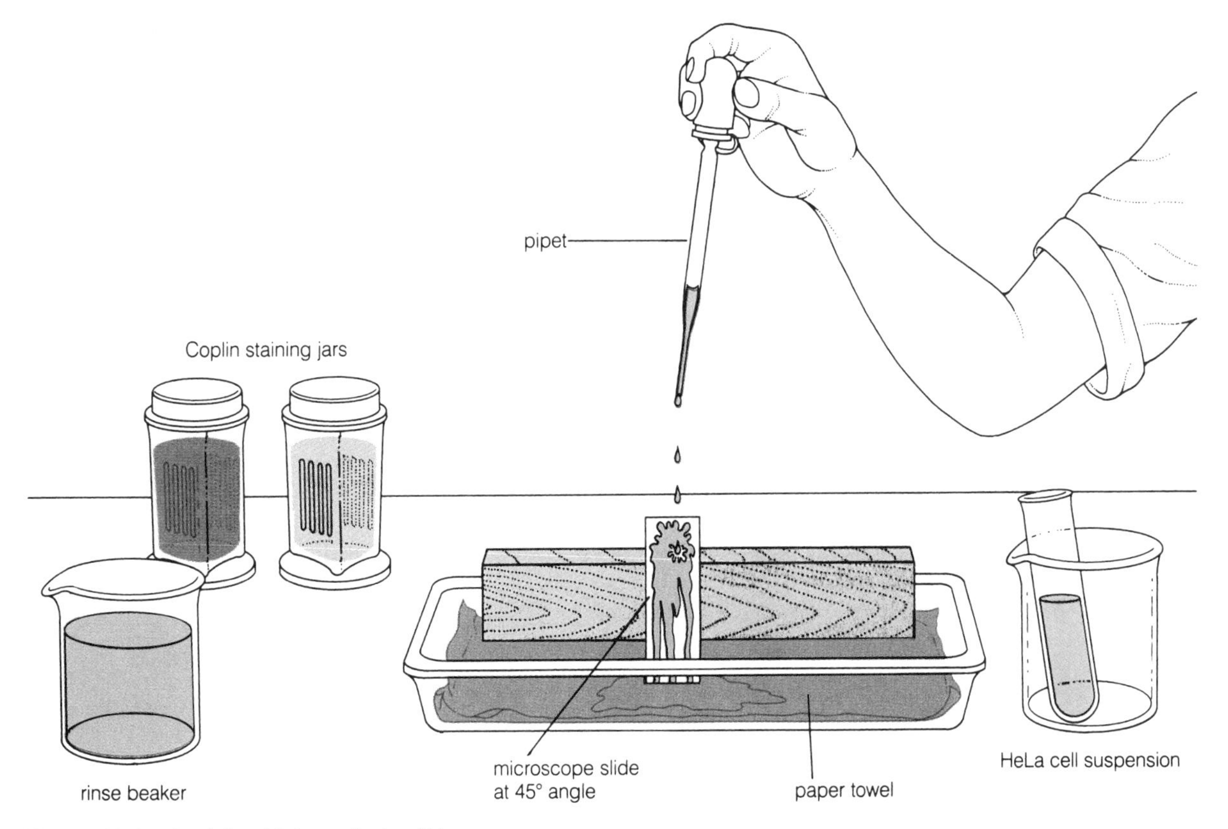

Figure 10-2 Applying HeLa cells to slide.

7. Rinse the slide in dH_2O by swishing it gently back and forth in the beaker.
8. Allow the slides to *air-dry completely*.
9. If making a permanent preparation, place 2 drops of Permount mounting medium on the slide in the region of the stained cells. Using the technique illustrated in Figure 3-9 (page 35), place a coverslip on the slide. (*Note:* Of course, you *do not* add the drop of water shown in Figure 3-9.) Squeeze out excess Permount by applying gentle pressure to the coverslip with the blunt eraser end of a pencil.

> **Caution**
>
> ***Allow Permount to dry for an hour before attempting to view if using an oil-immersion objective.***

A nonpermanent preparation can be made by adding a drop of water to the preparation and making a traditional wet mount slide. Viewing may then proceed immediately.

10. Place the slide on the microscope stage and observe your chromosome spread, focusing first with the medium-power objective and then with the high-dry objective. Locate cells that appear to have burst and have the chromosomes spread out. The number of good spreads will be low; so careful observation of many cells is necessary.

Note: **If your microscope has an oil-immersion objective, proceed to step 11. If not, skip to step 12.**

11. Once a good spread has been located, rotate the high-dry objective out of the light path and place a drop of immersion oil on that spot. Rotate the oil-immersion objective into the light path.
12. Observe the structure of the chromosomes, identifying sister chromatids and centromeres. Draw several of the chromosomes in Figure 10-3, labeling these parts.

Figure 10-3 Drawing of human chromosomes from HeLa cells. (______×)
Labels: chromosome, sister chromatid, centromere

13. Select 10 cells in which the chromosomes are visible and count the number of chromosomes you observe in each cell, inserting your results in Table 10-1. When you have counted the chromosomes in 10 individual cells, calculate the average number of chromosomes per cell.

TABLE 10-1 Chromosome Numbers in HeLa Cells

Cell	1	2	3	4	5	6	7	8	9	10	Average

Now examine Figure 10-4, an electron micrograph of a human chromosome. Label the two chromatids, the centromere, and the duplicated chromosome.

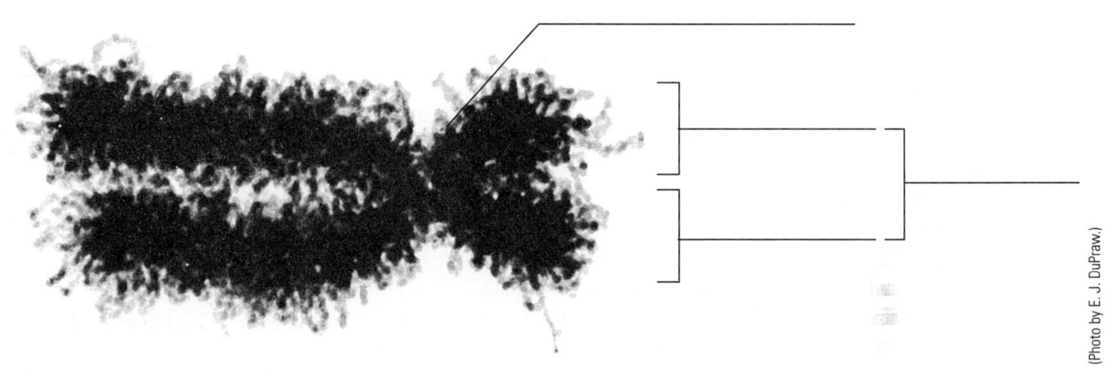

Figure 10-4 Electron micrograph of a chromosome (26,000×).
Labels: chromatid (2), centromere, duplicated chromosome

Note: **Use illustrations in your textbook to aid you in the following study.**

10.2 The Cell Cycle in Plant Cells: Onion Roots *(About 45 min.)*

Nuclear and cell divisions in plants are, for the most part, localized in specialized regions called meristems. **Meristems** are regions of active growth. A meristem contains cells that have the capability to divide repeatedly.

Plants have two types of meristems: apical and lateral. Apical meristems are found at the tips of plant organs (shoots and roots) and increase length. Lateral meristems, located beneath the bark of woody plants, increase girth. In this section, you will examine structures related to the cell cycle in the apical meristem of onion roots.

MATERIALS

Per student:

- prepared slide of onion, *Allium,* root tip mitosis
- compound microscope

PROCEDURE

Obtain a prepared slide of a longitudinal section (l.s.) of an *Allium* (onion) root tip. This slide has been prepared from the terminal several millimeters of an actively growing root. It was "fixed" (killed) by chemicals to preserve the cellular structure and stained with dyes that have an affinity for the structures involved in nuclear division.

Focus first with the low-power objective of your compound microscope to get an overall impression of the root's morphology.

Concentrate your study in the region about 1 mm behind the actual tip. This region is the apical meristem of the root (Figure 10-5). Keep in mind that the cell cycle is a continuous cycle; we separate events into different stages as a convenience to aid in their study, but it is often difficult to say definitively when one phase begins and another ends.

1. **Interphase.** Use the medium-power objective to scan the apical meristem. Note that most of the nuclei are in interphase.

   Switch to the high-dry objective, focusing on a single interphase cell. Note the distinct **nucleus,** with one or more **nucleoli,** and the **chromatin** dispersed within the bounds of the **nuclear envelope.** Label these features in cell 1 of Figure 10-6.
2. **Mitosis.**
   (a) *Prophase.* During **prophase** the chromatin condenses, rendering the duplicated chromosomes visible as threadlike structures. At the same time, microtubules outside the nucleus are beginning to assemble into **spindle fibers.** Collectively, the spindle fibers make up the **spindle,** a three-dimensional structure widest in the middle and tapering to a point at the two **poles** (opposite ends of the cell). *You will not see the spindle during prophase.*

   Find a nucleus in prophase. Draw and label a prophase nucleus in cell 2 of Figure 10-6.

   The transition from prophase to metaphase is marked by the fragmentation and disappearance of the nuclear envelope. At about the same time, the nucleoli disappear.

   (b) *Metaphase.* When the nuclear envelope is no longer distinct, the cell is in **metaphase.** Identify a metaphase cell by locating a cell with the duplicated chromosomes, each consisting of two **sister chromatids,** lined up midway between the two poles. This imaginary midline is called the **spindle equator.** (You will not be able to distinguish the chromatids.) The spindle has moved into the space the nucleus once occupied. The microtubules have become attached to the chromosomes at the **kinetochores,** groups of proteins that form the outer faces of the centromeres. Find a cell in metaphase. Label cell 3 of Figure 10-6.

   (c) *Anaphase.* During **anaphase** sister chromatids of each chromosome separate, each chromatid moving toward an opposite pole.

   Find an early anaphase cell, recognizable by the slightly separated chromatids. Notice that the chromatids begin separating at the centromere. The last point of contact before separation is complete is at the ends of the "arms" of each chromatid. Although incompletely understood, the mechanism of chromatid separation is based on action of the spindle-fiber microtubules. Once separated, each chromatid is referred to as an individual daughter chromosome. Note that now the chromosome consists of a *single* chromatid.

   Find a late anaphase cell and draw it in cell 4 of Figure 10-6.

   (d) *Telophase.* When the daughter chromosomes arrive at opposite poles, the cell is in **telophase.** The spindle disorganizes. The chromosomes expand again into chromatin form, and a nuclear envelope re-forms around each newly formed daughter nucleus.

   Find a telophase cell and label individual chromosomes, nuclei, and nuclear envelopes on cell 5 of Figure 10-6.

Figure 10-5 Root tip, l.s. (20×).

B. Cytokinesis in Onion Cells

Cytokinesis, division of the cytoplasm, usually follows mitosis. In fact, it often overlaps with telophase. Find a cell undergoing cytokinesis in the onion root tip. In plants, cytokinesis takes place by **cell plate formation** (Figure 10-7). During this process, Golgi body–derived vesicles migrate to the spindle equator, where they fuse. Their contents contribute to the formation of a new cell wall, and their membranes make up the new plasma membranes. In most plants, cell plate formation starts in the *middle* of the cell.

1. Examine Figure 10-7, an electron micrograph showing cell plate formation. Note the microtubules that are part of the spindle apparatus.
2. Find a cell undergoing cytokinesis on the prepared slide of onion root tips. With your light microscope, the developing **cell plate** appears as a line running horizontally between the two newly formed nuclei. Return to cell 5 of Figure 10-6 and label the developing cell plate.

Recently divided cells are often easy to distinguish by their square, boxy appearance. Find two recently divided **daughter cells;** then draw and label their contents in cell 6 of Figure 10-6. Include cytoplasm, nuclei, nucleoli, nuclear envelopes, and chromatin.

What is the difference between chromatin and chromosomes?

____

____

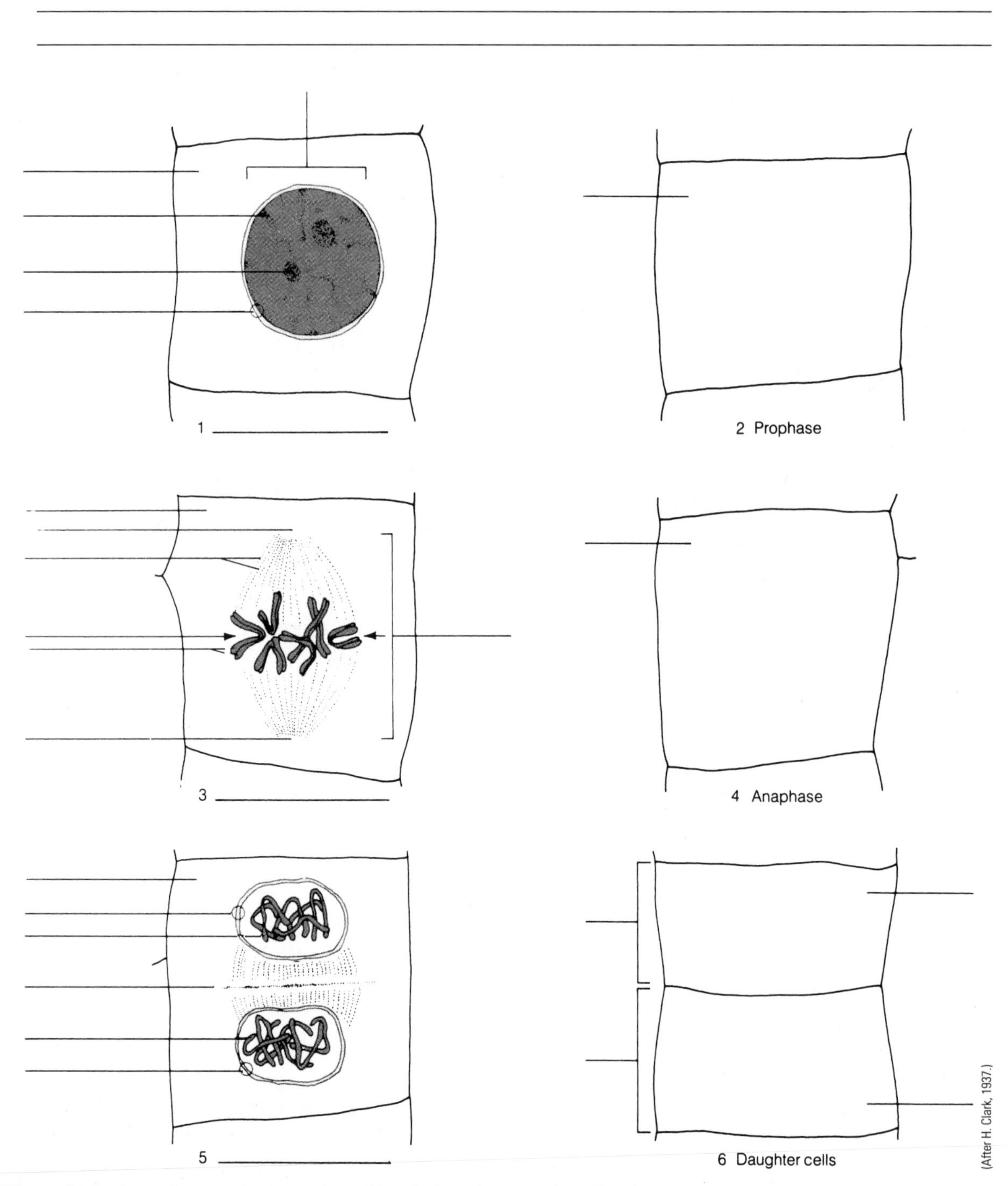

Figure 10-6 Interphase, mitosis, and cytokinesis in onion root tip cells.
Labels: interphase, cytoplasm, nucleus, nucleolus, chromatin, nuclear envelope, metaphase, spindle fibers, spindle, pole, spindle equator (between arrows), sister chromatids, telophase and cell plate formation, chromosome, cell plate, daughter cell
(*Note:* Some terms are used more than once.)

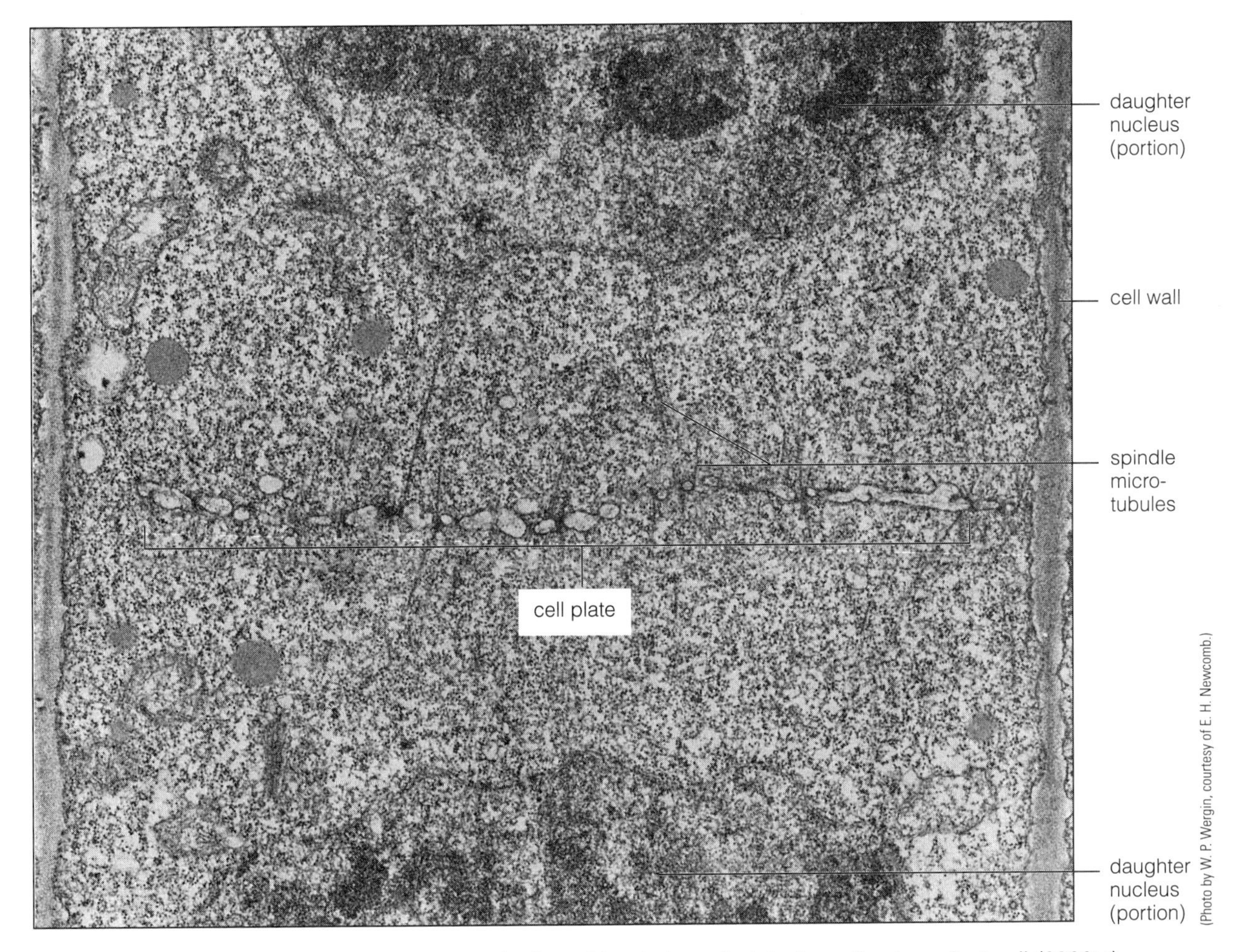

Figure 10-7 Transmission electron micrograph of cytokinesis by cell plate formation in a plant cell (2000×).

Following cytokinesis, the cell undergoes a period of growth and enlargement, during which time the nucleus is in interphase. Interphase may be followed by another mitosis and cytokinesis, or in some cells interphase may persist for the rest of a cell's life.

C. Duration of Phases of the Cell Cycle

An onion root tip preparation can be thought of as a snapshot capturing cells in various phases of the cell cycle at a particular moment in time. The frequency of occurrence of a cell cycle phase is directly proportional to the length of a phase. You can therefore estimate the amount of time each phase takes by tallying the proportions of cells in each phase.

The length of the cell cycle for cells in actively dividing onion root tips is approximately 24 hours, with mitosis lasting for about 90 minutes.

1. Examine the meristem region of an onion root tip slide. Count the number of cells in each of the stages of mitosis plus interphase in one field of view. Repeat this procedure for other fields of view until you count 100 cells. Record your data in Table 10-2.
2. Now calculate the time spent in each stage based on a 24-hour cell cycle by dividing the number of cells in each stage by the total number of cells counted to determine percent of total cells in each stage.
3. Multiply the fraction obtained in step 2 by 24 to determine duration.

   Although your calculations are only a rough approximation of the time spent in each stage, they do illustrate the differences in duration of each stage in the cell cycle.

TABLE 10-2 Determining Duration of Cell Cycle Phases			
Phase	**Number Seen**	**% of Total**	**Duration (hrs)**
Interphase			
Prophase			
Metaphase			
Anaphase			
Telophase			
Total			

10.3 The Cell Cycle in Animal Cells: Whitefish Blastula *(About 20 min.)*

Fertilization of an ovum by a sperm produces a zygote. In animal cells, the zygote undergoes a special type of cell division, *cleavage*, in which no increase in cytoplasm occurs between divisions. A ball of cells called a blastula is produced by cleavage. Within the blastula, repeated nuclear and cytoplasmic divisions take place; consequently, the whitefish blastula is an excellent example in which to observe the cell cycle of an animal.

Note a key difference between plants and animals: Whereas plants have meristems where divisions continually take place, animals do not have specialized regions to which mitosis and cytokinesis are limited. Indeed, divisions occur continually throughout many tissues of an animal's body, replacing worn-out or damaged cells.

With several important exceptions, mitosis in animals is remarkably like that in plants. These exceptions will be pointed out as we go through the cell cycle.

MATERIALS

Per student:

- prepared slide of whitefish blastula mitosis
- compound microscope

PROCEDURE

Obtain a slide labeled "whitefish blastula." Scan it with the low-power objective and then at medium power. This slide has numerous thin sections of a blastula. Select one section (Figure 10-8) and then switch to the high-dry objective for detailed observation.

As you examine the slides, draw the cells to show the correct sequence of events in the cell cycle of whitefish blastula.

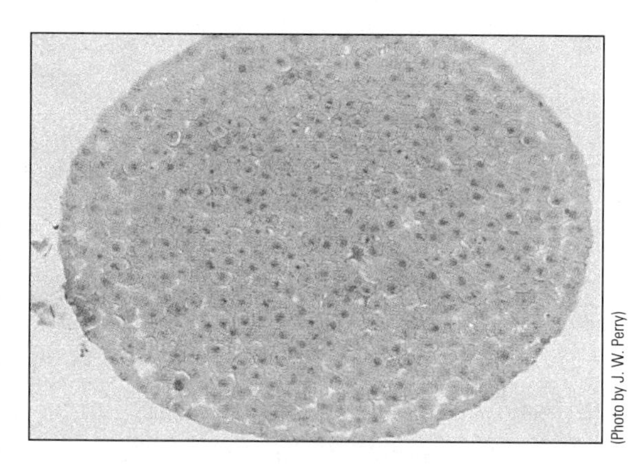

Figure 10-8 Section of a whitefish blastula (75×).

A. Interphase and Mitosis

1. **Interphase.** Locate a cell in **interphase**. As you observed in the onion root tip, note the presence of the nucleus and chromatin within it. Note also the absence of a cell wall.

   Draw an interphase cell above the word "Interphase" in Figure 10-9 and label the cytoplasm, nucleus, and plasma membrane.
2. **Mitosis.**
   (a) *Prophase.* The first obvious difference between mitosis in plants and animals is found in **prophase.** Unlike the onion cells, those of whitefish contain **centrioles** (Figure 10-10). As seen with the electron microscope, centrioles are barrel-shaped structures consisting of nine radially arranged triplets of microtubules.

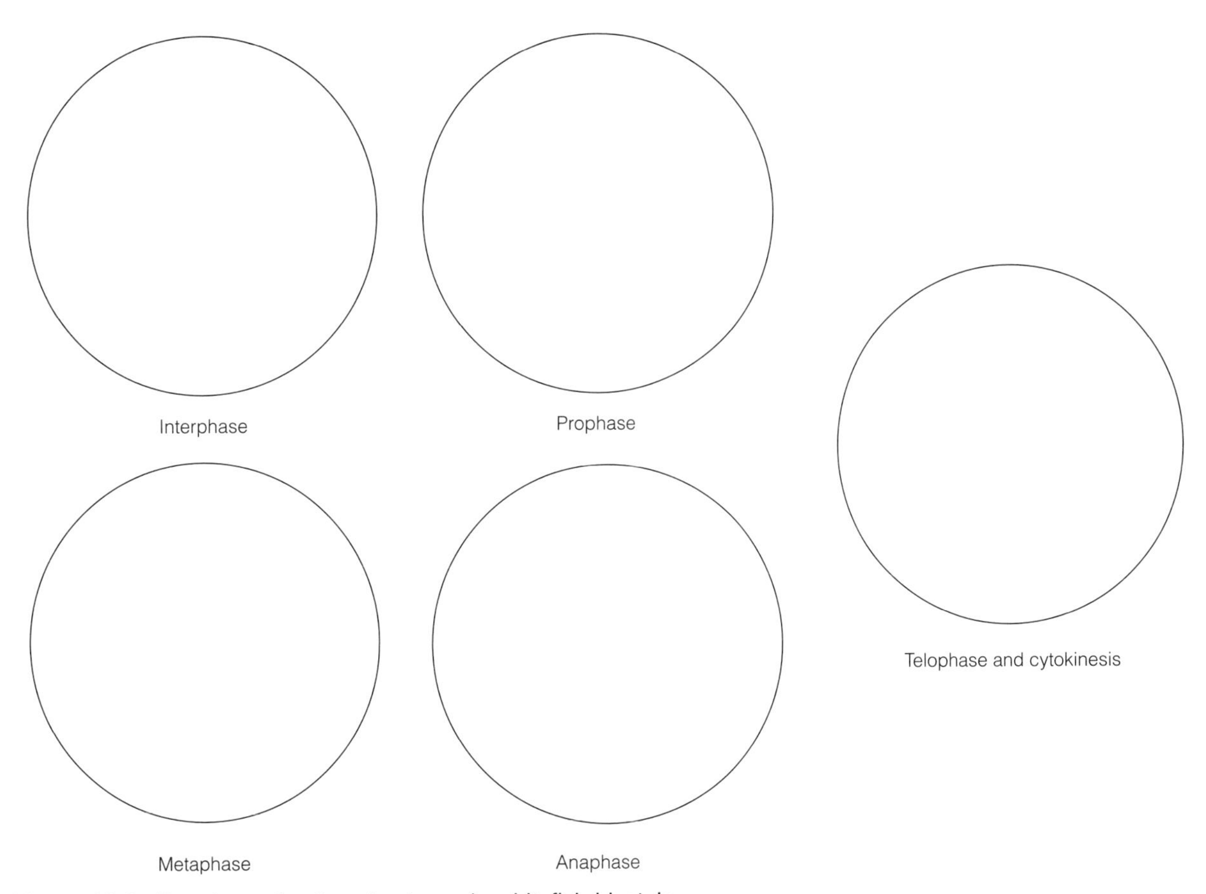

Figure 10-9 Drawings of cell cycle stages in whitefish blastula.
Labels: cytoplasm, nucleus, plasma membrane, spindle, chromosomes, spindle equator, sister chromatids, daughter nuclei, chromatin, furrow
(*Note:* Some terms are used more than once.)

One pair of centrioles was present in the cytoplasm in the G1 stage of interphase. These centrioles duplicated during the S stage of interphase. Subsequently, one new and one old centriole migrated to each pole.

Although the centrioles are too small to be resolved with your light microscope, you can see a starburst pattern of spindle fibers that appear to radiate from the centrioles. Other microtubules extend between the centrioles, forming the **spindle** (Figure 10-11). The chromosomes become visible as the chromatin condenses.

Find a prophase cell, identifying the spindle and starburst cluster of fibers about the centriole.

Draw the prophase cell in the proper location on Figure 10-9. Label the spindle, chromosomes, cytoplasm, and the position of the plasma membrane.

(b) *Metaphase.* As was the case in plant cells, during **metaphase** the spindle fiber microtubules become attached to the **kinetechore** of each centromere region, and the duplicated chromosomes (each consisting of two **sister chromatids**) line up on the **spindle equator.** Locate a metaphase cell.

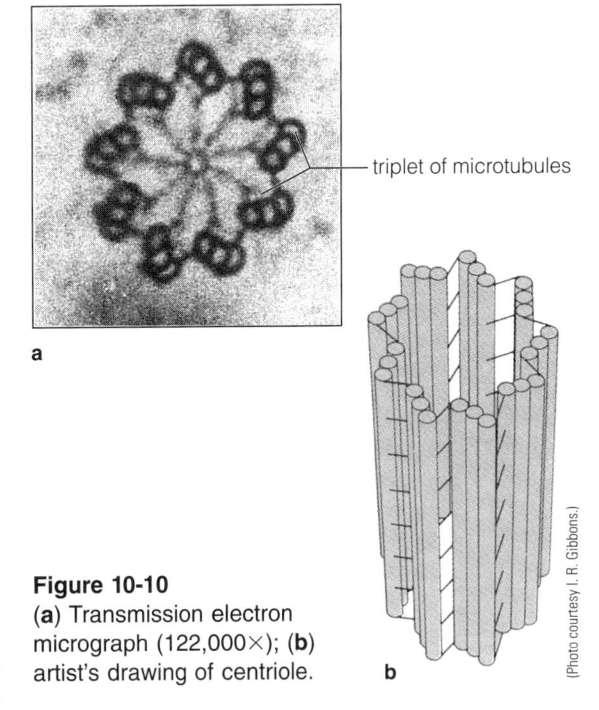

Figure 10-10
(**a**) Transmission electron micrograph (122,000×); (**b**) artist's drawing of centriole.

Draw the metaphase cell in the proper location on Figure 10-9. Label the chromosomes on the spindle equator, spindle, and plasma membrane.

(c) *Anaphase.* Again similar to that observed in plant cells, **anaphase** begins with the separation of sister chromatids into individual (daughter) chromosomes. Observe a blastula cell in anaphase.

Draw the anaphase cell in the proper location on Figure 10-9. Label the separating sister chromatids, spindle, cytoplasm, and plasma membrane.

(d) *Telophase.* **Telophase** is characterized by the arrival of the individual (daughter) chromosomes at the poles. A nuclear envelope forms around each daughter nucleus. Find a telophase cell.

Is the spindle still visible? _______

Is there any evidence of a nuclear envelope forming around the chromosomes?______

Draw the telophase cell in Figure 10-9. Label daughter nuclei, chromatin, cytoplasm, and plasma membrane.

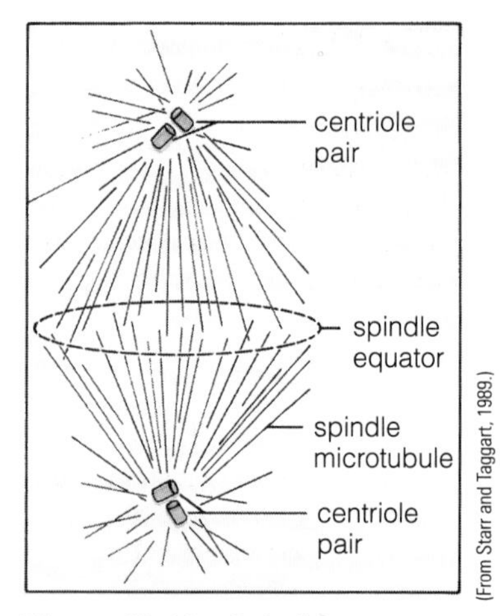

Figure 10-11 Spindle apparatus in animal cell.

B. Cytokinesis in Animal Cells

A second major distinction between cell division in plants and animals occurs during cytoplasmic division. Cell plates are absent in animal cells. Instead, cytokinesis takes place by **furrowing.**

To visualize how furrowing takes place, imagine wrapping a string around a balloon and slowly tightening the string until the balloon has been pinched in two. In life, the animal cell is pinched in two, forming two discrete cytoplasmic entities, each with a single nucleus. Figure 10-12 illustrates the cleavage furrow in animal cell.

Find a cell in the blastula undergoing cytokinesis. The telophase cell that you drew in Figure 10-9 may also show an early stage of cytokinesis. Label the cleavage furrow if it does.

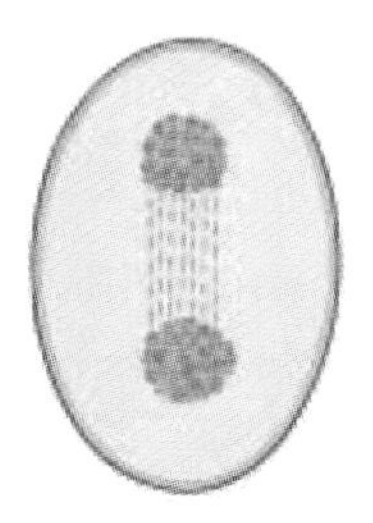

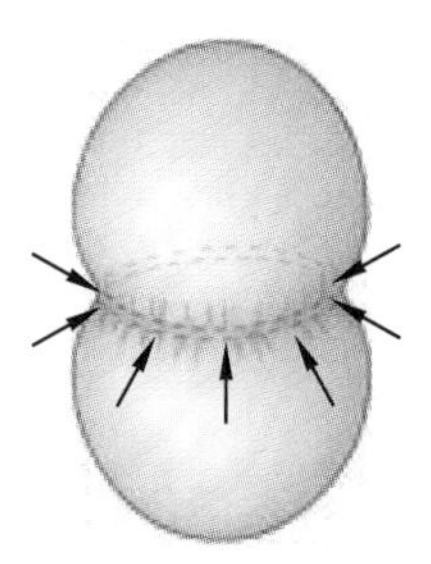

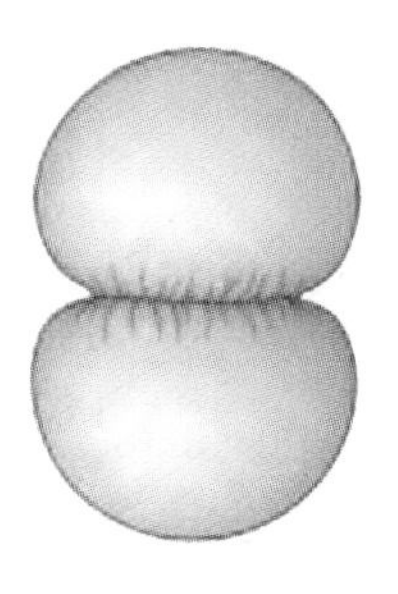

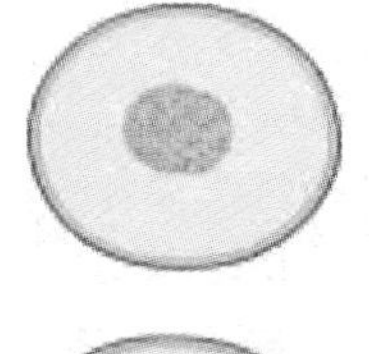

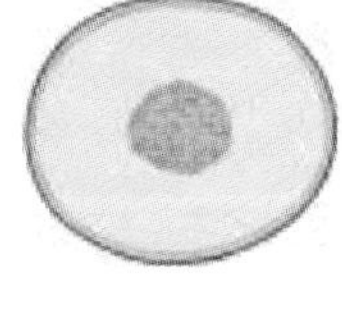

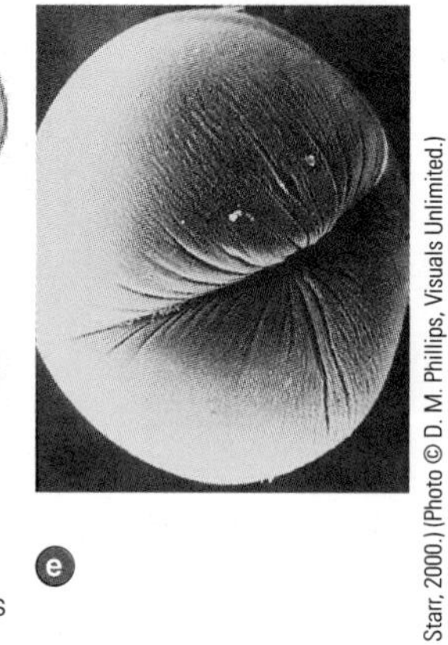

Figure 10-12 (a–d) Cytoplasmic division of an animal cell. **(e)** Scanning electron micrograph of the cleavage furrow at the plane of the former spindle equator.

10.4 Chromosome Squashes *(About 30 min.)*

You can make your own chromosome squash preparation quite simply. Cytologists and taxonomists do this routinely to count chromosomes. Observing whole sets of chromosomes is useful for studying chromosomal abnormalities and for determining if two organisms are different species.

MATERIALS

Per student:
- onion or daffodil root tips
- sharp razor blade
- 2 dissecting needles
- microscope slide and coverslip
- compound microscope

Per student pair:
- acetocarmine stain in dropping bottle
- iron alum in dropping bottle
- burner and matches

PROCEDURE

You will use onion or daffodil root tips that have been fixed, preserved, and softened to make squashes.

1. Obtain a single root and place it on a clean microscope slide. Notice that the terminal 2 mm or so is opaque white. This is the apical meristem.
2. With a sharp razor blade, separate the apical meristem from the rest of the root. Discard all *but* this meristem region.
3. Add a drop of acetocarmine stain and tease the tissue apart with dissecting needles.
4. Add a drop of iron alum. The iron intensifies the staining of the chromosomes.
5. Place a coverslip over the root tip. Spread the cells out by gently pressing down on the coverslip with your finger or a pencil eraser. Gently heat the slide over a flame.

Caution

Be careful. You don't want to cause the fluid to boil away!

6. Examine the preparation with your light microscope. Identify all stages of the cell cycle that have been described above.

   What do you notice about the shape of the cells after this preparation?

   ____

   ____

10.5 Simulating Mitosis *(About 20 min.)*

Understanding chromosome movements is crucial to understanding mitosis. You can simulate mitosis with a variety of materials. This is a simple activity, but a valuable one. It will be especially helpful when comparing the events of mitosis with those of meiosis in the next exercise.

MATERIALS

Per student pair:
- 44 pop beads each of two colors
- 8 magnetic centromeres

PROCEDURE

1. Build the components for two pairs of chromosomes by assembling strings of pop beads as follows:
   - **(a)** Assemble two strands of pop beads with 8 pop beads of one color on each arm, with a magnetic centromere connecting the two arms.
   - **(b)** Repeat step a, but use pop beads of the second color.
   - **(c)** Assemble two more strands of pop beads, but with 3 pop beads of one color on each arm.
   - **(d)** Repeat step c, using pop beads of the second color.

   You should have four long strings, two of each color, and four short strings, with two of each color. Each pop bead string should have a magnetic centromere at its midpoint. Note that pop bead strings can attach to each other at the magnetic centromere. Each pop bead string represents a single molecule of DNA plus proteins.
2. Place **one** of each kind of strand in the center of your workspace, which represents the nucleus. You have created a nucleus with four "chromosomes," two long and two short.
3. Manipulate these model chromosomes through the phases of the cell cycle, beginning in the G1 phase of interphase and proceeding through the rest of interphase, mitosis, and cytokinesis.

PRE-LAB QUESTIONS

_____ **1.** Reproduction in prokaryotes occurs primarily through the process known as
(a) mitosis
(b) cytokinesis
(c) furrowing
(d) fission

_____ **2.** The genetic material (DNA) of eukaryotes is organized into
(a) centrioles
(b) spindles
(c) chromosomes
(d) microtubules

_____ **3.** The process of cytoplasmic division is known as
(a) meiosis
(b) cytokinesis
(c) mitosis
(d) fission

_____ **4.** The product of chromosome duplication is
(a) two chromatids
(b) two nuclei
(c) two daughter cells
(d) two spindles

_____ **5.** The correct sequence of stages in *mitosis* is
(a) interphase, prophase, metaphase, anaphase, telophase
(b) prophase, metaphase, anaphase, telophase
(c) metaphase, anaphase, prophase, telophase
(d) prophase, telophase, anaphase, interphase

_____ **6.** During prophase, duplicated chromosomes
(a) consist of chromatids
(b) contain centromeres
(c) consist of nucleoproteins
(d) contain all of the above

_____ **7.** During the S period of interphase
(a) cell growth takes place
(b) nothing occurs because this is a resting period
(c) chromosomes divide
(d) synthesis (or replication) of the nucleoproteins takes place

_____ **8.** Chromatids separate during
(a) prophase
(b) telophase
(c) cytokinesis
(d) anaphase

_____ **9.** Cell plate formation
(a) occurs in plant cells but not in animal cells
(b) usually begins during telophase
(c) is a result of fusion of Golgi vesicles
(d) is all of the above

_____ **10.** Centrioles and a starburst cluster of spindle fibers would be found in
(a) both plant and animal cells
(b) only plant cells
(c) only animal cells
(d) none of the above

Name ______________________________________ Section Number ______________________________

EXERCISE 10

Mitosis and Cytokinesis: Nuclear and Cytoplasmic Division

POST-LAB QUESTIONS

Introduction

1. Distinguish among interphase, mitosis, and cytokinesis.

10.1 Chromosomal Structure

2. Distinguish between the structure of a duplicated chromosome before mitosis and the chromosome produced by separation of two chromatids during mitosis.

10.2 The Cell Cycle in Plant Cells: Onion Roots

3. If the chromosome number of a typical onion root tip cell is 16 before mitosis, what is the chromosome number of each newly formed nucleus after nuclear division has taken place?

4. In plants, what name is given to a region where mitosis occurs most frequently?

5. The cells in the following photomicrographs have been stained to show microtubules comprising the spindle apparatus. Identify the stage of mitosis in each and label the region indicated on (**b**). (Photos by Andrew S. Bajer.)

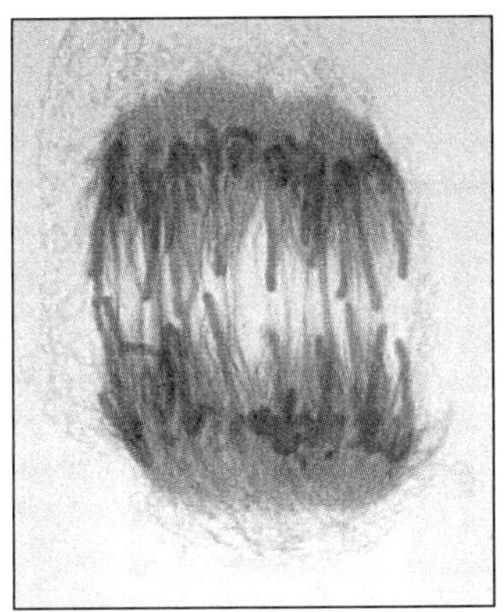

a stage? ____________________

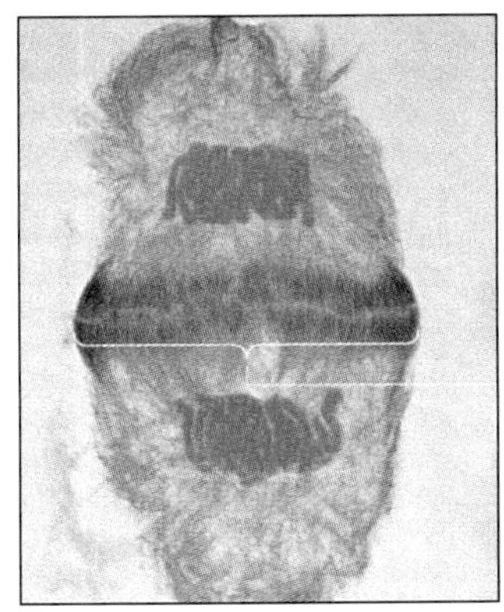

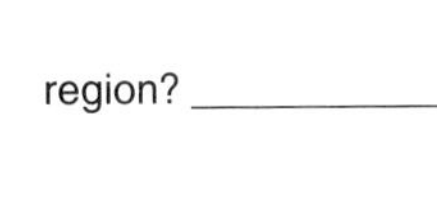

region? ____________

b stage? ____________________

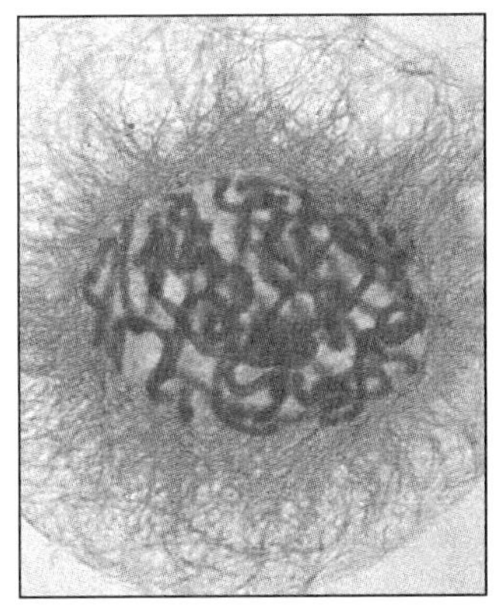

c stage? ____________________

10.3 The Cell Cycle in Animal Cells: Whitefish Blastula

6. Name two features of animal cell mitosis and cytokinesis you can use to distinguish these processes from those occurring in plant cells.

   a.

   b.

Food for Thought

7. Observe photomicrographs (**a**) and (**b**) below. Is (**a**) from a plant or an animal?

   Note the double nature of the blue "threads." Each individual component of the doublet is called a ________________________________. Is (**b**) from a plant or an animal? (Photos by Andrew S. Bajer.)

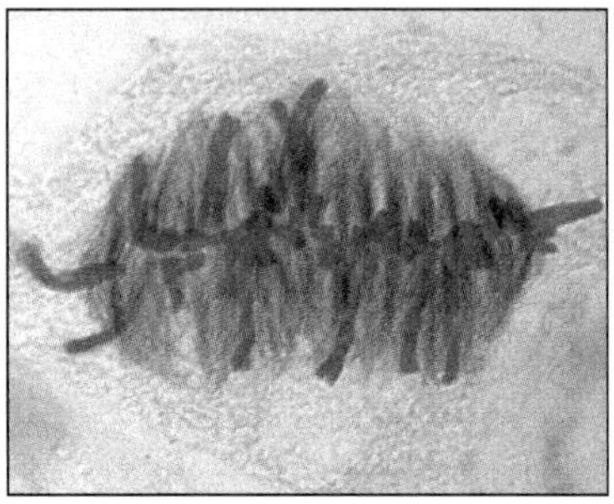

a plant or animal?

____

structure?

____

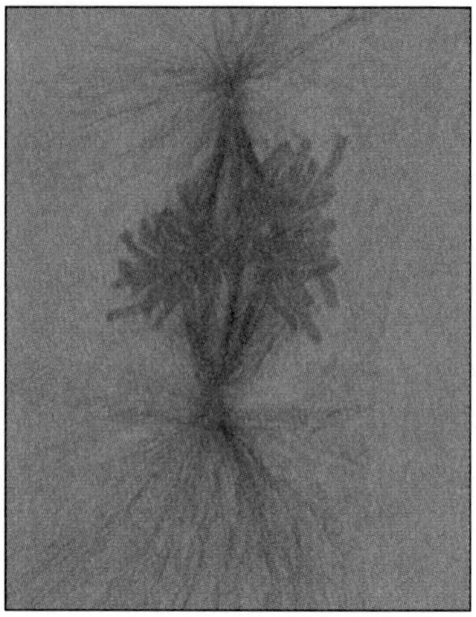

b plant or animal?

____

8. Why do you suppose cytokinesis generally occurs in the cell's midplane?

9. Why must the DNA be duplicated during the S phase of the cell cycle, prior to mitosis?

10. What would happen if a cell underwent mitosis but not cytokinesis?

EXERCISE 11

Meiosis: Basis of Sexual Reproduction

OBJECTIVES

After completing this exercise, you will be able to

1. define *meiosis, homologue (homologous chromosome), diploid, haploid, gene, gene pair, allele, gamete, ovum, sperm, fertilization, locus, synapsis, zygote, genotype, nondisjunction;*
2. indicate the differences and similarities between meiosis and mitosis;
3. describe the basic differences between the life cycles of higher plants and higher animals;
4. describe the process of meiosis, and recognize events that occur during each stage;
5. discuss the significance of crossing over, segregation, and independent assortment;
6. identify the meiotic products in male and female animals.;
7. describe the process of nondisjunction and chromosome number abnormalities in resulting gametes and zygotes.

INTRODUCTION

Like mitosis, meiosis is a process of nuclear division. During mitosis, the number of chromosomes in the daughter nuclei remains the same as that in the parental nucleus. In meiosis, however, the genetic complement is halved, resulting in daughter nuclei containing only one-half the number of chromosomes as the parental nucleus. Thus, while mitosis is sometimes referred to as an *equational division,* meiosis is often called *reduction division.* Moreover, while mitosis is completed after a single nuclear division, two divisions, called meiosis I and meiosis II, occur during meiosis. Table 11-1 summarizes the differences between mitosis and meiosis.

TABLE 11-1 Comparison of Mitosis and Meiosis

Mitosis	Meiosis
Equational division: Amount of genetic material remains constant	Reduction division: Amount of genetic material is halved
Completed in one division cycle	Requires two division cycles for completion
Produces two genetically identical nuclei	Produces two to four genetically different nuclei
Generally produces cells not directly involved in sexual reproduction	Ultimately produces cells used for sexual reproduction

In the body cells of most eukaryotes, chromosomes exist in pairs called homologues (homologous chromosomes); that is, there are two chromosomes that are physically similar and contain genetic information for the same traits. To visualize this, press your palms together, lining up your fingers. Each "finger pair" represents one pair of homologues.

When both homologues are in the *same* nucleus, the nucleus is **diploid** (2n); when only one of the homologues is present, the nucleus is **haploid** (n). If the parental nucleus normally contains the diploid (2n) chromosome number before meiosis, all four daughter nuclei contain the haploid (n) number at the completion of meiosis.

The reduction in chromosome number is the basis for sexual reproduction. In animals, the cells containing the daughter nuclei produced by meiosis are called **gametes: ova** (singular is *ovum*) if the parent is female, **sperm** cells if male. As you probably know, gametes are produced in the gonads—ovaries and testes, respectively. In fact, this is the *only* place where meiosis occurs in higher animals. Figure 11-1 shows where meiosis occurs in humans, while Figure 11-2 shows the life cycle of a higher animal.

Note when meiosis occurs—during gamete production. During **fertilization** (the fusion of a sperm nucleus with an ovum nucleus), the diploid chromosome number is restored as the two haploid gamete nuclei fuse to form the **zygote,** the first cell of the new diploid generation.

What about plants? Do plants have sex? Indeed they do. However, the plant life cycle is a bit more complex than that of animals. Plants of a single species have two completely different body forms. The primary function of one is the production of gametes. This plant is called a *gametophyte* ("gamete-producing plant") and it is haploid. Because the entire plant is haploid, gametes are produced in specialized organs by mitosis. The other body form, a *sporophyte,* is diploid. This diploid sporophyte has specialized organs in which meiosis occurs, producing haploid spores (hence the name *sporophyte,* "spore-producing plant"). When spores germinate and produce more cells by mitosis, they grow into haploid gametophytes, completing the life cycle.

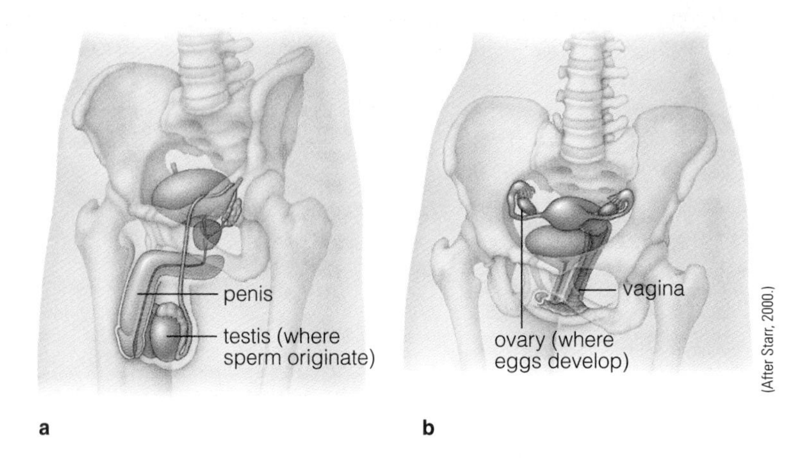

Figure 11-1 Gamete-producing structures in humans. (**a**) Human male, (**b**) human female.

Figure 11-3 shows the structures in a typical flower that produce sperm and eggs.

Examine Figure 11-4, which shows the gametophyte and sporophyte of a fern plant. Remember, the gametophyte and sporophyte are different, free-living stages of the *same* species of fern. Now look at Figure 11-5, which diagrams a typical plant life cycle. Again, note the consequence of meiosis. In plants, it results in the production of **spores,** not gametes.

You should understand an important concept from these diagrams: *Meiosis always halves the chromosome number. The diploid chromosome number is eventually restored when two haploid nuclei fuse during fertilization.*

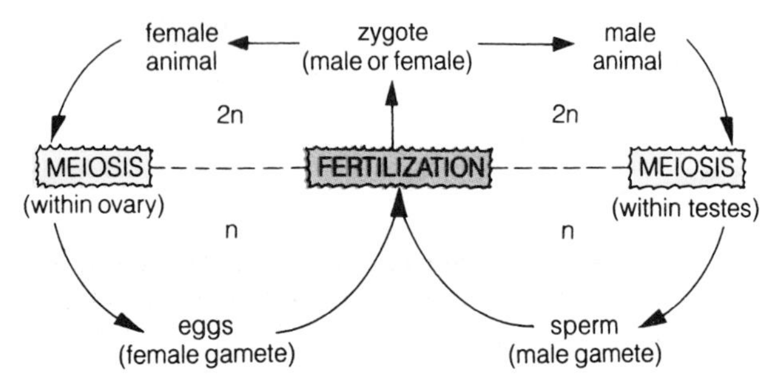

Figure 11-2 Life cycle of higher animals.

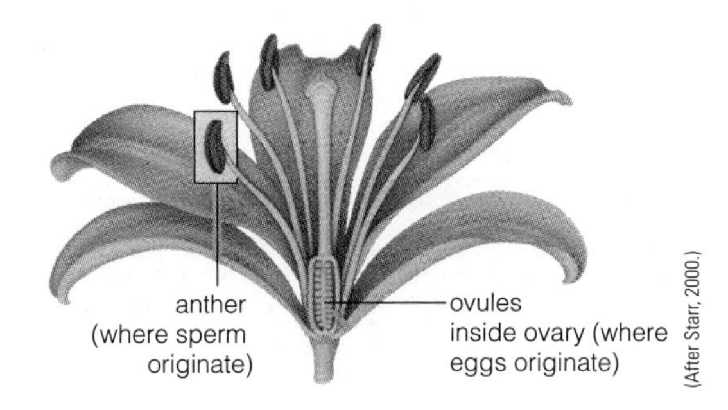

Figure 11-3 Gamete-producing structures in a flowering plant.

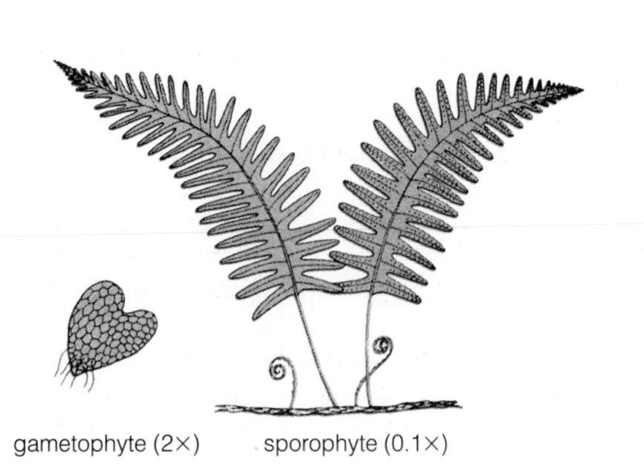

Figure 11-4 Gametophyte and sporophyte phases of the same fern species. Note size differences.

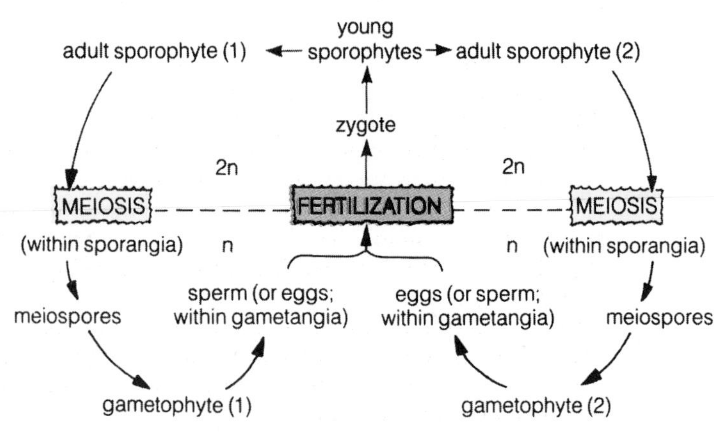

Figure 11-5 Life cycle of a plant.

Understanding meiosis is an absolute necessity for understanding the patterns of inheritance in Mendelian genetics. Gregor Mendel, an Austrian monk, spent years deciphering the complexity of simple genetics. Although he knew nothing of genes and chromosomes, he noted certain patterns of inheritance and formulated three principles, now known as Mendel's principles of recombination, segregation, and independent assortment. The following activities will demonstrate the events of meiosis and the genetic basis for Mendel's principles.

11.1 Demonstrations of Meiosis Using Pop Beads *(About 75 min.)*

MATERIALS

Per student pair:

- 44 pop beads each of two colors (red and yellow, for example)
- 8 magnetic centromeres
- marking pens
- 8 pieces of string, each 40 cm long
- meiotic diagram cards similar to those used here
- colored pencils

Per student group (table):

- bottle of 95% ethanol to remove marking ink
- tissues

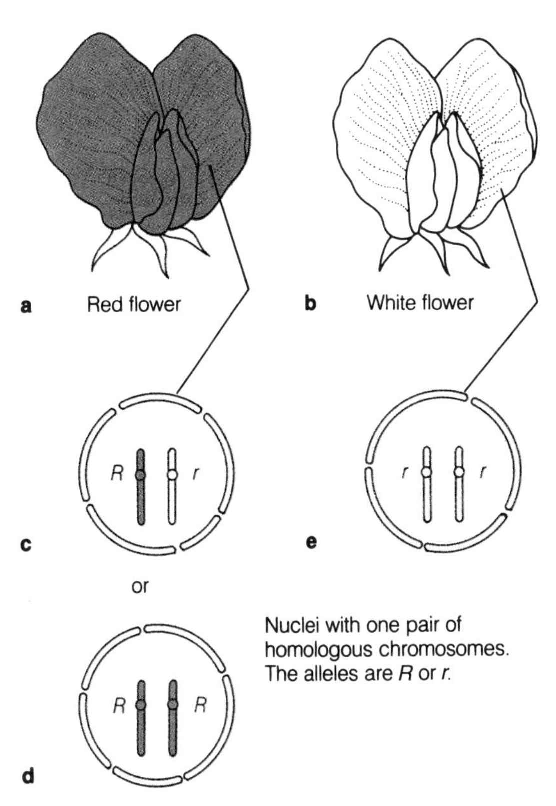

Figure 11-6 Chromosomal control of flower color. (**c–e**) show a nucleus with one pair of homologous chromosomes. The alleles for flower color are *R* or *r*.

PROCEDURE

Work in pairs.

Within the nucleus of an organism, each chromosome bears **genes,** which are units of inheritance. Genes may exist in two or more alternative forms called **alleles.** Each homologue bears *genes* for the same traits; these are the **gene pairs.** However, the homologues may or may not have the same *alleles.* An example will help here.

Suppose the trait in question is flower color and that a flower has only two possible colors, red or white (Figure 11-6a, b). The gene is coding (providing the information) for flower color. There are two homologues in the same nucleus, so each bears the gene for flower color. *But,* on one homologue, the *allele* might code for red flowers, while the allele on the other homologue might code for white flowers (Figure 11-6c). There are two other possibilities. The alleles on *both* homologues might be coding for red flowers (Figure 11-6d), or they *both* might be coding for white flowers (Figure 11-6e). Note that these three possibilities are mutually exclusive.

1. Build the components for two pairs of homologous chromosomes by assembling strings of pop beads as follows:
   - **(a)** Assemble two strands of pop beads with 8 pop beads of one color on each arm, with a magnetic centromere connecting the two arms.
   - **(b)** Repeat step a, but use pop beads of the second color.
   - **(c)** Assemble two more strands of pop beads, but with 3 pop beads of one color on each arm.
   - **(d)** Repeat step c, using pop beads of the second color.

   You should have four long strings, two of each color, and two short strings, also two of each color. Each pop-bead string should have a magnetic centromere at its midpoint by which pop-bead strings can attach to each other. Each pop-bead string represents a single molecule of DNA plus proteins, with each bead representing a gene.
2. Place **one** of each kind of strand in the center of your workspace, which represents the interphase nucleus of a cell that will undergo meiosis. You have created a nucleus with four "chromosomes," two long and two short. The long strands represent one homologous pair, and the short strands represent a second homologous pair of chromosomes.

We start by assuming that these chromosomes represent the diploid condition. The two colors represent the origin of the chromosomes: One homologue (color ________________) came from the male parent, and the other homologue (color ________________) came from the female parent.

3. The four single-stranded chromosomes represent four unduplicated chromosomes. Now simulate DNA duplication during the S-phase of interphase (Figure 10-1), whereby each DNA molecule and its associated proteins are copied exactly. The two copies, called sister chromatids, remain attached to each other at their centromeres (Figure 11-7). During chromosome replication, the genes also duplicate. Thus, alleles on sister chromatids are identical.

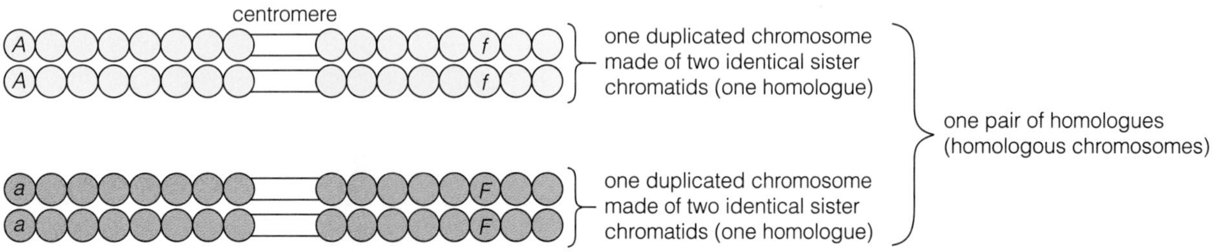

centromere

A f

A f

one duplicated chromosome made of two identical sister chromatids (one homologue)

one pair of homologues (homologous chromosomes)

a F

a F

one duplicated chromosome made of two identical sister chromatids (one homologue)

Figure 11-7 One pair of homologous pop-bead chromosomes.

How many sister chromatids are there in a duplicated chromosome? _______

How many chromosomes are represented by four sister chromatids? _______ By eight? _______

What is the diploid number of the starting (parental) nucleus? (*Hint:* Count the number of homologues to obtain the diploid number.) _______

4. As mentioned previously, genes may exist in two or more alternative forms, called alleles. The location of an allele on a chromosome is its **locus** (plural: *loci*). Using the marking pen, mark two loci on each long chromatid with letters to indicate alleles for a common trait. Suppose the long pair of homologous chromosomes codes for two traits, skin pigmentation and the presence of attached earlobes in humans. We'll let the capital letter *A* represent the allele for normal pigmentation and a lowercase *a* the allele for albinism (the absence of skin pigmentation); *F* will represent free earlobes and *f* attached earlobes. A suggested marking sequence is illustrated in Figure 11-7.
5. Let's assign a gene to our second homologous pair of chromosomes, the short pair. We'll suppose this gene codes for the production of an enzyme necessary for metabolism. On one homologue (consisting of two chromatids) mark the letter *E*, representing the allele causing enzyme production. On the other homologue, *e* represents the allele that interferes with normal enzyme production.
6. Obtain a meiotic diagram card like the one in Figure 11-8. Manipulate your model chromosomes through the stages of meiosis described below, locating the chromosomes in the correct diagram circles (representing nuclei) as you go along. Reference to Figure 11-8 will be made at the proper steps. *DO NOT* draw on the meiotic diagram cards.

A. Meiosis Without Crossing Over

Although crossing over is a nearly universal event during meiosis, we will first work with a simplified model to illustrate chromosomal movements and separations during meiosis. Refer to Figure 11-9 as you manipulate your model.

1. **Late interphase.** During interphase, the nuclear envelope is intact and the chromosomes are randomly distributed throughout the nucleoplasm (semifluid substance within the nucleus). All duplicated chromosomes (eight chromatids) should be in the parental nucleus, indicating that DNA duplication has taken place. The sister chromatids of each homologue should be attached by their magnetic centromeres, but the four homologues should be separate. Your model nucleus contains a diploid number 2n = 4.

   The pop-bead chromosomes should appear during interphase in the parental nucleus as shown in Figure 11-8. Be sure to mark the location of the alleles. Use different pencil or pen colors to differentiate the homologues on your drawings.
2. **Meiosis I.** During meiosis I, homologues are separated from each other into different nuclei. Daughter nuclei created are thus haploid.
   (a) *Prophase I.* During the first prophase, the parental nucleus contains four duplicated homologous chromosomes, each comprised of two sister chromatids joined at their centromeres. The chromatin condenses to form discrete, visible chromosomes. The homologues pair with each other. This pairing is called *synapsis.* Slide the two homologues together.

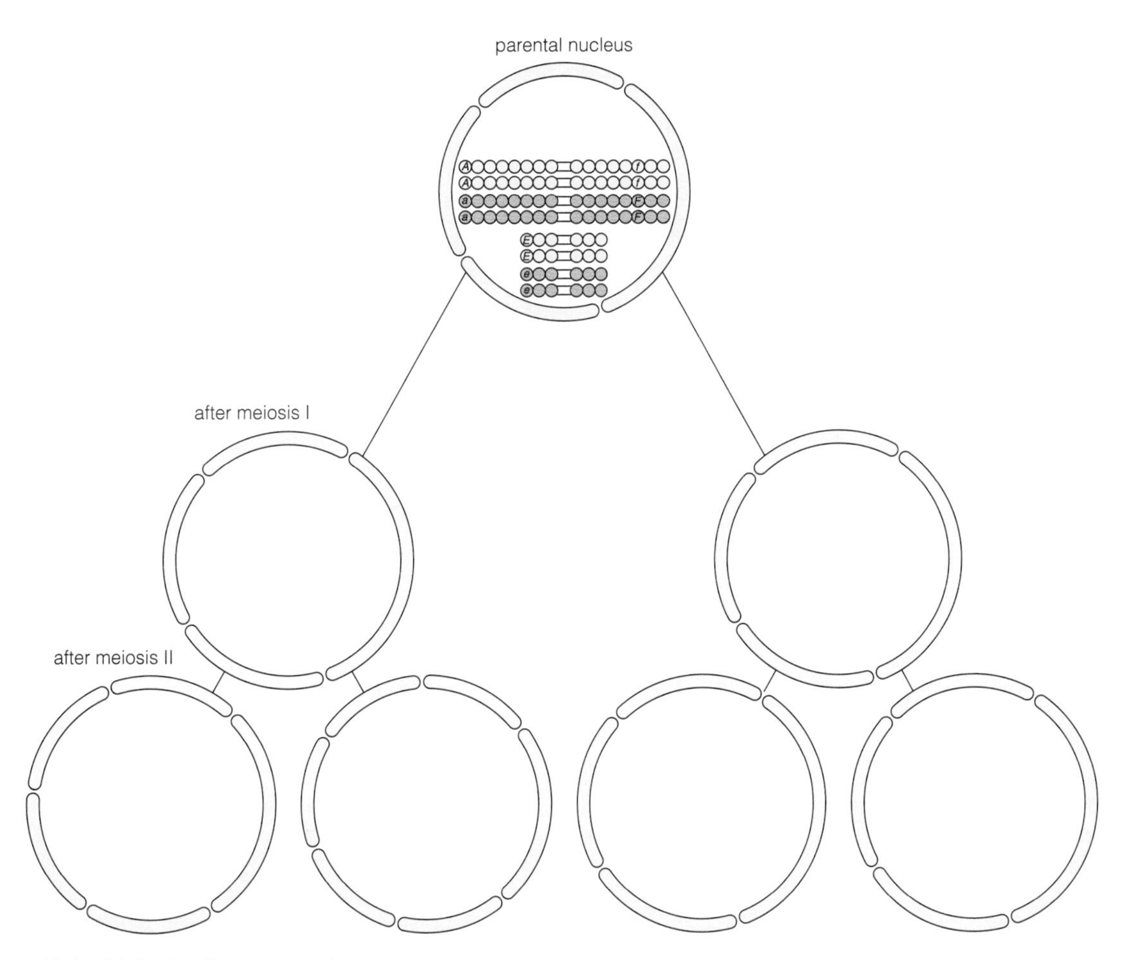

Figure 11-8 Meiosis without crossing over.

Twist the chromatids about one another to simulate synapsis.
The nuclear envelope disorganizes at the end of prophase I.

(b) *Metaphase I.* Homologous chromosomes now move toward the spindle equator, the centromeres of each homologue coming to lie *on either side of the equator.* Spindle fibers, consisting of aggregations of microtubules, attach to the centromeres. One homologue attaches to microtubules extending from one pole, and the other homologue attaches to microtubules extending from the opposite spindle pole.

To simulate the spindle fibers, attach one piece of string to each centromere. Then lay the free ends of strings from two homologues toward one spindle pole and the ends of the other homologues toward the opposite pole.

(c) *Anaphase I.* During anaphase I, the homologous chromosomes separate, one homologue moving toward one pole, the other toward the opposite pole. The movement of the chromosomes is apparently the result of shortening of some spindle fibers and lengthening of others. Each homologue is still in the duplicated form, consisting of two sister chromatids.

Pull the two strings of one homologous pair toward its spindle pole and the other toward the opposite spindle pole, separating the homologues from one another. Repeat with the second pair of homologues.

(d) *Telophase I.* Continue pulling the string spindle fibers until each homologue is now at its respective pole. The first meiotic division is now complete. You should have two nuclei, each containing two chromosomes (one long and one short) consisting of two sister chromatids.

Draw your pop-bead chromosomes as they appear after meiosis I on the two nuclei labeled "after meiosis I" of Figure 11-8. Depending on the organism involved, an interphase (interkinesis) and cytokinesis may precede the second meiotic division, *or* each nucleus may enter directly into meiosis II. The chromosomes decondense into chromatin form.

It is important to note here that DNA synthesis *does not* occur following telophase I (between meiosis I and meiosis II).

Before meiosis II, the spindle is rearranged into two spindles, one for each nucleus.

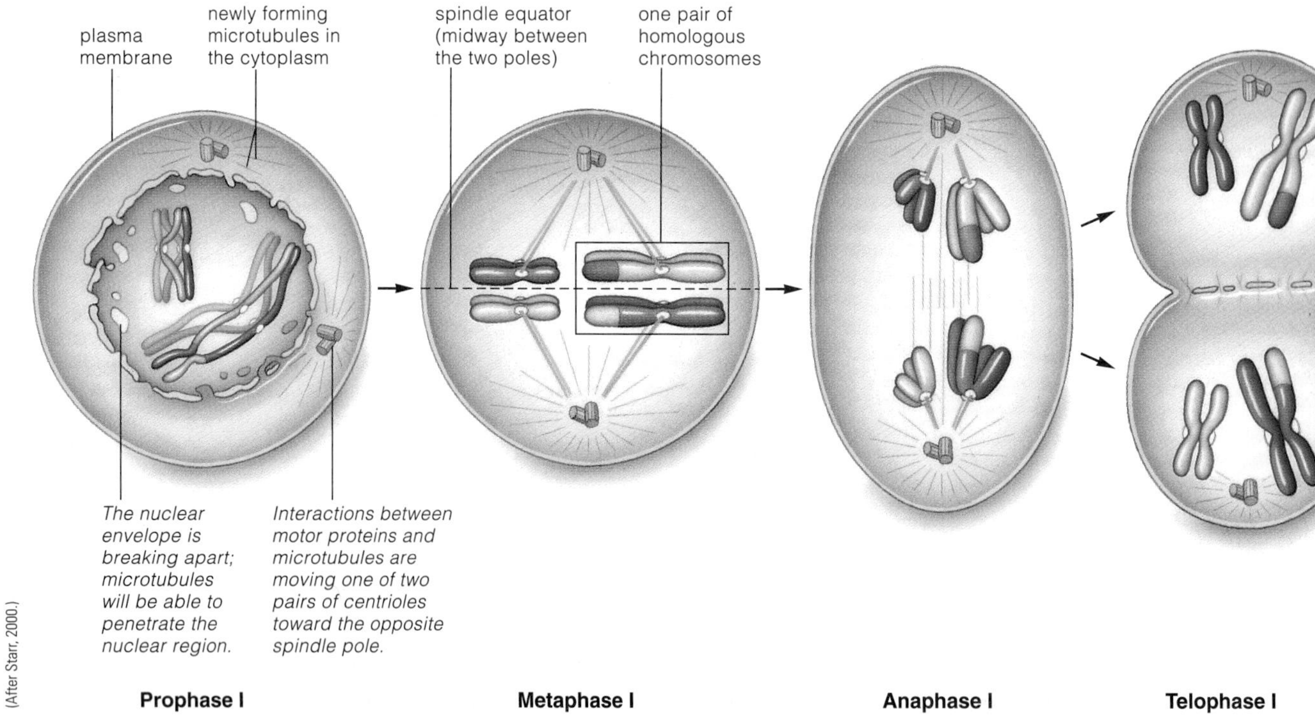

Figure 11-9 Meiosis in a generalized animal germ cell. Two pairs of chromosomes are shown. Maternal chromosomes are shaded purple. Paternal chromosomes are shaded light blue.

3. **Meiosis II.** During meiosis II, sister chromatids are separated into different daughter nuclei. The result is four haploid nuclei.
   (a) *Prophase II.* At the beginning of the second meiotic division, the sister chromatids are still attached by their centromeres. During prophase II, the nuclear envelope disorganizes, and the chromatin recondenses.
   (b) *Metaphase II.* Within each nucleus, the duplicated chromosomes align with the equator, the centromeres lying *on the equator.* Spindle fiber microtubules attach the centromeres of each chromatid to opposite spindle poles.

   Your string spindle fibers should be positioned so that the two spindle fiber strings from sister chromatids lie toward opposite poles. Note that each nucleus contains only *two* duplicated chromosomes (one long and one short) consisting of *two* sister chromatids each.
   (c) *Anaphase II.* The sister chromatids separate, moving to opposite poles. Pull on the string until the two sister chromatids separate. After the sister chromatids separate, each is an individual (not duplicated) daughter chromosome.
   (d) *Telophase II.* Continue pulling on the string spindle fibers until the two daughter chromosomes are at opposite poles. The nuclear envelope re-forms around each chromosome and the chromosomes decondense back into chromatin form. Four daughter nuclei now exist. Note that each nucleus contains two individual unduplicated chromosomes (each formerly a chromatid) originally present within the parental nucleus. These nuclei and the cells they're in generally undergo a differentiation and maturation process to become gametes (in animals) or spores (in plants).

   Draw your pop-bead chromosomes as they appear after meiosis II in the "gamete nuclei" of Figure 11-8. Your diagram should indicate the genetic (chromatid) complement *before* meiosis and *after* each meiotic division, *not* the stages of each division.

   Remember that meiosis takes place in both male and female organisms. (See Figure 11-2.)

   If the parental nucleus was from a male, what is the gamete called? ____________________

   If female? ____________________

   Is the parental nucleus diploid or haploid? ____________________

   Are the nuclei produced after the *first* meiotic division diploid or haploid? ____________________

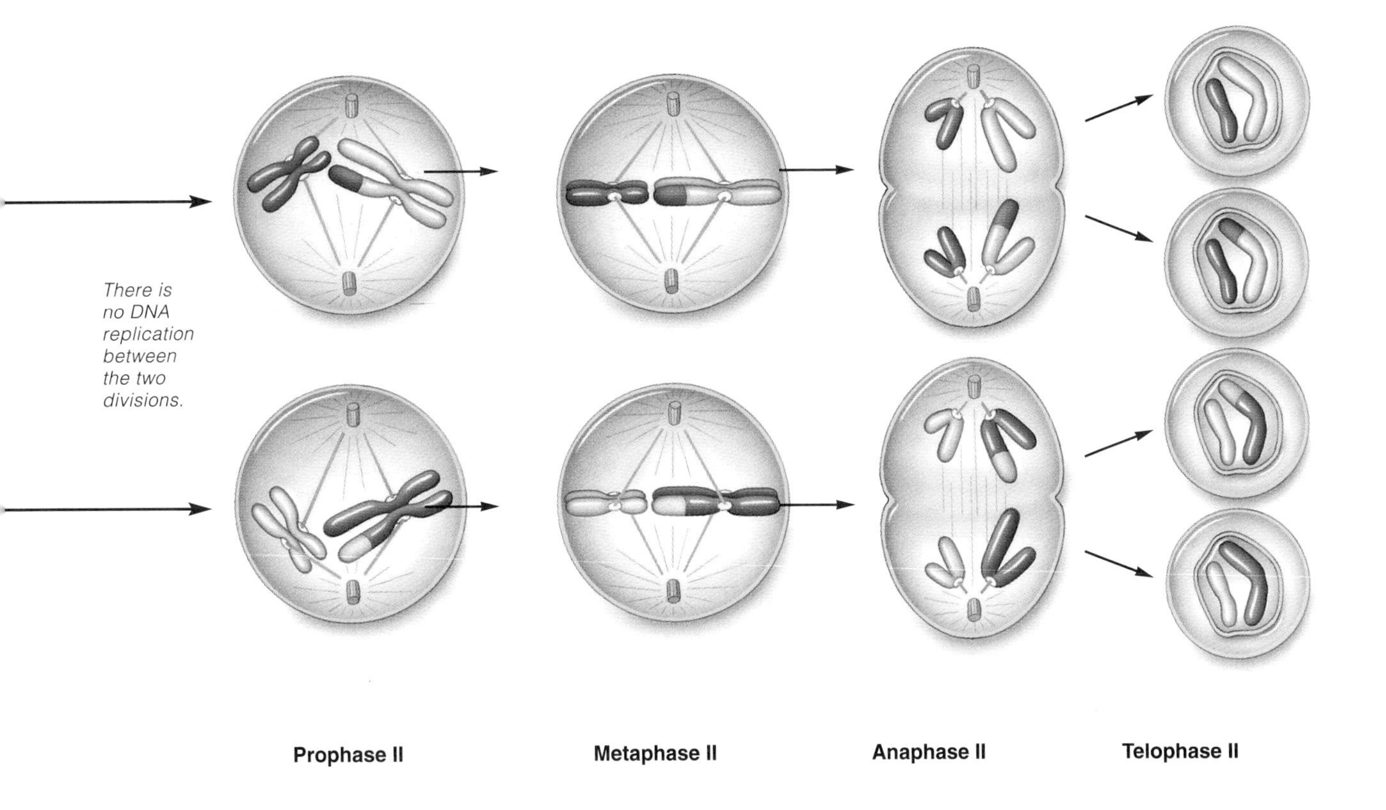

Are the nuclei of the gametes diploid or haploid? ______________________

What is the **genotype** of each gamete nucleus after meiosis II? (The genotype is the genetic composition of an organism, or the alleles present. Another way to ask this question is, What alleles are present in each gamete nucleus? Write these in the format: *AFE*, *afe*, and so on.)

____

If you answered the preceding questions correctly, you might logically ask, "If the chromosome number of the gametes is the same as that produced after the first meiotic division, why bother to have two separate divisions? After all, the genes present are the same in both gametes and first-division nuclei."

There are two answers to this apparent paradox. The first, and perhaps the most obvious, is that the second meiotic division ensures that a *single* chromatid (nonduplicated chromosome) is contained within each gamete. After gametes fuse, producing a zygote, the genetic material duplicates prior to the zygote's undergoing mitosis. If gametes contained two chromatids, the zygote would have four, and duplication prior to zygote division would produce eight, twice as many as the organism should have. If DNA duplication within the zygote were not necessary for the onset of mitosis, this problem would not exist. Alas, DNA synthesis apparently is a necessity to initiate mitosis.

You can discover the second answer for yourself by continuing with the exercise, for although you have simulated meiosis, you have done so without showing what happens in *real* life. That's the next step . . .

B. Meiosis with Crossing Over

A very important event that results in a reshuffling of alleles on the chromatids occurs during prophase I. Recall that synapsis results in pairing of the homologues. During synapsis, the chromatids break, and portions of chromatids bearing genes for the same characteristic (but perhaps *different* alleles) are exchanged between *nonsister* chromatids. This event is called **crossing over,** and it results in recombination (shuffling) of alleles.

1. Look again at Figure 11-7. Distinguish between sister and nonsister chromatids. Now look at Figure 11-10, which demonstrates crossing over in one pair of homologues.
2. Return your chromosome models to the nucleus format with two pairs of homologues entering prophase I.

3. To simulate crossing over, break four beads from the arms of two nonsister chromatids in the long homologue pair, exchanging bead color between the two arms. During actual crossing over, the chromosomes may break anywhere within the arms.

   Crossing over is virtually a universal event in meiosis. Each pair of homologues may cross over in several places simultaneously during prophase I.

4. Manipulate your model chromosomes through meiosis I and II again and watch what happens to the distribution of the alleles as a consequence of the crossing over. Fill in Figure 11-11 as you did before, but this time show the effects of crossing over. Again, use different colors in your sketches.

   What are the genotypes of the gamete nuclei?

   Is the distribution of alleles present in the gamete nuclei after crossing over the same as that which was present without crossing over?

   Is the distribution of alleles present in the gamete nuclei after crossing over the same as that in the nuclei after the first meiotic division?

   Crossing over provides for genetic recombination, resulting in increased variety. How many different genetic *types* of daughter chromosomes are present in the gamete nuclei without crossing over (Figure 11-8)?

   How many different types are present with crossing over (Figure 11-11)?

   We think you would agree that a greater number of *types* of daughter chromosomes indicates greater *variety.*

   Recall that the parental nucleus contained a pair of homologues, each homologue consisting of two sister chromatids. Because sister chromatids are identical in all respects, they have the same alleles of a gene (see Figure 11-7). As your models showed, the alleles on nonsister chromatids may not (or may) be identical; they bear the same genes but may have different alleles, different forms of some genes.

   What is the difference between a gene and an allele?

   Let's look at a single set of alleles on your model chromosomes that are, say, the alleles for pigmentation, A and a. Both alleles were present in the parental nucleus. How many are present in the gametes?

a A pair of duplicated homologous chromosomes.

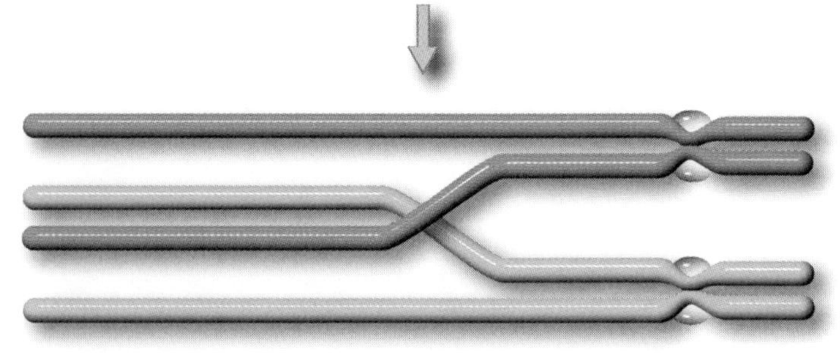

b Crossover between nonsister chromatids of the two chromosomes.

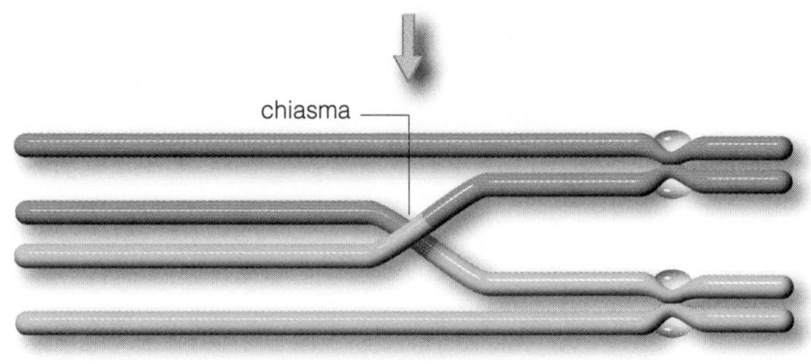

c Nonsister chromatids exchange segments.

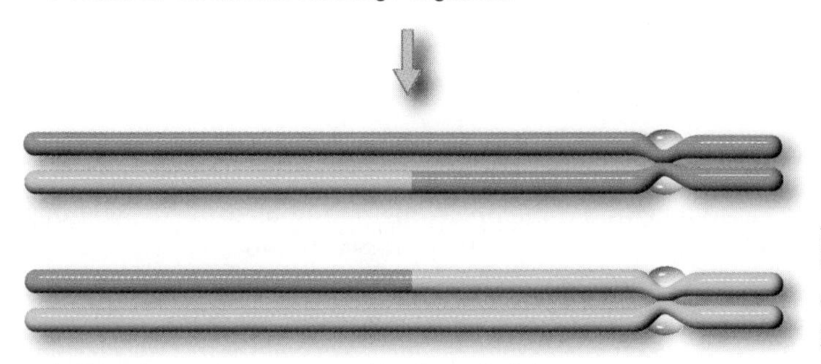

d Homologues have new combinations of alleles.

(After Starr, 2000.)

Figure 11-10 Crossing over in one pair of homologues. Maternal chromosomes are shaded purple. Paternal chromosomes are shaded light blue.

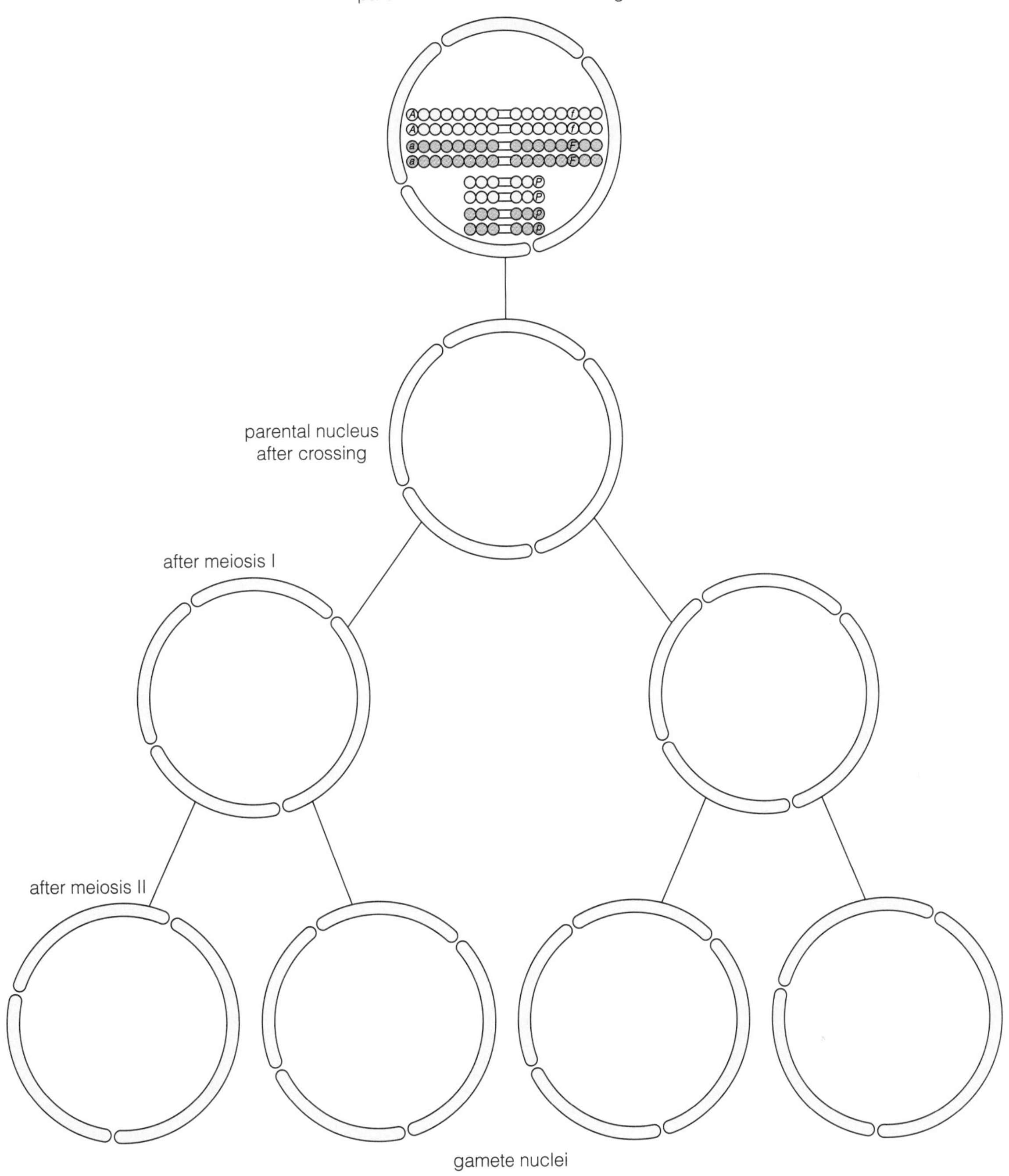

Figure 11-11 Meiosis with crossing over.

This illustrates Mendel's first principle, segregation. Segregation means that during gamete formation, pairs of alleles are separated (segregated) from each other and end up in different gametes.

C. Demonstrating Independent Assortment

Manipulate your model chromosomes again through meiosis with crossing over (Figure 11-11), searching for different possibilities in chromosome distribution that would make the gametes genetically different.

Does the distribution of the alleles for enzyme production to different gametes on the second set of homologues have any bearing on the distribution of the alleles on the first set (alleles for skin pigmentation and earlobe condition)? ____________________

This distribution demonstrates the principle of independent assortment, which states that segregation of alleles into gametes is independent of the segregation of alleles for other traits, *as long as the genes are on different sets of homologous chromosomes.* Genes that are on different (nonhomologous) chromosomes are said to be **nonlinked.** By contrast, genes for different traits that are on the same chromosome are **linked.**

Because the genes for enzyme production and those for skin pigmentation and earlobe attachment are on different homologous chromosomes, these genes are ___________, while the genes for skin pigmentation and earlobe attachment are __________________ because they are on the same chromosome.

In reality, most organisms have many more than two sets of chromosomes. Humans have 23 pairs (2n = 46), while some plants literally have hundreds!

A thorough understanding of meiosis is necessary to understand genetics. With this foundation, you'll find that problems involving Mendelian genetics are easy and fun to do. Without an understanding of meiosis, Mendelian genetics will be hopelessly confusing.

D. Nondisjunction and the Production of Gametes with Abnormal Chromosome Number

Errors in the process of meiosis can occur in many ways. Perhaps the best understood error process is that of **nondisjunction,** when one or more pairs of chromosomes fail to separate in anaphase. The result is gamete nuclei with too few or too many chromosomes.

1. Begin to manipulate your model chromosomes to show meiosis without crossing over (page 149). In modeling events at metaphase I, however, arrange the spindle fiber threads for the long pair of homologues so that they all extend to the same pole.
2. Model anaphase I, pulling the chromosomes toward their respective poles. Nondisjunction occurs in the long pair of homologues, with both duplicated chromosomes being pulled to the same pole. See Figure 11-12.

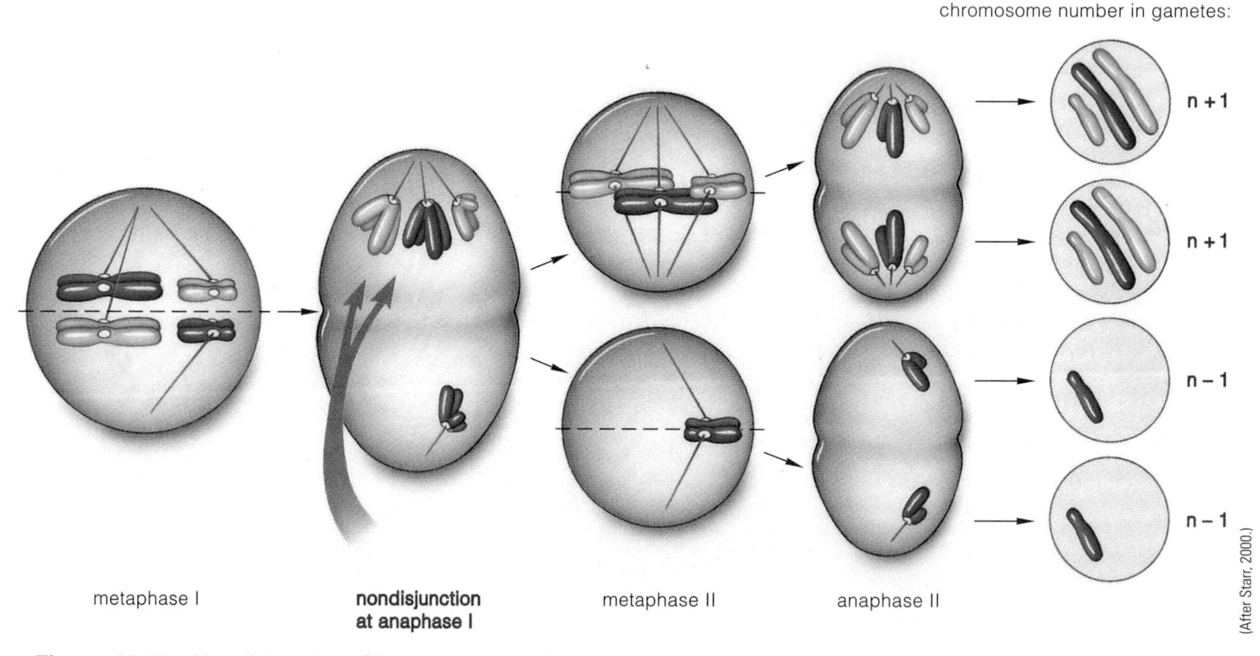

Figure 11-12 Nondisjunction. Of the two pairs of homologous chromosomes shown, one pair fails to separate at anaphase I of meiosis. The chromosome number changes in the gametes.

3. Continue to manipulate the model chromosomes through the remainder of the meiotic process.

   How many chromosomes are found in gamete nuclei? ______________________

   How does this compare to the chromosome number in normal gametes? ______________________

   Recall that each chromosome bears a unique set of genes and speculate about the effect of nondisjunction on the resulting zygotes formed from fertilization with such a gamete.

   ____

   ____

   ____

One of the most common human genetic disorders arises from nondisjunction during gamete (usually ovum) formation. Down syndrome results from nondisjunction in one of the 23 pairs of human chromosomes, chromosome 21. An individual with Down syndrome has three copies of chromosome 21 instead of the normal two copies. While symptoms of this genetic disorder vary greatly, most individuals show moderate to severe mental impairment and a host of associated physical defects. Relatively few other human genetic disorders arise from nondisjunction, probably because the consequences of abnormal chromosome number are often lethal.

Note: **Remove marking ink from pop beads with 95% ethanol and tissues.**

11.2 Meiosis in Animal and Plant Cells *(About 40 min.)*

Now that you have a conceptual understanding of meiosis, let's see the actual divisions as they occur in living organisms.

MATERIALS

Per lab room:

- set of demonstration slides of meiosis in animal testes and lily anther
- set of demonstration slides of meiosis in mammalian ovary

Per student pair:

- scissors
- tape or glue

PROCEDURE

A. Spermatogenesis in Male Animals

In animals, as mentioned previously, meiosis results in the production of gametes—ova in females and sperm in males.

1. Examine Figure 11-13, which depicts sperm formation in a seminiferous tubule within human testes. A diploid reproductive cell, the *spermatogonium,* first enlarges into a *primary spermatocyte.* The primary spermatocyte undergoes meiosis I to form two haploid *secondary spermatocytes.* After meiosis II, four haploid *spermatids* are produced, which develop flagella during differentiation into four *sperm cells.* This process is called *spermatogenesis.*
2. Examine the demonstration slide of spermatogenesis in animal testes. Under low power, note the many circular structures. These are the seminiferous tubules, where spermatogenesis takes place. See Figure 11-14.

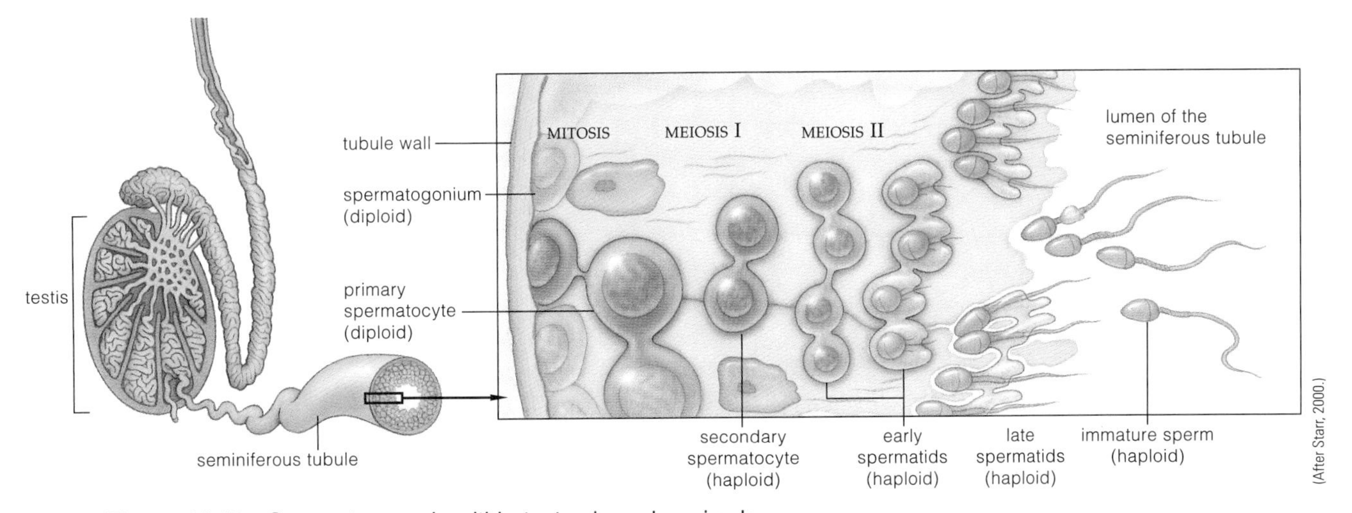

Figure 11-13 Spermatogenesis within testes in male animals.

3. Switch to high power and focus on one seminiferous tubule for closer observation. Mature sperm appear as fine dark lines in the center of the tubule. Cells in progressively earlier stages of meiosis are seen as you move toward the outer wall of the tubule. Adjacent to the tubule wall, you can see diploid primary spermatocytes and spermatogonia. See Figure 11-15.

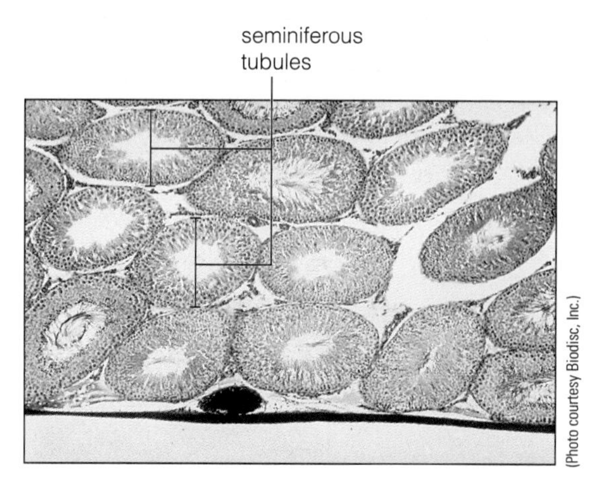

Figure 11-14 Section of mammalian testis (90×).

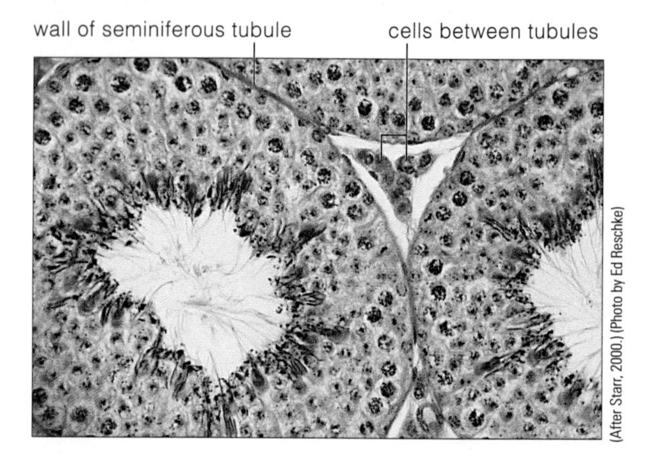

Figure 11-15 Light micrograph of cells inside three adjacent seminiferous tubules.

Can you identify any cells that appear to be undergoing meiotic divisions? __________

B. Oogenesis in Female Animals

In the ovaries of female animals, *ova* (eggs) are produced by meiosis during the process called oogenesis. Unlike spermatogenesis, only one of the meiotic products becomes a gamete.

1. Examine Figure 11-16, which depicts oogenesis in female animals.

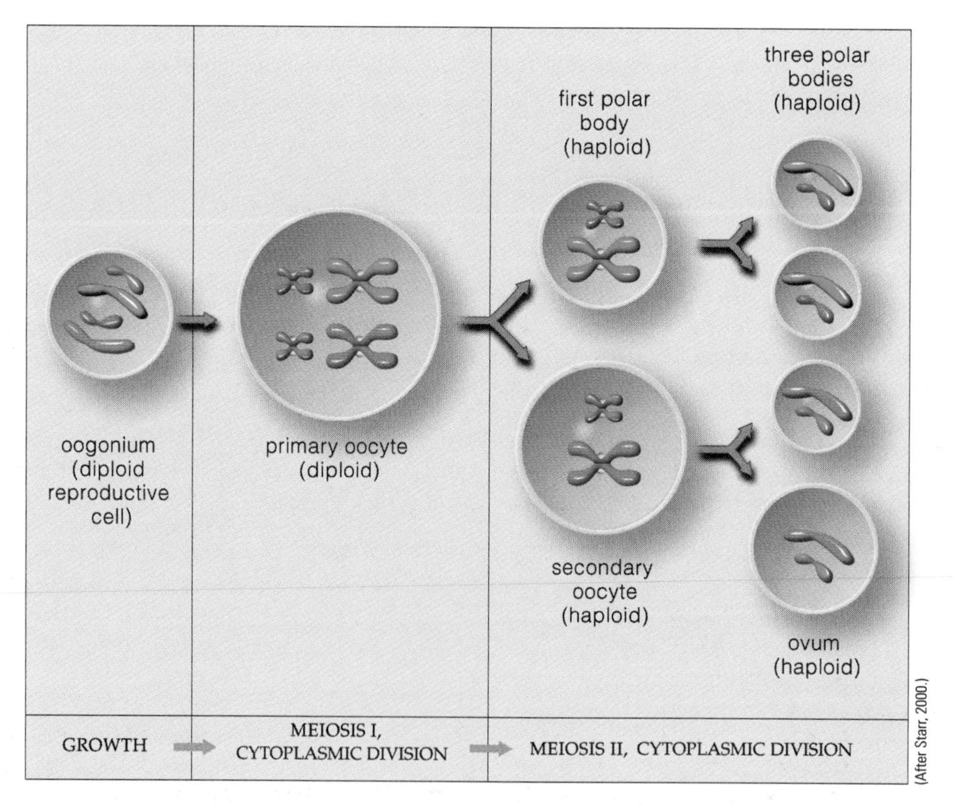

Figure 11-16 Egg formation in female animals. Unlike the sketch, the three polar bodies that actually form in meiosis are extremely small compared to the egg.

The diploid reproductive cell, called an *oogonium,* grows into a *primary oocyte.* The primary oocyte undergoes meiosis I, one product being the *secondary oocyte,* the other a *polar body.* Notice the difference in size of the secondary oocyte and the polar body. This is because the secondary oocyte ends up with nearly all of the cytoplasm after meiosis I.

In humans and other mammals, secondary oocytes are released from the ovary. If fertilization occurs, a sperm penetrates the secondary oocyte, which then continues through meiosis II. Following meiosis II, only the secondary oocyte becomes a mature, haploid *ovum;* depending on the species, the polar body may or may not undergo meiosis II. In any case, the polar bodies are extremely small and do not function as gametes.

2. Examine the demonstration slides of oogenesis in an animal. Identify the follicles within which oogenesis begins. Also identify oocytes that may be in various stages of development. (See Figures 11-17 and 11-18.)

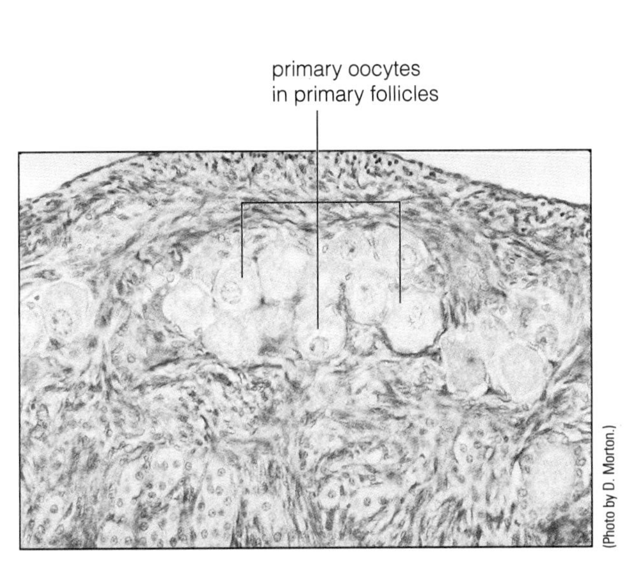

Figure 11-17 Section of mammalian ovary with primary follicles (220×).

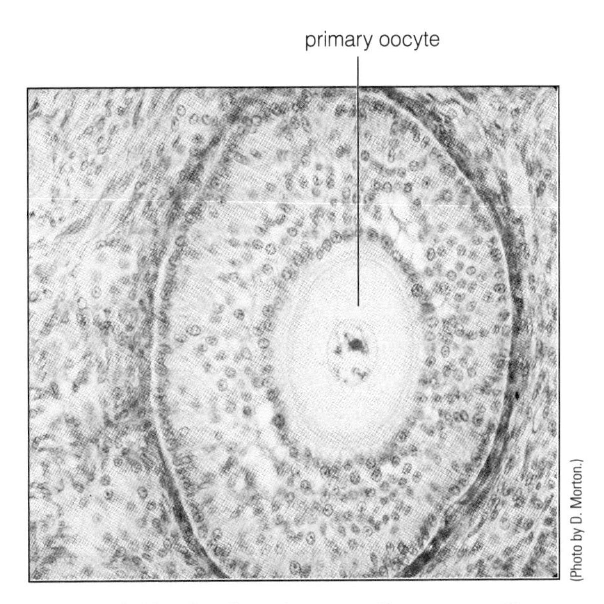

Figure 11-18 Section of mammalian ovary with primary oocyte (376×).

C. Meiosis in Plants

For the sake of brevity, we will only examine meiosis in the male reproductive structure of flowering plants. Recall from our earlier discussion that meiosis in plants results in spore production, not directly into gametes.

1. Examine the demonstration slides of meiosis beginning with the diploid *microsporocytes.* Microsporocytes are the cells within a flower that undergo meiosis to produce haploid *microspores.* Eventually these microspores develop into pollen grains, which in turn produce sperm.
2. As you examine the slides, cut out the photomicrographs on pages 165 and 167 and arrange them on Figure 11-19 to depict the meiotic events leading to microspore formation. Label each photo with the terms provided.
   - **(a)** *Interphase.* During interphase, the *nucleus* of each diploid *microsporocyte* is distinct, containing granular-appearing chromatin. The cells are compactly arranged.
   - **(b)** *Early prophase I.* Now the chromatin has begun to condense into discrete *chromosomes,* which have the appearance of fine threads within the nucleus.
   - **(c)** *Mid-prophase I.* Additional condensation of the *chromosomes* has taken place. Pairing of homologous chromosomes is taking place.
   - **(d)** *Late prophase I.* The chromosomes have condensed into short, rather fat structures. Synapsis and crossing over are taking place. Note that the nuclear envelope has disorganized.
   - **(e)** *Metaphase I.* The homologous chromosomes lie in the region of the *spindle equator.* The *spindle,* composed of *spindle fibers,* can be discerned as fine lines running toward the *poles.* (Note the absence of centrioles in plant cells.)
   - **(f)** *Early anaphase I.* Separation of homologous chromosomes is beginning to take place.
   - **(g)** *Later anaphase I.* Homologous chromosomes have nearly reached the opposite poles. Reduction division has occurred.

Figure 11-19 Meiosis and microsporogenesis in the anther. Cut out the photomicrographs on pages 165 and 167 and arrange them in proper sequence here. (Photos by J. W. Perry.)

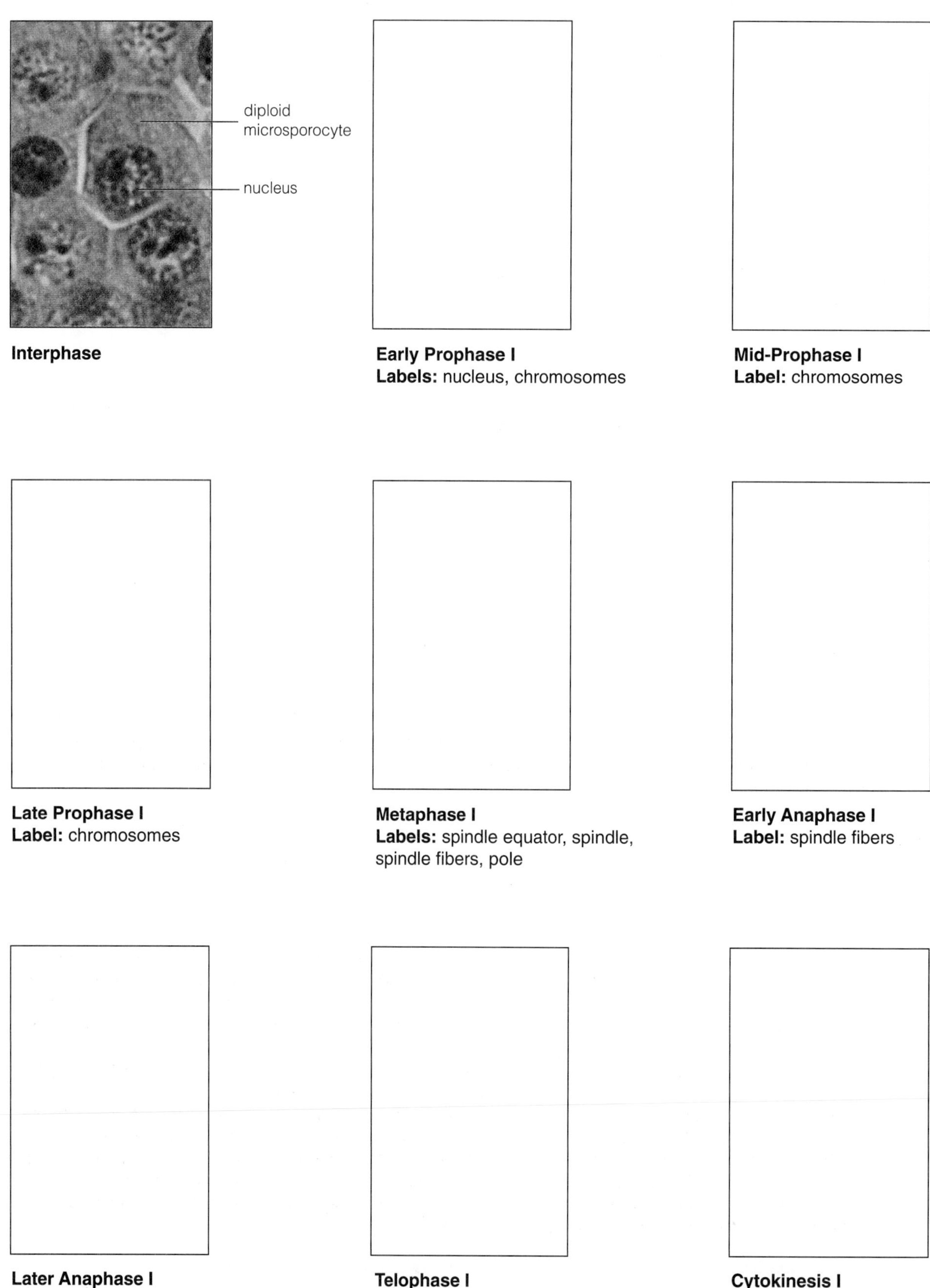

Interkinesis
Labels: nuclei, cell wall, daughter cells

Prophase II
Labels: daughter cells, nuclei

Metaphase II
Labels: chromosomes, spindle equator

Anaphase II
Labels: sister chromatids (unduplicated chromosomes)

Telophase II and Cytokinesis
Labels: cell plate, nuclei

Four haploid cells
Labels nuclei

(h) *Telophase I.* The homologous chromosomes have aggregated at opposite poles. The spindle remains visible.

(i) *Cytokinesis I.* The *cell plate* is forming in the midplane of the cell. Spindle fibers, which are aggregations of microtubules, are visible running perpendicularly through the cell plate. The microtubules are directing the movement of Golgi vesicles, which contain the materials that form the cell plate.

A nuclear envelope has re-formed about the chromosomes, resulting in a well-defined nucleus in each *daughter cell.*

(j) *Interkinesis.* In these plant cells, a short stage exists between meiosis I and II. Distinct nuclei are apparent in the two daughter cells. A cell wall has formed across the entirety of the midplane.

(k) *Prophase II.* The chromosomes in each nucleus of the two daughter cells condense again into distinct, threadlike bodies. As was the case at the end of prophase I, the nuclear envelope disorganizes.

(l) *Metaphase II.* Chromosomes consisting of sister chromatids line up on the spindle equator in both cells. (The photomicrograph shows the very early stages of separation of the chromatids.)

(m) *Anaphase II.* The sister chromatids (now more appropriately considered *unduplicated chromosomes*) are being drawn to their respective poles in each cell.

Before anaphase II begins, sister chromatids are attached to each other along their length. Shortening of the spindle fibers, which are attached to the chromatids at kinetochores within their centromeres, causes the chromatids to separate, beginning in the region of the centromere. This causes a V-shaped configuration of the chromosomes.

(n) *Telophase II and cytokinesis.* Nuclear envelopes are now re-forming around each of the four sets of chromosomes. Cell plate formation is occurring perpendicular to the cell wall that was formed after telophase I.

Four haploid cells. After cell wall formation is complete, the four haploid cells (microspores) will separate. Subsequently, each will develop into a pollen grain inside which sperm cells will form.

PRE-LAB QUESTIONS

_____ **1.** In meiosis, the number of chromosomes _____, while in mitosis, it _____.
(a) is halved/is doubled
(b) is halved/remains the same
(c) is doubled/is halved
(d) remains the same/is halved

_____ **2.** The term "2n" means
(a) the diploid chromosome number is present
(b) the haploid chromosome number is present
(c) chromosomes within a single nucleus exist in homologous pairs
(d) both a and c

_____ **3.** In higher animals, meiosis results in the production of
(a) egg cells (ova)
(b) gametes
(c) sperm cells
(d) all of the above

_____ **4.** Recombination of alleles on nonsister chromatids occurs during
(a) anaphase I
(b) meiosis II
(c) telophase II
(d) crossing over

_____ **5.** Alternative forms of genes are called
(a) homologues
(b) locus
(c) loci
(d) alleles

_____ **6.** If both homologous chromosomes of each pair exist in the same nucleus, that nucleus is
(a) diploid
(b) unable to undergo meiosis
(c) haploid
(d) none of the above

_____ **7.** DNA duplication occurs during
(a) interphase
(b) prophase I
(c) prophase II
(d) interkinesis

_____ **8.** Nondisjunction
(a) results in gametes with abnormal chromosome numbers
(b) occurs at anaphase
(c) results when homologues fail to separate properly in meiosis
(d) is all of the above

_____ **9.** Humans
(a) don't undergo meiosis
(b) have 46 chromosomes
(c) produce gametes by mitosis
(d) have all of the above characteristics

_____ **10.** Gametogenesis in male animals results in
(a) four sperm
(b) one gamete and three polar bodies
(c) four functional ova
(d) a haploid ovum and three diploid polar bodies

Name ______________________________ Section Number ______________________

EXERCISE 11

Meiosis: Basis of Sexual Reproduction

POST-LAB QUESTIONS

Introduction

1. If a cell of an organism has 46 chromosomes before meiosis, how many chromosomes will exist in each nucleus after meiosis?

2. What basic difference exists between the life cycles of higher plants and higher animals?

3. In animals, meiosis results directly in gamete production, while in plants meiospores are produced. Where do the gametes come from in the life cycle of a plant?

4. How would you argue that meiosis is the basis for sexual reproduction in plants, even though the *direct* result is a spore rather than a gamete?

11.1 Demonstrations of Meiosis Using Pop Beads

5. Suppose one sister chromatid of a chromosome has the allele *H*. What allele will the other sister chromatid have? (Assume crossing over has not taken place.) ______________________

6. Suppose that two alleles on one homologous chromosome are *A* and *B*, and the other homologous chromosome's alleles are *a* and *b*.
   a. How many different genetic types of gametes would be produced *without* crossing over? ________
   b. What are the genotypes of the gametes? ______________________
   c. If crossing over were to occur, how many different genetic types of gametes could occur? ________
   d. List them. ______________________

7. Assume that you have built a homologous pair of *duplicated* chromosomes, one chromosome red and the other yellow. Describe or draw the appearance of two nonsister chromatids after crossing over.

8. Examine the meiotic diagram at right. Describe in detail what's wrong with it.

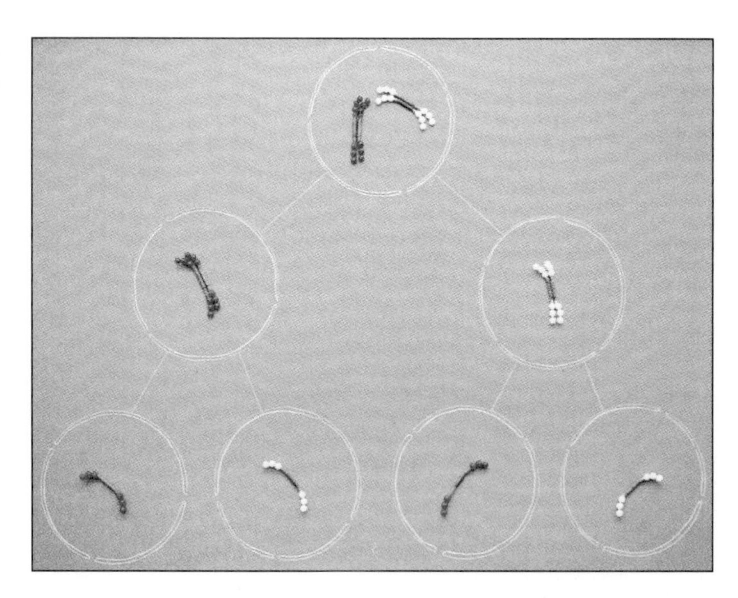

11.2 Meiosis in Animal and Plant Cells

9. Observe the photo at right, which shows a stage of meiosis occurring in a flower anther. Are the cells shown haploid or diploid?

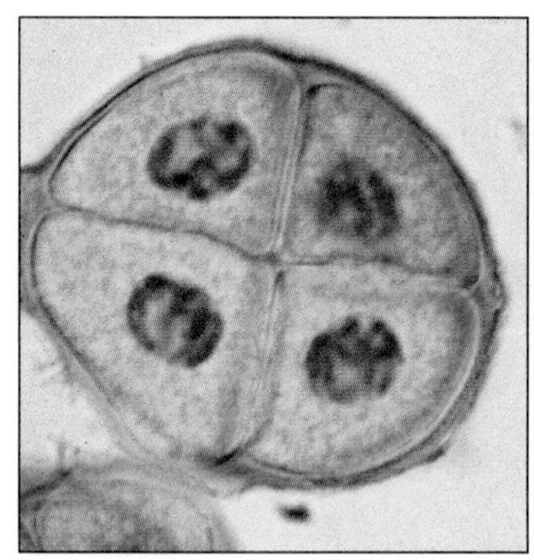

(Photo by J. W. Perry.)

Food for Thought

10. From a genetic viewpoint, of what significance is fertilization?

Photomicrographs for Figure 11-19 Meiosis in anther of a flowering plant. Cut from this page and arrange in proper sequence in Figure 11-19. (1200×) Continues on next page. (Photos by J. W. Perry.)

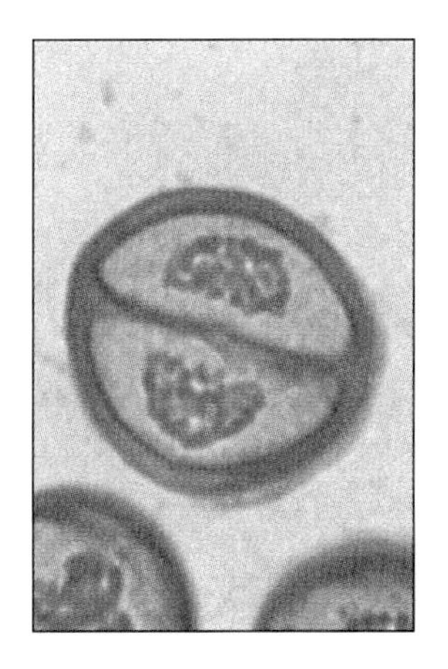
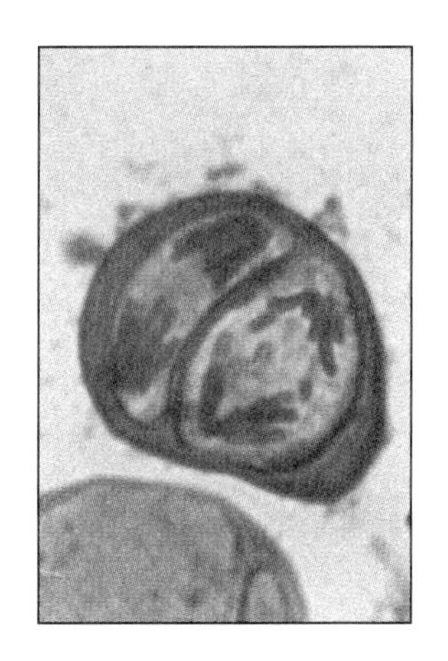
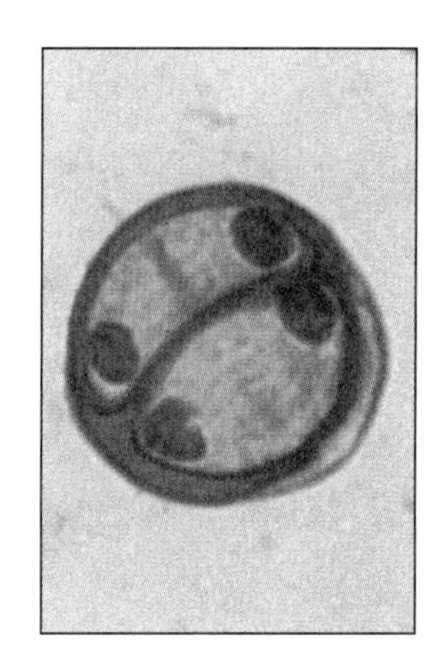
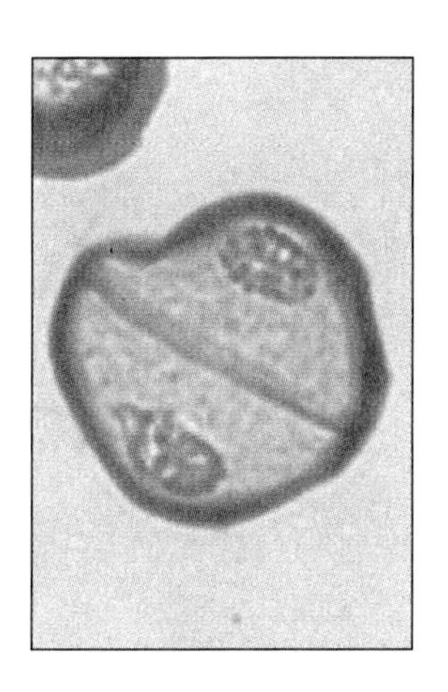
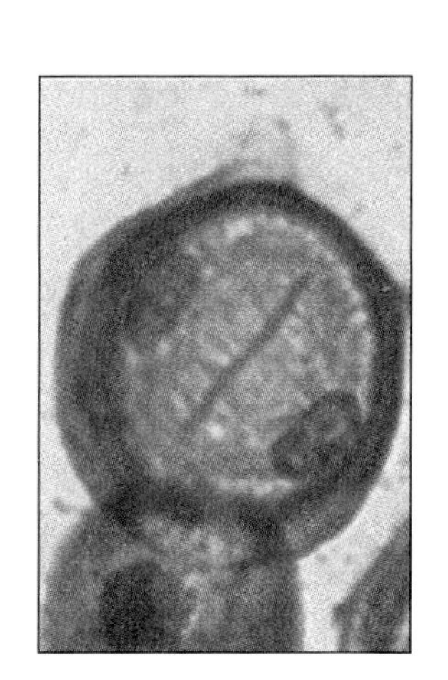
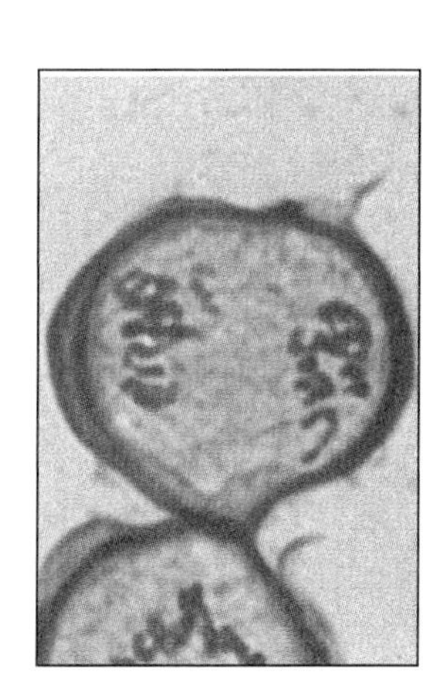
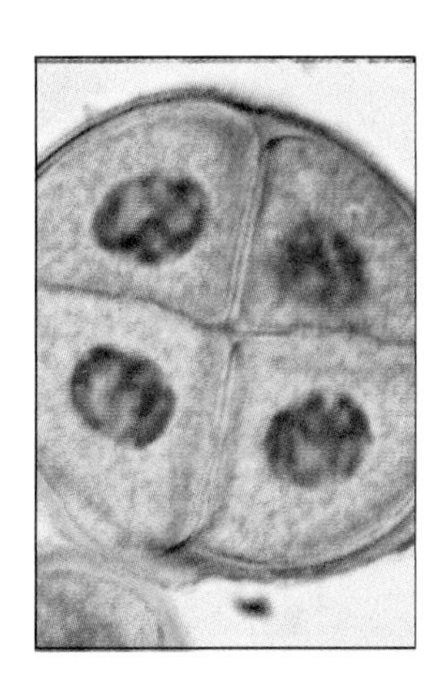
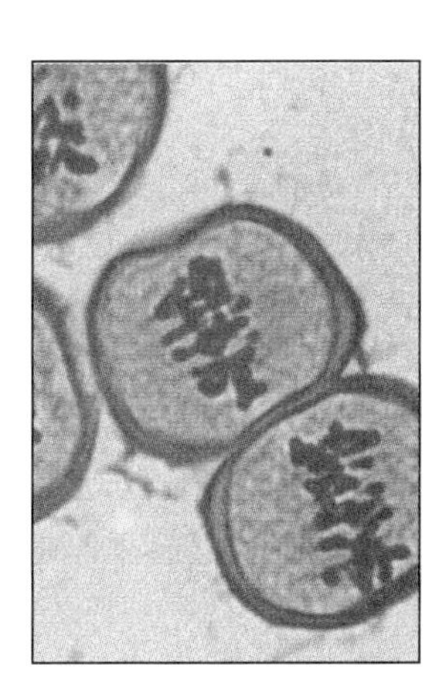
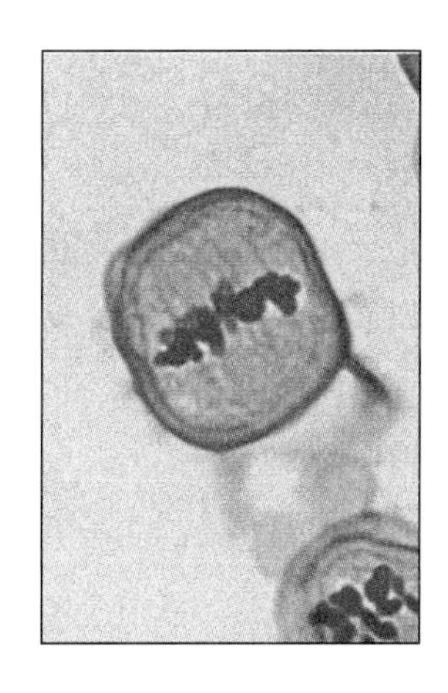

Photomicrographs for Figure 11-19 (continued) Meiosis in anther of a flowering plant. Cut from this page and arrange in proper sequence in Figure 11-19. (1200×) (Photos by J. W. Perry.)

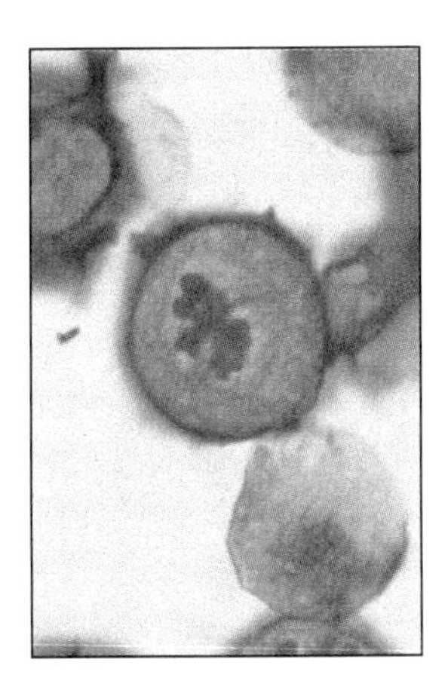

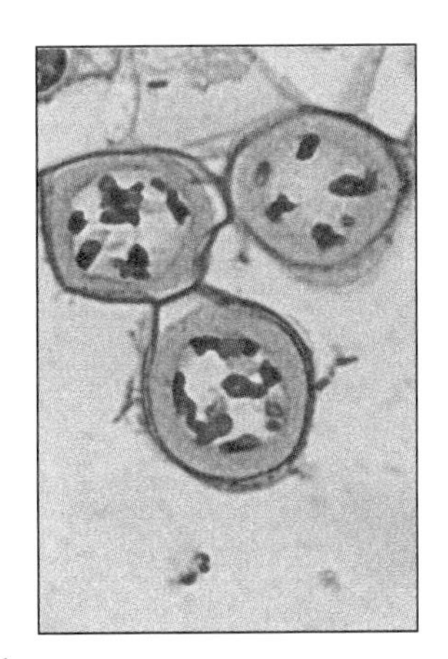

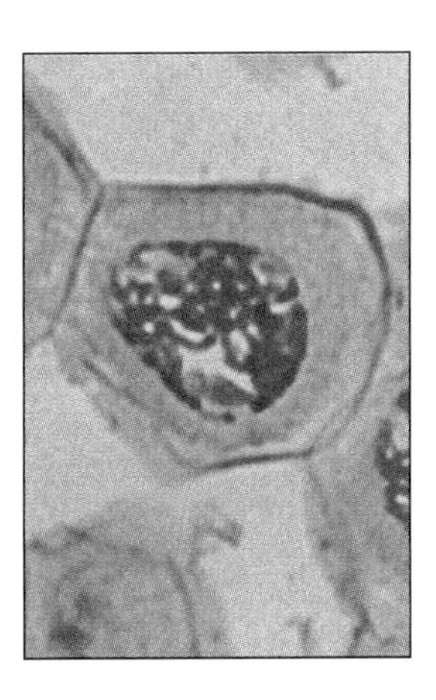

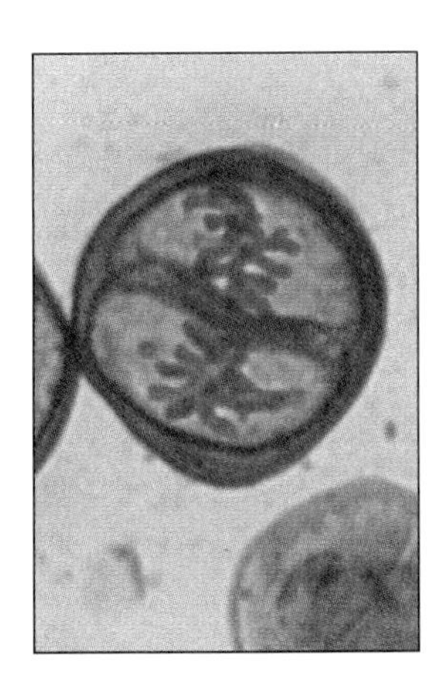

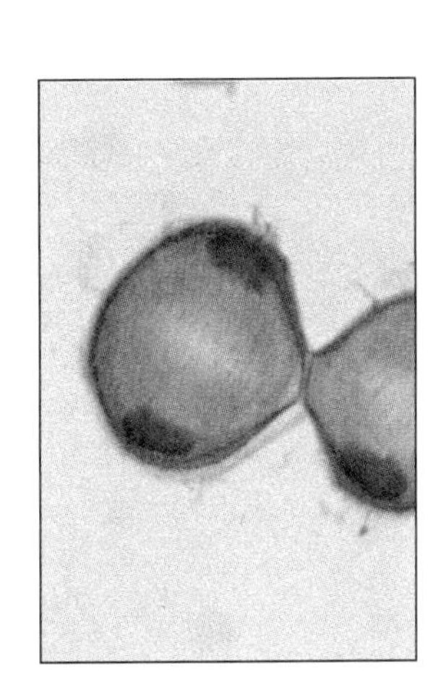

EXERCISE 12

Heredity

OBJECTIVES

After completing this exercise, you will be able to

1. define *true-breeding, hybrid, monohybrid cross, law of segregation, diploid, haploid, genotype, phenotype, dominant, recessive, complete dominance, homozygous, heterozygous, dihybrid cross, probability, chi-square test;*
2. solve monohybrid and dihybrid cross problems;
3. use sampling to determine phenotypic ratios of a visible trait in the gametophytes of an F_1 C-fern hybrid;
4. observe sperm release and fertilization events that lead to an F_2 C-fern sporophyte generation;
5. form hypotheses about genotypic and phenotypic ratios in the F_2 C-fern sporophyte generation;
6. use a chi-square test to determine whether observed results are consistent with expected results;
7. determine your phenotype and give your probable genotype for some common traits.

INTRODUCTION

In 1866, an Austrian monk, Gregor Mendel, presented the results of painstaking experiments on the inheritance of the garden pea, but the scientific community ignored them, possibly because they didn't understand their significance. Now, more than a century later, Mendel's work seems elementary to modern-day geneticists, but its importance cannot be overstated. The principles generated by Mendel's pioneering experimentation are the foundation for the genetic counseling so important today to families with genetically based health disorders. They are also the framework for the modern research that is making inroads into treating diseases previously believed to be incurable. In this era of genetic engineering—the incorporation of foreign DNA into chromosomes of other species—it's easy to lose sight of the concepts underlying the processes that make it all possible. These experiments and genetics problems should give you a good basic understanding of these processes.

12.1 Monohybrid Crosses

Garden peas have both male and female parts in the same flower and are able to self-fertilize. For his experiments, Mendel chose parental plants that were **true-breeding,** meaning that all self-fertilized offspring displayed the same form of a trait as their parent. For example, if a true-breeding purple-flowered plant self-fertilizes, all of its offspring will have purple flowers.

When parents that are true-breeding for *different* forms of a trait are crossed—for example, purple flowers and white flowers—the offspring are called **hybrids.** When only one trait is being studied, the cross is a **monohybrid cross.** We'll look first at monohybrid problems and crosses.

A. Monohybrid Problems with Complete Dominance (About 20 min.)

MATERIALS

Optional:

- pop beads used in Exercise 12
- bottle of 70% ethanol
- simulated chromosomes, consisting of pop beads with magnetic centromeres, and meiotic diagram cards (page 159, Exercise 12)

PROCEDURE

1. Most organisms are diploid; that is, they contain homologous chromosomes with genes for the same traits. The location of a gene on a chromosome is its *locus* (plural: *loci*). Two genes at homologous loci are called a

gene pair. Chromosomes have numerous genes, as shown in Figure 12-1. Genes exist in different forms, called *alleles*. Let's consider one gene pair at the *F* locus. There are three possibilities for the allelic makeup at the *F* locus.

Both alleles are *FF*:

portion of one chromosome — allele *F*
. . . its homologue — allele *F*

Both alleles are *ff*:

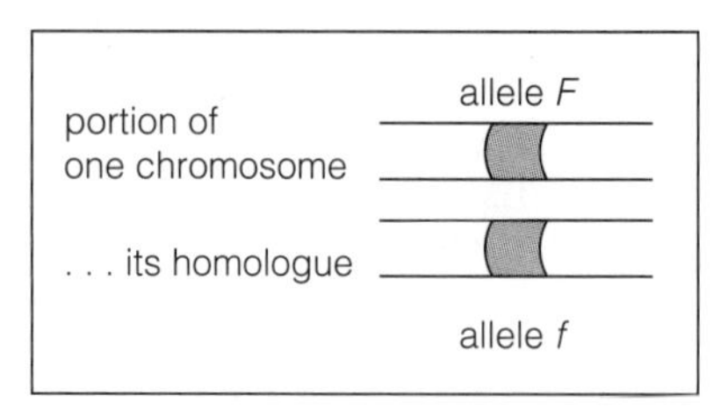

One allele is *F*, and the other is *f*:

portion of one chromosome — allele *F*
. . . its homologue — allele *f*

Gametes, on the other hand, are haploid; they contain only one of the two homologues and thus only one of the two alleles for a specific trait. According to Mendel's first law of inheritance, the **law of segregation,** each organism contains two alleles for each trait, and the alleles segregate (separate) during the formation of gametes during meiosis. Each gamete then contains only one allele of the pair.

The **genotype** of an organism represents its genetic constitution—that is, the alleles present, either for each locus, or taken cumulatively as the genotype of the entire organism.

For each of these diploid genotypes, indicate all possible genotypes of the gametes that can be produced by the organism:

Diploid Genotype	Potential Gamete Genotype(s)
FF	______
ff	______
Ff	______, ______

If you don't understand the process that gives rise to the gamete genotypes, manipulate the pop-bead models that you used in Exercise 12. Using a marking pen, label one bead of each chromosome and go through the meiotic divisions that give rise to the gametes. *It is imperative that you understand meiosis before you attempt to do genetics problems.*

2. During fertilization, two haploid gamete nuclei fuse, and the diploid condition is restored. Give the diploid genotype produced by fusion of the following gamete genotypes.

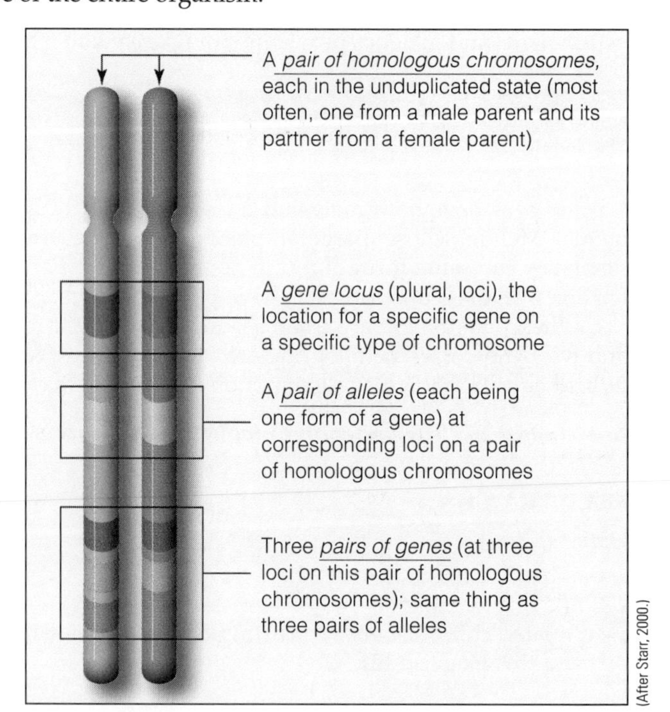

Figure 12-1 A few genetic terms illustrated.

Gamete Genotype	X	Gamete Genotype	⟶	Diploid Genotype
F		*F*		________
F		*f*		________
f		*f*		________

3. Now let's attach some meaning to genotypes. As you see, the genotype is the actual genetic makeup of the organism. The **phenotype** is the outward expression of the genotype—that is, what the organism looks like because of its genotype, as well as its physiological traits and behavior. (Although phenotype is determined primarily by genotype, in many instances environmental factors can modify phenotype.)

   Human earlobes are either attached or free (Figure 12-2). This trait is determined by a single gene consisting of two alleles, *F* and *f*. An individual whose genotype is *FF* or *Ff* has free earlobes. This is the **dominant** condition. Note that the presence of one *or* two *F* alleles results in the dominant phenotype, free earlobes. The allele *F* is said to be dominant over its allelic partner, *f*. The **recessive** phenotype, attached earlobes, occurs only when the genotype is *ff*. In the case of **complete dominance,** the dominant allele completely masks the expression or effect of the recessive allele.

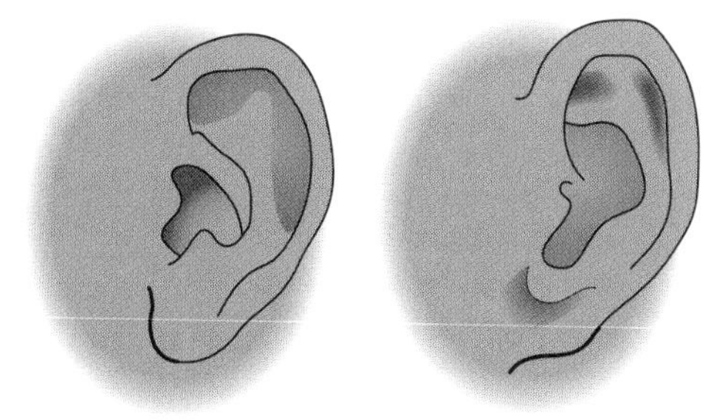

Figure 12-2 Free and attached earlobes in humans.

   When both alleles in a nucleus are identical, the nucleus is **homozygous.** Those with both dominant alleles are homozygous dominant.

   When both recessives are present in the same nucleus, the individual is said to be *homozygous recessive* for the trait.

   When both the dominant and recessive alleles are present in a single nucleus, the individual is **heterozygous** for that trait.

   A man has the genotype *FF*. What is the genotype of his gamete (sperm) nuclei? ________

   A woman has attached earlobes. What is her genotype? ________

   What allele(s) does her gametes (ova) carry? ________

   These two individuals produce a child. Show the genotype of the child by doing the cross:

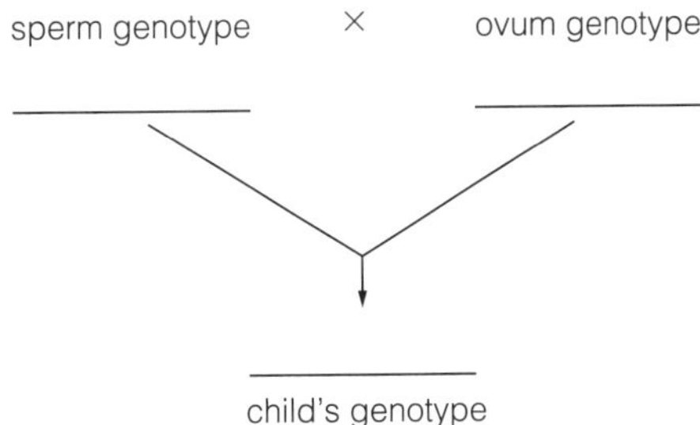

   What is the phenotype of the child? (That is, does this child have attached or free earlobes?) ______

4. In garden peas, purple flowers are dominant over white flowers. Let *A* represent the allele for purple flowers, *a* the allele for white flowers.

   (a) What is the phenotype (color) of the flowers with the following genotypes?

Genotype	Phenotype
AA	
aa	
Aa	

Note: **Always distinguish clearly between upper- and lowercase letters.**

A white-flowered garden pea is crossed with a homozygous dominant purple-flowered plant.

(b) Name the genotype(s) of the gametes of the white-flowered plant. ____________

(c) Name the genotype(s) of the gametes of the purple-flowered plant. ____________

(d) Name the genotype(s) of the plants produced by the cross. ____________

(e) Name the phenotype(s) of the plants produced by the cross. ____________

(f) The Punnett square is a convenient way to perform the mechanics of a cross. The circles along the top and side of the Punnett square represent the possible gamete nuclei. Insert the proper letters indicating the genotypes of the possible gamete nuclei for the white/purple flowered cross in the circles, then fill in the Punnett square for all the possible genetic outcomes.

gametes of white-flowered plant

gametes of purple-flowered plant

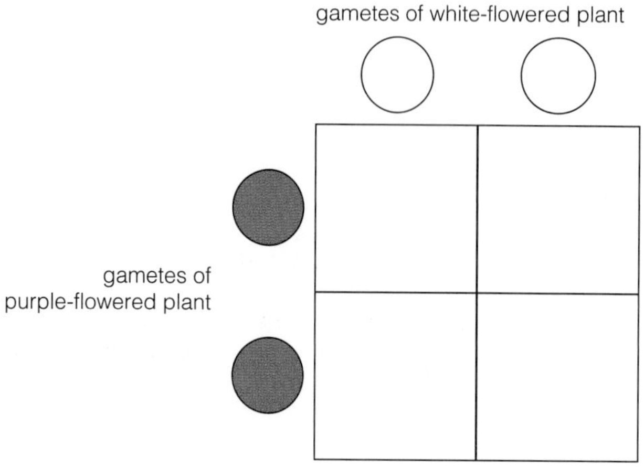

(g) A heterozygous plant is crossed with a white-flowered plant. Fill in the Punnett square, then give the genotypes and phenotypes of all the possible genetic outcomes.

gametes of white-flowered plant

gametes of heterozygote

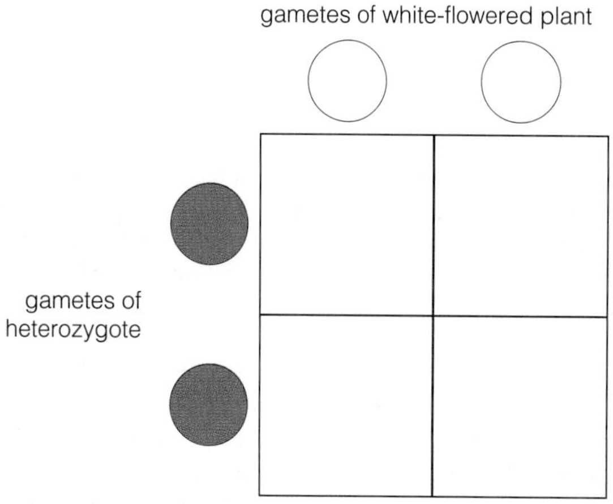

Possible genotypes: ____________

Possible phenotypes: ____________

Draw a line from each possible genotype to its respective phenotype.

(h) It's unlikely that every cross between two pea plants will produce four seeds that will in turn grow into four offspring every time. Rather, one of the most useful facets of problems such as these is that they allow you to *predict* the chances of a particular genetic outcome occurring. Genetics is really a matter of **probability,** the likelihood of the occurrence of a particular outcome.

To take a simple example, consider that the probability of coming up with heads in a single toss of a coin is one chance in two, or ½. Now let's apply this idea to the probability that offspring will have a certain genotype. Look at your Punnett square in part (g). The probability of having a genotype is the sum of all occurrences of that genotype. For example, the genotype *Aa* occurs in two of the four boxes. The probability that the genotype *Aa* will be produced from that particular cross is thus $2/4$, or 50%.

(i) What is the probability of an individual from part (g) having the genotype *aa*? ____________

B. An Observable Monohybrid Cross *(About 15 min.)*

MATERIALS

Per student group (table):

- genetic corn ears illustrating monohybrid cross
- hand lens (optional)

Examine the monohybrid genetic corn demonstration. This illustrates a monohybrid cross between plants producing purple kernels and ones producing yellow kernels. By convention, P stands for the parental generation. The offspring are called the *first filial generation*, abbreviated F_1. If these F_1 offspring are crossed, their offspring are the *second filial generation*, designated F_2:

PROCEDURE

1. Note that all the first-generation kernels (F_1) in the genetic corn demonstration are purple, while the second-generation ear (F_2) has both purple and yellow kernels. Count the purple kernels and then the yellow ones in the F_2 ear.
   ________ purple; ________ yellow. When reduced to the lowest common denominator, is this ratio closest to 1:1, 2:1, 3:1, or 4:1? ________. This is called the *phenotypic ratio.*
2. A corncob with kernels represents the products of multiple instances of sexual reproduction. Each kernel represents a single instance; fertilization of one egg by one sperm produced *each* kernel. Thus each kernel represents a different cross.

Let the letter *P* represent the gene for kernel color.

(a) What genotypes produce a purple phenotype? ____________
(b) Which allele is dominant? ____________
(c) What is the genotype of the yellow kernels on the F_2 ear? ____________
(d) You are given an ear with purple kernels. How do you determine its genotype with a single cross?

____

____

C. Experiment: Monohybrid Heredity in a Fern

A significant limitation of carrying out genetics experiments in the biology lab is that most take several months or years to collect relevant data. Even though Mendel could raise two generations of peas in a growing season, the experiments conducted in his garden plot often lasted several years. Two growing seasons were required to produce the corn monohybrid cross studied above. Fortunately, we can now look at inheritance in organisms with a much shorter life cycle. We'll use C-ferns to investigate a monohybrid cross.

Like all ferns, C-ferns have two independent life cycle phases: a structurally simple, haploid gametophyte and a more complex diploid sporophyte. (See Figures 12-4 and 12-5, page 158, to review the generalized fern life cycle.) A mature C-fern plant produces haploid spores via the process of meiosis. The spores germinate under suitable environmental conditions, and begin to divide mitotically to produce the gametophyte phase (Figure 12-3). This haploid phase develops very rapidly, with gametophytes maturing within 2 weeks.

At maturity, the gametophyte consists of a small (2 mm), simple, photoautrophic flattened structure with sex organs that produce *by mitosis* eggs in structures called archegonia and/or sperm in structures called antheridia. In the presence of water, flagellated sperm are discharged. The sperm are attracted to substances produced by the archegonia and swim toward the egg. Eventually one sperm fertilizes the egg, producing the first cell of the diploid sporophyte generation, the zygote.

The photoautotrophic sporophyte also develops rapidly, with roots and leaves visible within 1–2 weeks. The C-fern sporophytes reaches an ultimate height of 10–40+ cm. Spores are produced by meiosis in structures on the leaves, completing the life cycle.

C.1. Week 1—Observation of F_1 Hybrid Gametophytes *(About 30 min.)*

MATERIALS

Per student:

- 2-week-old C-fern culture in petri dish
- dissecting microscope
- sterile dH_2O
- sterile pipet
- marking pen
- calculator (optional)

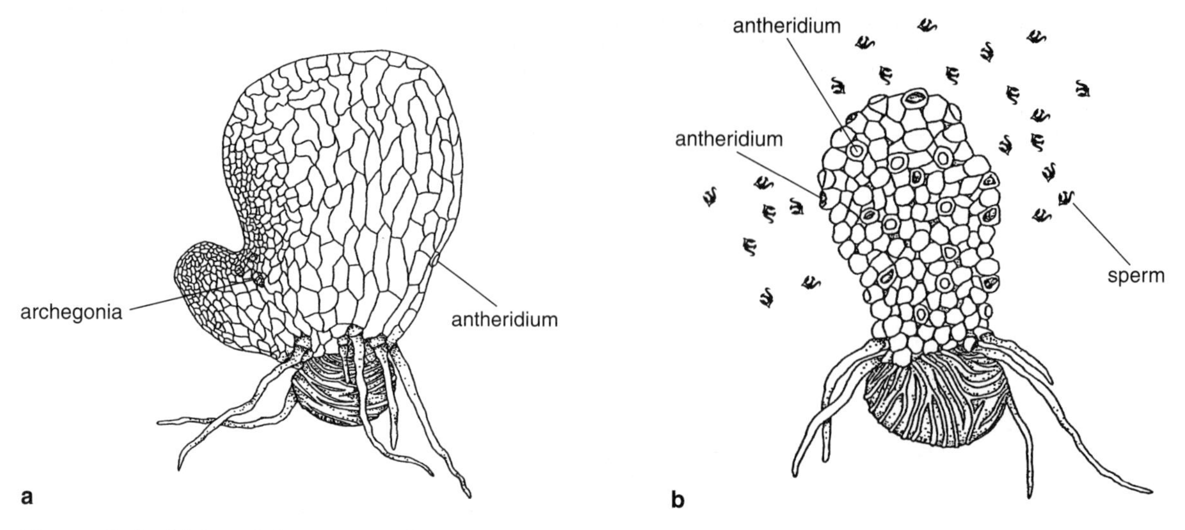

Figure 12-3 Mature C-fern gametophytes and gametes. (**a**) Hermaphroditic gametophytes produce both eggs and sperm and are somewhat heart-shaped. (**b**) Male gametophytes produce only sperm and appear tongue-shaped. (After University of Tennessee Research Foundation, 1998.)

Prior to this class, petri dishes with nutrient medium were inoculated with spores from an F_1 hybrid C-fern plant. The spores germinated and have grown into mature haploid gametophytes.

PROCEDURE

1. Observe the cultures under the dissecting microscope with the highest magnification possible and transmitted light (from below.) You can prevent your cultures from drying out from the heat of the microscope by leaving the lid on the culture as much as possible and by turning the light off or removing the culture to the lab bench when it is not being observed.
2. While you are observing the culture, tilt the lid up and use a sterile pipet to add 1–2 mL sterile distilled water. Lower the lid and tilt the plate back and forth to cover all of the gametophytes with water. Observe the release of swimming sperm from antheridia and their attempts to find and fertilize mature eggs within archegonia.

   Do all the gametophytes have the same phenotype? Describe any differences you observe.

   ____

   ____

   ____

   For this experiment, we will focus only on the larger, heart-shaped hermaphroditic gametophytes. Which of the phenotypes would you designate as a mutant? Why?

   ____

   ____

3. Take a random sample of the hermaphroditic gametophyte population by counting up to 50 individuals and tallying their phenotype.
   (a) Why is it important to take a random sample from the cultures?

   ____

   (b) What is a suitable method of collecting data that would ensure a random sample?

   ____

   ____

4. Record your data in Table 12-1 and in the location designated by your instructor for the class data. Leave the lid on during this procedure if possible. If it becomes fogged, quickly exchange the fogged lid for a clean lid from an unused dish. After scoring, replace the old lid over the culture.
5. When you've finished your observations, remove any excess water from the culture by lifting the lid slightly and pouring off the excess. Place the culture in the location designated by your instructor. Be sure the petri dish lid is in place.
   (a) Are the plants in the culture dish haploid or diploid? ______________________________
   (b) What products will result from the fertilization events in the culture? ______________________________
   Will they be haploid or diploid? ______________________________

TABLE 12-1 Gametophyte Phenotypes

Description of Phenotypes	Number of Gametophytes	Class Total

(c) If the F_1 sporophyte is heterozygous for a single mutant trait, what genotypes will be present in the spores? ____________

What genotypes will be present in the gametophytes? ____________

(d) What is the expected ratio of genotypes? ____________

(e) What is the approximate phenotypic ratio of the gametophytes? ____________

(f) Can you determine the dominance relationships from the data in Table 12-1?

____

____

Biologists have assigned the designation *CP* to the single gene responsible for the two different phenotypes seen in this experiment. The dominant allele is thus designated *CP*, and the recessive allele *cp*.

(g) Predict the genetic outcome of the fertilizations taking place in the cultures by formulating a hypothesis to explain the inheritance of the trait. Indicate expected ratios of both the gametophyte and the F_2 generations.

____

____

C.2. Data Analysis *(About 15 min.)*

Gametes from true-breeding gametophyte parents (the P generation) combine to produce hybrid F_1 sporophyte fern plants. Meiosis within the F_1 plants produces spores. Each gametophyte you observe resulted from mitotic divisions of one of those spores. Their gametes will combine to produce the F_2 generation sporophytes.

In the space below, diagram the crosses involved in the F_1 and F_2 generations, indicating which generations/structures are haploid, and which are diploid.

You can now test your hypothesis concerning the method of trait inheritance to determine whether the data you collected support or do not support your model. Geneticists typically use the chi-square (χ^2) statistical test to determine whether experimentally obtained data are a satisfactory approximation of the expected data. In short, this test expresses the difference between expected (hypothetical) and observed (collected) numbers as a single value, χ^2. If the difference between observed and expected results is large, a large χ^2 results, while a small difference results in a small χ^2. Chi-square values are calculated according to the formula

$$\chi^2 = \sum \frac{(O - E)^2}{E}$$

where O = *observed* number of individuals

E = *expected* number of individuals

Σ = the sum of all values of $(O - E)^2/E$ for the various categories of phenotypes

Let's use this formula in our garden-pea monohybrid cross, question 4 from page 181. Suppose 81 flowers are counted in a cross. Our hypothesis (expectation) is that three-fourths of them will be purple:

$$\frac{3}{4} \times 81 = 60.75$$

Similarly, we expect one-fourth to be white:

$$\frac{1}{4} \times 81 = 20.25$$

Suppose we actually count 64 purple flowers and 17 white flowers. Examine Table 12-2, noting how these values are used.

$$\chi^2 = \sum(0.174 + 0.522) = 0.696$$

TABLE 12-2 Calculations of Chi Square for Garden-Pea Monohybrid Cross

Phenotype	Genotype	*O*	*E*	$(O-E)$	$(O-E)^2$	$(O-E)^2/E$
Purple	*P_*	64	60.75	3.25	10.56	0.174
White	*pp*	17	20.25	–3.25	10.56	0.522
Total		81	81	0		0.696

Now, how do we interpret the χ^2 value we found? Suppose the expected and observed values were identical. Then $\chi^2 = 0$. You might guess that a number very close to zero indicates close agreement between observed and expected and a large χ^2 value suggests that "something unusual" is taking place. The problem is that chance alone almost always causes small deviations between observed and expected results, *even when the hypothesis being tested is correct.*

When does the χ^2 value indicate that chance alone cannot explain the deviation? Geneticists generally agree on a probability value of 1 in 20 (or 5% = .05) as the lowest acceptable value derived from the χ^2 test. This number indicates that if the experiment is repeated many times, the deviations expected due to chance alone will be as large as or larger than those observed only about 5% or less of the time. Probabilities equal to or greater than .05 are considered to support the hypothesis, while probabilities lower than .05 do not support the hypothesis. Here we must consult a table of χ^2 values to make our decision (Table 12-3).

In our example, the χ^2 value is 0.696. Since this is a monohybrid problem with only two categories of possible outcomes (purple or white flowers), the number of degrees of freedom (*n* in the left-hand column of Table 12-3) is 1. Read across the table until you come to .05 and find the χ^2 value 3.84. Because 0.696, our calculated χ^2 value, is less than 3.841, it is likely that the variation in the observed and expected is the result of chance, and that our hypothesized outcome is correct. A value *greater than* 3.841, however, would indicate that chance alone cannot explain the deviation between observed and expected, and we would reject our hypothesis.

The term *degrees of freedom* requires further explanation. The number of degrees of freedom is always 1 *less* than the number of categories of possible outcomes. Thus, if you are dealing with a dihybrid problem with a ratio of 9:3:3:1 (four possible phenotypes), $n = 3$.

TABLE 12-3 Distribution of χ^2

Degrees of Freedom, *n*	Probability of Obtaining a χ^2 Value as Large or Larger			
	.10	.05	.01	.001
1	2.71	3.84	6.63	10.83
2	4.61	5.99	9.21	13.82
3	6.25	7.82	11.35	16.27
4	7.78	9.49	13.28	18.47

1. Transfer your individual or group data from the Totals columns in Table 12-1 to Table 12-4 and calculate χ^2.

TABLE 12-4 χ^2 Calculation from Gametophyte Data

Phenotype	Observed (*O*)	Expected (*E*)[a]	(*O* – *E*)	$(O - E)^2$	$(O - E)^2/E$
Totals					χ^2=

[a] This number should be based on the hypothesis you developed in week 1 observations (page 175).

2. Use Table 12-3 to determine the probability of obtaining this χ^2 value for the gametophyte data in Table 12-4. How many degrees of freedom are there? ________________

Is your hypothesis supported or not supported? ________________ If not, what might be changed in your hypothesis or in the experimental design?

___

___

C.3. Week 3—Observation of F_2 Sporophytes (About 45 min.)

MATERIALS

Per student:

- 4-week-old C-fern culture in petri dish
- dissecting microscope
- dissecting needle or toothpicks
- calculator (optional)

PROCEDURE

1. Examine your cultures with the dissecting microscope. Mutant and wild-type phenotypes are best observed using reflected light from the top or the side. Carefully observe the oldest leaves. Can you see mutant and wild-type phenotypes? ____________

   Are the young sporophytes haploid or diploid? Why?

   ___

   ___

2. Sketch what you are observing, and label it with the following terms: gametophyte, sporophyte leaf, sporophyte root.
3. Take a random sample of the sporophyte population in a dish by counting up to 50 individuals and identifying their phenotype. You can remove the lid from the culture to do this. It may be easier to score the phenotype after gently and randomly pulling up individual sporophytes with a dissecting needle or toothpick and laying them out in a row on empty areas of the culture plate. Observe the largest leaf on each sporophyte and examine the differences carefully before recording data in Table 12-5 and in the location designated for class data.
4. Following scoring of phenotypes, place the lid back on the plate and return the culture to the designated location, or take it home so that you can observe it over the next several weeks to determine whether the phenotype of older sporophytes is apparent without use of a microscope.

TABLE 12-5 Sporophyte Phenotypes

Description of Phenotypes	Number of Sporophytes	Class Total

5. Restate your hypothesis regarding the inheritance of the mutant and wild-type alleles, and your prediction of the genetic outcome in the F_2 sporophytes.

____

____

____

6. Transfer your individual or class total data to Table 12-6 to calculate χ^2.

TABLE 12-6 χ^2 Calculation from Sporophyte Data

Phenotype	Observed (O)	Expected (E)[a]	($O - E$)	($O - E$)^2	($O - E$)$^{2/E}$
Totals					χ^2=

[a]This number should be based on the hypothesis you developed in this week's observations.

7. Use Table 12-3 to determine the probability of obtaining this χ^2 value for the sporophyte data in Table 12-6.
   How many degrees of freedom are there? _______________
   What is the approximate probability? _______________
   Is your hypothesis supported or not supported? _______________
   Which allele is the dominant allele? _______________ Which is recessive? _______________
   If gametophytes had not expressed the phenotype, would you be able to form a hypothesis from observations of the gametophyte generation? _______________ Why or why not? _______________

____

____

12.2 Dihybrid Inheritance

All the problems and experiments so far have involved the inheritance of only one trait; that is, they are monohybrid problems. Now we'll examine cases in which two traits are involved: **dihybrid problems.**

Note: **We will assume that the genes for these traits are carried on different (nonhomologous) chromosomes.**

A. Dihybrid Problems (*About 15 min.*)

1. Let's consider these two traits:
   - In humans, a pigment in the front part of the eye masks a blue layer at the back of the iris. The dominant allele P causes production of this pigment. Those who are homozygous recessive (pp) lack the pigment, and the back of the iris shows through, resulting in blue eyes. (Other genes determine the color of the pigment, but in this problem we'll consider only the presence or absence of *any* pigment at the front of the eye.)
   - Dimpled chins (D = allele for dimpling) are dominant over undimpled chins (d = allele for lack of dimple).

   (a) List all possible genotypes for an individual with pigmented iris and dimpled chin. _______________
   (b) List the possible genotypes for an individual with pigmented iris but lacking a dimpled chin. _______________
   (c) List the possible genotypes of a blue-eyed, dimple-chinned individual. _______________
   (d) List the possible genotypes of a blue-eyed individual lacking a dimpled chin. _______________

2. An individual is heterozygous for both traits (eye pigmentation and chin form).
   (a) What is the genotype of such an individual? _______________
   (b) What are the possible genotypes of that individual's gametes? _______________
   If determining the answer for question 2 was difficult, recall from Exercise 12 that the principle of independent assortment states that genes on different (nonhomologous) chromosomes are separated

out independently of one another during meiosis. That is, the occurrence of an allele for eye pigmentation in a gamete has no bearing on which allele for chin form will occur in that same gamete.

There is a useful method for determining possible gamete genotypes produced during meiosis from a given parental genotype. Using the genotype *PpDd* as an example, here's the method:

Follow the four arrows to determine the four gamete genotypes.

(c) Two individuals heterozygous for both eye pigmentation and chin form have children. What are the possible genotypes of those F_1 offspring?

____

You can set up a Punnett square to do dihybrid problems just as you did with monohybrid problems. However, depending on the parental genotypes, the square may have as many as 16 boxes, rather than just 4. Insert the possible genotypes of the gametes from one parent in the top circles and the gamete genotypes of the other parent in the circles to the left of the box.

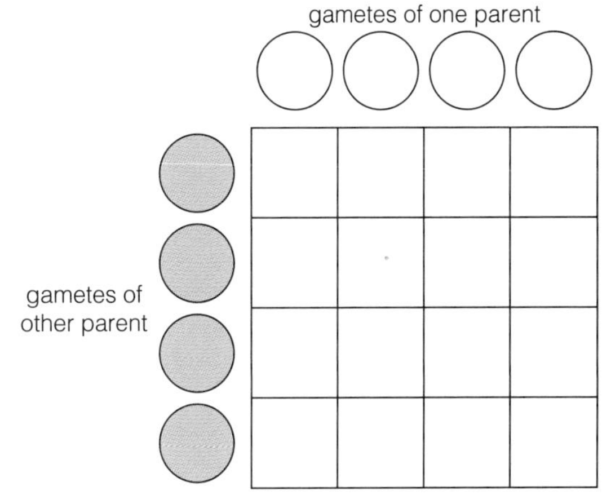

(d) Possible genotypes of children produced by two parents heterozygous for both eye pigmentation and chin form:

____

What is the ratio of the genotypes?__________

What is the phenotypic ratio? __________

(e) Recalling the discussion of probability in Section 12.1, state the probability of a child from part (d) having the following genotypes.

ppDD ______________________

PpDd ______________________

PPDd ______________________

To extend the probability discussion, let's reconsider flipping a coin by asking the question, What is the probability of flipping heads twice in a row? The chance of flipping heads the first time is $\frac{1}{2}$. The same is true for the second flip. The chance (probability) that we'll flip heads twice in a row is $\frac{1}{2} \times \frac{1}{2} = \frac{1}{4}$. The probability that we could flip heads 3 times in a row is $\frac{1}{2} \times \frac{1}{2} \times \frac{1}{2} = \frac{1}{8}$.

(f) State the probability that three children born to the parents in part (d) will have the genotype *ppdd*.

____

What is the probability that three children born to these parents will have dimpled chins and pigmented eyes? ______________________

(g) What is the genotype of the F_1 generation when the father is homozygous for both pigmented eyes and dimpled chin, but the mother has blue eyes and no dimple? ______________________

What is the phenotype of this individual? ______________________

B. An Observable Dihybrid Cross *(About 20 min.)*

MATERIALS

Per student group (table):

- genetic corn ears illustrating a dihybrid cross

PROCEDURE

1. Examine the demonstration of dihybrid inheritance in corn. Notice that not only are the kernels two different colors (one trait), but they are also differently shaped (second trait). Kernels with starchy endosperm (the carbohydrate-storing tissue) are smooth, while those with sweet endosperm are shriveled. Notice that all *four* possible phenotypic combinations of color and shape are present in the F_2 generation.

The P gene is involved in pigment production, with two alleles P and p. The S gene determines carbohydrate (sugar) storage, with two alleles S and s.

Which genotypes of the parents produced the F_2 generation kernels? ____________

2. Set up a Punnett square of this dihybrid cross:

   What is the predicted phenotypic ratio? ________

3. Count the number of kernels of each possible phenotype and record in Table 12-7. To increase your sample size, count three ears.

   Which traits seem dominant?

   ____

   Which traits seem recessive?

   ____

4. Calculate the actual phenotypic ratio you observed:

   ____

   Do your observed results differ from the expected results? ____________

gametes of one parent

gametes of other parent

	○	○	○	○
●				
●				
●				
●				

TABLE 12-7 Phenotypes in Dihybrid Corn Cross

	Number of Kernels with Phenotypes			
Ear	**Yellow Smooth**	**Yellow Shriveled**	**Purple Smooth**	**Purple Shriveled**
1				
2				
3				
Totals				

5. Use the chi-square test to determine if the deviation from the expected results can be accounted for by chance alone.

   Chi-square test results: ____________

   ____

12.3 Some Readily Observable Human Traits *(About 15 min.)*

In the preceding pages, we examined several human traits that are fairly simple and that follow the Mendelian pattern of inheritance. Most of our traits are much more complex, involving many genes or interactions between genes. For example, hair color is determined by at least four genes, each one coding for the production of melanin, a brown pigment. Because the effect of these genes is cumulative, hair color can range from blond (little melanin) to very dark brown (much melanin).

Clearly, human traits are of great interest to us. Table 12-8 lists a number of traits that exhibit Mendelian inheritance. For each trait, work with a lab partner to determine your phenotype, then fill in Table 12-8. List your possible genotype(s) for each trait. When convenient, examine your parents' phenotypes and attempt to determine your actual genotype.

1. *Mid-digital hair* (Figure 12-4a). Examine the joint of your fingers for the presence of hair, the dominant condition (*MM*, *Mm*). Complete absence of hair is due to the homozygous-recessive condition (*mm*). You may need a hand lens to determine your phenotype. Even the slightest amount of hair indicates the dominant phenotype.
2. *Tongue rolling* (Figure 12-4b). The ability to roll one's tongue is due to a dominant allele, *T*. The homozygous-recessive condition, *tt*, results in inability to roll the tongue (Figure 12-5).
3. *Widow's peak* (Figure 12-4c). Widow's peak describes a distinct downward point in the frontal hairline and is due to the dominant allele, *W*. The recessive allele, *w*, results in a continuous hairline. (Omit study of this trait if baldness is affecting the hairline.)
4. *Earlobe attachment* (Figure 12-4d). Most individuals have free (unattached) earlobes (*FF*, *Ff*). Homozygous recessives (*ff*) have earlobes attached directly to the head.

TABLE 12-8 Summary of My Mendelian Traits							
			Mom's		Dad's		
Trait	My Phenotype	My Possible Genotypes	Phenotype	Possible Genotype	Phenotype	Possible Genotype	My Possible/ Probable Genotype
Mid-digital hair							
Tongue rolling							
Widow's peak							
Earlobe attachment							
Hitchhiker's thumb							
Relative finger length							

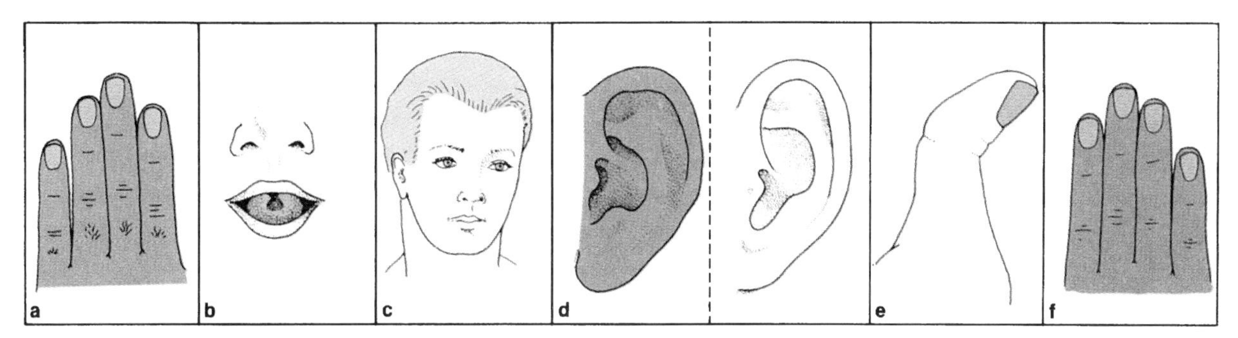

Figure 12-4 Some readily observable human Mendelian traits.

5. *Hitchhiker's thumb* (Figure 12-4e). Although considerable variation exists in this trait, we'll consider those individuals who *cannot* extend their thumbs backward to approximately 45° to be carrying the dominant allele, *H*. Homozygous-recessive persons (*hh*) can bend their thumbs at least 45°, if not farther.
6. *Relative finger length* (Figure 12-4f). An interesting sex-influenced (*not* sex-linked) trait relates to the relative lengths of the index and ring finger. In males, the allele for a short index finger (*S*) is dominant. In females, it's recessive. In rare cases, each hand is different. If one or both index fingers are greater than or equal to the length of the ring finger, the recessive genotype is present in males, and the dominant one present in females.

(Photo courtesy Evan Cerasoli.)

Figure 12-5 A student at San Diego State University exhibiting the tongue-rolling trait for the benefit of a tongue-roll-challenged student.

Genetics Problems. One of the best ways to solidify your understanding of different patterns of inheritance is to work genetics problems. Your instructor may assign you portions of Appendix 2, which is a collection of these useful and interesting problems.

PRE-LAB QUESTIONS

_____ **1.** In a monohybrid cross
(a) only one trait is being considered
(b) the parents are always dominant
(c) the parents are always heterozygous
(d) no hybrid is produced

_____ **2.** The genetic makeup of an organism is its
(a) phenotype
(b) genotype
(c) locus
(d) gamete

_____ **3.** An allele whose expression is completely masked by the expression or effect of its allelic partner is
(a) homologous
(b) homozygous
(c) dominant
(d) recessive

_____ **4.** The physical appearance and physiology of an organism, resulting from interactions of its genetic makeup and its environment, is its
(a) phenotype
(b) hybrid vigor
(c) dominance
(d) genotype

_____ **5.** When both dominant and recessive alleles are present within a single nucleus, the organism is ____ for the trait.
(a) diploid
(b) haploid
(c) homozygous
(d) heterozygous

_____ **6.** A Punnett square is used to determine
(a) probable gamete genotypes
(b) possible parental phenotypes
(c) possible parental genotypes
(d) possible genetic outcomes of a cross

_____ **7.** The gametophyte of a fern is
(a) haploid
(b) photoautotrophic
(c) a structure that produces eggs and/or sperm
(d) all of the above

_____ **8.** A chi-square test is used to
(a) determine if experimental data adequately matches what was expected
(b) analyze a Punnett square
(c) determine parental genotypes producing a given offspring genotype
(d) determine if a trait is dominant or recessive

_____ **9.** Possible gamete genotypes produced by an individual of genotype *PpDd* are
(a) *Pp* and *Dd*
(b) all *PpDd*
(c) *PD* and *pd*
(d) *PD*, *Pd*, *pD*, and *pd*

_____ **10.** If you can roll your tongue.
(a) you have at least one copy of the dominant allele *T*
(b) you have two copies of the recessive allele *t*
(c) you must be male
(d) you are haploid

Name ______________________________ Section Number ______________________

EXERCISE 12

Heredity

POST-LAB QUESTIONS

12.1 Monohybrid Crosses

1. Explain the implications of Mendel's law of segregation as it applies to the distribution of alleles in gametes.

2. Assume that production of hairs on a plant's leaves is controlled by a single gene with two alleles *H* (dominant) and *h* (recessive). Hairy leaves are dominant to smooth (nonhairy) leaves.
   a. Name the genotype(s) of a smooth-leaved plant. __________
   b. Name the genotype(s) of a hairy-leaved plant. __________
   c. What are the possible genotypes of gametes produced by the smooth-leaved plant? __________
   d. What are the possible genotypes of gametes produced by the hairy-leaved plant? __________

3. *Non*-true-breeding hairy-leaved plants are crossed with smooth-leaved plants.
   a. What genotypic and phenotypic ratios would you expect for the potential offspring? __________
   b. Suppose you perform such a cross, collect data, and do a chi-square test to aid in data analysis. How many degrees of freedom would there be? __________
   c. Suppose your chi-square value is very large (>25). What does this indicate about your experiment and/or hypothesis?

4. What genotypic ratio would you expect in the gametophyte generation of C-ferns produced by F_1 spores if two traits on separate chromosomes were being followed?

5. Were dominant and recessive traits observed equally in both gametophytes and sporophytes of C-ferns? How did you determine which character was dominant and which was recessive?

12.2 *Dihybrid Inheritance*

6. Suppose you have two traits controlled by genes on separate chromosomes. If sexual reproduction occurs between two heterozygous parents, what is the genotypic ratio of all possible gametes?

Food for Thought

7. Explain the usefulness of the chi-square test.

8. Suppose students in previous semesters had removed some of the corn kernels from the genetic corn ears before you counted them. What effect would this have on your results?

9. Assume that one allele is completely dominant over the other for the following questions.
   a. Two individuals heterozygous for a *single* trait have children. What is the expected phenotypic ratio of the possible offspring? _____________
   b. Two individuals heterozygous for *two* traits have children. What would be the expected phenotypic ratio of the possible offspring? _____________
   c. Crossing two individuals heterozygous for two traits results in the same phenotypic ratio as for a single trait. Are the genes for these two traits on separate chromosomes or on the same chromosome? Explain your answer. (Remember that the gene for each trait is located at a locus, a physical region on the chromosome.)

10. How does probability differ from actuality?

EXERCISE 13

Nucleic Acids: Blueprints for Life

OBJECTIVES

After completing this exercise, you will be able to

1. define *DNA, RNA, purine, pyrimidine, principle of base pairing, replication, transcription, translation, codon, anticodon, peptide bond, gene, genetic engineering, recombinant DNA, plasmid, bacterial conjugation;*
2. identify the components of deoxyribonucleotides and ribonucleotides;
3. distinguish between DNA and RNA according to their structure and function;
4. describe DNA replication, transcription, and translation;
5. give the base sequence of DNA or RNA when presented with the complementary strand;
6. identify a codon and anticodon on RNA models and describe the location and function of each;
7. give the base sequence of an anticodon when presented with that of a codon, and vice versa;
8. describe what is meant by the *one-gene, one-polypeptide hypothesis;*
9. describe the process of DNA recombination by bacterial conjugation;
10. explain the difference between DNA recombination by bacterial conjugation and the technique by which eukaryotic gene products are produced by bacteria.

INTRODUCTION

By 1900, Gregor Mendel had demonstrated patterns of inheritance, based solely on careful experimentation and observation. Mendel had no clear idea how the traits he observed were passed from generation to generation, although the seeds of that knowledge had been sown as early as 1869, when the physician-chemist Friedrich Miescher isolated the chemical substance of the nucleus. Miescher found the substance to be an acid with a large phosphorus content and named it "nuclein." Subsequently, nuclein was identified as **DNA,** short for **deoxyribonucleic acid.** Some 75 years would pass before the significance of DNA would be revealed.

Few would argue that the demonstration of DNA as the genetic material and the subsequent determination of its molecular structure are among the most significant discoveries of the twentieth century. Since the early 1950s, when James Watson and Francis Crick built on discoveries of others before them to construct their first model of DNA, tremendous advances in molecular biology have occurred, many of them based on the structure of DNA. Today we speak of gene therapy and genetic engineering in household conversations. In the minds of some, these topics raise hopes for curing or preventing many of the diseases plaguing humanity. For others, thoughts turn to "playing with nature," undoing the deeds of God, or creating monstrosities that will wipe humanity off the face of the earth.

This exercise will familiarize you with the basic structure of nucleic acids and their role in the cell. Understanding the function of nucleic acids—both DNA and **RNA (ribonucleic acid)**—is central to understanding life itself. We hope you will gain an understanding that will allow you to form educated opinions concerning what science should do with its newfound technology.

13.1 Isolation and Identification of Nucleic Acids *(About 30 min.)*

At the beginning of the twenty-first century, no scientific endeavor holds more promise for humanity than the Human Genome Project. This initiative by the National Institutes of Health and private companies has created a complete map of every gene in our chromosomes. With this knowledge, researchers are beginning to learn the function of each gene. This huge project, biology's "moon shot," has the potential to unlock the secrets of human life.

The first step in this undertaking is the isolation of DNA. In the following section, you'll isolate and identify a nucleic acid component of *Halobacterium salinarum*, a bacterium that grows in habitats with extremely high salt (NaCl) concentrations.

Halobacterium is able to live in its specialized environment because of its cell wall, which differs from that of most other bacteria by maintaining its rodlike shape *only* at high salt concentrations. As NaCl levels drop, the cell shape first becomes irregular and finally spherical. At still lower concentrations, the cell ruptures because of osmotic effects (Exercise 6). We will take advantage of this response to allow us to release the cells' contents, including the nucleic acids, for isolation.

MATERIALS

Per student pair:

- culture tube of *Halobacterium salinarum*
- cotton applicator stick
- 10 mL of 95% ethanol in a test tube
- glass rod
- inoculating needle
- 5- or 10-mL sterile pipette
- two 10-mL graduated cylinders
- 2 test tubes
- test tube rack
- china marker
- agarose gel plate with methylene blue stain
- dropping pipette

Per lab bench:

- TBE buffer solution bottle and in dropping bottle
- DNA standard solution in dropper bottle
- 1% albumin solution in dropper bottle
- paper towels

Per lab room:

- raw egg white in beaker
- source of dH_2O
- white light transilluminator or other source of white light

PROCEDURE

Work in pairs.

A. Isolation of Nucleic Acids

1. With the 10-mL graduated cylinder, measure 1.5 mL of distilled water into a clean test tube.
2. Remove the cap from the slant culture of *H. salinarum* and insert the cotton swab applicator stick into the culture tube. Gently swab the entire surface of the culture by carefully rotating the cotton swab over the pink bacterial colonies. Try to pick up as much of the bacterial colony as possible on the swab. Remove the cotton swab from the culture tube and replace the cap.
3. Transfer the cotton swab applicator stick to the test tube containing the distilled water. Release all the adhering cells by vigorously swirling the cotton swab in the distilled water and occasionally pressing the swab against the wall of the tube.

The bacterial cells rupture from osmotic shock, since the cell wall cannot withstand the change from conditions of extremely high salt concentration to those of salt absence.

4. Withdraw the swab from the test tube, pressing it against the tube wall to squeeze out as much fluid as possible. *The gelatinous fluid adhering to the swab's surface after wetting will contain a large concentration of nucleic acids. This fluid must be left in the test tube.* Discard the swab.
5. Wipe the surface of the glass rod with a piece of paper towel moistened with 95% ethanol. Insert the clean rod into the test tube containing the cell suspension and stir vigorously. This action will assure total cell lysis. Remove the glass rod.
6. Using the sterile pipet, add 3 mL of 95% ethanol one drop at a time down the *side* of the test tube containing the cell suspension. Any material adhering to the tube wall should be washed into the suspension. The alcohol should form a layer on top of the aqueous cell suspension; *be careful not to mix the water and alcohol layers.*
7. Clean the inoculating wire with a paper towel moistened in 95% ethanol. Insert the wire into the culture tube so that the hook is at the cell suspension–alcohol interface and rotate the wire in a circular motion. The rotation should mix the contents only at the partition layer between the alcohol and cell suspension.

Nucleic acids precipitate and are extracted at this boundary between alcohol and water. Notice that strands of material adhere to the wire and trail off into the solution. The long linear-chain molecules of DNA appear stringy and form a cottony, viscous cloud around the wire.

The nucleic acids you have extracted are not pure; they contain cellular debris as well as adhering proteins. Observe the appearance and texture of the raw egg white on demonstration. Egg white has a very high concentration of the protein albumin. You will test your extracted precipitate to identify its major component.

B. Identification—Test for DNA

1. Use a 10-mL graduated cylinder to measure 3 mL of TBE buffer solution into a clean test tube.
2. Place the nucleic acid from the wire winder into the test tube. Swirl the wire winder in the buffer solution to thoroughly dissolve the viscous material in the buffer solution.
3. Obtain an agarose plate with methylene blue stain. (Agarose is an inert gel-like substance; methylene blue stain binds specifically with DNA, causing a visible purplish color.) Turn the unopened plate over and use the marking pen to draw four small, widely separated circles on the *underside* of the plate. Label the circles P, D, A, and C. These will be visible when looking at the plate from above to mark the locations where different test substances will be applied.
4. Open the plate, and with a dropping pipette apply a drop of the precipitated bacterial material dissolved in TBE buffer solution onto the area marked P.
5. Similarly, apply a drop of DNA standard solution onto the area marked D.
6. Apply a drop of 1% albumin protein solution onto the area marked A.
7. Apply a drop of TBE buffer onto the area marked C.
8. Set the plate aside for 20–30 minutes to allow time for color development.
9. View the petri plate on the white light transilluminator. A purple color indicates the presence of DNA. (See Figure 13-1.) Record your results in Table 13-1.

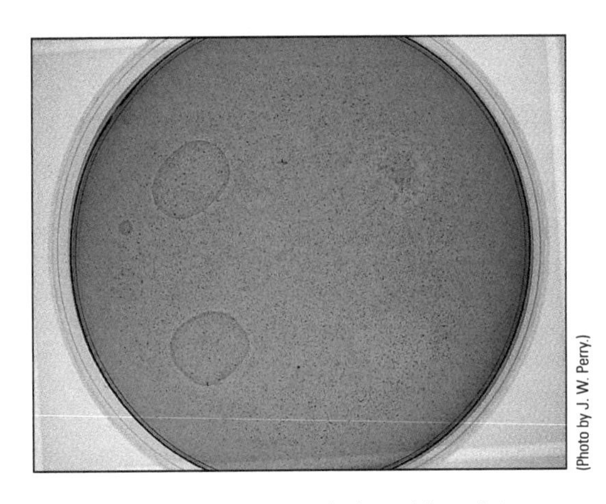

(Photo by J. W. Perry.)

Figure 13-1 Agarose-methylene blue plate on white light transilluminator showing positive (left) and negative (right) tests for DNA.

TABLE 13-1 Identification of Contents Extracted from *Halobacterium*

Droplet Code	Droplet Contents	Color
P	Precipitate in TBE buffer	
D	DNA standard	
A	1% albumin	
C	TBE buffer	

Name the substance(s) present in the material you isolated from *Halobacterium*.

____

What is the purpose of the DNA standard droplet?

____

What is the purpose of the TBE buffer droplet?

____

13.2 Modeling the Structure and Function of Nucleic Acids and Their Products *(About 90 min.)*

MATERIALS

Per student pair or group:
- DNA puzzle kit

Per lab room:
- DNA model

PROCEDURE

Work in pairs or groups.

Note: Clear your work surface of everything except your lab manual and the DNA puzzle kit.

In this section, we are concerned with three processes: *replication, transcription,* and *translation.* But before we study these three *per se,* let's formulate an idea of the structure of DNA itself.

A. Nucleic Acid Structure

1. Obtain a DNA puzzle kit. It should contain the following parts:
   - 18 deoxyribose sugars
   - 9 ribose sugars
   - 18 phosphate groups
   - 4 adenine bases
   - 6 guanine bases
   - 6 cytosine bases
   - 4 thymine bases
   - 2 uracil bases
   - 3 transfer RNA (tRNA)
   - 3 amino acids
   - 3 activating units
   - ribosome template sheet

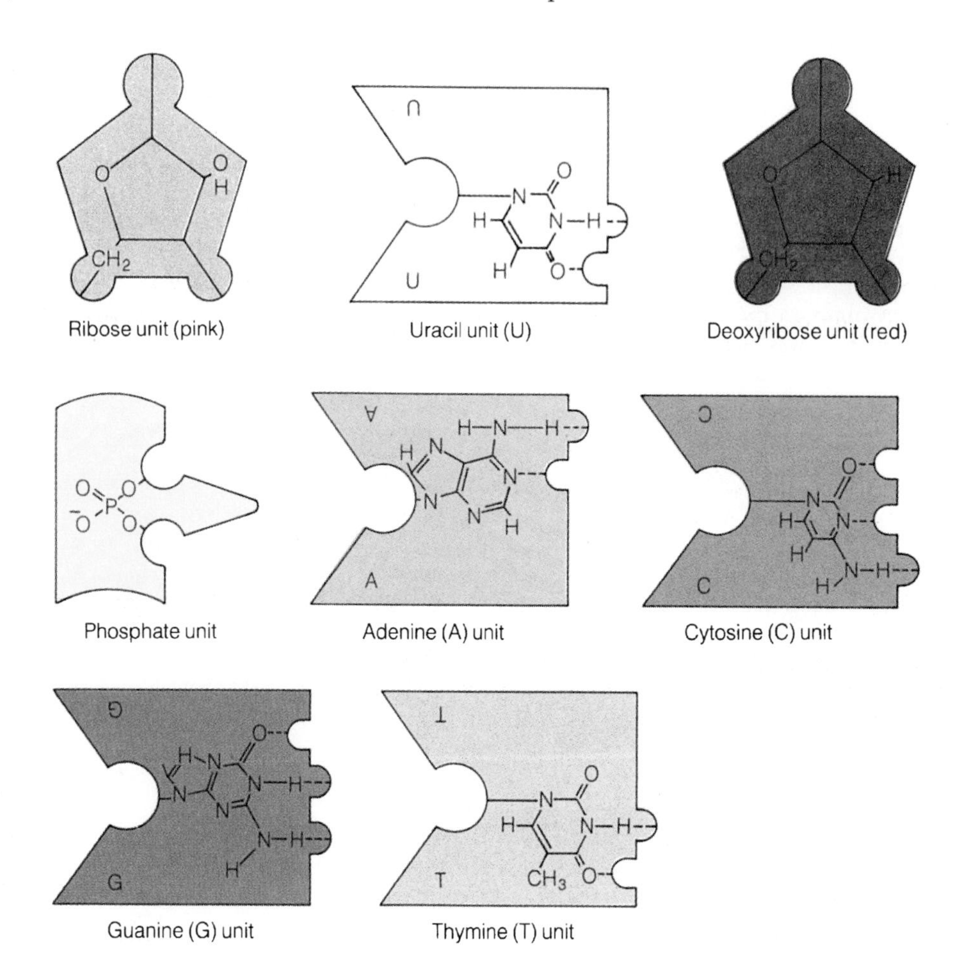

2. Group the components into separate stacks. Select a single deoxyribose sugar, an adenine base (labeled A), and a phosphate, fitting them together as shown in Figure 13-2. This is a single nucleotide (specifically a *deoxy*ribonucleotide), a unit consisting of a sugar (deoxyribose), a phosphate group, and a nitrogen-containing base (adenine).

Let's examine each component of the nucleotide.

Deoxyribose (Figure 13-3) is a sugar compound containing five carbon atoms. Four of the five are joined by covalent bonds into a ring. Each carbon is given a number, indicating its position in the ring. (These numbers are read "1-prime, 2-prime," and so on. "Prime" is used to distinguish the carbon atoms from the position of atoms that are sometimes numbered in the nitrogen-containing bases.) This structure is usually drawn in a simplified manner, without actually showing the carbon atoms within the ring (Figure 13-4).

There are four kinds of nitrogen-containing bases in DNA. Two are **purines** and are double-ring structures. Specifically, the two purines are *adenine* and *guanine* (abbreviated A and G, respectively; Figure 13-5).

The other two nitrogen-containing bases are **pyrimidines,** specifically *cytosine* and *thymine* (abbreviated C and T, respectively). Pyrimidines are single-ring compounds, as shown in Figure 13-6.

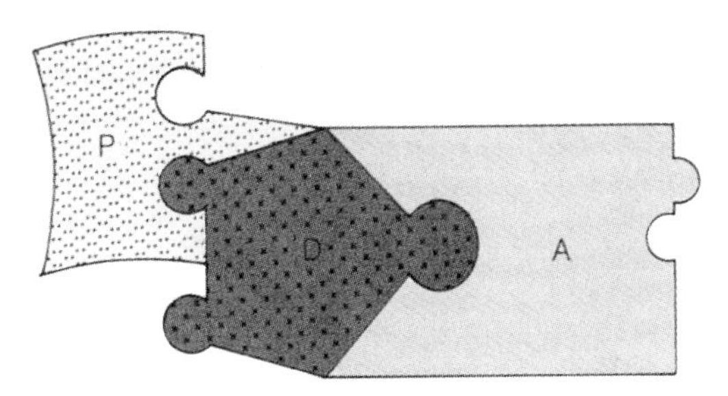

Figure 13-2 One deoxyribonucleotide.

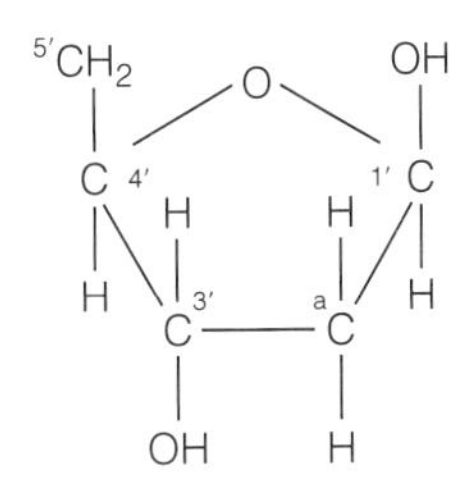

Figure 13-3 Deoxyribose.

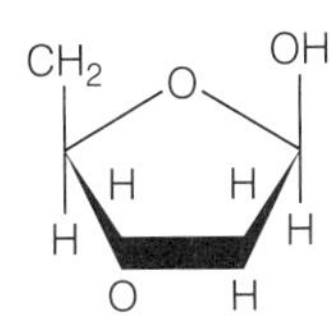

Figure 13-4 Simplified representation of deoxyribose.

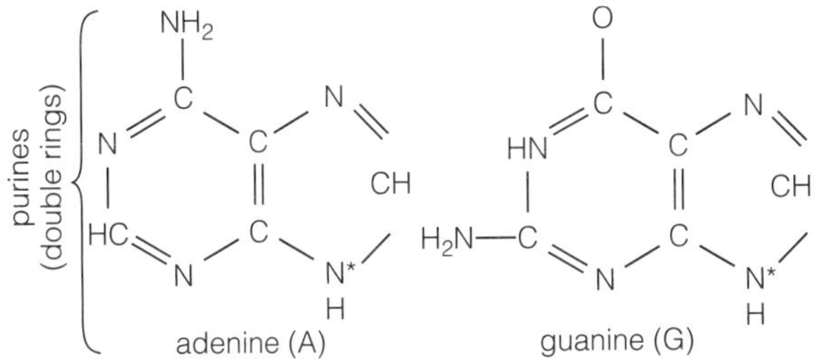

purines (double rings)

adenine (A) guanine (G)

Figure 13-5 Double-ringed purines found in DNA.

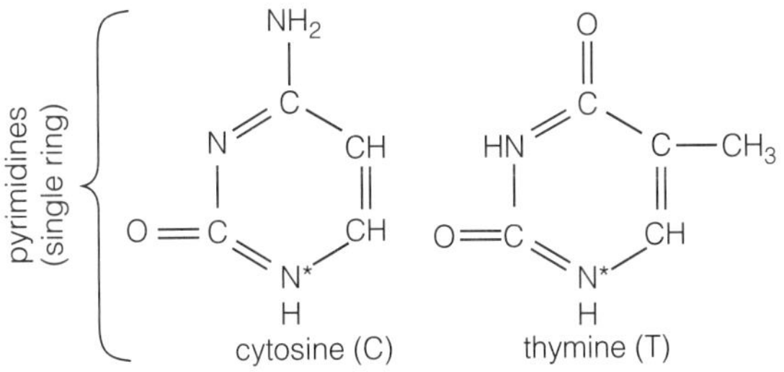

pyrimidines (single ring)

cytosine (C) thymine (T)

Figure 13-6 Pyrimidines found in DNA.

The symbol * indicates where a bond forms between each nitrogen-containing base and the 1′ carbon atom of the sugar ring structure. Although deoxyribose and the nitrogen-containing bases are organic compounds (they contain carbon), the phosphate group is an inorganic compound, with the structural formula shown in Figure 13-7.

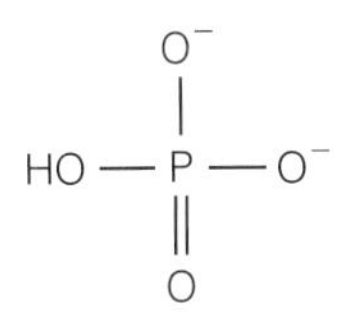

Figure 13-7 Phosphate group found in nucleic acids.

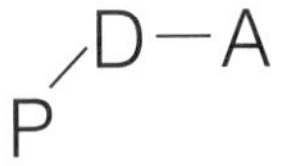

The phosphate end of the deoxyribonucleotide is referred to as the 5′ end, because the phosphate group bonds to the 5′ carbon atom.

There are four kinds of deoxyribonucleotides, each differing only in the type of base it possesses. Construct the other three kinds of deoxyribonucleotides, then draw them in Figures 13-8b–d. Rather than drawing the somewhat complex shape of the model, in this and other drawings, just give the correct position and letters. Use D for deoxyribose, P for a phosphate group, and A, C, G, and T for the different bases (as shown in Figure 13-8a).

Note the small notches and projections in the nitrogen-containing bases. Will the notches of adenine and thymine fit together? ______________________

Will guanine and cytosine? ______________________

Will adenine and cytosine? ______________________

Will thymine and guanine? ______________________

The notches and projections represent bonding sites. Make a prediction about which bases will bond with one another.

____

Will a purine base bond with another purine? ____________

Will a purine base bond with both types of pyrimidines?

____

P /D—A

Deoxyribonucleotide containing adenine **a**	Deoxyribonucleotide containing guanine **b**
Deoxyribonucleotide containing cytosine **c**	Deoxyribonucleotide containing thymine **d**

Figure 13-8 Drawings of deoxyribonucleotides containing guanine, cytosine, and thymine.

3. Assemble the three additional deoxyribonucleotides, linking them with the adenine-containing unit, to form a nucleotide strand of DNA. Note that the sugar backbone is bonded together by phosphate groups. Your strand should appear like that shown in Figure 13-9.
4. Now assemble a second four-nucleotide strand, similar to that of Figure 13-9. However, this time make the base sequence T-A-C-G, from bottom to top. DNA molecules consist of *two* strands of nucleotides, each strand the *complement* of the other.

5. Assemble the two strands by attaching (bonding) the nitrogen bases of complementary strands. Note that the adenine of one nucleotide always pairs with the thymine of its complement; similarly, guanine always pairs with cytosine. This phenomenon is called the **principle of base pairing.** On Figure 13-10, attach letters to the model pieces indicating the composition of your double-stranded DNA model.

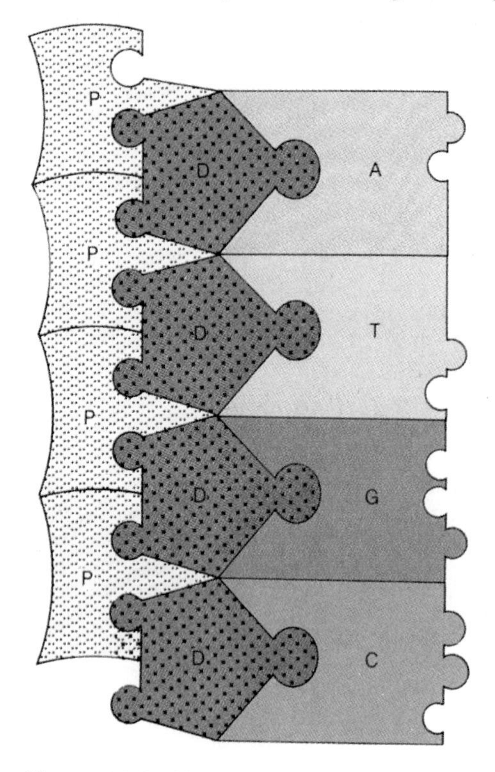

Figure 13-9 Four-nucleotide strand of DNA.

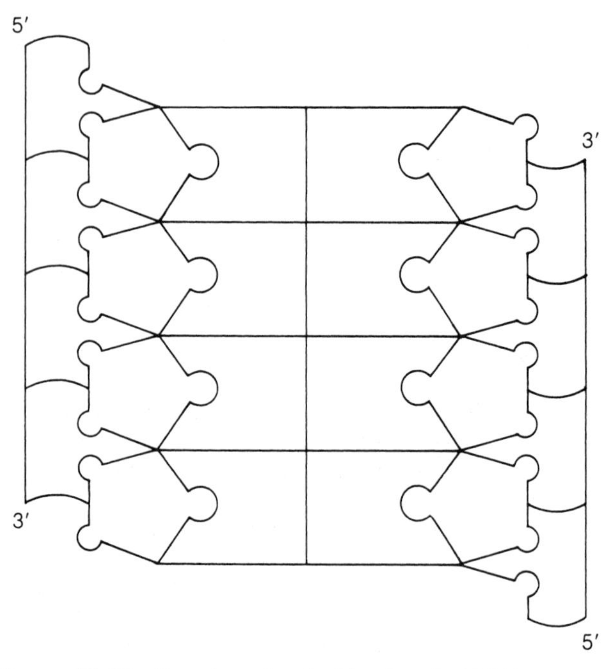

Figure 13-10 Drawing of a double strand of DNA. **Labels:** A, T, G, C, D, P (all used more than once)

What do you notice about the *direction* in which each strand is running? (That is, are both 5′ carbons at the same end of the strands?)

____

____

____

(Does the second strand of your drawing show this? It should.)

In life, the purines and pyrimidines are joined together by hydrogen bonds. Note again that the sugar backbone is linked by phosphate groups. Your model illustrates only a very small portion of a DNA molecule. The entire molecule may be tens of thousands of nucleotides in length!

6. Slide your DNA segment aside for the moment.
7. Examine the three-dimensional model of DNA on display in the laboratory (Figure 13-11). Notice that the two strands of DNA are twisted into a spiral-staircaselike pattern. This is why DNA is known as a *double helix.* Identify the deoxyribose sugar, nitrogen-containing bases, hydrogen bonds linking the bases, and the phosphate groups.

The second type of nucleic acid is RNA, short for ribonucleic acid. There are three important differences between DNA and RNA:

(a) RNA is a *single strand* of nucleotides.
(b) The sugar of RNA is **ribose.**
(c) RNA lacks the nucleotide that contains thymine. Instead, it has one containing the pyrimidine uracil (U) (see Figure 13-12).

Compare the structural formulas of ribose (Figure 13-13) and deoxyribose (Figure 13-4). How do they differ?

____

Why is the sugar of DNA called *deoxy*ribose?

____

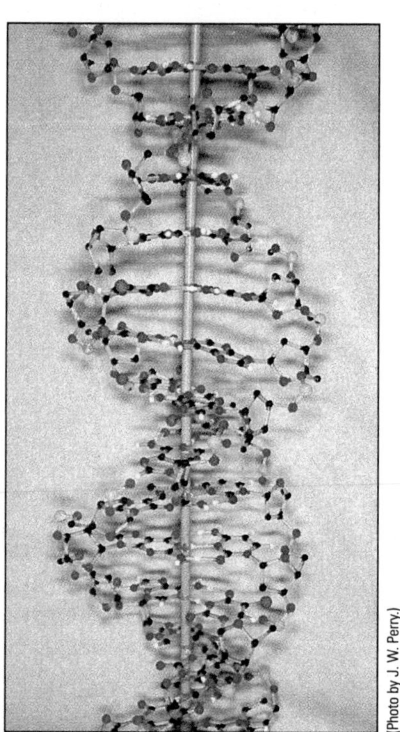

(Photo by J. W. Perry.)

Figure 13-11 Three-dimensional model of DNA.

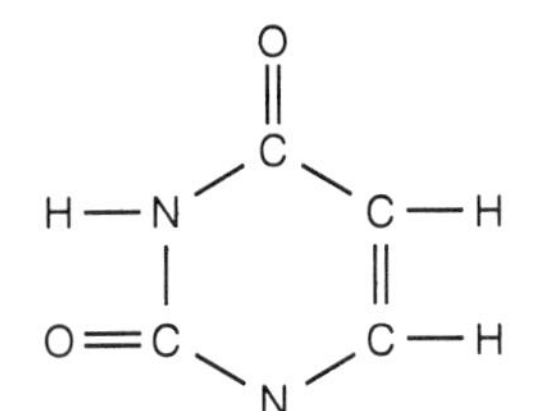

Figure 13-12 The pyrimidine uracil.

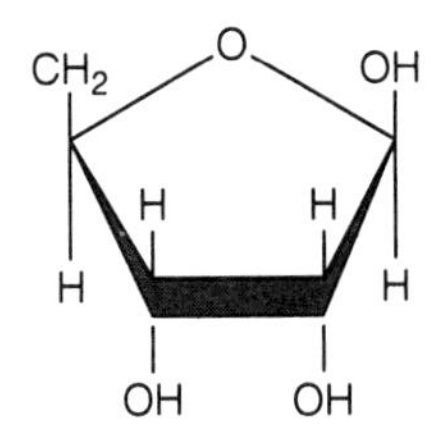

Figure 13-13 Ribose.

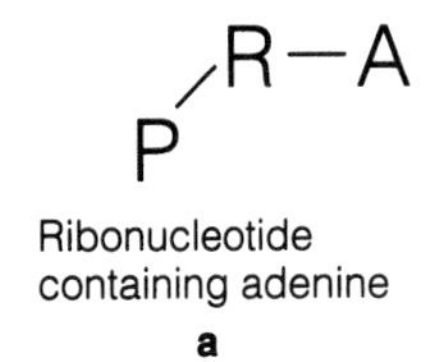

Ribonucleotide containing adenine
a

Ribonucleotide containing guanine
b

Ribonucleotide containing cytosine
c

Ribonucleotide containing uracil
d

Figure 13-14 Drawings of four possible ribonucleotides.

8. From the remaining pieces of your model kit, select four ribose sugars, an adenine, uracil, guanine, and cytosine, and four phosphate groups. Assemble the four **ribonucleotides** and draw each in Figure 13-14. (Use the convention illustrated in Figure 13-8 rather than drawing the actual shapes.)

Disassemble the RNA models after completing your drawing.

B. Modeling DNA Replication

DNA **replication** takes place during the S-stage of interphase of the cell cycle (see Exercise 10). Recall that the DNA is aggregated into chromosomes. Before mitosis, the chromosomes duplicate themselves so that the daughter nuclei formed by mitosis will have the same number of chromosomes (and hence the same amount of DNA) as did the parent cell.

Replication begins when hydrogen bonds between nitrogen bases break and the two DNA strands "unzip." Free nucleotides within the nucleus bond to the exposed bases, thus creating *two* new strands of DNA (as described below). The process of replication is controlled by enzymes called **DNA polymerases.**

1. Construct eight more deoxyribonucleotides (two of each kind) but don't link them into strands.
2. Now return to the double-stranded DNA segment you constructed earlier. Separate the two strands, imagining the zipperlike fashion in which this occurs within the nucleus.
3. Link the free deoxyribonucleotides to each of the "old" strands. When you are finished, you should have two double-stranded segments.

Note that one strand of each is the parental ("old") strand and the other is newly synthesized from free nucleotides. This illustrates the *semiconservative* nature of DNA replication. Each of the parent strands remains intact—it is *conserved*—and a new complementary strand is formed on it. Two "half-old, half-new" DNA molecules result.

Figure 13-15 Drawing of two replicated DNA segments, illustrating their semiconservative nature.

4. Draw the two replicated DNA molecules in Figure 13-15, labeling the old and new strands. (Once again, use the convention shown in Figure 13-8.)

C. Transcription: DNA to RNA

DNA is an "information molecule" residing *within* the nucleus. The information it provides is for assembling proteins *outside* the nucleus, within the cytoplasm. The information does not go directly from the DNA to the cytoplasm. Instead, RNA serves as an intermediary, carrying the information from DNA to the cytoplasm.

Synthesis of RNA takes place within the nucleus by **transcription.** During transcription, the DNA double helix unwinds and unzips, and a single strand of RNA, designated **messenger RNA (mRNA),** is assembled using the nucleotide sequence of *one* of the DNA strands as a pattern (template). Let's see how this happens.

1. Disassemble the replicated DNA strands into their component deoxyribonucleotides.
2. Construct a new DNA strand consisting of nine deoxyribonucleotides. With the purines and pyrimidines pointing away from you, lay the strand out horizontally in the following base sequence: T-G-C-A-C-C-T-G-C
3. Now assemble RNA ribonucleotides complementary to the exposed nitrogen bases of the DNA strand. Don't forget to substitute the pyrimidine uracil for thymine.

What is the sequence from left to right of nitrogen bases on the mRNA strand?

____

After the mRNA is synthesized within the nucleus, the hydrogen bonds between the nitrogen bases of the deoxyribonucleotides and ribonucleotides break.

4. Separate your mRNA strand from the DNA strand. (You can disassemble the deoxyribonucleotides now.) At this point, the mRNA moves out of the nucleus and into the cytoplasm.

By what avenue do you suppose the mRNA exits the nucleus? (*Hint:* Reexamine the structure of the nuclear membrane, as described in Exercise 5.)

____

To *transcribe* means to "make a copy of." Is transcription of RNA from DNA the formation of an *exact* copy? __________ Explain.

____

____

You will use this strand of mRNA in the next section. Keep it close at hand.

D. Translation—RNA to Polypeptides

Once in the cytoplasm, mRNA strands attach to *ribosomes,* on which translation occurs. To *translate* means to change from one language to another. In the biological sense, **translation** is the conversion of the linear message encoded on mRNA to a linear strand of amino acids to form a polypeptide. (A *peptide* is two or more amino acids linked by a peptide bond.)

Translation is accomplished by the interaction of mRNA, ribosomes, and **transfer RNA (tRNA),** another type of RNA. The tRNA molecule is formed into a four-cornered loop. You can think of tRNA as a baggage-carrying molecule. Within the cytoplasm, tRNA attaches to specific free amino acids. This occurs with the aid of activating enzymes, represented in your model kit by the pieces labeled "glycine activating" or "alanine activating." The amino acid–carrying tRNA then positions itself on ribosomes where the amino acids become linked together to form polypeptides.

1. Obtain three tRNA pieces, three amino acid units, and three activating units.
2. Join the amino acids first to the activating units and then to the tRNA. Will a particular tRNA bond with *any* amino acid, or is each tRNA specific? _______________
3. Now let's do some translating. In the space below, list the sequence of bases on the *messenger* RNA strand, starting at the left.
   (left, 3′ end)_______________(right, 5′ end)

Translation occurs when a *three*-base sequence on mRNA is "read" by tRNA. This three-base sequence on mRNA is called a **codon.** Think of a codon as a three-letter word, read right (5′) end to left (3′) end. What is the order of the rightmost (first) mRNA codon? (Remember to list the letters in the *reverse* order of that in the mRNA sequence.)

The first codon on the mRNA model is (5′ end)_______________ (3′ end)

4. Slide the mRNA strand onto the ribosome template sheet, with the first codon at the 5′ end.
5. Find the tRNA–amino acid complex that complements (will fit with) the first codon. The complementary three-base sequence on the tRNA is the **anticodon.** Binding between codons and anticodons begins at the P site of the 40s subunit (the smaller subunit) of the ribosome. The tRNA–amino acid complex with the correct anticodon positions itself on the P site.
6. Move the tRNA–amino acid complex onto the P site on the ribosome template sheet and fit the codon and anticodon together. In the boxes below, indicate the codon, anticodon, and the specific amino acid attached to the tRNA.

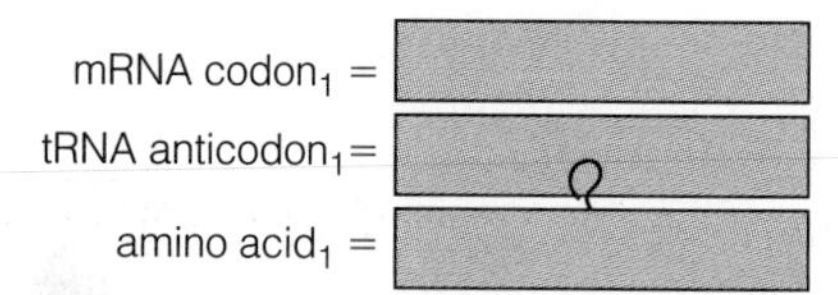

7. Now identify the second mRNA codon and fill in the boxes.

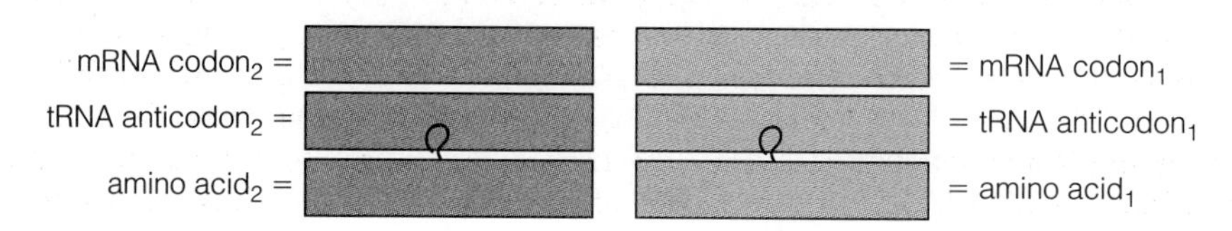

8. The second tRNA–amino acid complex moves onto the A site of the 40s subunit. Position this complex on the A site. An enzyme now catalyzes a condensation reaction, forming a **peptide bond** and linking the two amino acids into a dipeptide. (Water, HOH, is released by this condensation reaction.)
9. Separate amino acid$_1$ from its tRNA and link it to amino acid$_2$. (In reality, separation occurs somewhat later, but the puzzle doesn't allow this to be shown accurately; see below for correct timing.)

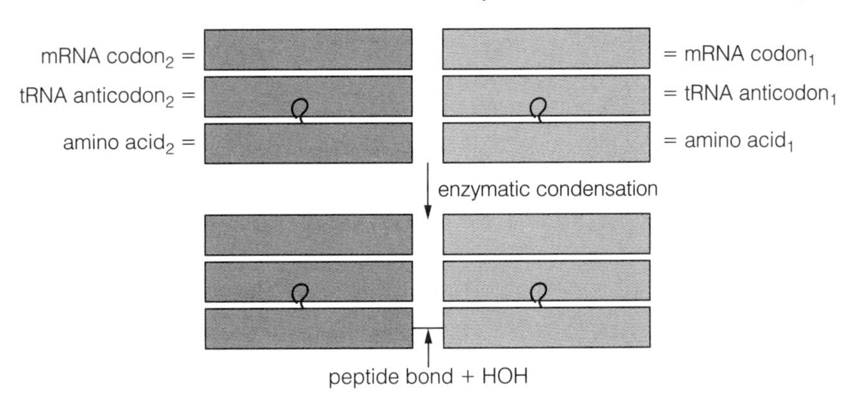

One tRNA–amino acid complex remains. It must occupy the A site of the ribosome in order to bind with its codon. Consequently, the dipeptide must move to the right.

10. Slide the mRNA to the right (so that tRNA$_2$ is on the P site) and fit the third mRNA codon and tRNA anticodon to form a peptide bond, creating a model of a tripeptide. At about the same time that the second peptide bond is forming, the first tRNA is released from both the mRNA and the first amino acid. Eventually, it will pick up another specific amino acid.

What amino acid will tRNA$_1$ pick up? ____________.

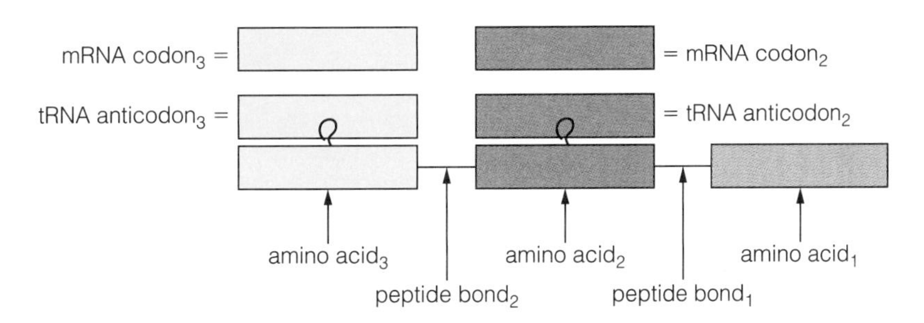

Record the tripeptide that you have just modeled. __________

You have created a short polypeptide. Polypeptides may be thousands of amino acids in length. As you see, the amino acid sequence is ultimately determined by DNA, because it was the original source of information.

Finally, let's turn our attention to the concept of a gene. A **gene** is a unit of inheritance. Our current understanding of a gene is that a gene codes for one polypeptide. This is appropriately called the **one-gene, one-polypeptide hypothesis.** Given this concept, do you think a gene consists of one, several, or many deoxyribonucleotides?

A gene probably consists of __________ deoxyribonucleotides.

***Note:* Please disassemble your models and return them to the proper location.**

13.3 Principles of Genetic Engineering: Recombination of DNA *(About 15 min.)*

People suffering from Type 1 diabetes are unable to produce enough insulin, a hormone that is synthesized by the pancreas and that is instrumental in regulating the amount of blood sugar. Therapy for severe diabetes includes daily injections of insulin. Until recently, that insulin was extracted from the pancreas of slaughtered pigs and cows. With the advent of techniques commonly referred to as genetic engineering, human insulin is now produced by bacteria. These organisms grow and reproduce rapidly, hence producing quantities of insulin en masse.

Genetic engineering is a convenient phrase to describe what is more properly called methods in recombinant DNA. **Recombinant DNA** is DNA into which a set of "foreign" nucleotides has been inserted. In the case of insulin production, researchers first located on human chromosomes the gene (set of nucleotides) that codes for insulin production. Once identified, the nucleotides were removed from the human DNA and inserted into the DNA within a bacterium. As this bacterial cell reproduces, each new generation contains the gene coding for human insulin. The bacteria produce the hormone, which is harvested and purified. Thus these recombinant bacteria are "insulin factories."

Bacteria have been exchanging genes with each other for millennia. In the process, new genetic strains of bacteria may be produced. The following demonstration will familiarize you with genetic recombination in bacteria; these principles are the basis for genetic engineering.

Two strains of the bacterium *Escherichia coli* will be used in this experiment:

- Strain 1 carries a chromosomal gene that causes it to be resistant to the antibiotic drug streptomycin; it is susceptible to (killed by) another antibiotic, ampicillin. (See Figure 13-16a.)
- Strain 2 is resistant to the antibiotic drug ampicillin but susceptible to streptomycin; the gene for resistance to ampicillin is located on a small extrachromosomal (that is, not on its chromosome) loop of DNA called a **plasmid.** (See Figure 13-16b.)

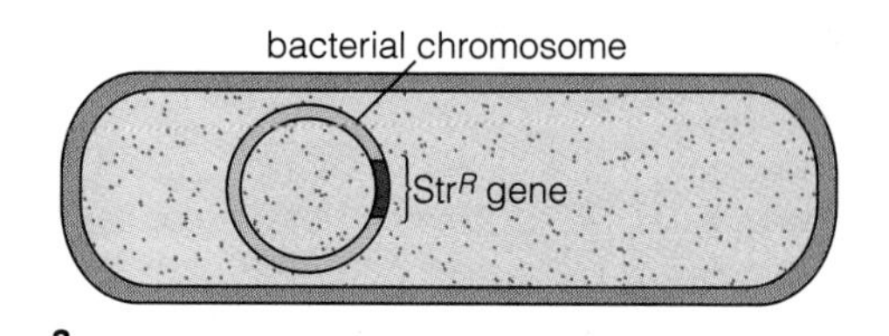

a

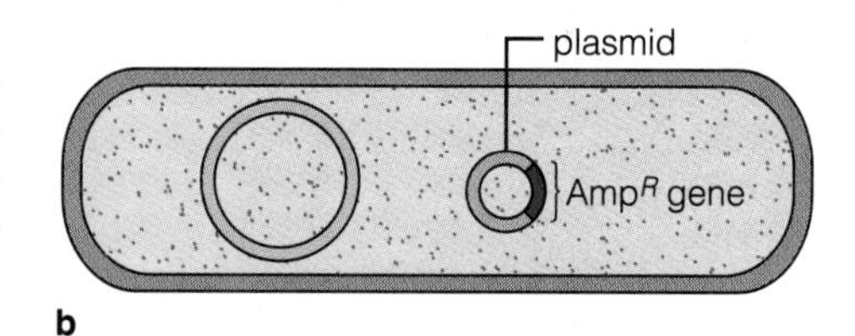

b

Figure 13-16 Genetic components of two strains of *Escherichia coli* bacteria. (**a**) Strain 1 has a gene for resistance to the antibiotic streptomycin (designated Str^R) on its chromosome. (**b**) Strain 2 has a gene for resistance to the antibiotic ampicillin (designated Amp^R) on a plasmid within the cell.

Plasmids contain relatively few genes compared to the bacterial chromosome. Like chromosomal DNA, plasmids can replicate. Insertion of a nonbacterial DNA segment (set of nucleotides) results in the formation of a hybrid plasmid that can replicate the foreign DNA as well. This is the basis for human insulin production within bacteria, as mentioned above. Genes for resistance to various antibiotics are commonly found on plasmids as well.

Plasmids may also be transferred from a host (donor) bacterium to a recipient bacterial cell by a process called **bacterial conjugation.** Thus, the plasmid acts both as a carrier of foreign DNA and as an agent (vector) for the introduction of that DNA into the recipient cell. Once plasmid DNA is transferred to the recipient, the recipient bears the genes (and hence makes the gene products) formerly in the host.

Plasmid transfer between host and recipient (in this case, two bacterial cells) occurs through a bridge formed by the host cell that connects it to the recipient. Figure 13-17 illustrates bacterial conjugation. Note that genes on the bacterial chromosome are not transferred between cells.

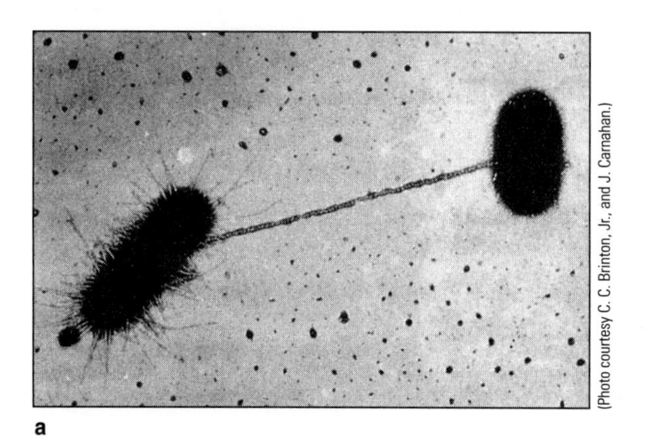

(Photo courtesy C. C. Brinton, Jr., and J. Carnahan.)

a

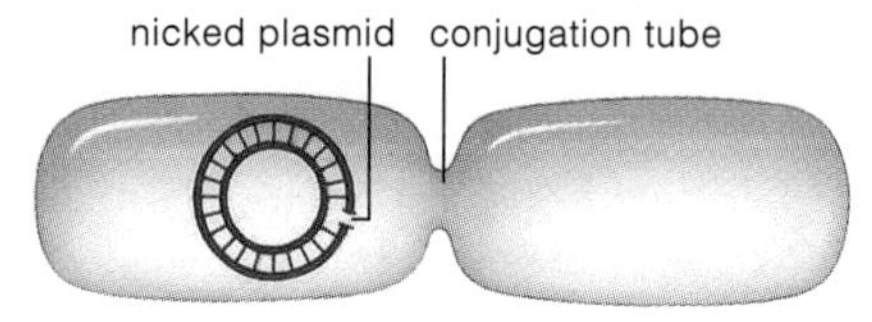

1 A conjugation tube has already formed between a donor and a recipient cell. An enzyme has nicked the donor's plasmid.

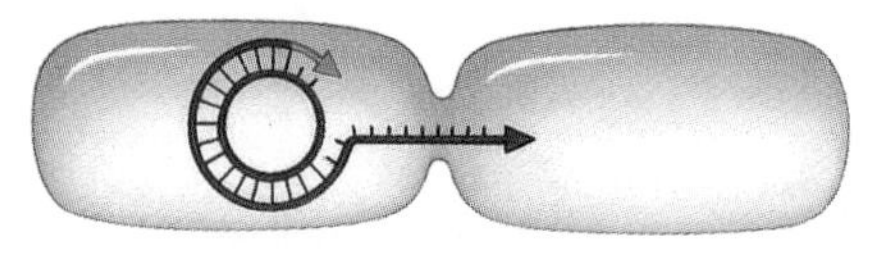

2 DNA replication starts on the nicked plasmid. The displaced DNA strand moves through the tube and enters the recipient cell.

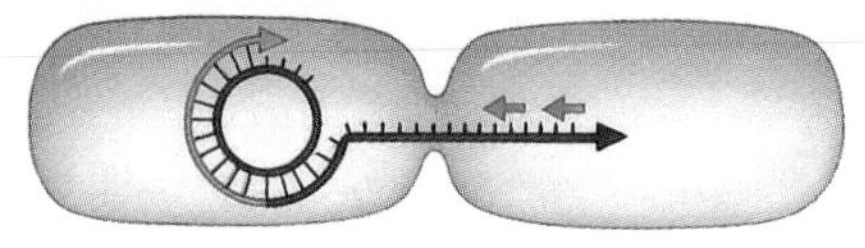

3 In the recipient cell, replication starts on the transferred DNA.

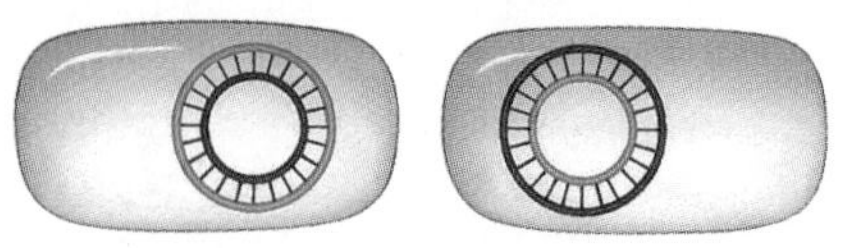

4 The cells separate from each other; the plasmids circularize.

b

(After Starr, 2000.)

Figure 13-17 (**a**) Conjugation between two bacteria. (**b**) Plasmid gene transfer during bacterial conjugation. The bacterial chromosome is not shown.

MATERIALS

Per student group (4):

- nutrient agar plate containing ampicillin, incubated with *E. coli* Strain 1 on one half, and *E. coli* Strain 2 on the other half
- nutrient agar plate containing streptomycin, incubated with *E. coli* Strain 1 on one half, and *E. coli* Strain 2 on the other half
- nutrient agar plate containing *both* ampicillin and streptomycin, incubated with a mating solution of both *E. coli* Strain 1 and *E. coli* Strain 2 spread across the plate (The mating solution is designed to allow bacterial conjugation to occur.)
- demonstration nutrient agar plates showing growth of *E. coli* Strain 1 and *E. coli* Strain 2

PROCEDURE

1. Examine the plates for growth. (Your instructor will provide demonstration plates for you to examine so that you can recognize bacterial growth.)
2. Record your observations in Table 13-2, using a + to indicate the growth of bacteria, a – to indicate absence of growth. A bacterium that is sensitive to (killed by) an antibiotic will be unable to grow on nutrient media containing that specific antibiotic.

TABLE 13-2 Bacterial Growth on Antibiotic-Containing Plates

Growth of Bacteria on Nutrient Agar Containing			
Strain of *E. coli*	**Ampicillin**	**Streptomycin**	**Ampicillin plus Streptomycin**
Strain 2 (donor)			
Strain 1 (recipient)			
"Mating mixture"			

3. Discard your plates in the designated location.

Was Strain 2 susceptible *or* resistant to ampicillin? ______________________

To streptomycin? ______________________

Was Strain 1 susceptible *or* resistant to ampicillin? ______________________

To streptomycin? ______________________

Make a conclusion about the presence and location of a gene in each of the two strains of *E. coli* for resistance to each of the antibiotics.

Strain 2

____

____

Strain 1

____

____

Make a conclusion about what happened when the two strains were mixed together. Incorporate your observations concerning antibiotic resistance into your conclusion.

____

____

____

PRE-LAB QUESTIONS

_____ **1.** The individuals responsible for constructing the first model of DNA structure were
- (a) Wallace and Watson
- (b) Lamarck and Darwin
- (c) Mendel and Meischer
- (d) Crick and Watson

_____ **2.** Deoxyribose is
- (a) a five-carbon sugar
- (b) present in RNA
- (c) a nitrogen-containing base
- (d) one type of purine

_____ **3.** A nucleotide may consist of
- (a) deoxyribose or ribose
- (b) purines or pyrimidines
- (c) phosphate groups
- (d) all of the above

_____ **4.** Which of the following is consistent with the principle of base pairing?
- (a) purine-purine
- (b) pyrimidine-pyrimidine
- (c) adenine-thymine
- (d) guanine-thymine

_____ **5.** Nitrogen-containing bases between two complementary DNA strands are joined by
- (a) polar covalent bonds
- (b) hydrogen bonds
- (c) phosphate groups
- (d) deoxyribose sugars

_____ **6.** The difference between deoxyribose and ribose is that ribose
- (a) is a six-carbon sugar
- (b) bonds only to thymine, not uracil
- (c) has one more oxygen atom than deoxyribose has
- (d) is all of the above

_____ **7.** Replication of DNA
- (a) takes place during interphase
- (b) results in two double helices from one
- (c) is semiconservative
- (d) is all of the above

_____ **8.** Transcription of DNA
- (a) results in formation of a complementary strand of RNA
- (b) produces two new strands of DNA
- (c) occurs on the surface of the ribosome
- (d) is semiconservative

_____ **9.** An anticodon
- (a) is a three-base sequence of nucleotides on tRNA
- (b) is produced by translation of RNA
- (c) has the same base sequence as does the codon
- (d) is the same as a gene

_____ **10.** Bacterial plasmids
- (a) are the only genetic material in bacteria
- (b) may carry genes for antibiotic resistance
- (c) may be transferred between bacteria during the process of conjugation
- (d) are both b and c

Name ______________________________ Section Number ______________________

EXERCISE 13

Nucleic Acids: Blueprints for Life

POST-LAB QUESTIONS

13.2 Modeling the Structure and Function of Nucleic Acids and Their Products

1. The following diagram represents some of the puzzle pieces used in this section.
   a. Assembled in this form, do they represent a(an) amino acid, base, portion of messenger RNA, or deoxyribonucleotide?

   b. Justify your answer.

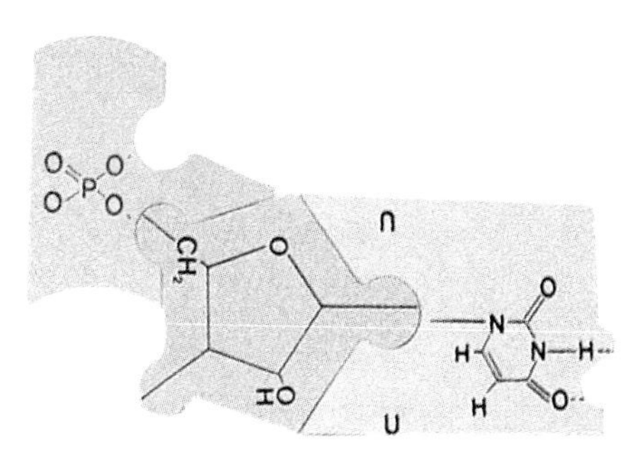

2. Why is DNA often called a *double helix?*

3. State the following ratios.
   a. Guanine to cytosine in a double-stranded DNA molecule: __________
   b. Adenine to thymine: __________

4. Define the following terms.
   a. replication

   b. transcription

   c. translation

   d. codon

   e. anticodon

5. What does it mean to say that DNA replication is *semiconservative*?

6. a. If the base sequence on one DNA strand is ATGGCCTAG, what is the sequence on the other strand of the helix?

   b. If the original strand serves as the template for transcription, what is the sequence on the newly formed RNA strand?

7. a. What amino acid would be produced if *transcription* takes place from a nucleotide with the three-base sequence ATA? _______________

   b. A genetic mistake takes place during *replication* and the new DNA strand has the sequence ATG. What is the three-base sequence on an RNA strand transcribed from this series of nucleotides?

   ____

   c. Which amino acid results from this codon?

13.3 Principles of Genetic Engineering: Recombination of DNA

8. a. What is a plasmid?

   b. How are plasmids used in genetic engineering?

9. How does bacterial conjugation differ from the process by which eukaryotic gene products are produced by bacteria?

Food for Thought

10. What kinds of molecules are present in bacterial cells in addition to the DNA you might have isolated in Section 13.1?

EXERCISE 14

Evidences of Evolution

OBJECTIVES

After completing this exercise, you will be able to

1. define *evolution, fossil, natural selection, population, species, fitness, mass extinction, adaptive radiation, anthropologist, hominid;*
2. explain how natural selection operates to alter the genetic makeup of a population;
3. describe the general sequence of evolution of life forms over geologic time;
4. recognize primitive and advanced characteristics of skull structure of human ancestors and relatives;
5. describe the evolutionary relationships among *Dryopithecus, Australopithecus afarensis,* and other australopith species, and the various *Homo* species.

INTRODUCTION

Have you ever marveled over the diversity of animals in a zoo? Lions, antelopes, zebras, giraffes, elephants, and chimpanzees are all mammals, as are we humans. All mammals living today, and many forms that are extinct, descended from a mammalian ancestor who lived more than 200 million years ago. The process by which this incredible diversity of mammals (and all the other living and extinct species) came to exist is called **evolution,** the focus of this exercise.

Evolution, the process that results in changes in the genetic makeup of populations of organisms through time, is the unifying framework for the whole of biology. Less than 200 years ago, it seemed obvious to most people that living organisms had not changed over time—that dandelions looked like dandelions and humans looked like humans, year after year, generation after generation, without change. As scientists studied the natural world more closely, however, evidence of change and the relatedness of all living organisms emerged from geology and the fossil record, as well as from comparative morphology, developmental patterns, and biochemistry.

Charles Darwin (and, independently, Alfred Russel Wallace) postulated the major mechanism of evolution to be **natural selection**—the difference in survival and reproduction that occurs among individuals of a population that differ in one or more alleles. (See Exercise 11 for the definition of *allele.*) A **population** is a group of individuals of the same species occupying a given area. A **species** is one or more populations that closely resemble each other, interbreed under natural conditions, and produce fertile offspring.

Individuals in a population that have particular combinations of alleles better adapted to their environment survive and produce more offspring than individuals with different genetic makeups. The offspring have received alleles that also make their chances for survival and reproduction greater, and so it goes, generation after generation. The population evolves as some traits become more common and others decrease or disappear over time.

Many people think of evolution in historic terms—as something that produced the dinosaurs or the Galapagos finches Darwin studied but that no longer operates in today's world. Remember, though, that *the process of genetic change over time continues today* and is a dominant force shaping the living organisms of our planet.

The scientific evidence for evolution is overwhelming, although scientists continue to debate the exact mechanisms by which natural selection and other evolutionary agents change allelic frequencies. In this exercise, you will observe the effects of natural selection in a living plant population; consider the time over which evolution has occurred; and examine the fossil record for evidence of large-scale trends and change among our human ancestors and relatives.

14.1 Experiment: Natural Selection*

Natural selection discriminates among individuals with respect to their ability to produce offspring. Those individuals that survive and reproduce will perpetuate more of their alleles in the population. These individuals are said to exhibit greater **fitness** than those who leave no or fewer offspring.

You will examine individuals of two different populations of dandelions. Dandelions are interesting because, despite their showy yellow flowers, they reproduce primarily asexually. Instead of the sexual reproduction that typically occurs in flowers, dandelions form seeds without fertilization. Thus, each dandelion is essentially genetically (and physically) identical to its parent plant.

One set of dandelion seedlings in your laboratory was grown from seeds collected from an area that has endured frequent, close mowing for many years. The second group of seedlings was grown from seeds collected from plants mowed infrequently, if at all. Your instructor will describe in more detail the conditions under which the parent plants of your seedlings have grown.

This experiment tests the hypothesis that *dandelion populations grown for generations under different mowing regimes will have different growth forms.* Write a prediction regarding this experiment in Table 14-2. What would you predict the dandelion plants grown from seeds produced in mowed and unmowed areas to look like?

MATERIALS

Per student group (4):

- flat of dandelion seedlings, labeled Mowed or Unmowed
- trowel or large spoon
- metric ruler
- knife or single-edged razor blade
- calculator

Per lab room:

- several dishpans half-filled with water
- paper towels

PROCEDURE

1. Observe the flats of young dandelion plants. Note and describe any general differences in appearance between those grown from seeds of unmowed plants versus those grown from seeds of mowed plants.

____

____

____

2. Take one flat of plants to your lab bench. Remove the plants carefully from the flat with a trowel or spoon. Be careful to remove all the root system but do not damage or lose root material. Wash each plant gently in the dishpan provided, *not in the sink,* to remove the residual soil. Blot the plants dry with the paper towels.
3. Cut each plant at the region separating the root from the leafy shoot, keeping each root portion with its corresponding shoot. See Figure 14-1.
4. Measure shoot length to the nearest millimeter as the length between the cut surface and the tip of the longest leaf. Record this measurement in Table 14-1.
5. Measure root length to the nearest millimeter as the length between the cut surface and the tip of the longest tap root. Sometimes it's hard to tell where the tap root ends, but try to avoid measuring branch roots. See Figure 14-1. Record this data in Table 14-1.
6. For each plant, divide the shoot length by the root length to calculate the shoot-to-root ratio. Record the ratio in Table 14-1.

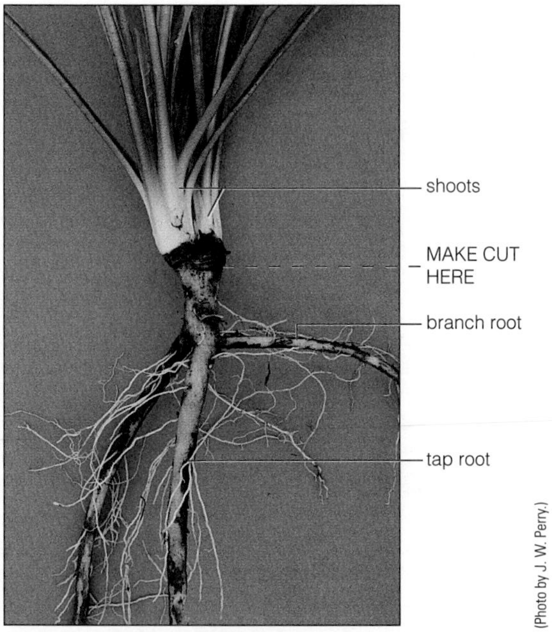

shoots

MAKE CUT HERE

branch root

tap root

(Photo by J. W. Perry.)

Figure 14-1 Dandelion plant.

*Adapted from an exercise by Thomas Hilbish and Minnie Goodwin, University of South Carolina.

TABLE 14-1 Measurements of Dandelions (Mowed or Unmowed? ____________)			
Plant	Shoot Length (mm)	Root Length (mm)	Shoot-to-Root Ratio
1			
2			
3			
4			
5			
Total number of plants =			Average =

7. Graph the distribution of the shoot-to-root ratio of each plant in your flat in Figure 14-2.

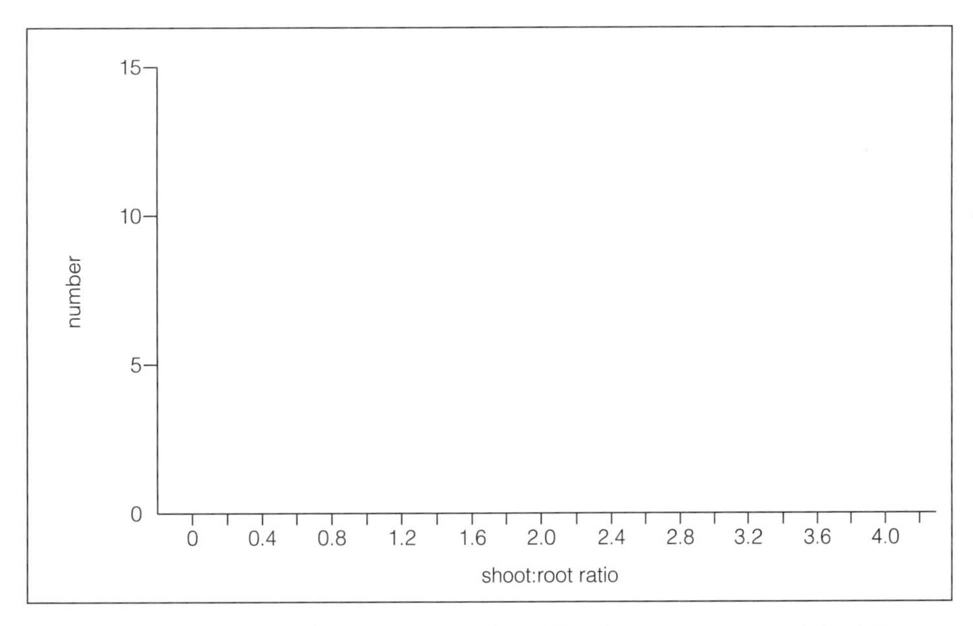

Figure 14-2 Distribution of shoot-to-root ratios of length measurements of dandelions.

8. Calculate the average shoot-to-root ratio of your plants by adding the shoot-to-root ratios of all the individual plants and dividing that number by the total number of plants.

9. Your instructor has drawn two graphs like that in Figure 14-2, one for mowed plants and one for unmowed plants. Pool your data with those of your classmates who have measured plants of the same type and enter that pooled data in Table 14-2 and on the corresponding graph. Calculate the average shoot-to-root ratio of the pooled data for each set of plants.

Are there noticeable differences in the distribution of shoot-to-root ratios between descendants of dandelions from intensely mowed versus unmowed areas? ___

If so, what are those differences?

___

___

___

Do the results support your hypothesis? Write a conclusion accepting or rejecting the hypothesis in Table 14-2.

TABLE 14-2 Class Shoot-to-Root Ratios for Mowed and Unmowed Plants

Prediction:

Group	**Mowed**	**Unmowed**
1		
2		
3		
4		
5		
6		
7		
8		
Average		

Conclusion:

If you failed to detect a difference between seeds grown from plants of mowed versus unmowed populations, speculate on the reason(s) for this failure. Does this failure necessarily mean that natural selection is not occurring?

___

___

Explain the results of this demonstration in terms of the effects of natural selection on the genetic makeup of the two dandelion populations.

___

___

14.2 Geologic Time *(About 30 min.)*

Our earth is an ancient planet that formed from a cloud of dust and gas approximately 4.6 billion years ago.

Life formed relatively quickly on the young planet, and the first cells emerged in the seas by 3.8 billion years ago. For more than a billion years, the only living organisms on earth were prokaryotic bacteria and bacterialike cells. Eventually, though, more complex, eukaryotic organisms evolved. The seas were colonized by single-celled organisms first, followed by multicelled and colonial plants and animals (Figure 14-3). Only much later did life move onto the land.

Throughout the history of life on earth, there have been many episodes of **mass extinctions** (catastrophic global events in which major groups of species are wiped out) followed by **adaptive radiation** (in which a lineage fills a wide range of habitats in a burst of evolutionary activity). For example, scientists postulate that a huge asteroid impacted the earth about 65 million years ago. The resulting environmental destruction caused a mass extinction in which nearly all the dinosaurs and marine organisms became extinct. After this extinction, the early mammals underwent an adaptive radiation and filled the habitats that had previously been occupied by dinosaurs.

Geologic time, "deep time," is difficult for humans, with our recent evolution and 75-year life spans to grasp. In this section, you will construct a geologic time line to help you gain perspective on the almost incomprehensible sweep of time over which evolution has operated.

MATERIALS

Per student pair:

- one 4.6-m rope or string
- meter stick and/or metric ruler
- masking tape
- calculator

PROCEDURE

1. Obtain a 4.6-m length of rope or string. The string represents the entire length of time (4.6 billion years) since the earth was formed. On this time line, you will attach masking-tape labels to mark the points when the events below occurred. Measure from the starting point with the meter stick or metric ruler.
2. Locate and mark with tape on the time line the following events (bya = billion years ago, mya = million years ago, ya = years ago):

4.6 bya	Formation of earth
3.8 bya	First living prokaryotic cells
2.5 bya	Oxygen-releasing photosynthetic pathway
1.2 bya	Origin of eukaryotic cells
580 mya	Origin of multicellular marine plants and animals
500 mya	First vertebrate animals (marine fishes)
435 mya	First vascular land plants and land animals
345 mya	Great coal forests; amphibians and insects undergo great adaptive radiation
240 mya	Mass extinction of nearly all living organisms
225 mya	Origin of mammals, dinosaurs; seed plants dominate
135 mya	Dinosaurs reach peak; flowering plants arise
65 mya	Mass extinction of dinosaurs and most marine organisms
63 mya	Flowering plants, mammals, birds, insects dominate
6 mya	First hominid (upright walking) human ancestors
2.5 mya	Human ancestors (*Homo*) using stone tools, forming social life
200,000 ya	Evolution of *Homo sapiens*
6400 ya	Egyptian pyramids constructed
2000 ya	Peak of Roman Empire
225 ya	Signing of the Declaration of Independence
In A.D. 1969	Humans step on earth's moon

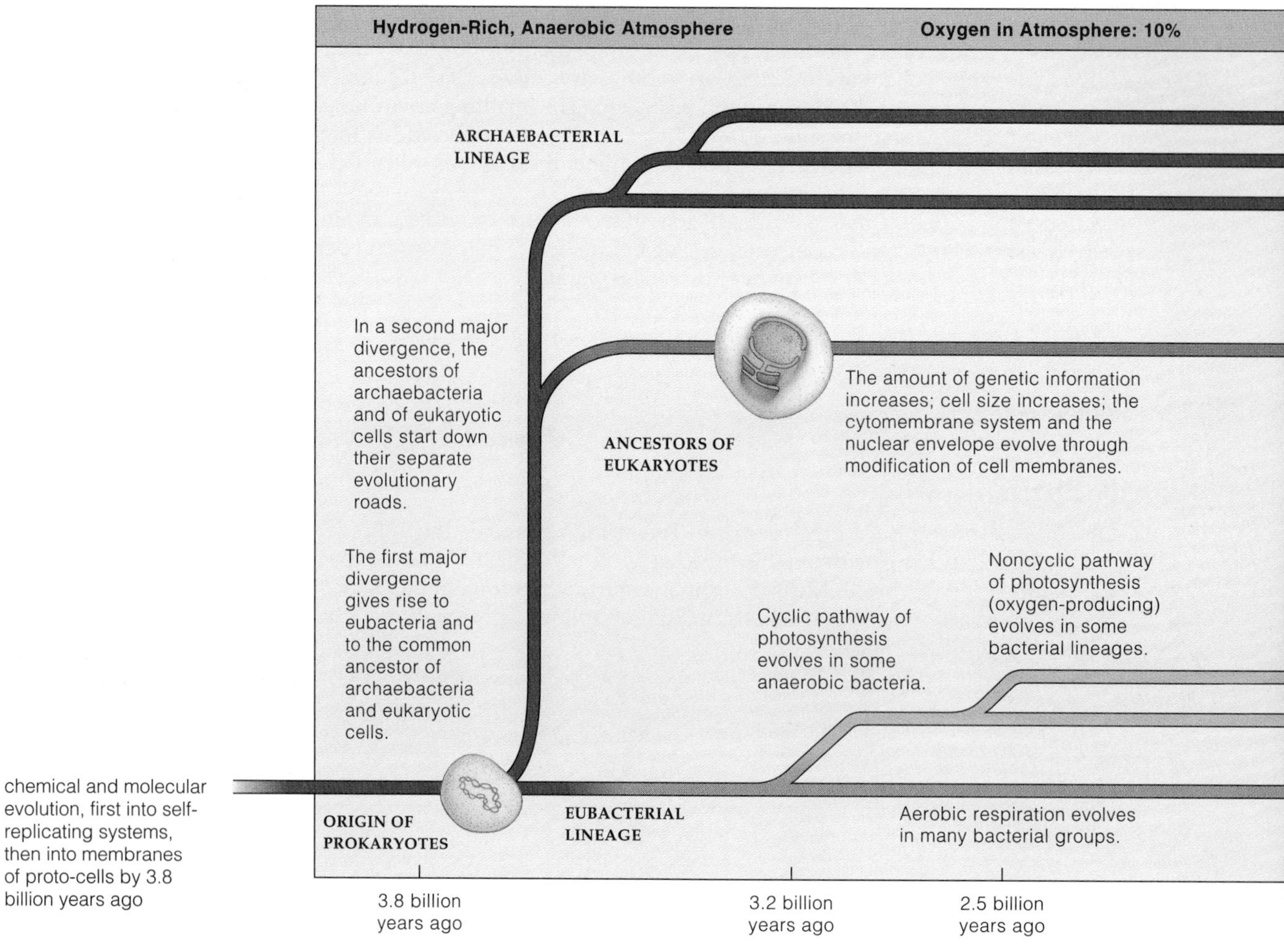

Figure 14-3 An evolutionary tree of life that reflects mainstream thinking about the connections among major lineages and origins of some eukaryotic organelles. (After Starr, 2000.)

Over what proportion of the earth's history were there only single-celled living organisms? __________
Over what proportion of the earth's history have multicelled organisms existed? __________
Over what proportion of the earth's history have mammals been a dominant part of the fauna? __________
Over what proportion of the earth's history have modern humans existed? __________

14.3 The Fossil Record and Human Evolution *(About 30 min.)*

The fossil record provides us with compelling evidence of evolution. Most fossils are parts of organisms, such as shells, teeth, and bones of animals, or stems and seeds of plants. These parts are replaced by minerals to form stone or are surrounded by hardened material that preserves the external form of the organism.

Anthropologists, scientists who study human origin and cultures, are piecing together the story of human evolution through the study of fossils, as well as through comparative biochemistry and anatomy. The fossil record of our human lineage is fragmentary at times and generates much discussion and differing interpretations. However, scientists agree that **hominids,** all species on the evolutionary branch leading to modern humans, arose in Africa between 10 and 5 million years ago from the same genetic line that also produced gorillas and chimpanzees.

Many anatomical changes occurred in the course of evolution from hominid ancestor to modern humans. Arms became shorter, feet flattened and then developed arches, and the big toe moved in line with the other toes. The legs moved more directly under the pelvis. These features allowed upright posture and bipedalism (walking on two legs), and they also allowed the use of hands for tasks other than locomotion.

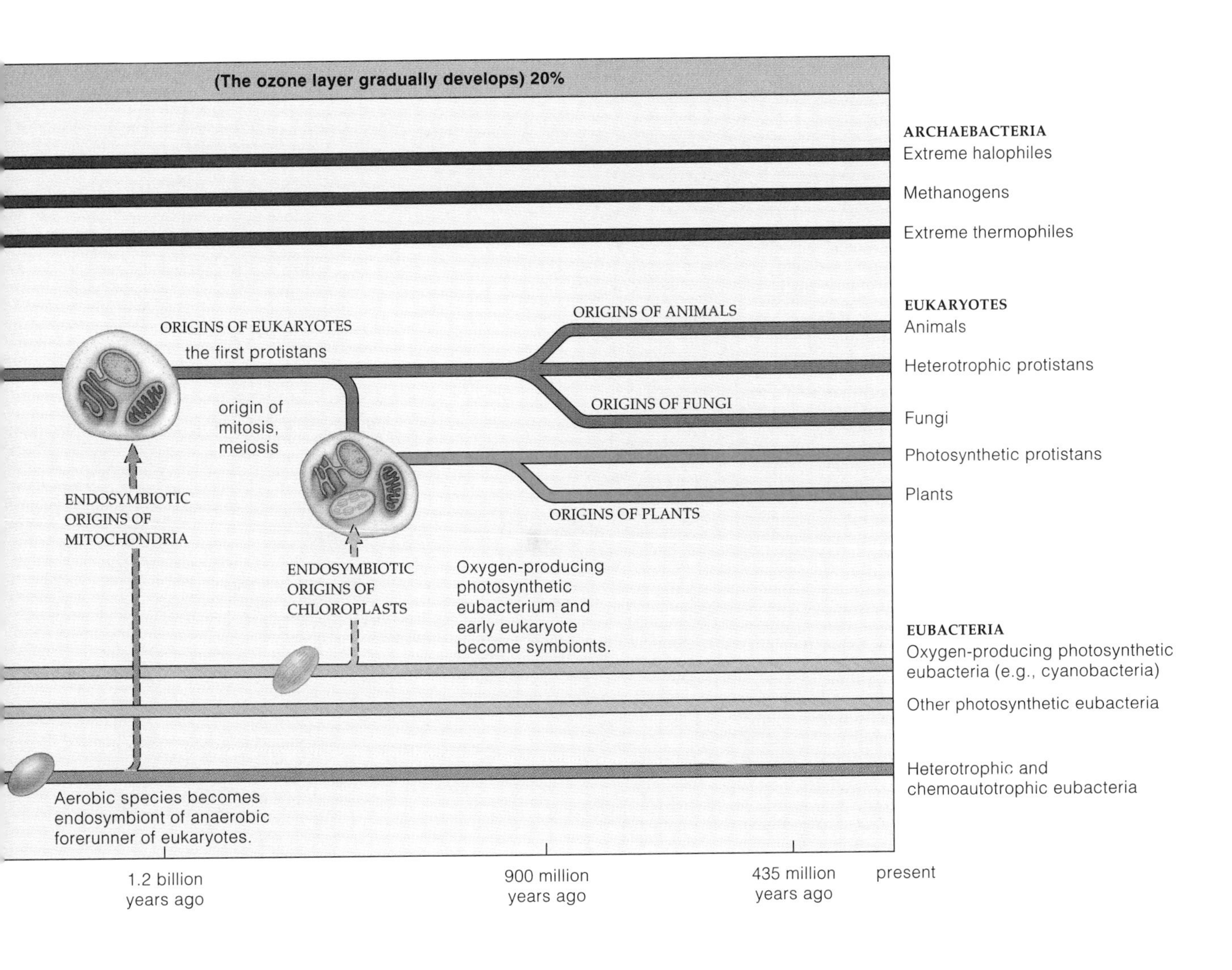

Still other anatomical changes were associated with the head. In this section, you will observe some of the evolutionary trends associated with skull structure by studying reproductions of the skulls of several ancestors and relatives of modern humans.

MATERIALS

Per laboratory room:

- 1 set of skull reproductions and replicas, including *Dryopithecus africanus, Australopithecus afarensis, Australopithecus africanus, Paranthropus robustus, Homo habilis, Homo erectus, Homo neandertalensis, Homo sapiens,* and chimpanzee (*Pan troglodytes*)
- collection of fossil plants and animals (optional)
- guide to fossils (optional)
- wall chart of human evolution (optional)

PROCEDURE

1. If they are available, examine the fossil collection and field guide to fossils. Note which fossils are forms that lived in the seas and which lived on land.
   (a) Do any of the fossils resemble organisms now living? If so, describe the similarities between fossil and current forms.

   ____

   ____

   ____

(b) Are there fossils in the collection that are unlike any organisms currently living? If so, describe the features that appear to be most unlike today's forms.

____

____

____

2. Examine the hominid skulls in the laboratory. Note that these skulls are not actual fossils themselves. Instead, each skull is a plaster or plastic restoration designed partly from fossilized remains and partly from reconstruction. The *Homo sapiens* and chimpanzee skulls are replicas of a modern specimen.

   The skulls are labeled with the scientific name of the organism, approximate date of origin, and cranial capacity. As you can see, some of these hominids appear to have lived at the same approximate time in history.

   Study the skulls and Figure 14-4. In Table 14-3, record your observations of the changes that occurred over time. Rate the following characteristics for each skull using this scale: (– – –) for the most primitive and (+ + +) for the most advanced.

Characteristic	Primitive	Advanced
Teeth	Many large, specialized	Fewer, smaller, less specialized
Jaw	Jaw and muzzle long	Jaw and muzzle short
Cranium (braincase)	Cranium small, forehead receding	Cranium large, prominent vertical forehead
Muscle attachments	Prominent eyebrow ridges and cranial keel (ridge)	Much reduced eyebrow ridges, no keel
Nose	Nose not protruding from face	Prominent nose with distinct bridge

3. Examine the skulls of *Dryopithecus* and the australopiths more closely. Scientists believe that apelike forms like *Dryopithecus* may have been probably the common ancestors of modern apes and humans.

The first true hominids found in the fossil record have been assigned to the genus *Australopithecus* (from the Greek *australis* for "southern" and *pithecus* for "ape," because the first specimens were found in southern Africa). Some of the oldest known fossils of *Australopithecus* are of the species *A. afarensis* (the most famous skeleton has been named Lucy). Lucy and her conspecifics (members of the same species) lived approximately 3.5 million years ago and may have been the ancestors to the later forms of *Australopithecus*. (See Figure 14-5.)

(a) Study the skull of *Dryopithecus* and the skull of *A. afarensis*. What resemblances do you see between the two specimens?

____

____

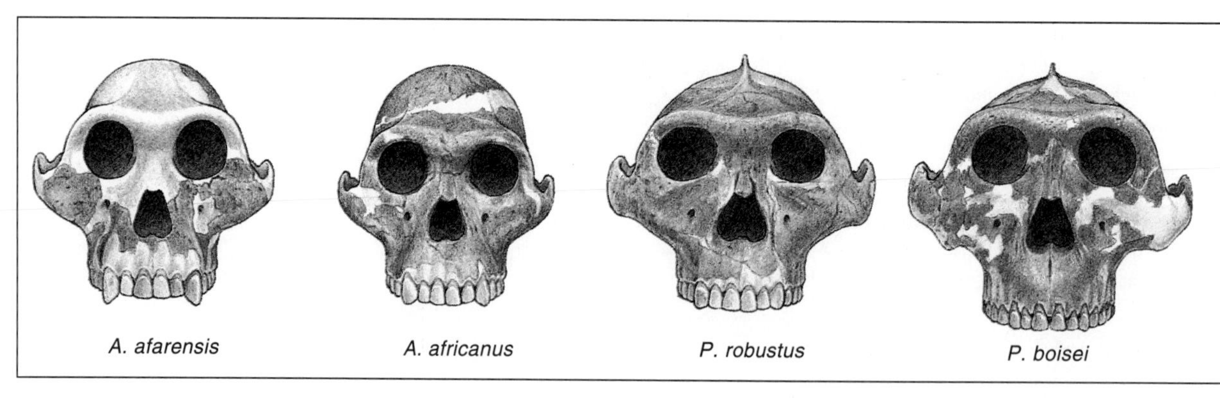

Figure 14-4 Comparison of skull shapes of human ancestors and relatives. (After Starr and Taggart, 1992.)

TABLE 14-3 Skull Characteristics of Human Ancestors and Relatives, from (– – –) = Most Primitive to (+ + +) = Most Advanced

Genus and Species	Teeth	Jaw	Structure	Cranium	Muscle	Attachments	Nose
Dryopithecus							
Australopithecus afarensis							
Australopithecus africanus							
Paranthropus robustus							
Homo habilis							
Homo erectus							
Homo neandertalensis							
Homo sapiens							
Pan troglodytes							

(b) What structural differences do you see between the two skull specimens?

____

____

(c) What advances do you see in skull characteristics between *Dryopithecus* and *Australopithecus africanus*?

____

____

(d) Between *Dryopithecus* and *Paranthropus robustus*?

____

____

The genetic line(s) of *Australopithecus africanus* and *P. robustus* became extinct approximately 2–1.5 million years ago. There is clear evidence, though, that these hominids were fully bipedal, though not as upright in stance as modern humans. These australopiths also probably used unworked stones and pieces of wood as tools.

4. Now turn your attention to the skulls of the *Homo* lineage. There is debate also about the lineage of our own genus, *Homo* ("Man"). *Homo habilis,* often dubbed simply "early *Homo,*" lived in eastern and southern Africa from about 2.5–1.6 million years ago. *Homo erectus* ("upright man") also arose in Africa. However, it was *Homo erectus* whose populations left Africa in waves between 2 million and 500,000 years ago. Our own species, *Homo sapiens* ("wise man"), arose from other *Homo* ancestors.

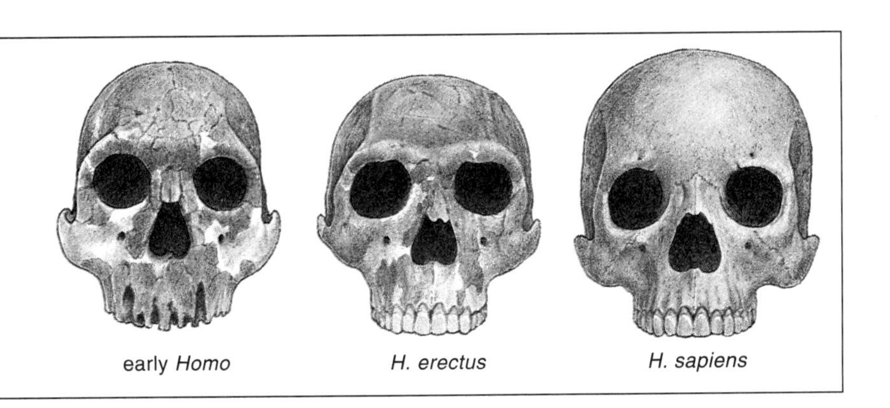

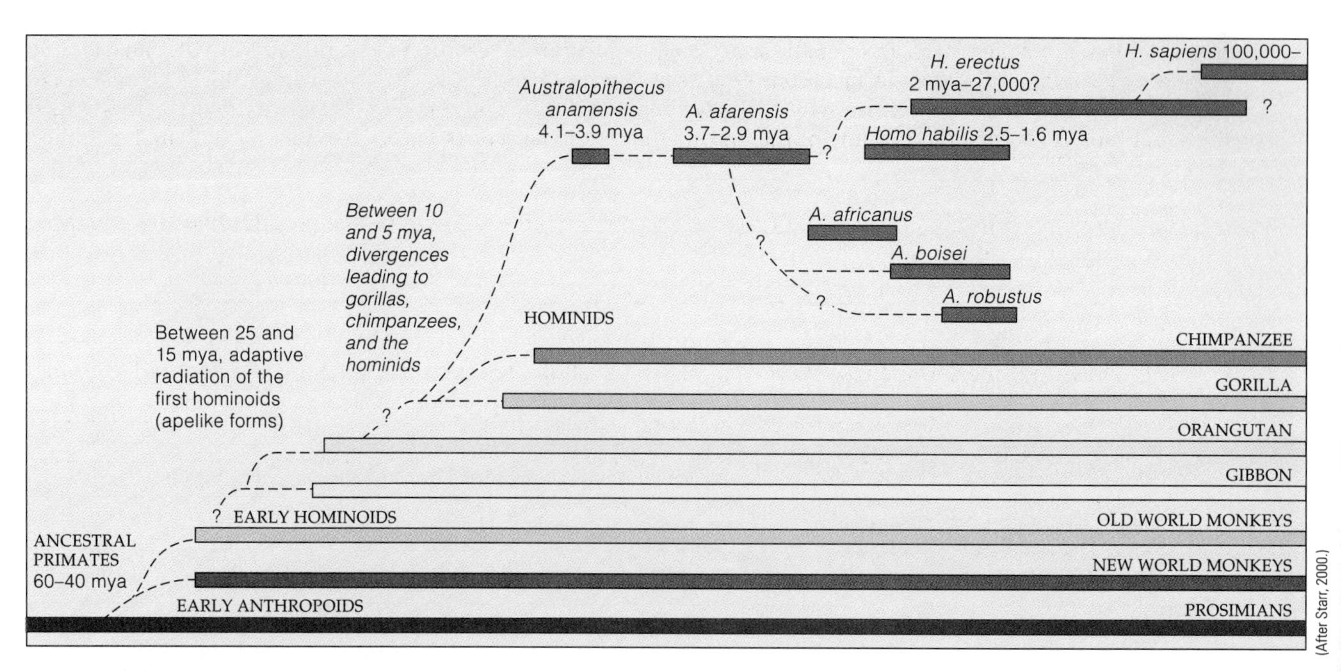

Figure 14-5 Timeline for appearance and extinction of major hominid species.

(a) What changes from primitive to advanced characteristics do you see in the evolutionary line from *A. afarensis* to *Homo sapiens*?

____

____

____

(b) What is the trend in cranial capacity between the ancestral *A. afarensis* and *Homo erectus*?

____

____

(c) Between *Homo erectus* and *Homo sapiens*?

____

____

Fossils of the Neandertals (*Homo neandertalensis*) are known from Europe, the Near East, and China. These humans hunted and gathered food, made many kinds of tools, and buried their dead, which has been interpreted as showing their capability for abstract and religious thought. The Neandertals disappeared mysteriously 40,000–35,000 years ago, soon after the modern form of *Homo sapiens* arrived from Africa in their range.

(d) What differences do you see between the skulls of Neandertal man and modern humans?

____

____

(e) How do the cranial capacities of Neandertal man and modern humans compare?

____

____

5. If one is available, study the wall chart in your lab room showing the course of human evolution. Pay particular attention to the conceptual drawings of the appearance of the hominids and humans, and to the evolution of tool making and cultural development.
6. Finally, study the skulls of the modern human form (*Homo sapiens*) and our nearest living relative, the chimpanzee (*Pan troglodytes*).

Fossil evidence, biochemistry, and genetic analyses indicate that chimpanzees and humans are the most closely related of all living primates. In fact, comparisons of amino acid sequences of proteins and DNA sequences show that chimpanzees and humans share about 99% of their genes! Other evidence shows that the separation from the hominid line of the lineage that led to chimpanzees occurred no more than 4–6 million years ago.

(a) How does the chimpanzee skull compare to the human skull with respect to primitive and advanced anatomical features?

____

____

(b) How does the chimpanzee skull compare to the skull of *Dryopithecus,* the apelike hominid ancestor?

____

____

PRE-LAB QUESTIONS

_____ 1. Evolution is
(a) the difference in survival and reproduction of a population
(b) the process that results in changes in the genetic makeup of a population over time
(c) a group of individuals of the same species occupying a given area
(d) change in an individual's genetic makeup over its lifetime

_____ 2. The major mechanism of evolution is
(a) adaptive radiation
(b) drift
(c) fitness
(d) natural selection

_____ 3. An organism whose genetic makeup allows it to produce more offspring than another of its species is said to have greater
(a) fitness
(b) evolution
(c) adaptive radiation
(d) selection

_____ 4. The earth was formed approximately
(a) 4600 years ago
(b) 1 million years ago
(c) 1 billion years ago
(d) 4.6 billion years ago

_____ 5. The first living organisms appeared on the earth approximately
(a) 4.6 billion years ago
(b) 3.8 billion years ago
(c) 1 billion years ago
(d) 6400 years ago

_____ 6. A catastrophic global event in which major groups of species disappear is called
(a) adaptive radiation
(b) natural selection
(c) mass extinction
(d) evolution

_____ 7. Fossils
(a) are remains of organisms
(b) are formed when organic materials are replaced with minerals
(c) provide evidence of evolution
(d) are all of the above

_____ 8. The organism believed to be the common ancestor of modern apes and humans is
(a) *Australopithecus africanus*
(b) *Homo habilis*
(c) the chimpanzee
(d) *Dryopithecus*

_____ 9. Hominids with evolutionarily advanced skull characteristics would have
(a) many large, specialized teeth
(b) a large cranium with prominent vertical forehead
(c) prominent eyebrow ridges
(d) a long jaw and muzzle

_____ 10. The human ancestor whose populations dispersed from Africa to other parts of the world was
(a) *Homo sapiens*
(b) *Australopithecus afarensis*
(c) *Dryopithecus*
(d) *Homo erectus*

Name ______________________________ Section Number ______________________

EXERCISE 14

Evidences of Evolution

POST-LAB QUESTIONS

Introduction

1. Define the following terms.
   a. evolution

   b. natural selection

14.1 Experiment: Natural Selection

2. Describe how natural selection would operate to change the genetic makeup of a dandelion population that had been growing in an unmowed area if that area became subject to frequent low mowing.

14.2 Geologic Time

3. Which event occurred first in the earth's history?
   a. Photosynthesis or eukaryotic cells? ______________________
   b. Vertebrate animals or flowering plants? ______________________
   c. Extinction of dinosaurs or origin of hominids? ______________________

14.3 The Fossil Record and Human Evolution

4. What primitive characteristics are visible in the skull pictured here?

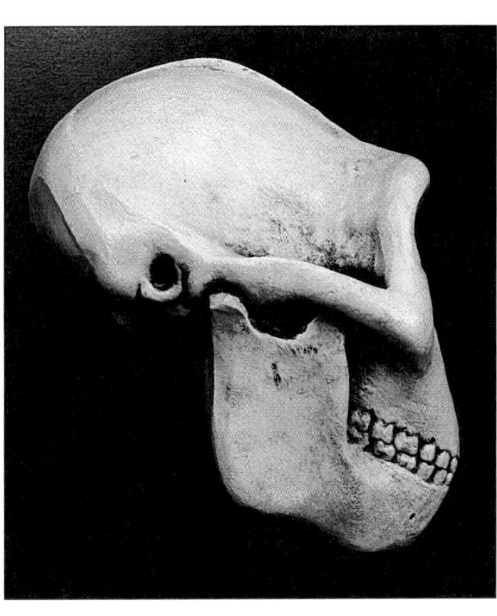

(Photo by J. W. Perry.)

5. Describe anatomical changes in the skull that occurred in human evolution between a *Dryopithecus*-like ancestor and *Homo sapiens*.

6. Compare the slope of the forehead of chimpanzees with that of modern humans.

7. Compare the teeth of *Australopithecus afarensis* and *Homo sapiens.*
   a. Describe similarities and differences between the two species.

   b. Write a hypothesis about dietary differences between the two species based on their teeth.

Food for Thought

8. Why was the development of bipedalism a major advancement in human evolution?

9. Do you believe humans are still evolving? If not, why not? If yes, explain in what ways humans may be evolving.

10. Humans have selectively bred many radically different domestic animals (for example, St. Bernard and chihuahua dog breeds). Does this activity result in evolution? Why or why not?

EXERCISE 15

Taxonomy: Classifying and Naming Organisms

OBJECTIVES

After completing this exercise, you will be able to

1. define *common name, scientific name, binomial, genus, specific epithet, species, taxonomy, phylogenetic system, dichotomous key, herbarium;*
2. distinguish common names from scientific names;
3. explain why scientific names are preferred over common names in biology;
4. identify the genus and specific epithet in a scientific binomial;
5. write out scientific binomials in the form appropriate to the Linnean system;
6. construct a dichotomous key;
7. explain the usefulness of an herbarium;
8. use a dichotomous key to identify plants, animals, or other organisms as provided by your instructor.

INTRODUCTION

We are all great classifiers. Every day, we consciously or unconsciously classify and categorize the objects around us. We recognize an organism as a cat or a dog, a pine tree or an oak tree. But there are numerous kinds of oaks, so we refine our classification, giving the trees distinguishing names such as "red oak," "white oak," or "bur oak." These are examples of **common names,** names with which you are probably most familiar.

Scientists are continually exchanging information about living organisms. But not all scientists speak the same language. The common name "white oak," familiar to an American, is probably not familiar to a Spanish biologist, even though the tree we know as white oak may exist in Spain as well as in our own backyard. Moreover, even within our own language, the same organism can have several common names. For example, within North America a gopher is also called a ground squirrel, a pocket mole, and a groundhog. On the other hand, the same common name may describe many different organisms; there are more than 300 different trees called "mahogany"! To circumvent the problems associated with common names, biologists use **scientific names** that are unique to each kind of organism and that are used throughout the world.

A scientific name is two-parted, a binomial. The first word of the binomial designates the group to which the organism belongs; this is the **genus** name (the plural of genus is *genera*). All oak trees belong to the genus *Quercus,* a word derived from Latin. Each kind of organism within a genus is given a **specific epithet.** Thus, the scientific name for white oak is *Quercus alba* (specific epithet is *alba*), while that of bur oak is *Quercus macrocarpa* (specific epithet is *macrocarpa*).

Notice that the genus name is always capitalized; the specific epithet usually is not capitalized (although it can be if it is the proper name of a person or place). The binomial is written in *italics* (since these are Latin names); if italics are not available, the genus name and specific epithet are underlined.

You will hear discussion of "species" of organisms. For example, on a field trip, you may be asked "What species is this tree?" Assuming you are looking at a white oak, your reply would be "*Quercus alba.*" The scientific name of the **species** includes *both* the genus name and specific epithet.

If a species is named more than once within textual material, it is accepted convention to write out the full genus name and specific epithet the first time and to abbreviate the genus name every time thereafter. For example, if white oak is being described, the first use is written *Quercus alba,* and each subsequent naming appears as *Q. alba.*

Similarly, when a number of species, all of the same genus, are being listed, the accepted convention is to write both the genus name and specific epithet for the first species and to abbreviate the genus name for each species listed thereafter. Thus, it is acceptable to list the scientific names for white oak and bur oak as *Quercus alba* and *Q. macrocarpa,* respectively.

Taxonomy is the science of classification (categorizing) and nomenclature (naming). Biologists prefer a system that indicates the evolutionary relationships among organisms. To this end, classification became a **phylogenetic system;** that is, one indicating the presumed evolutionary ancestry among organisms.

Current taxonomic thought separates all living organisms into six kingdoms:

- Kingdom Bacteria (prokaryotic cells that include pathogens)
- Kingdom Archaea (prokaryotic organisms that are evolutionarily closer to eukaryotes than bacteria)
- Kingdom Protista (euglenids, chrysophytes, diatoms, dinoflagellates, slime molds, and protozoans)
- Kingdom Fungi (fungi)
- Kingdom Plantae (plants)
- Kingdom Animalia (animals)

Let's consider the scientific system of classification, using ourselves as examples. All members of our species belong to

- Kingdom Animalia (animals)
- Phylum Chordata (animals with a notochord)
- Class Mammalia (animals with mammary glands)
- Order Primates (mammals that walk upright on two legs)
- Family Hominidae (human forms)
- Genus *Homo* (mankind)
- Specific epithet *sapiens* (wise)
- Species: *Homo sapiens*

The more closely related evolutionarily two organisms are, the more categories they share. You and I are different individuals of the same species. We share the same genus and specific epithet, *Homo* and *sapiens.* A creature believed to be our closest extinct ancestor walked the earth 1.5 million years ago. That creature shared our genus name but had a different specific epithet, *erectus.* Thus, *Homo sapiens* and *H. erectus* are *different* species.

Like all science, taxonomy is subject to change as new information becomes available. Modifications are made to reflect revised interpretations.

15.1 Constructing a Dichotomous Key *(About 45 min.)*

To classify organisms, you must first identify them. A *taxonomic key* is a device for identifying an object unknown to you but that someone else has described. The user chooses between alternative characteristics of the unknown object and, by making the correct choices, arrives at the name of the object.

Keys that are based on successive choices between two alternatives are known as **dichotomous keys** (*dichotomous* means "to fork into two equal parts"). When using a key, always read both choices, even though the first appears to describe the subject. Don't guess at measurements; use a ruler. Since living organisms vary in their characteristics, don't base your conclusion on a single specimen if more are available.

MATERIALS

Per lab room:

- several meter sticks or metric height charts taped to a wall

PROCEDURE

1. Suppose the geometric shapes below have unfamiliar names. Look at the dichotomous key following the figures. Notice there is a 1a and a 1b. Start with 1a. If the description in 1a fits the figure you are observing better than description 1b, then proceed to the choices listed under 2, as shown at the end of line 1a. If 1a does *not* describe the figure in question, 1b does. Looking at the end of line 1b, you see that the figure would be called an Elcric.
2. Using the key provided, determine the hypothetical name for each object. Write the name beneath the object and then check with your instructor to see if you have made the correct choices.

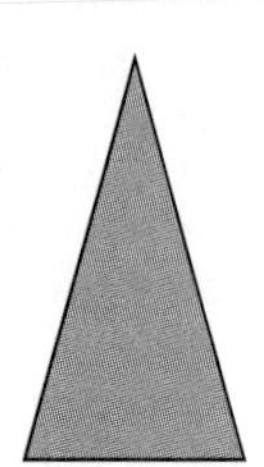

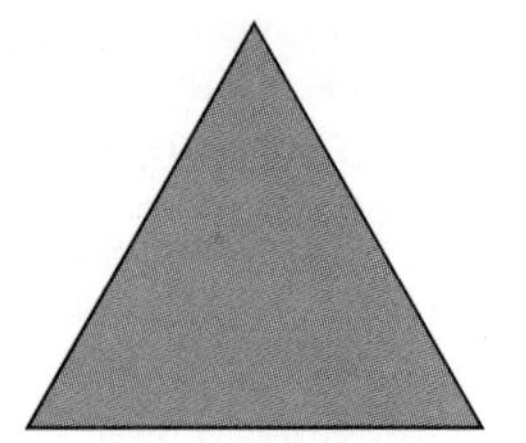
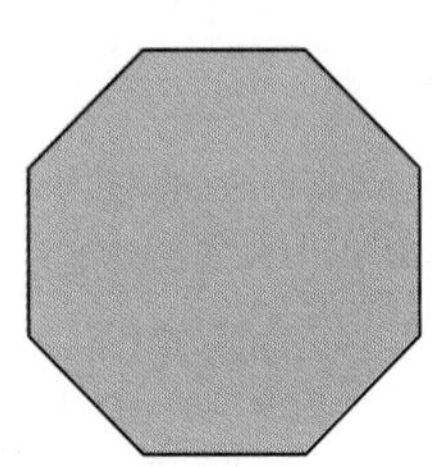

______ ______ ______ ______ ______

Key		
1a.	Figure with distinct corners	2
1b.	Figure without distinct corners	Elcric
2a.	Figure with 3 sides	3
2b.	Figure with 4 or more sides	4
3a.	All sides of equal length	Legnairt
3b.	Only 2 sides equal	Legnairtosi
4a.	Figure with only right angles	Eraqus
4b.	Figure with other than right angles	Nogatco

3. Now you will construct a dichotomous key, using your classmates as subjects. The class should divide up into groups of eight (or as evenly as the class size will allow). Working with the individuals in your group, fill in Table 15-1, measuring height with a metric ruler or the scale attached to the wall.
4. To see how you might plan a dichotomous key, examine the following branch diagram. If there are both men and women in a group, the most obvious first split is male/female (although other possibilities for the split could be chosen as well). Follow the course of splits for two of the men in the group.

   Note that each choice has *only* two alternatives. Thus, we split into "under 1.75 m" and "1.75 m or taller." Likewise, our next split is into "blue eyes" and "nonblue eyes" rather than all the possibilities.

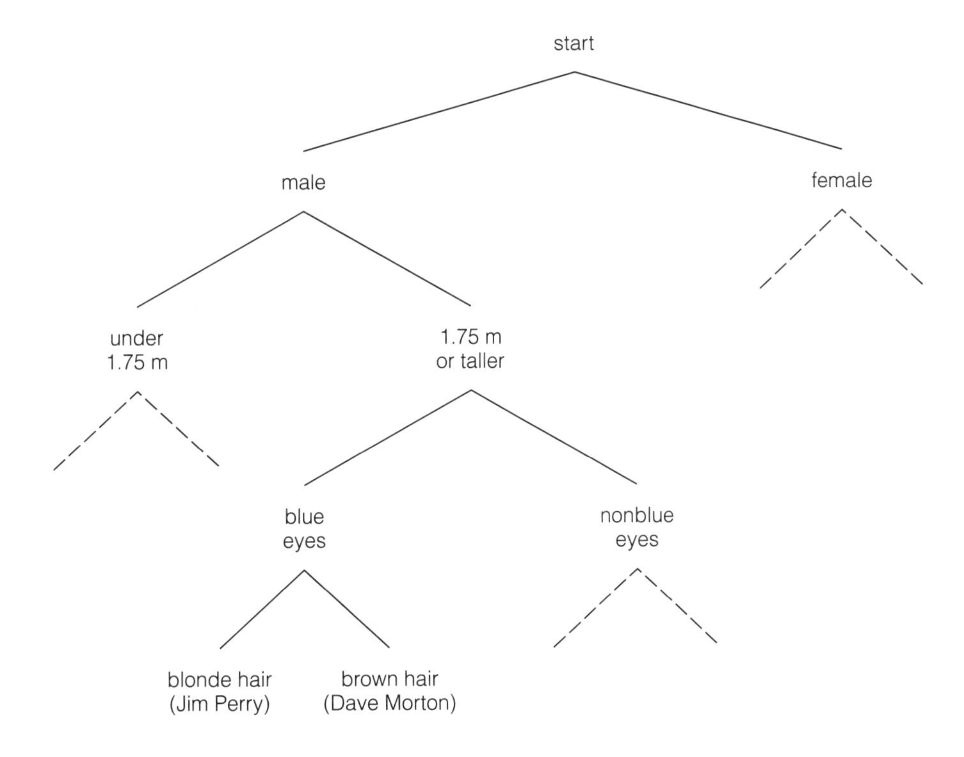

5. On a separate sheet of paper, construct a branch diagram for your group using the characteristics in Table 15-1 and then condense it into the dichotomous key that follows. When you have finished, exchange your key with that of an individual in another group. Key out the individuals in the other group without speaking until you believe you know the name of the individual you are examining. Ask that individual if you are correct. If not, go back to find out where you made a mistake, or possibly where the key was misleading. (Depending on how you construct your key, you may need more or fewer lines than have been provided.)

TABLE 15-1 Characteristics of Students

Student (name)	Sex (m/f)	Height (m)	Eye Color	Hair Color	Shoe Size
1.					
2.					
3.					
4.					
5.					
6.					
7.					
8.					

Key to Students in Group ____________________
1a. 1b.
2a. 2b.
3a. 3b.
4a. 4b.
5a. 5b.
6a. 6b.
7a. 7b.
8a. 8b.

15.2 Using a Taxonomic Key

A. Some Microscopic Members of the Freshwater Environment (About 30 min.)

Suppose you want to identify the specimens in some pond water. The easiest way is to key them out with a dichotomous key, now that you know how to use one. In this section, you will do just that.

MATERIALS

Per student:

- compound microscope
- microscope slide
- coverslip
- dissecting needle

Per student group (table):

- cultures of freshwater organisms
- 1 disposable plastic pipet per culture
- methylcellulose in dropping bottle

PROCEDURE

1. Obtain a clean glass microscope slide and clean coverslip.
2. Using a disposable plastic pipet or dissecting needle, withdraw a small amount of the culture provided.
3. Place *one* drop of the culture on the center of the slide.
4. Gently lower the coverslip onto the liquid.
5. Using your compound light microscope, observe your wet mount. Focus first with the low-power objective and then with the medium or high-dry objective, depending on the size of the organism in the field of view.
6. Concentrate your observation on a single specimen, keying out the specimen using the Key to Selected Freshwater Inhabitants that follows.
7. In the space provided, write the scientific name of each organism you identify. After each identification, have your instructor verify your conclusion.
8. Clean and reuse your slide and coverslip after each identification.

Key to Selected Freshwater Inhabitants

1a.	Filamentous organism consisting of green, chloroplast-bearing threads	2
1b.	Organism consisting of a single cell or nonfilamentous colony	4
2a.	Filament branched, each cell mostly filled with green chloroplast	*Cladophora*
2b.	Filament unbranched	3
3a.	Each cell of filament containing 1 or 2 spiral-shaped green chloroplasts	*Spirogyra*
3b.	Each cell of filament containing 2 star-shaped green chloroplasts	*Zygnema*
4a.	Organism consisting of a single cell	5
4b.	Organism composed of many cells aggregated into a colony	6
5a.	Motile, teardrop-shaped or spherical organism	*Chlamydomonas*
5b.	Nonmotile, elongate cell on either end; clear, granule-containing regions at ends	*Closterium*
6a.	Colony a hollow round ball of more than 500 cells; new colonies may be present inside larger colony	*Volvox*
6b.	Colony consisting of less than 50 cells	7
7a.	Organism composed of a number of tooth-shaped cells	*Pediastrum*
7b.	Colony a loose square or rectangle of 4–32 spherical cells	*Gonium*

Organism 1 is ______________________
Organism 2 is ______________________
Organism 3 is ______________________
Organism 4 is ______________________
Organism 5 is ______________________
Organism 6 is ______________________
Organism 7 is ______________________
Organism 8 is ______________________

B. Common Trees and Shrubs (About 1 hour)

Suppose you want to identify the trees growing on your campus or in your yard at home. Without having an expert present, you can now do that, because you know how to use a taxonomic key. But how can you be certain that you have keyed your specimen correctly?

Typically, scientists compare their tentative identifications against *reference specimens*—that is, preserved organisms that have been identified by an expert *taxonomist* (a person who names and classifies organisms). If you are identifying fishes or birds, the reference specimen might be a bottled or mounted specimen with the name on it. In the case of plants, reference specimens most frequently take the form of *herbarium mounts* (Figure 15-1) of the plants. An **herbarium** (plural, *herbaria*) is a repository, a museum of sorts, of preserved plants. The taxonomist flattens freshly collected specimens in a plant press. They are then dried and mounted on sheets of paper. Herbarium labels are affixed to the sheets, indicating the scientific name of the plant, the person who collected it, the location and date of collection, and often pertinent information about the habitat in which the plant was found.

It is likely that your school has an herbarium. If so, your instructor may show you the collection. To some, this endeavor may seem boring, but herbaria serve a critical function. The appearance or disappearance of plants from the landscape often gives a very good indication of environmental change. An herbarium records the diversity of plants in the area, at any point in history since the start of the collection.

Label indicates name of specimen, site and date of collection, associated species at same site, name(s) of collector(s)

(Photo by J. W. Perry)

Figure 15-1 A typical herbarium mount.

MATERIALS

Per student group (table):

- set of 8 tree twigs with leaves (fresh or herbarium specimens) *or*
- trees and shrubs in leafy condition (for an outdoor lab)

PROCEDURE

Use the appropriate following key to identify the tree and shrub specimens that have been provided in the lab or that you find on your campus. Refer to the *Glossary to Accompany Tree Key* (pages 224–225) and Figures 15-2 through 15-9 (pages 219–220) when you encounter an unfamiliar term. When you have finished keying a specimen, confirm your identification by checking the herbarium mounts or asking your instructor.

Note: **Some descriptions within the key have more characteristics than your specimen will exhibit. For example, the key may describe a fruit type when the specimen doesn't have a fruit on it. However, other specimen characteristics are described, and these should allow you to identify the specimen.**

Note: **The keys provided are for *selected* trees of your area. In nature, you will find many more genera than can be identified by these keys.**

Common names within parentheses follow the scientific name. A metric ruler is provided on page 220 for use where measurements are required.

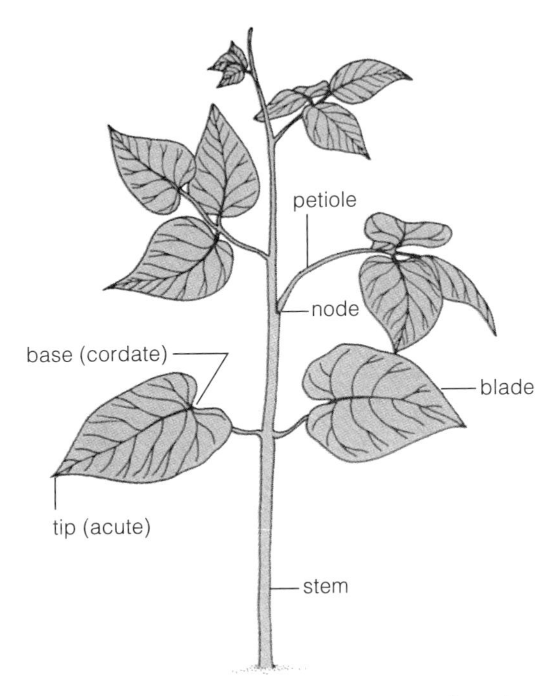

Figure 15-2 Structure of a typical plant (bean).

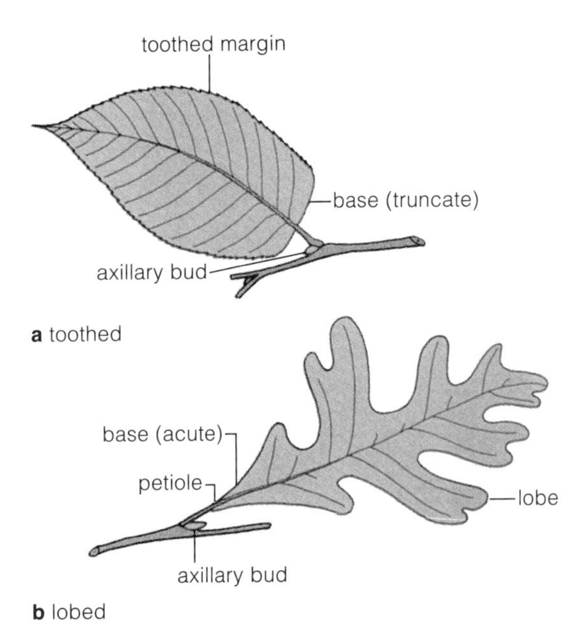

Figure 15-3 Simple leaves.

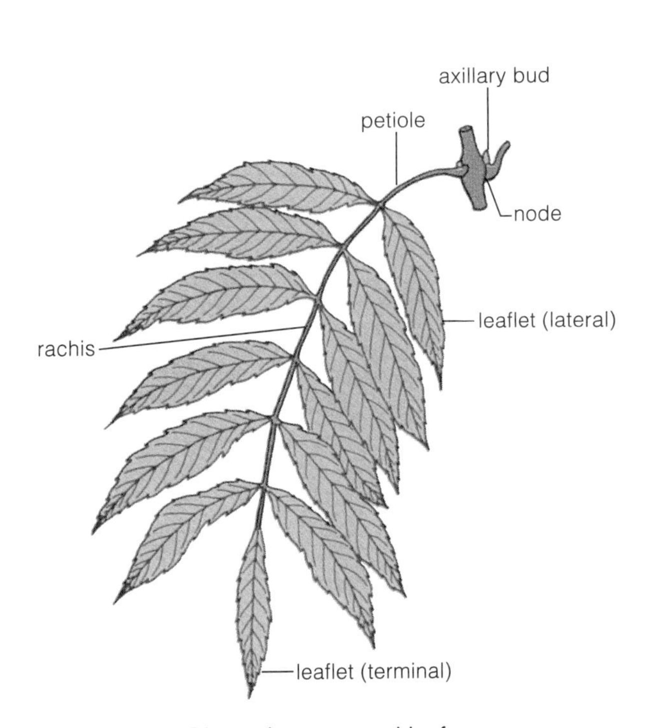

Figure 15-4 Pinnately compound leaf.

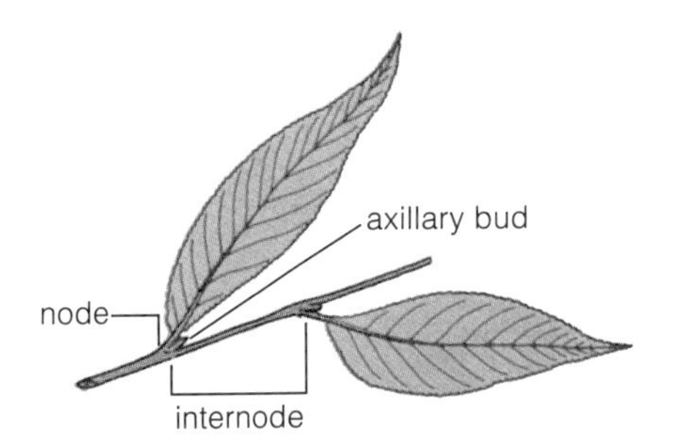

Figure 15-5 Simple leaves—alternating.

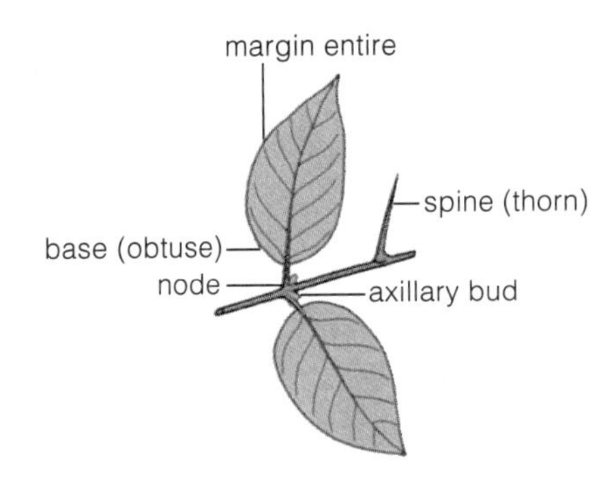

Figure 15-6 Simple leaves—opposite.

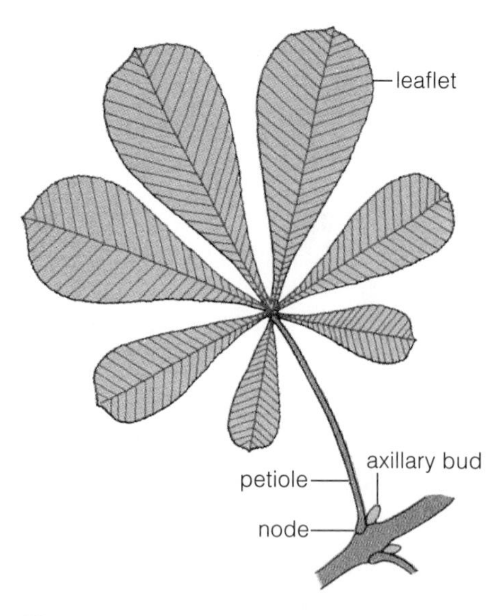

Figure 15-7 Palmately compound leaf.

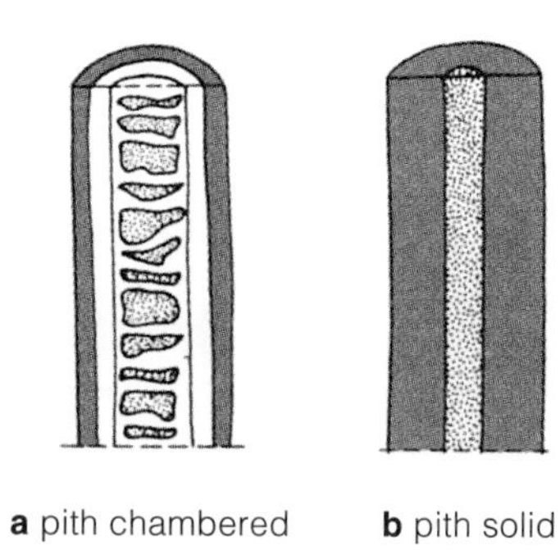

Figure 15-8 Pith types.

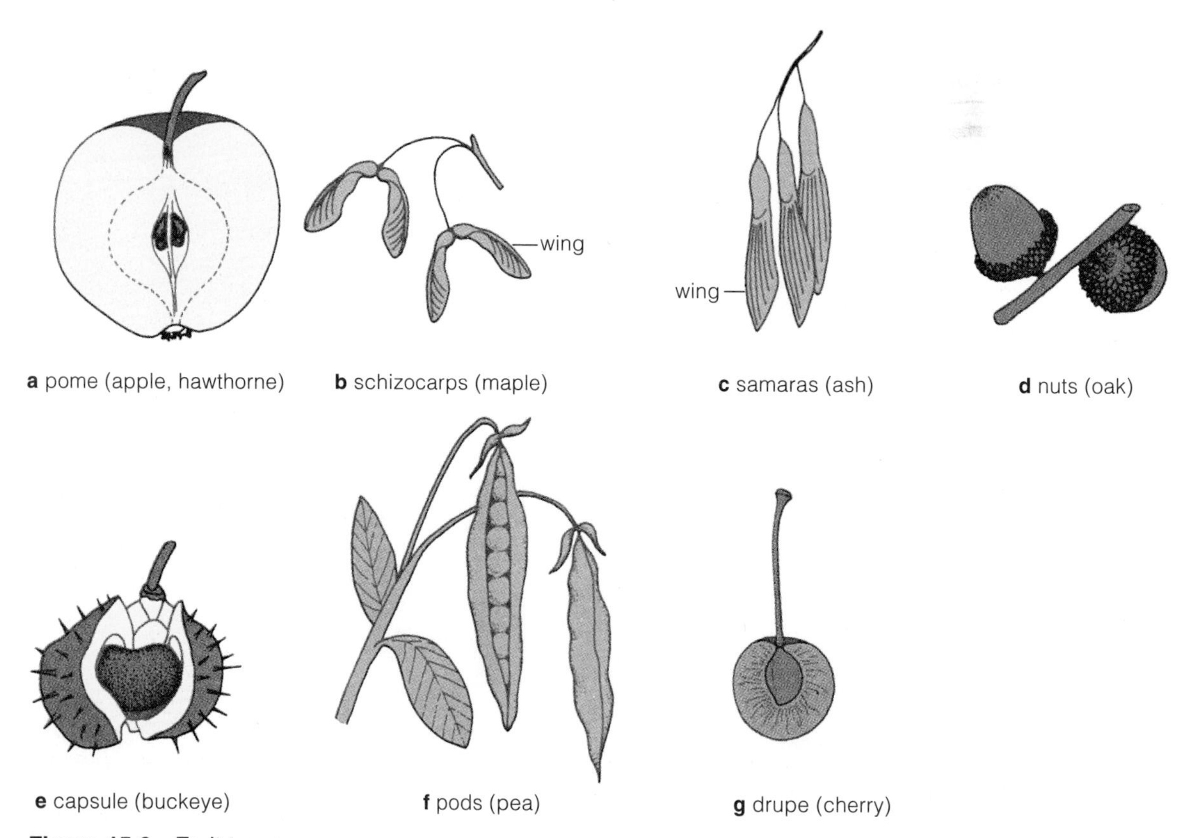

Figure 15-9 Fruit types.

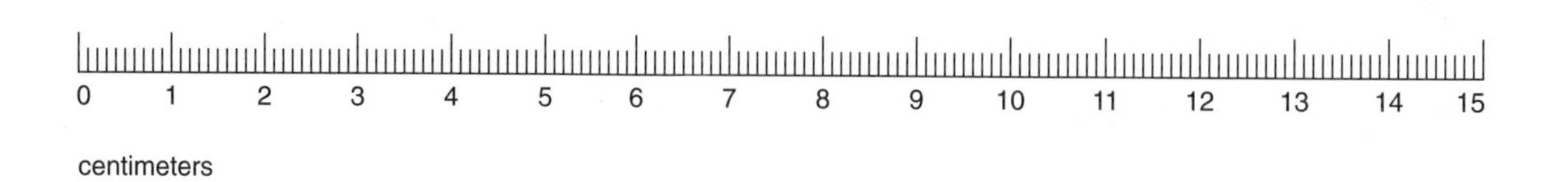

Key to Some Common Genera of Trees of the Midwestern and Eastern United States and Canada		
1a.	Leaves broad and flat; plants producing flowers and fruits (angiosperms)	2
1b.	Leaves needlelike or scalelike; plants producing cones, but no flowers or fruits (gymnosperms)	22
2a.	Leaves compound	3
2b.	Leaves simple	9
3a.	Leaves alternate	4
3b.	Leaves opposite	7
4a.	Leaflets short and stubby, less than twice as long as broad; branches armed with spines or thorns; fruit a beanlike pod	5
4b.	Leaflets long and narrow, more than twice as long as broad; trunk and branches unarmed; fruit a nut	6
5a.	Leaflet margin without teeth; terminal leaflet present; small deciduous spines at leaf base	*Robinia* (black locust)
5b.	Leaflet margin with fine teeth; terminal leaflet absent; large permanent thorns on trunk and branches	*Gleditsia* (honey locust)
6a.	Leaflets usually numbering less than 11; pith of twigs solid	Carya (hickory)
6b.	Leaflets numbering 1 or more, pith of twigs divided into chambers	*Juglans* (walnut, butternut)
7a.	Leaflets pinnately arranged; fruit a light-winged samara	8
7b.	Leaflets palmately arranged; fruit a heavy leathery spherical capsule	*Aesculus* (buckeye)
8a.	Leaflets numbering mostly 3–5; fruit a schizocarp with curved wings	*Acer* (box elder)
8b.	Leaflets numbering mostly more than 5; samaras borne singly, with straight wings	*Fraxinus* (ash)
9a.	Leaves alternate	10
9b.	Leaves opposite	21
10a.	Leaves very narrow, at least 3 times as long as broad; axillary buds flattened against stem	*Salix* (willow)
10b.	Leaves broader, less than 3 times as long as broad	11
11a.	Leaf margin without small, regular teeth	12
11b.	Leaf margin with small, regular teeth	13
12a.	Fruit a pod with downy seeds; leaf blade obtuse at base; petioles flattened, or if rounded, bark smooth	*Populus* (poplar, popple, aspen)
12b.	Fruit an acorn; leaf blade acute at the base; petioles rounded; bark rough	*Quercus* (oaks)
13a.	Leaves (at least some of them) with lobes or other indentations in addition to small, regular teeth	14
13b.	Leaves without lobes or other indentations except for small, regular teeth	16
14a.	Lobes asymmetrical, leaves often mitten-shaped	*Morus* (mulberry)
14b.	Lobes or other indentations fairly symmetrical	15

15a.	Branches thorny (armed); fruit a small applelike pome	*Crataegus* (hawthorne)
15b.	Branches unarmed	17
16a.	Bark smooth and waxy, often separating into thin layers; leaf base symmetrical	*Betula* (birch)
16b.	Bark rough and furrowed, leaf base asymmetrical	*Ulmus* (elm)
17a.	Leaf base asymmetrical, strongly heart-shaped, at least on one side	*Tilia* (basswood or linden)
17b.	Leaf base acute, truncate, or slightly cordate	18
18a.	Leaf base asymmetrical; bark on older stems (trunk) often warty	*Celtis* (hackberry)
18b.	Leaf base symmetrical	19
19a.	Leaf blade usually about twice as long as broad, generally acute at the base; fruit fleshy	20
19b.	Leaf not much longer than broad, generally truncate at base; fruit a dry pod	*Populus* (poplar, popple, aspen)
20a.	Leaf tapering to a pointed tip, glandular at base	*Prunus* (cherry)
20b.	Leaf spoon-shaped with a rounded tip, no glands at base	*Crataegus* (hawthorne)
21a.	Leaf margins with lobes and points, fruit a schizocarp	*Acer* (maple)
21b.	Leaf margins without lobes or points; fruit a long capsule	*Catalpa* (catalpa)
22a.	Leaves needlelike, with 2 or more needles in a cluster	23
22b.	Leaves needlelike or scalelike, occurring singly	24
23a.	Leaves more than 5 in a cluster, soft, deciduous, borne at the ends of conspicuous stubby branches	*Larix* (larch, tamarack)
23b.	Leaves 2–5 in a cluster	*Pinus* (pines)
24a.	Leaves soft, not sharp to the touch	25
24b.	Leaves stiff or sharp and often unpleasant to touch	27
25a.	Leaves about 0.2 cm and scalelike, overlapping	*Thuja* (white cedar, arbor vitae)
25b.	Leaves needlelike, appear to form two ranks on twig	26
26a.	Leaves with distinct petioles, 0.8–1.5 cm long; twigs rough; female cones drooping from branches	*Tsuga* (hemlock)
26b.	Leaves without distinct petioles, 1–3 cm long; twigs smooth; female cones erect on branches	*Abies* (firs)
27a.	Leaves appear triangular-shaped, about 0.5 cm, and tightly pressed to twig; cone blue, berrylike	*Juniperus* (juniper, Eastern red cedar)
27b.	Leaves elongated and needlelike	28
28a.	Tree; leaves 4-sided, protrude stiffly from twig; female cones droop from branch	*Picea* (spruces)
28b.	Shrub; leaves flattened, pressed close to twig at base; seed partially covered by a fleshy coat, usually red	*Taxus* (yew)

Key to Some Common Genera of Trees of the Pacific Region of the United States and Canada		
1a.	Leaves broad and flat; plants producing flowers and fruits (angiosperms)	2
1b.	Leaves needlelike or scalelike; plants producing cones, but no flowers or fruits (gymnosperms)	15
2a.	Leaves compound	3
2b.	Leaves simple	6
3a.	Leaves pinnately arranged	4
3b.	Leaves palmately arranged	5
4a.	Leaflets number 7, fruit a samara	*Fraxinus* (ash)
4b.	Leaflets 15–17, fruit a nut	*Juglans* (walnut)
5a.	Leaflets numbering 3, lobed; fruit a schizocarp	*Acer* (box elder)
5b.	Leaflets numbering more than 3; fruit a smooth or spiny capsule	*Aesculus* (buckeye)
6a.	Three or more equal-sized veins branching from leaf base	7
6b.	Leaf with single large central vein with other main veins branching from the central vein	9
7a.	Leaves opposite; fruit a schizocarp	*Acer* (maple)
7b.	Leaves alternate	8
8a.	Leaves nearly round in outline; fruit a pod	*Cercis* (redbud)
8b.	Leaves deeply lobed, very hairy beneath; fruit consisting of an aggregation of many 1-seeded nutlets surrounded by long hairs	*Platanus* (sycamore)
9a.	Leaf lobed; fruit a nut	*Quercus* (oak)
9b.	Leaf not lobed	10
10a.	Leaves opposite; fruit a drupe; halves of leaf remain attached by "threads" after blade has been creased and broken	*Cornus* (dogwood)
10b.	Leaves alternate	11
11a.	Upon crushing, blade gives off strong, penetrating odor	*Eucalyptus* (eucalyptus)
11b.	Blade not strongly odiferous upon crushing	12
12a.	Branch bark smooth, conspicuously red-brown; fruit a red ororange berry	*Arbutus* (madrone)
12b.	Branch rough, not colored red-brown	13
13a.	Undersurface of leaves golden-yellow; fruit a spiny, husked nut	*Catanopsis* (golden chinquapin)
13b.	Leaves green beneath	14
14a.	Petiole hairy; leathery blade with a stubby spine at end of each main vein; fruit a nut	*Lithocarpus* (tanoak)
14b.	Petiole and leaf lacking numerous hairs, leaves long and narrow, more than twice as long as wide	*Salix* (willow)

15a.	Leaves needlelike	16
15b.	Leaves scalelike	23
16a.	Leaves needlelike, with 2 or more needles in a cluster	17
16b.	Leaves needlelike, occurring singly	18
17a.	Needles 2–5 in a cluster	*Pinus* (pine)
17b.	Needles 6 or more per cluster, soft, deciduous, borne at the ends of stubby, conspicuous branches	*Larix* (larch)
18a.	Round scars on twigs where old needles have fallen off; twigs smooth; needles soft to the grasp; cones pointing upward with reference to stem	*Abies* (fir)
18b.	Twigs rough, with old needle petioles remaining	19
19a.	Needles angled, stiff, sharp, pointed, unpleasant to grasp; cones hanging downward from branch	*Picea* (spruce)
19b.	Needles soft, not sharp when grasped	20
20a.	Needles round in cross section, can be rolled easily between thumb and index finger; needles less than 1.3 cm long; cones small, less than 1.5 cm	*Tsuga* (hemlock)
20b.	Needles too flat to be rolled easily	21
21a.	Tips of needles blunt or rounded, undersurface with 2 white bands; cones with long, conspicuous, 3-lobed bracts	*Pseudotsuga* (Douglas fir)
21b.	Tips of needles pointed	22
22a.	Tops of needles grooved; woody seed cones broadly oblong in outline	*Sequoia* (redwood)
22b.	Tops of needles with ridges; lacking in cones, instead having a red, fleshy, cuplike seed covering	Taxus (yew)
23a.	Twig ends appear as if jointed	*Calocedrus* (incense cedar)
23b.	Tips of branches flattened, not jointed in appearance	24
24a.	Leaves glossy and fragrant	*Thuja* (Western red cedar)
24b.	Leaves awl-shaped, arranged spirally on twig	*Sequoiadendron* (giant sequoia)

Glossary to Tree Key

- *Acorn*—The fruit of an oak, consisting of a nut and its basally attached cup (Fig. 15-9d)
- *Acute*—Sharp-pointed (Fig. 15-2)
- *Alternate*—Describing the arrangement of leaves or other structures that occur singly at successive nodes or levels; not opposite or whorled (Fig. 15-5)
- *Angiosperm*—A flowering seed plant (e.g., bean plant, maple tree, grass)
- *Armed*—Possessing thorns or spines
- *Asymmetrical*—Not symmetrical
- *Axil*—The upper angle between a branch or leaf and the stem from which it grows
- *Axillary bud*—A bud occurring in the axil of a leaf (Figs. 15-3 through 15-7)
- *Basal*—At the base
- *Blade*—The expanded, more or less flat portion of a leaf (Fig. 15-2)
- *Bract*—A much reduced leaf
- *Capsule*—A dry fruit that splits open at maturity (e.g., buckeye; Fig. 15-9e)

- *Compound leaf*—Blade composed of 2 or more separate parts (leaflets) (Figs. 15-4, 15-7)
- *Cordate*—Heart-shaped (Fig. 15-2)
- *Deciduous*—Falling off at the end of a functional period (such as a growing season)
- *Drupe*—Fleshy fruit containing a single hard stone that encloses the seed (e.g., cherry, peach, or dogwood; Fig. 15-9g)
- *Fruit*—A ripened ovary, in some cases with associated floral parts (Figs. 15-9a–g)
- *Glandular*—Bearing secretory structures (glands)
- *Gymnosperm*—Seed plant lacking flowers and fruits (e.g., pine tree)
- *Lateral*—On or at the side (Fig. 15-4)
- *Leaflet*—One of the divisions of the blade of a compound leaf (Figs. 15-4, 15-7)
- *Lobed*—Separated by indentations (sinuses) into segments (lobes) larger than teeth (Fig. 15-3b)
- *Node*—Region on a stem where leaves or branches arise (Figs. 15-2 through 15-7)
- Nut—A hard, 1-seeded fruit that does not split open at maturity (e.g., acorn; Fig. 15-9d)
- *Obtuse*—Blunt (Fig. 15-6)
- *Opposite*—Describing the arrangement of leaves of other structures that occur 2 at a node, each separated from the other by half the circumference of the axis (Fig. 15-6)
- *Palmately compound*—With leaflets all arising at apex of petiole (Fig. 15-7)
- *Petiole*—Stalk of a leaf (Figs. 15-2, 15-3, 15-4, 15-7)
- *Pinnately compound*—A leaf constructed somewhat like a feather, with the leaflets arranged on both sides of the rachis (Fig. 15-4)
- *Pith*—Internally, the centermost region of a stem (Figs. 15-8a, b)
- *Pod*—A dehiscent, dry fruit; a rather general term sometimes used when no other more specific term is applicable (Fig. 15-9f)
- *Pome*—Fleshy fruit containing several seeds (e.g., apple or pear; Fig. 15-9a)
- *Rachis*—Central axis of a pinnately compound leaf (Fig. 15-4)
- *Samara*—Winged, 1-seeded, dry fruit (e.g., ash fruits; Fig. 15-9c)
- *Schizocarp*—Dry fruit that splits at maturity into two 1-seeded halves (Fig. 15-9b)
- *Simple leaf*—One with a single blade, not divided into leaflets (Figs. 15-3, 15-5, 15-6)
- *Spine*—Strong, stiff, sharp-pointed outgrowth on a stem or other organ (Fig. 15-6)
- *Symmetrical*—Capable of being divided longitudinally into similar halves
- *Terminal*—Last in a series (Fig. 15-4)
- *Thorn*—Sharp, woody, spinelike outgrowth from the wood of a stem; usually a reduced, modified branch
- *Tooth*—Small, sharp-pointed marginal lobe of a leaf (Fig. 15-3a)
- *Truncate*—Cut off squarely at end (Fig. 15-3a)
- *Unarmed*—Without thorns or spines
- *Whorl*—A group of 3 or more leaves or other structures at a node

15.3 What Species Is Your Christmas Tree? *(About 20 min.)*

Each year millions of "evergreen" trees become the center of attraction in human dwellings during the Christmas season. The process of selecting the all-important tree is the same whether you reside in the city where you buy your tree from a commercial grower, or whether you cut one off your "back forty." You ponder and evaluate each specimen until, with the utmost confidence, you bring home that perfect tree. Now that you have it, just what kind of tree stands in your home, looking somewhat like a cross between Old Glory and the Sistine Chapel? This key contains most of the trees that are used as Christmas trees; other gymnosperm trees are included, too. The common "Christmas trees" have an asterisk after their scientific name. Note that this key, unlike those in the preceding sections, indicates actual species designations.

1a.	Tree fragrant, boughs having supported (on clear moonlit nights) masses of glistening snow on their green needles; tree a product of nature	2
1b.	Tree not really a tree but rather a product of a cold and insensitive society; tree never giving life and never having life	17
2a.	Leaves persistent and green throughout the winter, needlelike, awl-shaped, or scalelike	3
2b.	Leaves deciduous; for this reason not a desirable Christmas tree	4
3a.	Leaves in clusters of 2–5, their bases within a sheath	5
3b.	Leaves borne singly, not in clusters	9
4a.	Cones 1.25–1.8 cm long, 12–15 scales making up cone	*Larix laricina* (tamarack)
4b.	Cones 1.8–3.5 cm long, 40–50 scales comprising cone	*Larix decidua* (larch)
5a.	Leaves 5 in a cluster, cones 10–25 cm long	*Pinus strobus** (white pine)
5b.	Leaves 2 in a cluster, cones less than 10 cm long	6
6a.	Leaves 2.5–7.5 cm long	7
6b.	Leaves 7.5–15 cm long	8
7a.	Leaves with a bluish cast; cones with a stout stalk, pointing away from the tip of the branch; bark orange in the upper part of the tree	*Pinus sylvestris** (scotch pine)
7b.	Leaves 1.25–3.75 cm long; cones stalkless, pointing forward toward the tip of branch	*Pinus banksiana* (jack pine)
8a.	Leaves slender, shiny; bark of trunk red-brown; cones 5–7.5 cm long; scales of cones without any spine at tip	*Pinus resinosa** (red pine)
8b.	Leaves thickened, dull; bark of trunk gray to nearly black; cones 5 to 7.5 cm long; scales of cone armed with short spine at tip	*Pinus nigra** (Austrian pine)
9a.	Leaves scalelike or awl-shaped	10
9b.	Leaves needlelike	11
10a.	Twigs flattened, leaves all of one kind, scalelike, extending down the twig below the point of attachment	*Thuja occidentalis* (white cedar, arbor vitae)
10b.	Twigs more or less circular in cross section; leaves of 2 kinds, either scalelike or awl-shaped, often both on same branch, not extending down the twig; coneless but may have a blue berrylike structure	*Juniperus virginiana* (red cedar)
11a.	Leaves with petioles	12
11b.	Leaves lacking petioles, leaf tip notched, needles longer than 1.25 cm	*Abies balsamea** (balsam fir)
12a.	Leaves angular, 4-sided in cross section, harsh to the touch; petiole adheres to twig	13
12b.	Leaves flattened	15
13a.	Leaves 3–10 mm long, blunt-pointed; twigs rusty and hairy	*Picea mariana** (black spruce)
13b.	Leaves 2 cm long, sharp-pointed; twigs smooth	14

14a.	Cones 2.5–5 cm long; leaves ill-scented when bruised or broken; smaller branches mostly horizontal	*Picea glauca** (white spruce)
14b.	Cones 7.5–15 cm long; scales comprising cone with finely toothed markings; leaves not ill-scented when bruised or broken; smaller branches drooping	*Picea abies** (Norway spruce)
15a.	Leaves pointed, over 1.25 cm long; red fleshy, berrylike structures present	16
15b.	Leaves rounded at tip, less than 1.25 cm long, with 2 white lines on underside	*Tsuga canadensis* (hemlock)
16a.	Leaves 2–2.5 cm long, dull dark green on top, with 2 broad yellow bands on undersurface; petiole yellowish	*Taxus cuspidata* (Japanese yew)
16b.	Leaves 1.25–2 cm long, without yellow bands on underside	*Taxus canadensis* (American yew)
17a.	Tree a glittering mass of structural aluminum, sometimes illuminated by multicolored floodlights	*Aluminous ersatzenbaum** (aluminum substitute)
17b.	Tree green, produced with petroleum products, increasing our dependence upon oil; used year after year; exactly like all others of its manufacture	*Plasticus perfectus** (plastic substitute)

PRE-LAB QUESTIONS

_____ **1.** The name "human" is an example of a
(a) common name
(b) scientific name
(c) binomial
(d) polynomial

_____ **2.** Current scientific thought places organisms in one of ___ kingdoms.
(a) two
(b) four
(c) five
(d) six

_____ **3.** The scientific name for the ruffed grouse is *Bonasa umbellus. Bonasa* is
(a) the family name
(b) the genus
(c) the specific epithet
(d) all of the above

_____ **4.** A binomial is always a
(a) genus
(b) specific epithet
(c) scientific name
(d) two-part name

_____ **5.** The science of classifying and naming organisms is known as
(a) taxonomy
(b) phylogeny
(c) morphology
(d) physiology

_____ **6.** Which scientific name for the wolf is presented correctly?
(a) Canis lupus
(b) canis lupus
(c) *Canis lupus*
(d) Canis Lupus

_____ **7.** A road that dichotomizes is
(a) an intersection of two crossroads
(b) a road that forks into two roads
(c) a road that has numerous entrances and exits
(d) a road that leads nowhere

_____ **8.** Most scientific names are derived from
(a) English
(b) Latin
(c) Italian
(d) French

_____ **9.** One objection to common names is that
(a) many organisms may have the same common name
(b) many common names may exist for the same organism
(c) the common name may not be familiar to an individual not speaking the language of the common name
(d) all of the above are true

_____ **10.** Phylogeny is the apparent
(a) name of an organism
(b) ancestry of an organism
(c) nomenclature
(d) dichotomy of a system of classification

Name ________________________ Section Number ________________________

EXERCISE 15

Taxonomy: Classifying and Naming Organisms

POST-LAB QUESTIONS

Introduction

1. If you were to use a binomial system to identify the members of your family (mother, father, sisters, brothers), how would you write their names so that your system would most closely approximate that used to designate species?

2. Describe several advantages of using scientific names instead of common names.

3. Based on the following classification scheme, which two organisms are most closely phylogenetically related? Why?

	Organism 1	Organism 2	Organism 3	Organism 4
Kingdom	Animalia	Animalia	Animalia	Animalia
Phylum	Arthopoda	Arthropoda	Arthropoda	Arthropoda
Class	Insecta	Insecta	Insecta	Insecta
Order	Coleoptera	Coleoptera	Coleoptera	Coleoptera
Genus	*Caulophilus*	*Sitophilus*	*Latheticus*	*Sitophilus*
Specific epithet	*oryzae*	*oryzae*	*oryzae*	*zeamaize*
Common name	Broadnosed grain weevil	Rice weevil	Longheaded flour beetle	Maize weevil

15.2 Using a Taxonomic Key

Consider the drawing of plants A and B in answering questions 4–6.

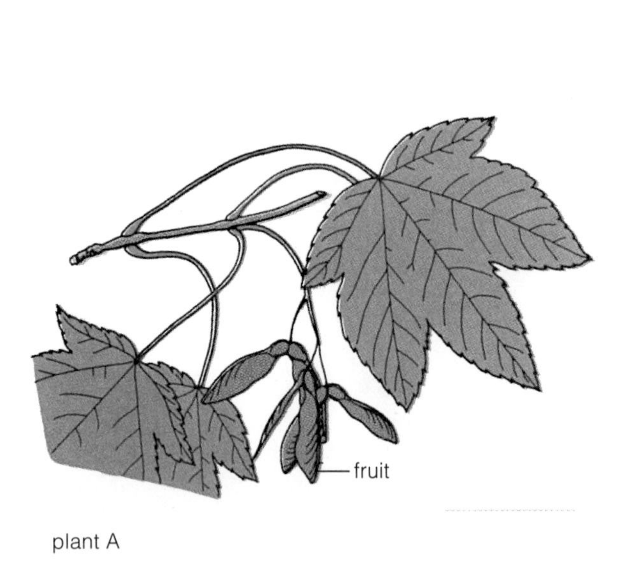

plant A

plant B

4. Using the taxonomic key in the exercise, identify the two plants as either angiosperms or gymnosperms.
   Plant A is a (an) ______________________.
   Plant B is a (an) ______________________.
5. To what genus does plant A belong? What is its common name?
   genus: ______________________
   common name: ______________________
6. To what genus does plant B belong? What is its common name?
   genus: ______________________
   common name: ______________________

Consider the drawing of plants C and D in answering questions 7–9.

plant C

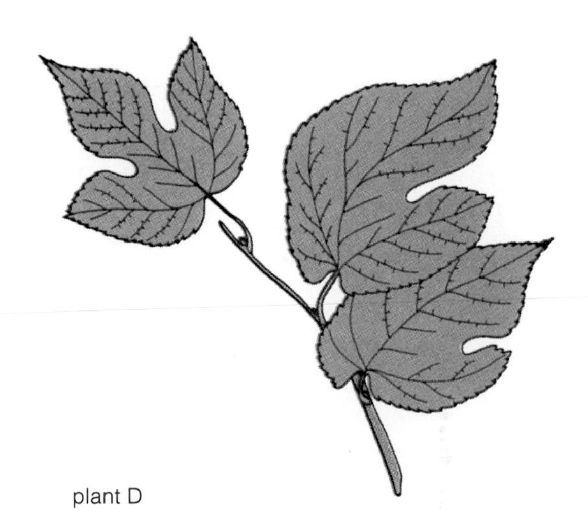

plant D

7. As completely as possible, describe the leaf of plant C.

8. To what genus does plant C belong? What is its common name?

genus: ______________________

common name: ______________________

9. Using the taxonomic key in the exercise, identify the genus of the organism below.

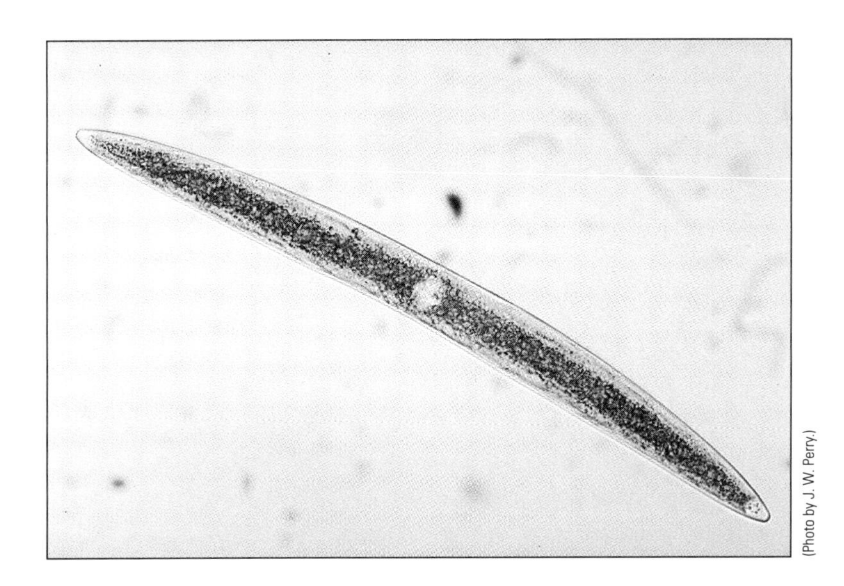
(Photo by J. W. Perry.)

genus: ______________________

Food for Thought

10. If you owned a large, varied music collection, how might you devise a key to keep track of all your different kinds of music?

Bacteria and Protists I

OBJECTIVES

After completing this exercise, you will be able to

1. define *prokaryotic, eukaryotic, pathogen, decomposer, coccus, bacillus, spirillum, Gram stain, antibiotic, symbiosis, parasitism, commensalism, mutualism, nitrogen fixation, monoecious, dioecious, plasmodium, obligate mutualism, gametangia, phagocytosis, vector;*
2. describe characteristics distinguishing bacteria from protists;
3. identify and classify the organisms studied in this exercise;
4. identify structures (those in **boldface** in the procedure sections) in the organisms studied;
5. distinguish Gram-positive and Gram-negative bacteria, indicating their susceptibility to certain antibiotics;
6. suggest measures that might be used to control malaria.

INTRODUCTION

The bacteria (Domain Bacteria, Kingdom Bacteria) and archaea (Domain Archaea, Kingdom Archaea) and protists (Domain Eukarya, Kingdom Protista) are among the simplest of living organisms. These have unicellular organisms within them, but that's where the similarity ends.

Bacteria and archaea are **prokaryotic** organisms, meaning that their DNA is free in the cytoplasm, unbounded by a membrane. They lack organelles (cytoplasmic structures surrounded by membranes). By contrast, the protists are **eukaryotic** organisms: The genetic material contained within the nucleus and many of their cellular components are compartmentalized into membrane-bound organelles.

Both bacteria and archaea, and unicellular protists, are at the base of the food chain. From an ecological standpoint, these simple organisms are among the most important organisms on our planet. Ecologically, they're much more important than we are.

16.1 Domain Bacteria, Kingdom Bacteria *(About 40 min.)*

The Kingdom Bacteria consists of bacteria and cyanobacteria. Most bacteria are heterotrophic, dependent upon an outside source for nutrition, while cyanobacteria are autotrophic (photosynthetic), able to produce their own carbohydrates.

Some heterotrophic bacteria are **pathogens,** causing plant and animal diseases, but most are **decomposers,** breaking down and recycling the waste products of life. Others are nitrogen fixers, capturing the gaseous nitrogen in the atmosphere and making it available to plants via a symbiotic association with their roots.

MATERIALS

Per student:

- bacteria type slide
- microscope slide
- coverslip
- dissecting needle
- compound microscope

Per lab room:

- Gram-stained bacteria (3 demonstration slides)
- *Oscillatoria*—living culture; disposable pipet
- *Azolla*—living plants

A. Bacteria: Are Bacteria Present in the Lab? *(About 15 min.)*

MATERIALS

Per student group (4):

- dH_2O in dropping bottle
- 4 nutrient agar culture plates
- sterile cotton swabs
- china marker
- 2 bottles labeled A and B (A contains tap water, B contains 70% ethyl alcohol)
- transparent adhesive tape
- paper towels

PROCEDURE

Work in groups of four.

1. Obtain four petri dishes containing sterile nutrient agar. Using a china marker, label one dish "Dish 1: Control." Label the others "Dish 2: Dry Swab," "Dish 3: Treatment A," and "Dish 4: Treatment B." Also include the names of your group members.
2. Run a sterile cotton swab over a surface within the lab. Some examples of things you might wish to sample include the surface of your lab bench, the floor, and the sink. Be creative!
3. Lift the lid on Dish 2 as little as is necessary to run the swab over the surface of the agar.

   Note: **Be careful that you don't break the agar surface.**

4. Tape the lid securely to the bottom half of the dish.
5. Soak one paper towel with liquid A and a second paper towel with liquid B.
6. Wipe down one-half of the surface you just sampled with liquid A (tap water), the other half with liquid B (70% ethyl alcohol). After the areas have dried, using dishes 3 and 4, repeat the procedures described in step 2. Place the cultures in a desk drawer to incubate until the next lab period. At that time, examine your culture for bacterial colonies, noting the color and texture of the bacterial growth.
7. Make a prediction of what you will find in the culture plates after the next lab period.

   ____

   ____

8. Describe what you see.
   Source of sample: ____________________

   Dish 1: Control. ____________________

   ____

   ____

   Dish 2: Dry Swab. ____________________

   ____

   ____

   Dish 3: Treatment A. ____________________

   ____

   ____

   Dish 4: Treatment B. ____

   ____

9. Make a conclusion about the usefulness of tap water and 70% ethyl alcohol as disinfectants.

   ____

   ____

Caution

Leave the lid on as you examine the cultures to prevent the spread of any potentially pathogenic (disease-causing) organisms. While the probability is small that pathogens are present, you should always err on the side of caution.

B. Bacteria: Bacterial Shape and Sensitivity to Antibiotics (About 10 min.)

MATERIALS

Per student:
- bacteria type slide
- compound microscope

Per lab room:
- Gram-stained bacteria (3 demonstration slides)

Work alone for this and the rest of these activities.

Bacteria exist in three shapes: **coccus** (plural, *cocci;* spherical), **bacillus** (plural, *bacilli;* rods), and **spirillum** (plural, *spirilla;* spirals).

PROCEDURE

1. Study a bacteria type slide illustrating these three shapes. You'll need to use the highest magnification available on your compound microscope. In Figure 16-1, draw the bacteria you are observing.

In addition to being differentiated by shape, bacteria can be separated according to how they react to a staining procedure called **Gram stain,** named in honor of a nineteenth-century microbiologist, Hans Gram. *Gram-positive* bacteria are purple after being stained by the Gram stain procedure, while *Gram-negative* bacteria appear pink. The Gram stain reaction is important to bacteriologists because it is one of the first steps in identifying an unknown bacterium. Furthermore, the Gram stain reaction indicates a bacterium's susceptibility or resistance to certain **antibiotics,** substances that inhibit the growth of bacteria.

2. Examine the *demonstration slides* illustrating Gram-stained bacteria. Gram-positive bacteria are susceptible to penicillin, while Gram-negative bacteria are not. In Table 16-1, list the species of bacteria that you have examined and their staining characteristics.

Figure 16-1 Drawings of the three bacterial shapes (____________×).

TABLE 16-1 Gram Stain Reaction of Various Bacteria

Bacterial Species	Gram Reactions (+ or –)

C. Cyanobacteria (Blue-Green Algae) (About 15 min.)

The cyanobacteria (sometimes called blue-green algae) are distinguished from the heterotrophic bacteria by being photosynthetic (photoautotrophic).

C.1. Oscillatoria

MATERIALS

Per student:

- microscope slide
- coverslip
- dissecting needle
- compound microscope

Per lab room:

- *Oscillatoria*—living culture; disposable pipet

PROCEDURE

1. Examine the culture provided of *Oscillatoria.* Describe what the culture looks like, including the texture and the color.

   ____

   ____

The color of the culture is a consequence of the photosynthetic accessory pigments that for the most part mask the chlorophyll.

(If you have immediate access to a greenhouse, you might walk through it now to see if you can find any *Oscillatoria* growing on the flowerpots or even the floor.)

2. Using a dissecting needle, scrape some of the culture from the surface and prepare a wet mount slide.
3. Examine your preparation with your compound microscope, starting with the medium-power objective and finally with the highest magnification available (oil immersion, if possible). Note that the individual cells are joined and so form the filament.

Do all the *Oscillatoria* cells look alike, or is there differentiation of certain cells within the filament?

____

____

4. Draw a portion of the filament in Figure 16-2.

Figure 16-2 Drawing of *Oscillatoria* (_______×).

C.2. Anabaena

Some cyanobacteria live as *symbionts* within other organisms. Literally, **symbiosis** means "living together." There are three types of symbiosis. In a parasitic symbiosis (**parasitism**), one organism lives at the expense of the other; that is, the parasite benefits while the host is harmed. A commensalistic symbiosis (**commensalism**) occurs when effects are positive for one species and neutral for the other. In a mutualistic symbiosis (**mutualism**) both organisms benefit from living together. In this section, you will examine one such symbiotic relationship.

MATERIALS

Per student:

- microscope slide
- coverslip
- dissecting needle
- compound microscope

Per lab room:

- *Azolla*—living plants

PROCEDURE

1. Place a leaf of the tiny water fern *Azolla* on a clean glass slide. Use a dissecting needle to crush the leaf into very small pieces. Now add a drop of water and a coverslip.
2. Scan your preparation with the medium-power objective of your compound microscope, looking for long chains (composed of numerous beadlike cells) of the filamentous cyanobacterium *Anabaena.* Switch to higher magnification when you find *Anabaena.*
3. Within the filament, locate the **heterocysts,** cells that are a bit larger than the other cells.
4. Draw what you see in Figure 16-3 and label it.

5. Examine the very high magnification electron micrograph of *Anabaena* in Figure 16-4. Identify the heterocyst with its distinctive *polar nodules* at either end of the cell. In the other cells, note the numerous wavy **thylakoids,** membranes on and in which the photosynthetic pigments are found. Identify the large electron-dense carbohydrate **storage granules** within the cytoplasm and the cell wall.

Is a nucleus present within the cells of *Anabaena?* Explain.

____

Based upon this electron micrograph, would you hypothesize that the heterocyst is photosynthetic?

____

6. Now look again at your microscope slide of *Anabaena.* Find the polar nodules, appearing as bright spots within the heterocysts.

Figure 16-3 Drawing of *Anabaena* that was within *Azolla* leaves (________×).
Label: heterocyst

Heterocysts convert nitrogen in the air (or water) to a form that the cyanobacterium can use for cellular metabolism. This process is called **nitrogen fixation.** Nitrogen-containing compounds accumulate in the water fern, some of it being used for its own metabolic needs. There is also an advantage to the *Anabaena*: It not only gets a place to live, but by being within the water fern leaf it is shielded from very high light levels that would otherwise do it damage.

Which type of symbiosis is the association between *Anabaena* and the water fern?

____

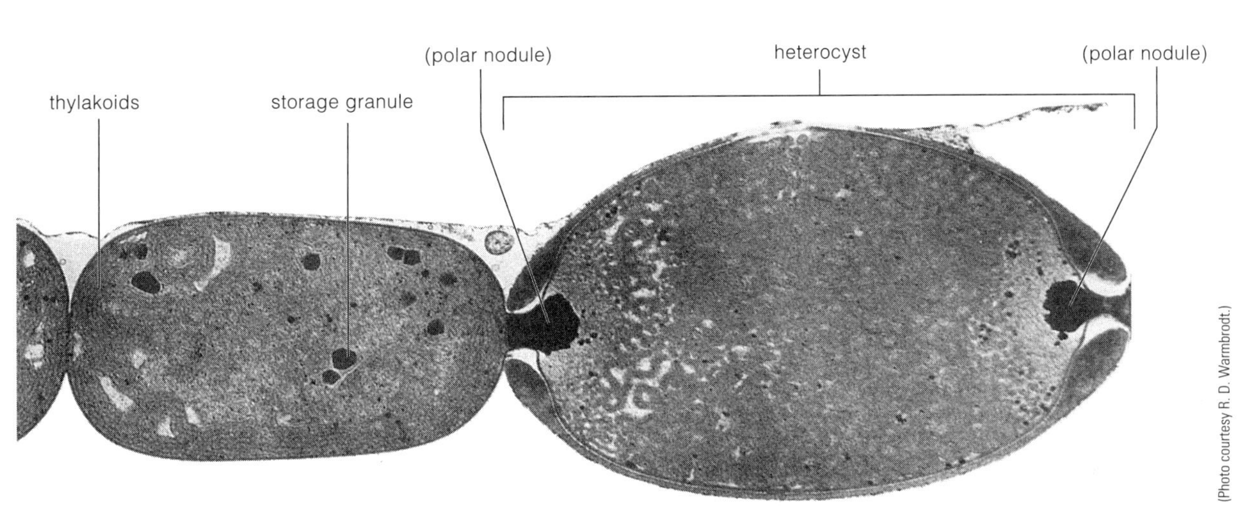

Figure 16-4 Electron micrograph of *Anabaena* (14,220×).

16.2 Domain Eukarya, Kingdom Protista *(About 2 hr.)*

The protists are a diverse assemblage of organisms, both green (photoautotrophic) and nongreen (heterotrophic). They are so diverse that different protists have previously been classified as fungi, animals, and plants! In these activities, you will examine the heterotrophic protists, saving the photoautotrophic protists for Exercise 17.

A. Phylum Stramenopila

A.1. Oomycotes: Water Molds *(About 30 min.)*

Oomycotes are commonly referred to as water molds because they are primarily aquatic organisms. While not all members of this division grow in freestanding water, they all rely on the presence of water to spread their asexual spores.

During sexual reproduction, all members produce large nonmotile female gametes, the eggs. These eggs are contained in sex organs (**gametangia;** singular, *gametangium*) called **oogonia.** By contrast, the male gametes are nothing more than *sperm nuclei* contained in **antheridia,** the male gametangia.

MATERIALS

Per student:

- culture of *Saprolegnia* or *Achlya*
- glass microscope slide
- compound microscope

A.1.a. Saprolegnia or Achlya

If you've ever kept goldfish, you may have seen a white, cottony mass growing on the sides of certain fish. This is a parasitic water mold that is easily controlled by the addition of chemicals to the aquarium.

PROCEDURE

1. Examine the water cultures of either *Saprolegnia* or *Achlya.* These fungi are chiefly saprophytes and are growing on a sterilized hemp seed that provides a carbohydrate source (Figure 16-5).
2. Examine its life cycle (Figure 16-6) as you study this organism.
3. Notice the numerous filamentous hyphae that radiate from the hemp seed. A **hypha** (plural, *hyphae*) is the basic unit of the fungal body. Collectively, all the hyphae constitute the **mycelium** (plural, *mycelia*).
4. Identify the hyphae and mycelium in Figures 16-5 and 16-6a.

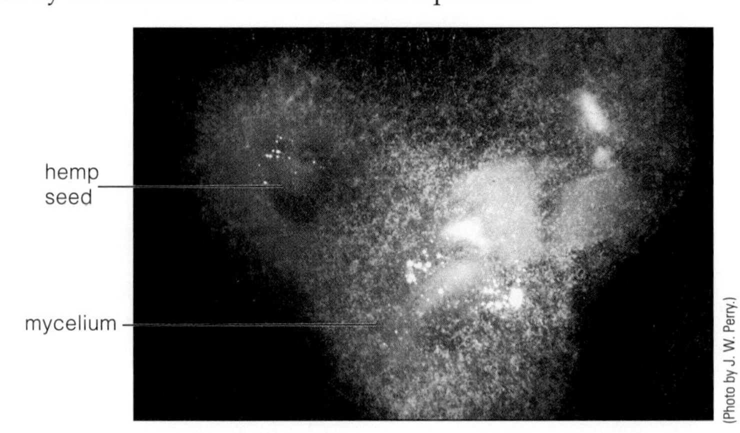

Figure 16-5 Mycelium of water mold growing on hemp seed (3×).

The mycelium of the water mold is **multinucleate,** meaning that there are many nuclei in each cell, not just one. Each nucleus is diploid (2n). Moreover, cross walls separating the mycelium into distinct cells are infrequent, forming only when reproductive organs are formed.

5. Place a glass slide on the stage of your compound microscope. This slide will serve as a platform for the culture dish that contains the fungal culture, allowing you to use the mechanical stage of the microscope to move the culture (if the microscope is so equipped).
6. Remove the lid from the culture dish, carefully place the culture dish on the platform, and examine with the low-power objective. Look first at the edge of the mycelium—that is, at the tips of the youngest hyphae.
7. Find the tips that appear denser (darker) than the rest of the hyphae. These dense tips are cells specialized for asexual reproduction. They are the **zoosporangia** (singular, *zoosporangium;* Figures 16-7, 16-6b), which produce biflagellate **zoospores** (Figure 16-6c). Note the cross wall separating the zoosporangium from the rest of the hyphae.
8. Examine Figure 16-6d. Zoospores are released from the zoosporangia, swim about for a period of time, lose their flagella, and encyst, meaning that they form a thick wall around the cytoplasm.
9. Look at Figure 16-6e. When the encysted zoospore germinates, it produces a second type of zoospore. This one also will eventually encyst, as shown in Figure 16-6f).
10. Look at Figure 16-6g. When the second cyst stage germinates, it produces a hypha that proliferates into a new mycelium (Figure 16-6a). As you see, this is asexual reproduction; no sex organs were involved in the formation of zoospores.

Sexual reproduction in the water molds occurs in the older portion of the mycelium. In your culture the older portion is located nearer the hemp seed.

11. Scan the colony to find the spherical female gametangia, the oogonia (singular, **oogonium;** Figures 16-7, 16-6h). Meiosis takes place within the oogonium to form haploid eggs.
12. Switch to the medium-power objective and study a single oogonium in greater detail. Depending on the stage of development, you will find either **eggs** (Figure 16-6i) or **zygotes** (fertilized eggs) (Figures 16-7, 16-6j) within the oogonium.

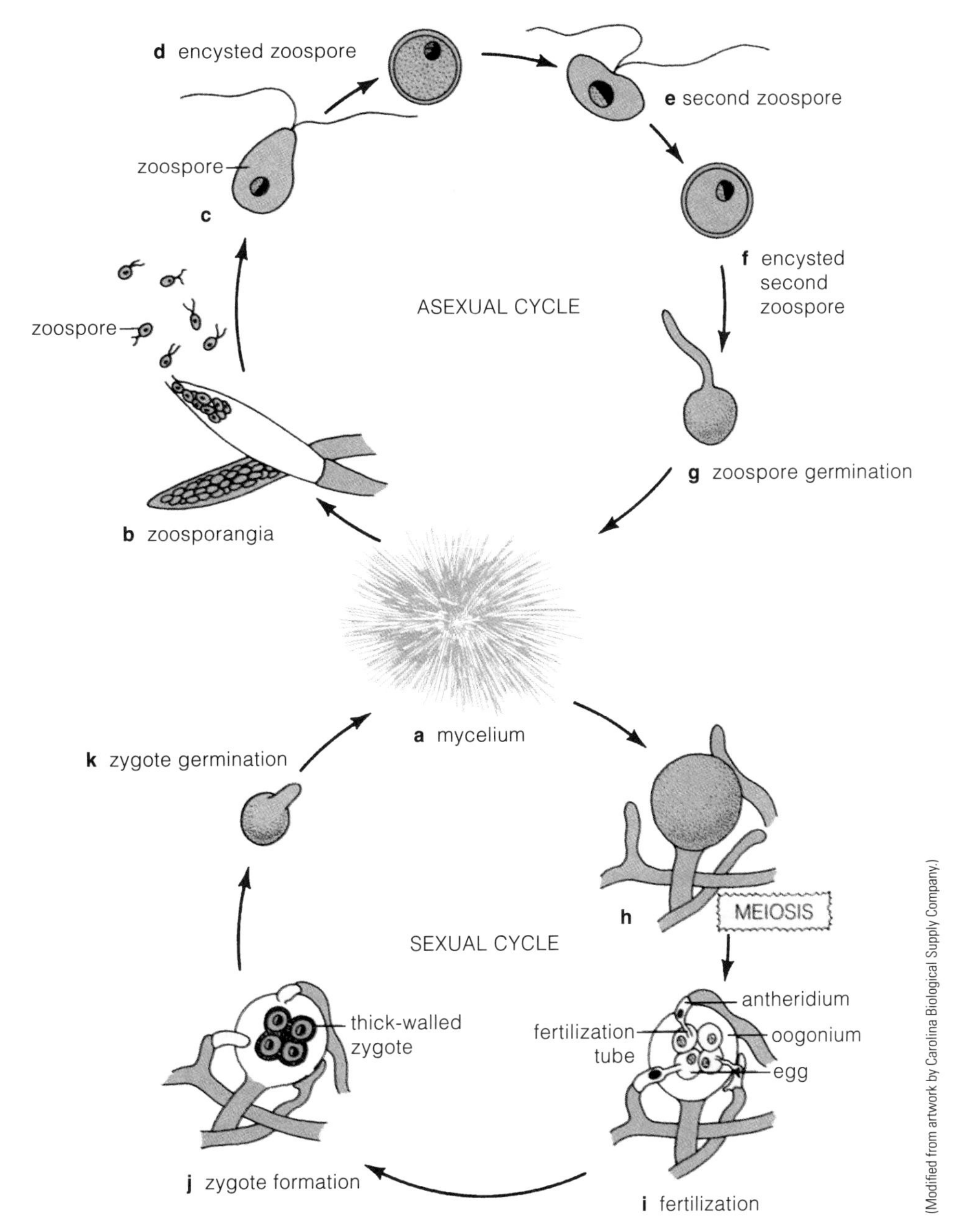

Figure 16-6 Life cycle of a water mold such as *Saprolegnia* or *Achlya.* (Structures colored green are 2n.)

13. Now you'll need to find the male gametangium, the **antheridium.** It's a short fingerlike hypha that attaches itself to the wall of the oogonium (Figures 16-8, 16-6i), much as you would wrap your finger around a baseball. Find an antheridium.

Because the nuclei of the antheridium have undergone meiosis, each nucleus is haploid.

14. Examine Figure 16-6i, which shows fertilization. Fertilization takes place when tiny fertilization tubes penetrate the wall of the oogonium. Rather than forming special male gametes, the haploid nuclei within the antheridium flow through the fertilization tubes to fuse with the egg nuclei.
15. Search your culture to see if you can locate any thick-walled zygotes within oogonia (Figures 16-9, 16-6j). Following a maturation period that can last several months, the zygote germinates by forming a germ tube (Figure 16-6k) that grows into a new mycelium (Figure 16-6a).

Notice that both male and female gametangia are produced on the same mycelium. The term describing this condition is **monoecious** (from the Greek words for "one house"). Organisms producing one type of sex organ on one body and the other type of sex organ on another body are **dioecious** (from the Greek words for "two" and "house").

Are humans monoecious or dioecious? ______________________

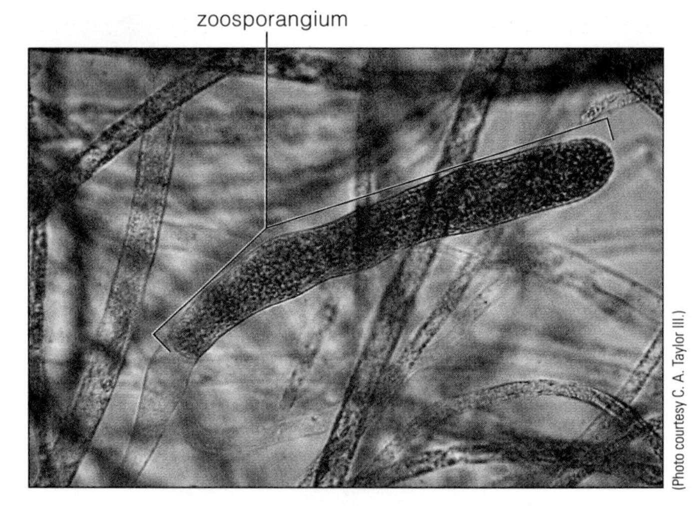

Figure 16-7 Water mold zoosporangium (272×).

A.1.b. Phytophthora

Some of the most notorious and historically important plant pathogens known to humans are water molds. Included in this group is *Phytophthora infestans,* the fungus that causes the disease known as late blight of potato. This disease spread through the potato fields of Ireland between 1845 and 1847. Most of the potato plants died, and 1 million Irish working-class citizens who had depended on potatoes as their primary food source starved to death. Another 2 million emigrated, many to the United States.

Perhaps no other plant pathogen so poignantly illustrates the importance of environmental factors in causing disease. While *Phytophthora infestans* had been present previous to 1845 in the potato-growing fields of Ireland, it was not until the region experienced several consecutive years of wet and especially cool growing seasons that late blight became a major problem. We can easily observe the importance of these climate-associated factors by studying another *Phytophthora* species, *P. cactorum.*

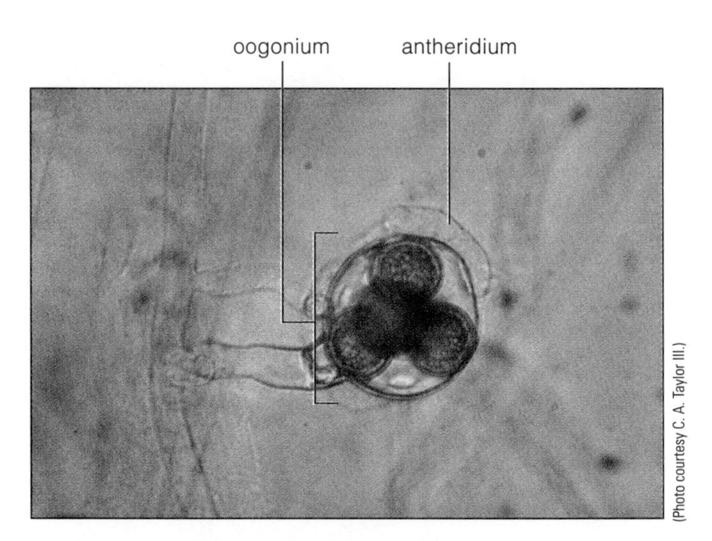

Figure 16-8 Gametangia of water mold (272×).

MATERIALS

Per student:

- culture of *Phytophthora cactorum*
- glass microscope slide
- dissecting needle
- compound microscope

Per lab room:

- refrigerator

  or

Per student group (4):

- ice bath

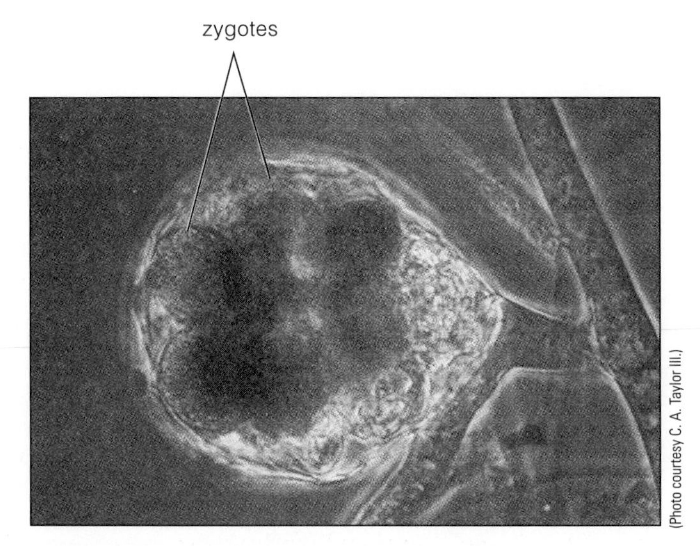

Figure 16-9 Oogonium with zygotes (720×).

PROCEDURE

1. Obtain a culture of *P. cactorum* that has been flooded with distilled water. Note that the agar has been removed from the edges of the petri dish and that the mycelium has grown from the agar edge into the water.
2. Place a glass slide on the stage of your compound microscope. This slide will serve as a platform for the culture, allowing you to use the mechanical stage of the microscope to move the culture (if the microscope is so equipped).

3. Remove the lid from the culture, carefully place the culture dish on the platform, and examine the culture with the low-power objective.
4. Search the surface of the mycelium, especially at the edges, until you find the rather *pear-shaped* **zoosporangia.** Switch to the medium-power objective for closer observation and then draw in Figure 16-10 a single zoosporangium.
5. Return your microscope to the low-power objective, remove the culture, replace the cover, and place it in a refrigerator or on ice for 15–30 minutes.
6. After the incubation time, again observe the zoosporangia microscopically. Find one in which the **zoospores** are escaping from the zoosporangium. Each zoospore has the potential to grow into an entirely new mycelium! Draw the zoospores in Figure 16-10.

Figure 16-10 Drawing of zoosporangium and zoospores of *Phytophthora cactorum.* (________×).
Labels: zoosporangium, zoospores

Like the other water molds, *Phytophthora* reproduces sexually. Your cultures contain the sexual structures as well as asexual zoosporangia.

7. Using a dissecting needle, cut a section about 1 cm square from the agar colony and *invert* it on a glass slide (so that the bottom side of the agar is now uppermost).
8. Place a coverslip on the agar block and observe with your compound microscope, first with the low-power objective, then with the medium-power, and finally with the high-dry objective.
9. Identify the spherical **oogonia** that contain **eggs** or thick-walled **zygotes** (depending on the stage of development). If present, the **antheridia** are club shaped and plastered to the wall of the oogonium.
10. In Figure 16-11, draw an oogonium, eggs (zygotes), and antheridia.

Figure 16-11 Drawing of the gametangia of *Phytophthora cactorum.* (________×).
Labels: oogonium, eggs (zygotes), antheridium

B. Phylum Amoebozoa: Slime Molds

The slime molds have both plantlike and animal-like characteristics. Because they engulf their food and lack a cell wall in their vegetative (nonreproductive) state, they are placed in the kingdom Amoebozoa. However, when they reproduce, they produce spores with a rigid cell wall, similar to plants.

B.1. Physarum: A Plasmodial Slime Mold (About 15 min.)

The vegetative (nonreproductive) body of the plasmodial slime molds consists of a naked multinucleate mass of protoplasm known as a **plasmodium.**

MATERIALS

Per student:

- petri dish culture of slime mold (*Physarum*)
- compound microscope

Per lab room:

- demonstration of slime mold (*Physarum*) sporangia

PROCEDURE

1. Obtain a petri dish culture of *Physarum* (Figure 16-12) and remove the cover. After examining it with your unaided eye, place the culture dish on the stage of your compound microscope and examine it with the low-power objective.
2. Watch the cytoplasm. The motion that you see within the plasmodium is cytoplasmic streaming (Exercise 6). Is the cytoplasmic streaming unidirectional, or does the flow reverse? ____________________

As you noticed, the *Physarum* culture was stored in the dark. That's because light (along with other factors) stimulates the plasmodium to switch to the reproductive phase.

3. Examine the spore-containing sporangia of *Physarum* that are on demonstration (Figure 16-13). Spores released from the sporangia germinate, producing a new plasmodium.

Because the plasmodium is multinucleate (whereas the spores are uninucleate), what event must occur following spore germination?

____

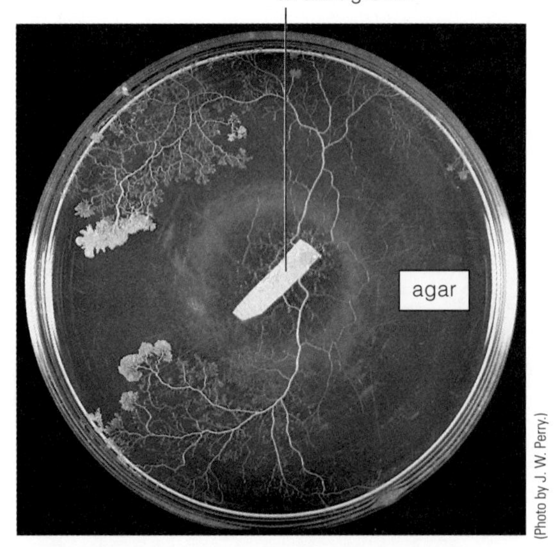

Figure 16-12 Culture dish containing the plasmodial slime mold, *Physarum* (0.7×).

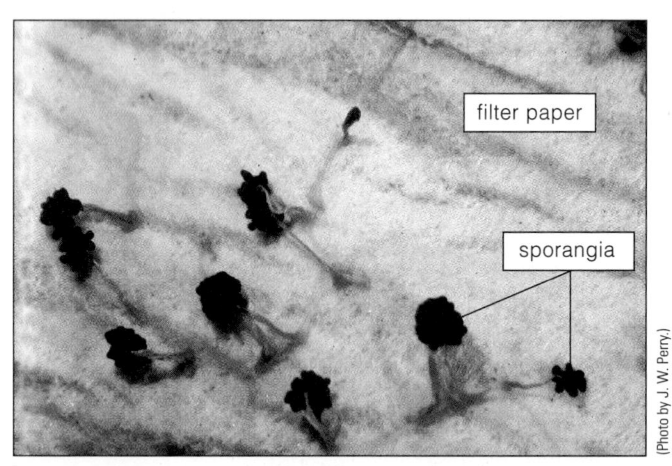

Figure 16-13 Sporangia of the plasmodial slime mold, *Physarum* (1.5×).

B.2. Amoeba *(About 20 min.)*

Amoebas continually change shape by forming projections called *pseudopodia* (singular, *pseudopodium,* "false foot"). One notorious human pathogen is *Entamoeba histolytica,* the cause of amoebic dysentery.

MATERIALS

Per student:

- depression slide
- coverslip
- dissecting needle
- compound microscope

Per group (table):

- carmine, in 2 screw-cap bottles
- tissue paper

Per lab room:

- *Amoeba*—living culture on demonstration at dissecting microscope; disposable pipet

PROCEDURE

1. Observe the *Amoeba*-containing culture on the stage of a demonstration dissecting microscope. (One species of amoeba has the scientific name *Amoeba.*) The microscope has been focused on the bottom of the culture dish, where the amoebas are located. Look for gray, irregularly shaped masses moving among the food particles in the culture.
2. Using a clean pipette, remove an amoeba and place it, along with some of the culture medium, in a depression slide.
3. Examine it with your compound microscope using the medium-power objective. You will need to adjust the diaphragm to increase the contrast (see Exercise 3) because *Amoeba* is nearly transparent.
4. Refer to Figure 16-14. Locate the **pseudopodia.** At the periphery of the cell, identify the **ectoplasm,** a thin, clear layer that

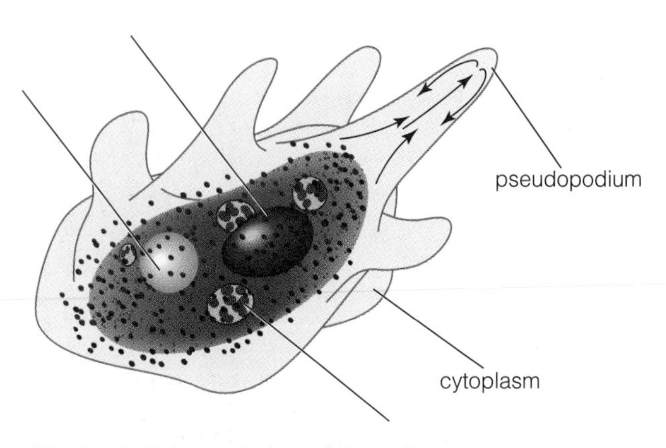

Figure 16-14 Artist's rendering of *Amoeba.*
Labels: ectoplasm, endoplasm, nucleus, contractile vacuole, food vacuole

surrounds the inner, granular **endoplasm.** Watch the organism as it changes shape. Which region of the endoplasm appears to stream, the outer or the inner? ______________________

This region, called the *plasmasol,* consists of a fluid matrix that can undergo phase changes with the semisolid *plasmagel,* the outer layer of the endoplasm. Pseudopodium formation occurs as the plasmasol flows into new environmental frontiers and then changes to plasmagel.

Numerous granules will be found within the endoplasm. Some of these are organelles; others are food granules.

5. Within the endoplasm, try to locate the **nucleus,** a densely granular, spherical structure around which the cytoplasm is streaming.
6. Find the clear, spherical **contractile vacuoles,** which regulate water balance within the cell. Watch for a minute or two to observe the action of contractile vacuoles. Label Figure 16-14.

Amoeba feeds by a process called **phagocytosis,** engulfing its food. Pseudopodia form around food particles, and then the pseudopodia fuse, creating a **food vacuole** within the cytoplasm. Enzymes are then emptied into the food vacuole, where the food particle is digested into a soluble form that can pass through the vacuolar membrane. You can stimulate feeding behavior by drawing carmine under the coverslip:

7. Place a drop of distilled water against one edge of the coverslip.
8. Pick up some carmine crystals by dipping a dissecting needle into the bottle; and deposit them into the water droplet.
9. Draw the suspension beneath the coverslip by holding a piece of absorbent tissue against the coverslip on the side *opposite* the carmine suspension.
10. Observe the *Amoeba* with your microscope again—you may catch it in the act of feeding.

C. Phylum Alveolata

C.1. Ciliata: Ciliated Protozoans (*About 20 min.*)

Most members of the phylum Ciliata are covered with numerous short locomotory structures called *cilia.* In this section, you will examine one of the largest ciliates, the predatory *Paramecium caudatum.*

Paramecium

MATERIALS

Per student:

- microscope slide
- coverslip
- compound microscope

Per group (table):

- methyl cellulose in 2 dropping bottles
- tissue paper
- acetocarmine stain in 2 dropping bottles
- box of toothpicks

Per lab room:

- *Paramecium caudatum*—living culture; disposable pipette
- Congo red—yeast mixture; disposable pipette

PROCEDURE

1. From the culture provided, prepare a wet mount of *Paramecium* on a clean microscope slide. Observe their rapid movement with the medium-power objective of your microscope. Look along the edge of the *Paramecium's* body to see evidence of the locomotory structures, the **cilia.** (Increasing the contrast by closing the microscope's iris diaphragm may help.)
2. Examine Figure 16-15, a scanning electron micrograph of *Paramecium.* Note the numerous **cilia** covering the body surface.
3. Figure 16-16 as an aid as you proceed with your observation of your living specimen.
4. Prepare another wet mount of *Paramecium,* but before you place a coverslip on the

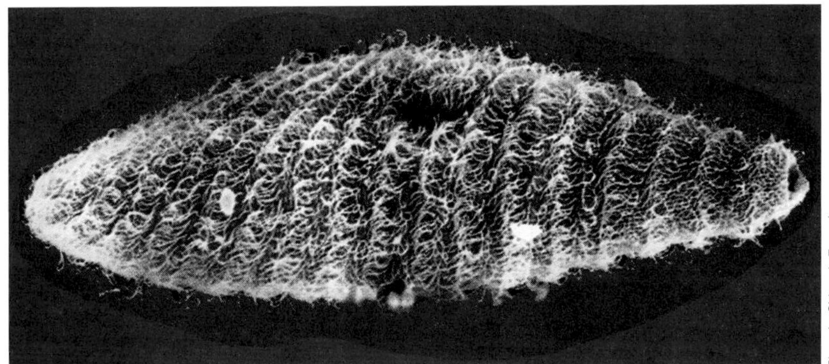

(Photo by Sidney L. Tamm.)

Figure 16-15 Scanning electron micrograph of *Paramecium caudatum.* (1170×).

slide, add a drop of aceto-carmne stain and a drop of methyl cellulose to slow the organisms' movement.

5. Find the large, centrally located **macronucleus.** It's likely that the second, a smaller **micronucleus adjacent** to the macronucleus, will be difficult to see. Surprise! This eukaryotic cell is *binucleate.* Each nucleus has its own distinctive functions.

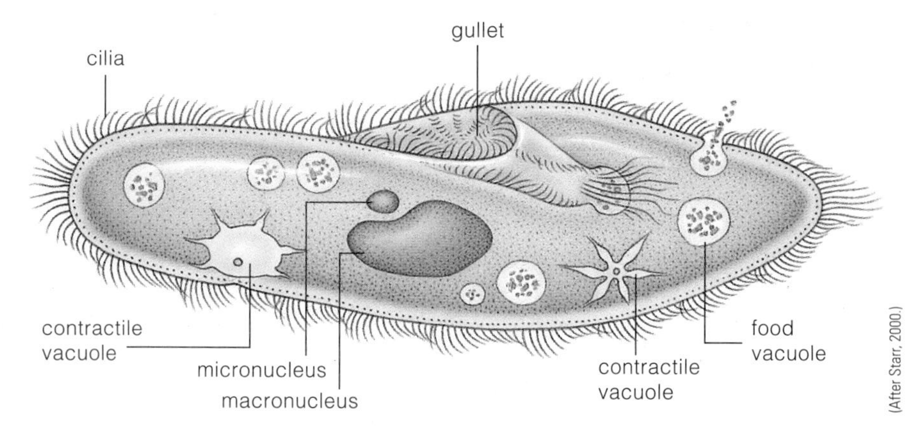

Figure 16-16 *Paramecium caudatum.*

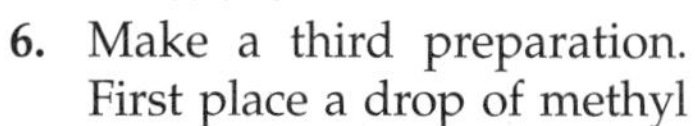

6. Make a third preparation. First place a drop of methyl cellulose on the slide, then a drop of Congo red/yeast mixture. Stir the mixture with a toothpick. Now add a drop of *Paramecium* culture. Place a coverslip on your wet mount and observe with the medium-power objective.
7. Locate the **oral groove,** a depression in the body that leads to the **gullet.**
8. Observe the yeast cells as they are eaten by *Paramecium.* The yeast will be taken into **food vacuoles,** where they are digested. Congo red is a pH indicator. As digestion occurs within the food vacuoles, the indicator will turn blue because of the increased acidity.
9. At either end of the organism, find the **contractile vacuoles,** which regulate water content within the organism. Watch them in operation.

*C.2. **Apicomplexans*** *(About 20 min.)*

All apicomplexans are parasites, infecting a wide range of animals, including humans. *Plasmodium vivax* causes one type of malaria in humans. In this section, you will study its life cycle with demonstration slides. Refer to Figure 16-17 as you proceed.

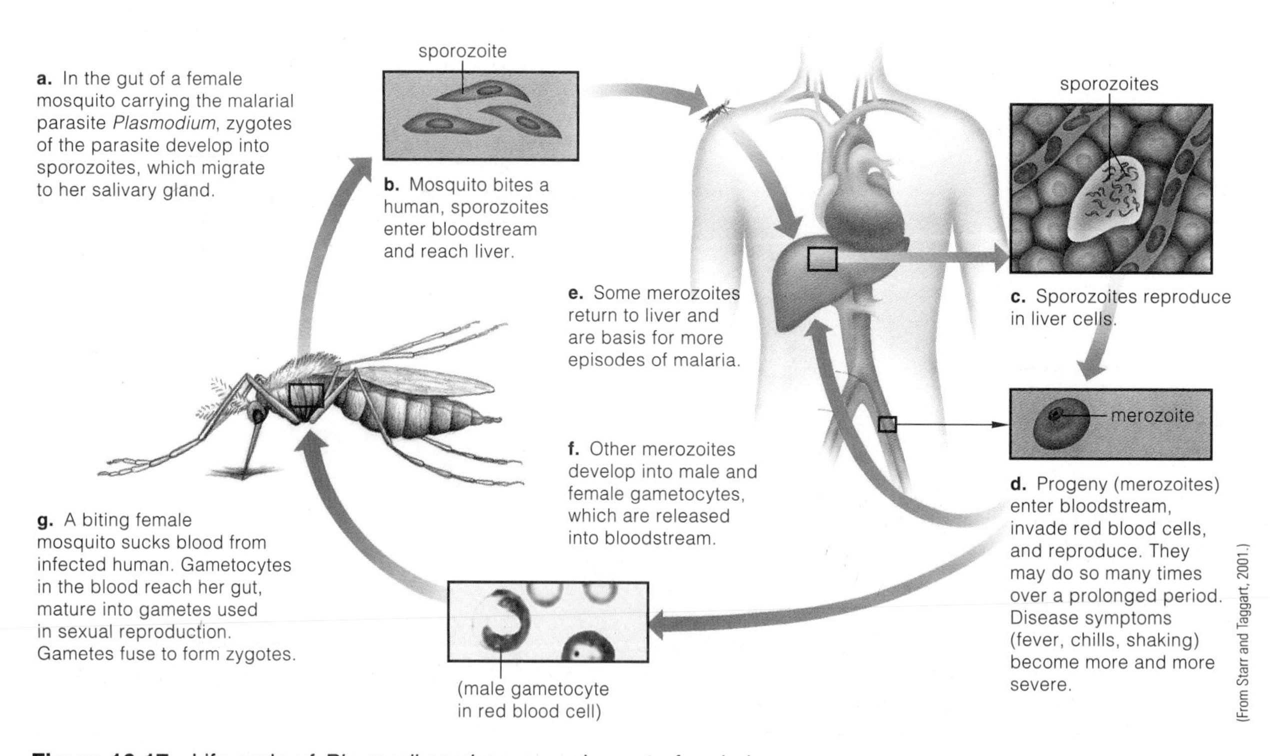

Figure 16-17 Life cycle of *Plasmodium vivax,* causal agent of malaria.

Plasmodium vivax

MATERIALS

Per lab room:

- demonstration slide of *Plasmodium vivax*, sporozoites
- demonstration slide of *P. vivax*, merozoites
- demonstration slide of *P. vivax*, immature gametocytes

PROCEDURE

1. Examine Figure 16-17a. *P. vivax* is transmitted to humans through the bite of an infected female *Anopheles* mosquito. The mosquito serves as a **vector,** a means of transmitting the organism from one host to another. Male mosquitos cannot serve as vectors, because they lack the mouth parts for piercing skin and sucking blood. If the mosquito carries the pathogen, **sporozoites** enter the host's bloodstream with the saliva of the mosquito.
2. Examine the demonstration slide of sporozoites (Figure 16-17b).
3. The sporozoites travel through the bloodstream to the liver, where they penetrate certain cells, grow, and multiply. When released from the liver cells, the parasite is in the form of a **merozoite** and infects the red blood cells. Figures 16-17c and 16-17d show this process.

(There are two intervening stages between sporozoites and merozoites. These are the *trophozoites* and *schizonts,* both developmental stages in red blood cells.)

4. Now examine the demonstration slide illustrating merozoites in red blood cells (Figure 16-17d).
5. Within the red blood cells, merozoites divide, increasing the merozoite population. At intervals of 48 or 72 hours, the infected red blood cells break down, releasing the merozoites. At this time, the infected individual has disease symptoms, including fever, chills, and shaking caused by the release of merozoites and metabolic wastes from the red blood cells. Some of these merozoites return to the liver cells, where they repeat the cycle and are responsible for recurrent episodes of malaria.
6. Merozoites within the bloodstream can develop into **gametocytes.** For development of a gametocyte to be completed, the gametocyte must enter the gut of the mosquito. This occurs when a mosquito feeds upon an infected (diseased) human.
7. Observe the demonstration slide of an **immature gametocyte** in a red blood cell (Figure 16-17f).
8. Within the gut of the mosquito, the gametocyte matures into a gamete (Figure 16-17g). When gametes fuse, they form a zygote that matures into an **oocyst.** Within each oocyst, sporozoites form, completing the life cycle of *Plasmodium vivax.* These sporozoites migrate to the mosquito's salivary glands to be injected into a new host.

D. Phylum Euglenozoa *(About 10 min.)*

These protistans have one or more flagella to provide motility. The group includes some fairly notorious human parasites that cause disease, including giardiasis (from drinking water contaminated with the causal protozoans), some sexually transmitted diseases (caused by *Trichomonas vaginalis*), and African sleeping sickness, caused by *Trypanosoma brucei.* (The photosynthetic Euglenozoa are considered in Ex. 17.)

D.1. *Trypanosoma*

D.2. *Trichonympha*

MATERIALS

Per student:

- prepared slide of *Trypanosoma* in blood smear
- prepared slide of *Trichonympha*
- compound microscope

PROCEDURE

1. Examine a prepared slide of human blood that contains the parasitic flagellate *Trypanosoma* (Figure 16-18), the cause of African sleeping sickness. This flagellate is transmitted from host to host by the bloodsucking tsetse fly. Note the **flagellum** arising from one end of the cell.
2. Another example of a flagellated protozoan is the termite-inhabiting *Trichonympha* (Figure 16-19). Study a prepared slide of these organisms. Examine the gut of the termite with the high-dry objective to find *Trichonympha.* Note the large number of **flagella** covering the upper portion of the cell, the more or less centrally located **nucleus** and wood fragments in the cytoplasm.

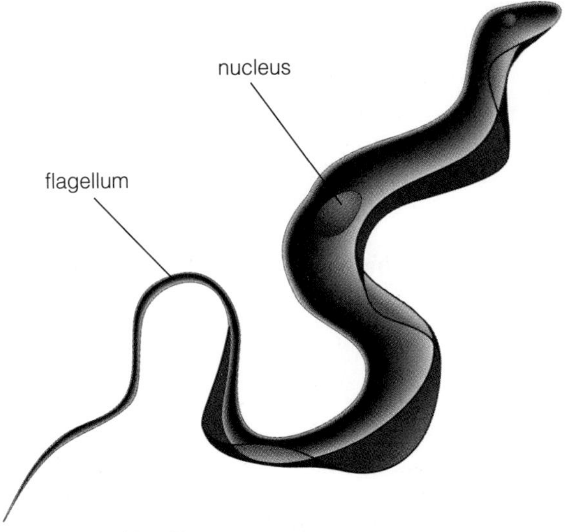

Figure 16-18 *Trypanosoma.*

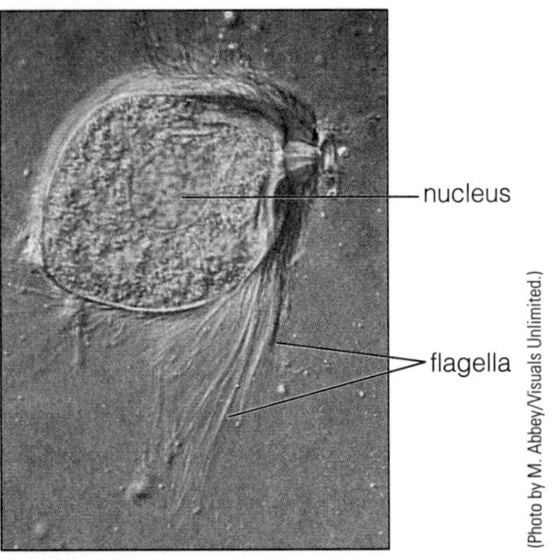

Figure 16-19 *Trichonympha* (900×).

The association of the termite and *Trichonympha* is an example of **obligate mutualism,** in which neither organism is capable of surviving without the other. Termites lack the enzymes to metabolize cellulose, a major component of wood. Wood particles ingested by termites are engulfed by *Trichonympha,* whose enzymes break the cellulose into soluble carbohydrates that are released for use by the termite.

PRE-LAB QUESTIONS

____ **1.** Some members of the Domain Bacteria have
(a) a nucleus
(b) membrane-bound organelles
(c) chloroplasts
(d) photosynthetic ability

____ **2.** A pathogen is
(a) a disease
(b) an organism that causes a disease
(c) a substance that kills bacteria
(d) the same as a heterocyst

____ **3.** Which of the following is true?
(a) All bacteria are autotrophic
(b) All protists are heterotrophic
(c) *Oscillatoria* is photoautotrophic
(d) *Trypanosoma* is photoautotrophic

____ **4.** Which organisms are characterized as decomposers?
(a) bacteria
(b) protozoans
(c) amoebas
(d) sporozoans

____ **5.** Organisms capable of nitrogen fixation
(a) include some bacteria
(b) include some cyanobacteria
(c) may live as symbionts with other organisms
(d) all of the above

____ **6.** A spherical bacterium is called a
(a) bacillus
(b) coccus
(c) spirillum
(d) none of the above

____ **7.** Gram stain is used to distinguish between different
(a) bacteria
(b) protistans
(c) dinoflagellates
(d) all of the above

____ **8.** Those organisms that are covered by numerous, tiny locomotory structures belong to the phylum
(a) Euglenozoa
(b) Stramenopila
(c) Amoebozoa
(d) Ciliata

____ **9.** A vector is
(a) an organism that causes disease
(b) a disease
(c) a substance that prevents disease
(d) an organism that transmits a disease causing organism

____ **10.** The organism that causes malaria is
(a) a pathogen
(b) *Plasmodium vivax*
(c) carried by a mosquito
(d) all of the above

Name ______________________ Section Number ______________________

EXERCISE 16

Bacteria and Protists I

POST-LAB QUESTIONS

Introduction

1. What major characteristic distinguishes bacteria from protists?

16.1A Bacteria

2. What form of bacterium is shown in this photomicrograph?

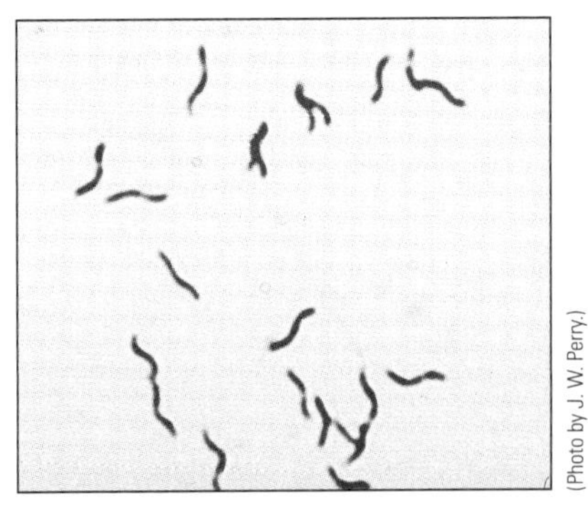
(Photo by J. W. Perry.)

(1100×).

16.1B Cyanobacteria

3. Examine the photomicrograph on the right, taken using an oil-immersion objective, the highest practical magnification of a light microscope. (The final magnification is 770×.)
   a. Based on your observation, identify the kingdom to which the organism belongs.
   b. Justify your answer.

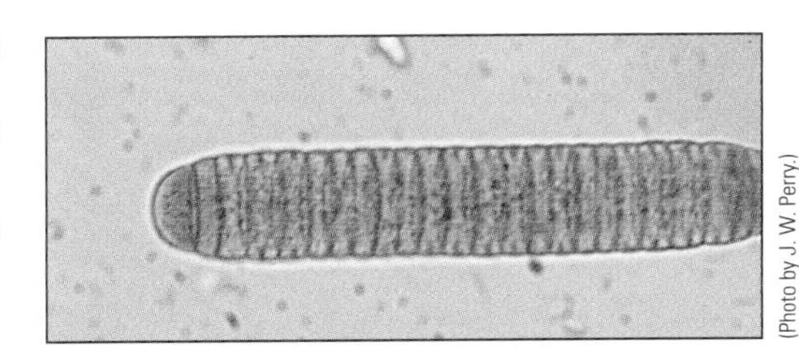
(Photo by J. W. Perry.)

(6000×).

4. Observe the photomicrograph on the right of an organism that was found growing symbiotically within the leaves of the water fern *Azolla*.
   a. What is the common name given an organism of this type?
   b. Give the name *and* function of the cell depicted by the line.

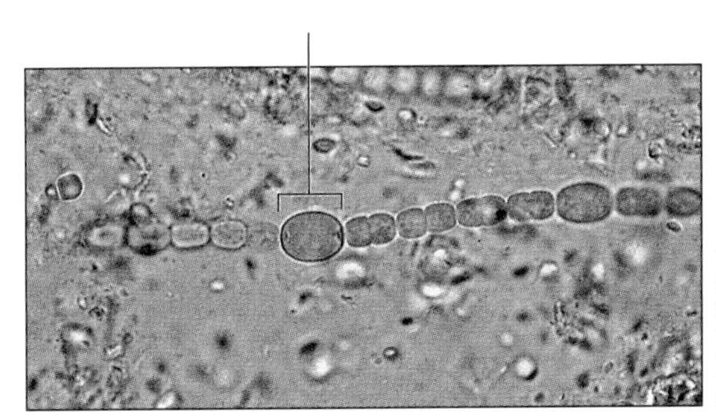

(Photo by J. W. Perry.)

(432×).

16.2A Phylum Stramenopila

5. You've never seen the protistan whose sexual structures appear here, but you have seen one very similar to it. What are the circular structures in photo?

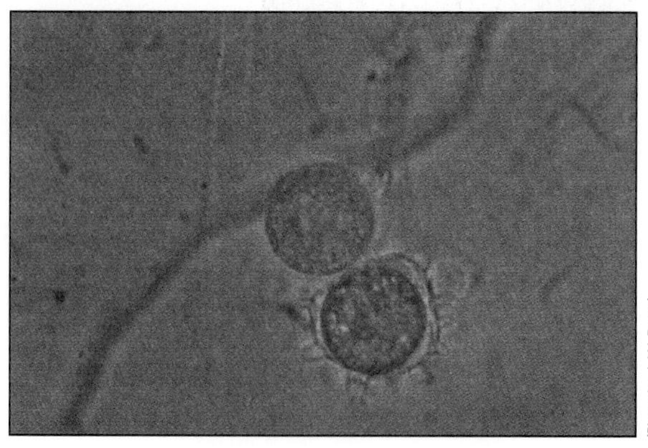

(287×).

(Photo by J. W. Perry.)

6. While the organism that resulted in the Irish potato famine of 1845–1847 had long been present, environmental conditions that occurred during this period resulted in the destructive explosion of disease. Indicate what those environmental conditions were and why they resulted in a major disease outbreak.

16.2B Phylum Amoebozoa

7. This photomicrograph was taken from a prepared slide of a stained specimen. You observed unstained living specimens in lab. Describe the mechanism by which it moves from place to place.

8. What is phagocytosis? What function does it serve?

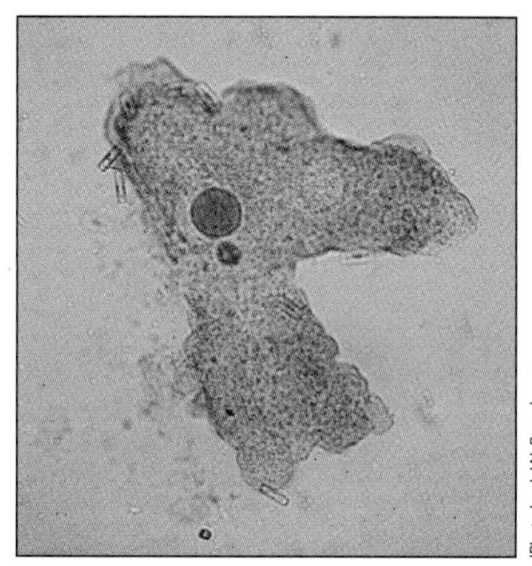

(425×).

(Photo by J. W. Perry.)

Food for Thought

9. If you were to travel to a region where rice is grown in paddies, you would see lots of the water fern *Azolla* growing in the water. A farmer would tell you this is done because *Azolla* is considered a "natural fertilizer." Explain why this is the case.

10. Based on your knowledge of the life history of *Plasmodium vivax,* suggest two methods for controlling malaria. Explain why each method would work.

a. ______________________________

b. ______________________________

EXERCISE 17

Protists II

OBJECTIVES

After completing this exercise, you will be able to

1. define *phytoplankton, phyt, phycobilin, agar, fucoxanthin, algin, kelp, gametangium, oogonia, antheridia;*
2. recognize and classify selected members of the seven phyla represented in this exercise;
3. distinguish among the structures associated with asexual and sexual reproduction described in this exercise;
4. identify the structures of the algae (those in **boldface** in the procedure sections).

INTRODUCTION

The organisms in this exercise make an enormous impact on the biosphere, both positively and negatively. Among their greatest contribution is the production of oxygen, because these are the photosynthetic protistans. Many are given the common name "algae."

Pond scum, frog spittle, seaweed, the stuff that clogs your aquarium if it's not cleaned routinely, the debris on an ocean beach after a storm at sea, the nuisance organisms of a lake—these are the images that pop into our minds when we first think about the organisms called algae. But many algae are also **phytoplankton,** the weakly swimming or floating algae, at the base of the aquatic food chain.

The following organisms are examined in this exercise:

Phylum	Common Name
Euglenozoa	Euglenoids
Chrysophyta	Yellow-green algae and diatoms
Dinoflagellata	Dinoflagellates
Rhodophyta	Red algae
Phaeophyta	Brown algae
Chlorophyta	Green algae
Charophyta	Desmids and stoneworts

Note the ending *-phyta* for all the phylum names. This derives from the Greek root word *phyt,* which means "plant." In the not too distant past, these organisms were considered plants because they were photosynthetic. While the endings remain, today they are considered to be members of the kingdom Protista.

17.1 Phylum Euglenozoa: Euglenoids *(About 15 min.)*

Euglenoids are motile, unicellular protists. Most species are photosynthetic, but there are also some heterotrophic species (see Ex. 16).

MATERIALS

Per student:

- microscope slide
- coverslip
- compound microscope

Per group (table):

- methylcellulose in 2 dropping bottles
- dH_2O in 2 dropping bottles

Per lab room:

- *Euglena*-living culture; disposable pipet

PROCEDURE

1. From the culture provided, prepare a wet mount slide of *Euglena.*
2. Observe with the medium-power objective of your compound microscope. Notice the motion of these green cells as they swim through the medium. If they're swimming too rapidly, prepare another slide, but add a drop of methylcellulose to the cell suspension before adding a coverslip.
3. Switch to the high-dry objective for more detailed observation. Figure 17-1a will serve as a guide in your study. This specimen was photographed by a special technique to give it a three-dimensional appearance.
4. Within the **cytoplasm,** identify the green **chloroplasts** and, if possible, the centrally located **nucleus.**
5. By closing the microscope's diaphragm to increase the contrast, you may be able to locate the **flagellum** at one end of the cell.
6. Search for the orange **eyespot,** a photoreceptive organelle located within the cytoplasm at the base of the flagellum.

In the world of microscopic swimmers, there are two types of flagella. One type, the *whiplash flagellum,* pushes the organism through the medium. An example of this is a human sperm cell. The other type, with which you're probably less familiar, is the *tinsel flagellum,* which pulls the organism through its watery environment. Tinsel flagella have tiny hairlike projections, visible only with the electron microscope.

7. Notice the direction of motion of the *Euglena* cells. Which type of flagellum does *Euglena* have?

____

Besides seeing the swimming motion caused by the flagellum, you may observe a contractionlike motion of the entire cell (*euglenoid movement*). *Euglena* is able to contort like this because it lacks a rigid cell wall. Instead, flexible helical interlocking proteinaceous strips within the cell membrane delimit the cytoplasm. These strips plus the cell membrane form the **pellicle.** Euglenoid movement provides a means of locomotion for mud-dwelling organisms.

8. Examine Figure 17-1b, a scanning electron micrograph showing the *flagellum* and the helical strips of the pellicle.
9. In Figure 17-2, make a series of sketches showing the different shapes *Euglena* takes during euglenoid movement.

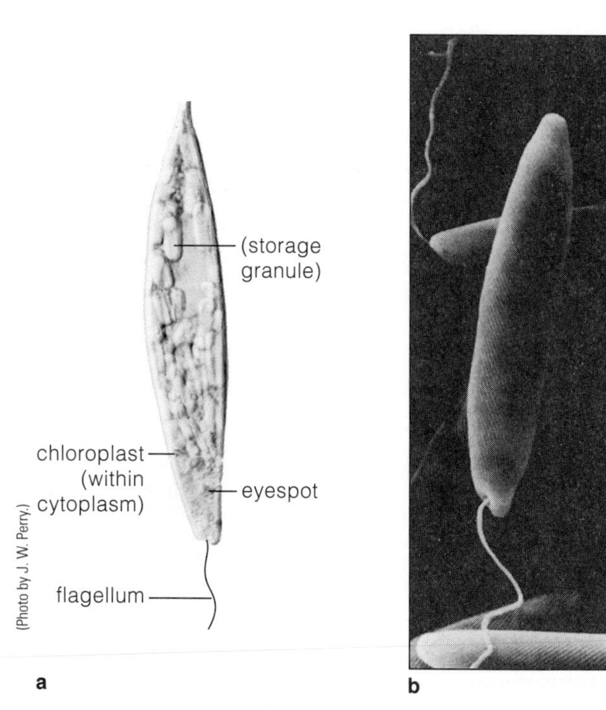

Figure 17-1 *Euglena.* (**a**) Light micrograph (550×). (**b**) Scanning electron micrograph (1,700×).

Figure 17-2 Different shapes possible in living *Euglena* exhibiting euglenoid movement.

17.2 Phylum Chrysophyta: Diatoms *(About 15 min.)*

The phylum Chrysophyta includes yellow green algae, and diatoms. You will examine only the diatoms here.

Diatoms are called the organisms that live in glass houses because their cell walls are composed largely of opaline *silica* ($SiO_2 \cdot nH_2O$). Diatoms are important as primary producers in the food chain of aquatic environments, and their cell walls are used for a wide variety of industrial purposes, ranging from the polishing agent in toothpaste to a reflective roadway paint additive. Massive deposits of cell walls of long-dead diatoms make up diatomaceous earth (Figure 17-3).

MATERIALS

Per student:

- microscope slide
- coverslip
- dissecting needle
- prepared slide of freshwater diatom
- compound microscope

Per group (table):

- diatomaceous earth
- dH_2O in 2 dropping bottles

Per lab room:

- diatoms—living culture; disposable pipet

(Photo courtesy Dan Williams.)

Figure 17-3 Diatomaceous earth quarry near Quincy, Washington.

PROCEDURE

1. Using a dissecting needle, scrape a small amount of diatomaceous earth onto a microscope slide and prepare a wet mount slide.
2. Examine your slide, starting with the low magnification objective, proceeding through the highest magnification available on your compound microscope. Close the iris diaphragm to increase the contrast.
3. You are looking at only the silica opaline shell. The cytoplasm of this organism disappeared hundreds of thousands, if not millions of years ago.
4. Note the tiny holes in the shells.
5. Now prepare a wet mount of living diatoms. Use the high-dry objective to note the golden brown chloroplasts within the cytoplasm and the numerous holes in the cell walls (shells).
6. In Figure 17-4, sketch several of the diatoms you are observing.
7. Now obtain a prepared slide of diatoms. These cells have been "cleaned," making the perforations in the cell wall especially obvious if you close the iris diaphragm on your microscope's condenser to increase the contrast. Study with the high-dry objective.

The pattern of the holes in the walls are characteristic of a given species. Before the advent of electronic techniques, microscopists observed diatom walls to assess the quality of microscope lenses. The resolving power (see discussion of resolving power in Exercise 3) could be determined if one knew the diameter of the holes under observation.

Figure 17-4 Drawing of diatoms (________×).

17.3 Phylum Dinoflagellata: Dinoflagellates *(About 15 min.)*

Dinoflagellates spin as they move through the water due to the position of their flagellum. On occasion, populations of certain dinoflagellates may increase dramatically, causing the seas to turn red or brown. These are the **red tides,** which can devastate fish populations because neurotoxins produced by the dinoflagellates poison fish.

MATERIALS

Per student:

- prepared slide of a dinoflagellate (for example, *Gymnodinium, Ceratium,* or *Peridinium*)
- compound microscope

Per lab room:

- dinoflagellate—living culture; disposable pipet (optional)

PROCEDURE

1. With the high-dry objective of your compound microscope, examine a prepared slide or living representative.
2. Compare what you are observing with Figures 17-5a and b.
3. Attempt to identify the stiff cellulosic plates encasing the cytoplasm.
4. Locate the two grooves formed by the junction of the plates. This is where the flagella are located. If you are examining living specimens, chloroplasts may be visible beneath the cellulose plates.
5. Examine the scanning electron micrograph of a dinoflagellate in Figure 17-6.

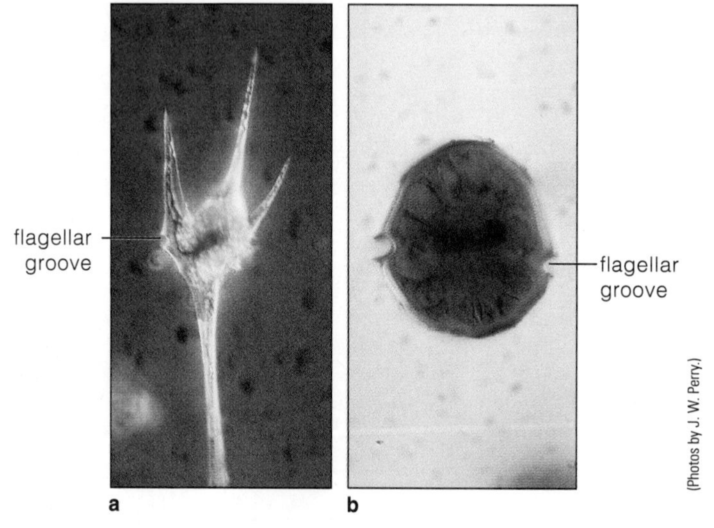

Figure 17-5 Two representative dinoflagellates. (**a**) *Ceratium* (240×). (**b**) *Peridinium* (640×).

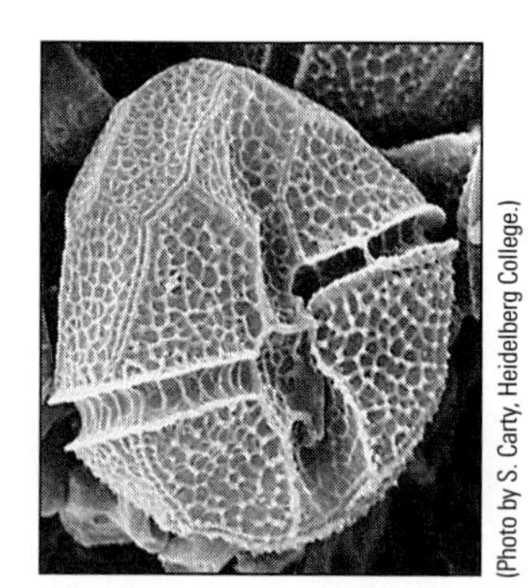

Figure 17-6 Scanning electron micrograph of a dinoflagellate (1000×).

17.4 Phylum Rhodophyta: Red Algae *(About 15 min.)*

Although commonly called red algae, members of the Rhodophyta vary in color from red to green to purple to greenish-black. The color depends on the quantity of their photosynthetic accessory pigments, the **phycobilins,** which are blue and red. These accessory pigments allow capture of light energy across the entire visible spectrum. This energy is passed on to chlorophyll for photosynthesis. One phycobilin, the red phycoerythrin, allows some red algae to live at great depths where red wavelengths, those of primary importance for green and brown algae, fail to penetrate.

Which wavelengths (colors) would be absorbed by a red pigment?

____

Most abundant in warm marine waters, red algae are the source of **agar** and carageenan substances extracted from their cell walls. Agar is the solidifying agent in microbiological media and carageenan is a thickener used in ice cream and bean.

MATERIALS

Per lab room:

- culture dish containing dried agar and petri dish of hydrated agar
- demonstration slide of *Porphyridium*
- demonstration specimen and slide of *Porphyra* (nori)

PROCEDURE

1. Observe the dried agar that is on demonstration.
2. Now observe the petri dish containing agar that has been hydrated, heated, and poured into the dish.
3. Examine these cells of the unicellular *Porphyridium* at the demonstration microscope, noting the reddish chloroplast.
4. In Figure 17-7, sketch a cell of *Porphyridium*.
5. Examine a portion of the multicellular membranous *Porphyra* specimen on demonstration. Draw what you see in Figure 17-8.
6. Compare the specimen with the photo of live *Porphyra* (Figure 17-9).
7. In the adjacent demonstration microscope, note the microscopic appearance of *Porphyra* (Figure 17-10). Note that the clear areas are actually the cell walls.

Figure 17-7 Drawing of *Porphyridium* (______×).

Porphyra is used extensively as a food substance in Asia, where it's commonly sold under the name *nori*. In Japan, nori production is valued at $20 million annually.

Figure 17-8 Drawing of the macroscopic appearance of *Porphyra* (nori) (______×).

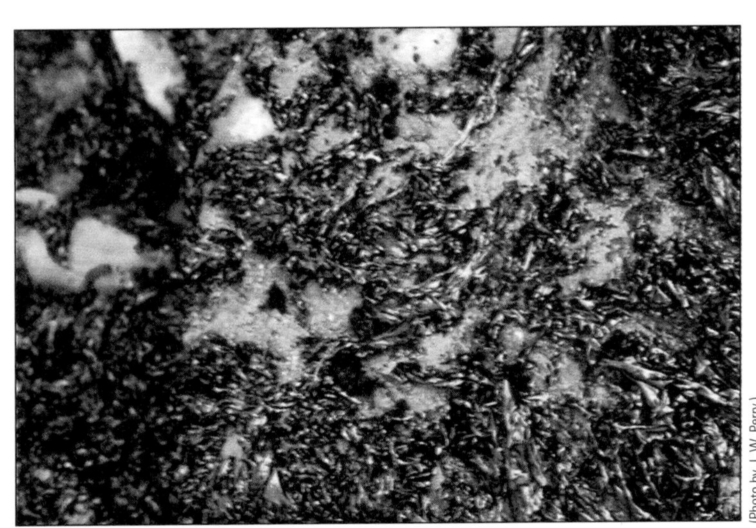

Figure 17-9 Living *Porphyra* (0.5×).

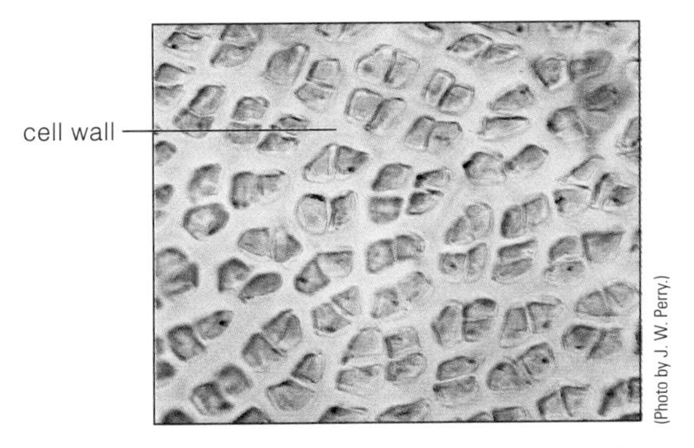

Figure 17-10 Microscopic appearance of *Porphyra* (nori) (278×).

17.5 Phylum Phaeophyta: Brown Algae *(About 10 min.)*

The vast majority of the brown algae are found in cold, marine environments. All members are multicellular, and most are macroscopic. Their color is due to the accessory pigment fucoxanthin, which is so abundant that it masks the green chlorophylls. Some species are used as food, while others are harvested for fertilizers. Of primary economic importance is **algin,** a cell wall component of brown algae that makes ice cream smooth, cosmetics soft, and paint uniform in consistency, among other uses.

A. *Kelps*—Laminaria and Macrocystis

Kelps are large (up to 100 m long), complex brown algae. They are common along the seashores in cold waters.

MATERIALS

Per lab room:

- demonstration specimen of *Laminaria*
- demonstration specimen of *Macrocystis*

PROCEDURE

Examine specimens of *Laminaria* (Figure 17-11) and *Macrocystis* (Figure 17-12). On each, identify the rootlike **holdfast** that anchors the alga to the substrate; the **stipe,** a stemlike structure; and the leaflike **blades.**

B. *Rockweed*—Fucus

Fucus is a common brown alga of the coastal shore, especially abundant attached to rocks where the plants are periodically wetted by splashing waves and the tides.

MATERIALS

Per lab room:

- demonstration specimen of *Fucus*

PROCEDURE

1. Examine demonstration specimens of *Fucus* (Figure 17-13), noting the branching nature of the body.
2. Locate the short **stipe** and **blade.**

Figure 17-11 *Laminaria.*

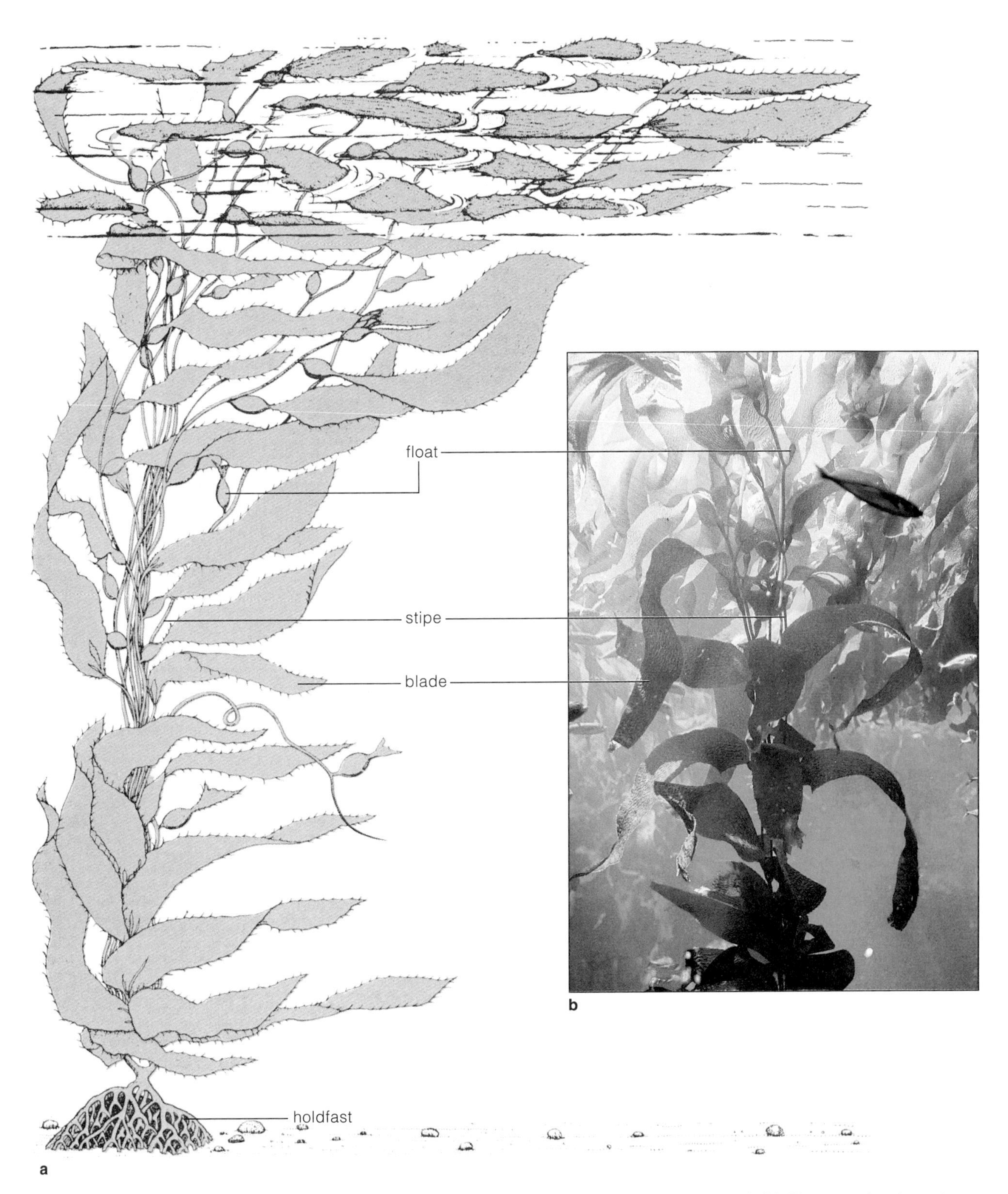

Figure 17-12 *Macrocystis.* (**a**) Artist conception. (After M. Neushul in Scagel et al., 1982.) (**b**) Photograph taken at Monterey Bay Aquarium, Monterey, CA. (Photo by J. W. Perry.) (Both 0.01×)

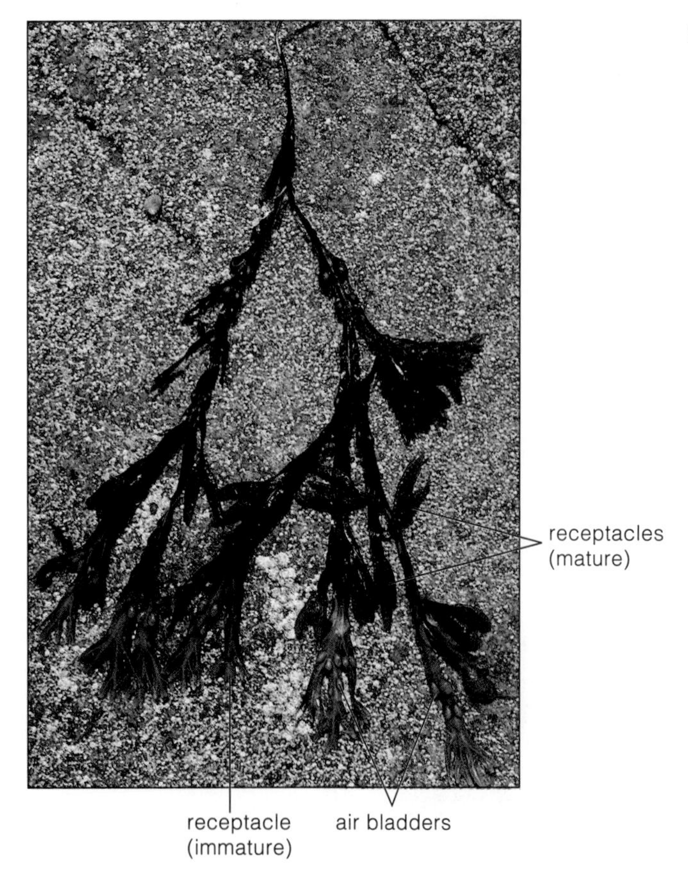

Figure 17-13 The rockweed, *Fucus* (0.60×). (Photo by J. W. Perry.)

3. Examine the swollen and inflated blade tips (Figure 17-14), housing the sex organs of the plant and apparently also serving to keep the plant buoyant at high tide. Notice the numerous tiny dots on the surface of these inflated ends. These are the openings through which motile sperm cells swim to fertilize the enclosed, nonmotile egg cells.

Figure 17-14 *Fucus* (colony, showing swollen tips that house the sex organs (0.35×).

17.6 Phylum Chlorophyta: Green Algae *(About 60 min.)*

The Chlorophyta is a diverse assemblage of green organisms, ranging from motile and nonmotile unicellular forms to colonial, filamentous, membranous, and multinucleate forms. Not only do they include the most species, they are also important phylogenetically because ancestral green algae gave rise to the land plants. As you would expect, the photosynthetic pigments — both primary and accessory — and the stored carbohydrates are identical in the green algae and land plants.

In this section, you will examine the green algae from a morphological standpoint, starting with unicellular forms. For some, you'll also study their modes of reproduction and identify their characteristic sex organs, the gametangia (the singular is *gametangium*).

A. *Chlamydomonas—A Motile Unicell*

MATERIALS

Per student:

- microscope slide
- coverslip
- compound microscope

Per student pair:

- tissue paper
- I_2KI in dropping bottle

Per lab room:

- living culture of *Chlamydomonas,* disposable pipet

PROCEDURE

1. Prepare a wet mount of *Chlamydomonas* cells from the culture provided.
2. First examine with the low-power objective of your compound microscope. Notice the numerous small cells swimming across the field of view.
3. It will be difficult to study the fast-swimming cells, so kill the cells by adding a drop of I_2KI to the edge of the coverslip; draw the I_2KI under the coverslip by touching a folded tissue to the opposite edge. To observe the cells, switch to the highest power objective available.

4. Identify the large, green cup-shaped **chloroplast** filling most of the cytoplasm.
5. *Chlamydomonas* stores its excess photosynthate as *starch grains,* which appear dark blue or black when stained with I_2KI. Locate the starch grains.
6. If the orientation of the cell is just right, you may be able to detect an orange **stigma** (eyespot) that serves as a light receptor.
7. Close the iris diaphragm to increase the contrast and find the two **flagella** at the anterior end of the cell.
8. Examine Figure 17-15, a transmission electron micrograph of *Chlamydomonas.* The electron microscope makes much more obvious the structures you can barely see with your light microscope. Notice the magnification.
9. The green algae have a specialized center for starch synthesis located within their chloroplast, the **pyrenoid.** Find the pyrenoid.

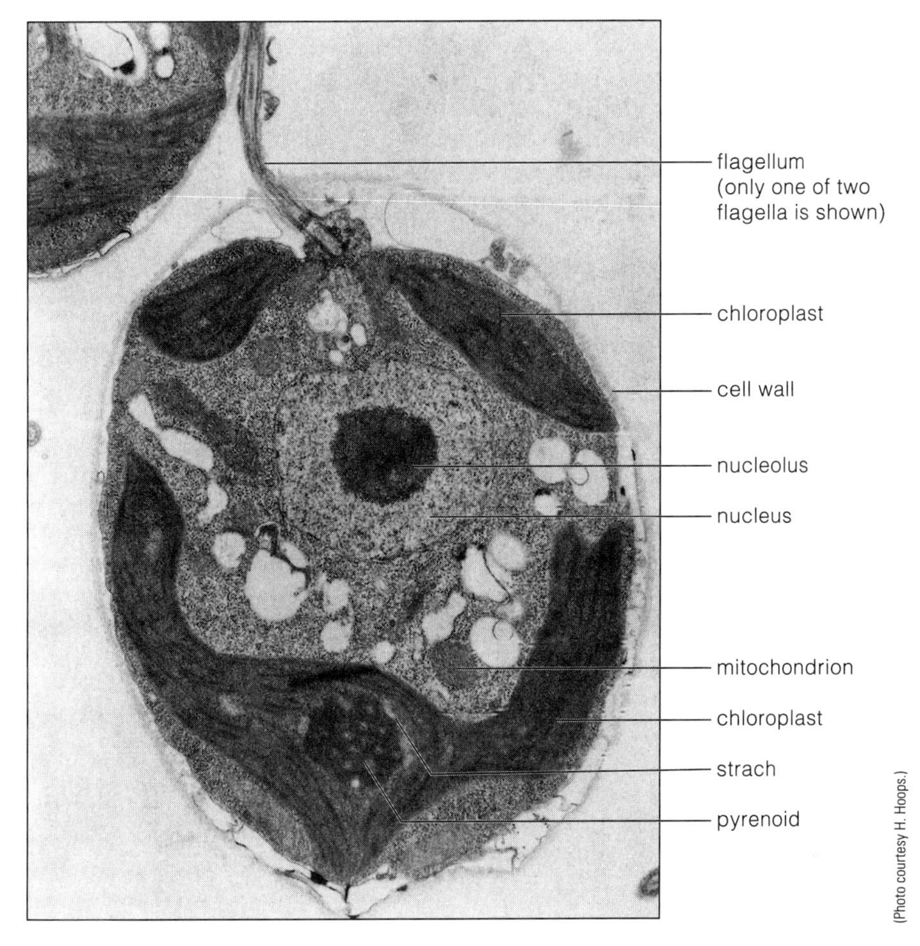

Figure 17-15 Transmission electron micrograph of *Chlamydomonas* (9750×).

B. Volvox*—A Motile Colony*

MATERIALS

Per student:

- depression slide
- compound microscope

Per lab room:

- living culture of *Volvox,* disposable pipet or prepared slide of *Volvox*

PROCEDURE

1. From the culture provided, place a drop of *Volvox*-containing culture solution on a depression slide. (A prepared slide may be substituted if living specimens are not available.)
2. Observe the large motile **colonies** (Figure 17-16), first with the microscope's low- and medium-power objectives. The colony is a hollow cluster of mostly identical, *Chlamydomonas*-like cells that are held together by a **gelatinous matrix.**
3. Identify the gelatinous matrix, which appears as the transparent region between individual cells.

Each cell possesses flagella. As the flagella beat, the entire colony rolls through the water. (The scientific name, *Volvox,* comes from the Latin word *volvere,* which means "to roll.")

4. Asexual reproduction takes place by **autocolony (daughter colony) formation.** Certain cells within the colony divide and then round up into a sphere called an **autocolony.** Find the autocolonies within your specimen.

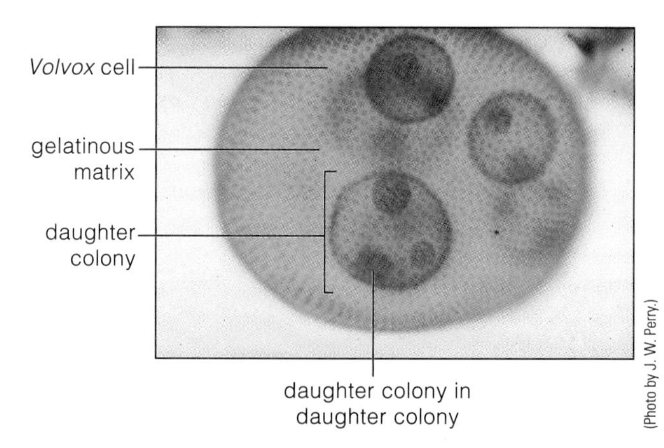

Figure 17-16 *Volvox,* with autocolonies (56×).

C. Oedogonium—*A Nonmotile Filament*

Oedogonium (Figure 17-17) is a filamentous, zoospore-producing green alga that is found commonly attached to sides of aquaria and slow-moving freshwater streams.

MATERIALS

Per student:

- microscope slide
- coverslip
- prepared slide of *Oedogonium*
- compound microscope

Per lab room:

- living culture of *Oedogonium*

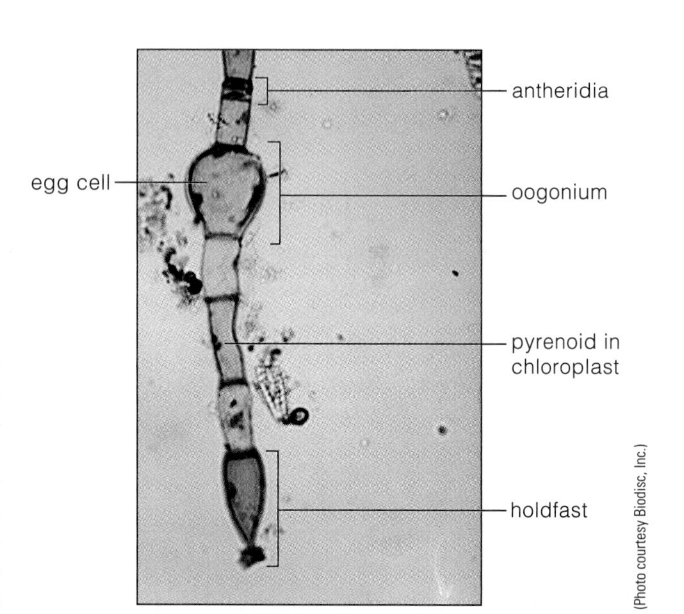

Figure 17-17 *Oedogonium,* with gametangia (250×).

PROCEDURE

1. Prepare a wet mount slide from the living culture provided and observe with the medium-power and high-dry objectives of your compound microscope.
2. Within each cell, locate the single, netlike **chloroplast** with its many **pyrenoids.** The cell that attaches the filament to the substrate is specialized as a **holdfast.** Scan your specimen to determine whether any holdfasts are present at the ends of the filaments. Each cell in the filament contains a single nucleus and a large central vacuole (both of which are difficult to distinguish).

Oedogonium reproduces both asexually and sexually. Asexual reproduction takes place by zoospore production. Sexual reproduction is easy to observe on prepared slides, and illustrates features commonly found in many other algae, land plants, and animals. These feature include the presence of a very large non-motile egg and motile sperm.

3. Obtain a prepared slide of a bisexual species of *Oedogonium,* which bears both male and female gametangia on the same filament. Examine with the medium-power objective of your compound microscope.
4. Find the large, spherical **oogonia.** These are the female sex organs. Within each oogonium locate a single, large **egg cell,** which virtually fills the oogonium.

Now find the male sex organs, the **antheridia,** which appear as short, boxlike cells. Each antheridium produces two *sperm.*

When the egg cell is fertilized by a sperm, a *zygote* is formed. The zygote develops a thick, heavy wall. Some oogonia on your slide may contain zygotes. When the zygote germinates, the diploid nucleus undergoes meiosis, producing motile spores. When the spore settles down, mitosis and cell division occur, producing the haploid filament you have examined.

Eventually, the two nuclei of each gamete fuse to form a **zygote,** which develops a thick wall. This thick-walled zygote serves as an overwintering structure. In the spring, the zygote nucleus undergoes meiosis. Three of the four nuclei die, leaving one functional, haploid nucleus. Germination of this haploid cell results in the formation of a haploid filament.

D. Ulva—A Membranous Form

A final representative of the green algae illustrates the fourth morphological form in the group, those having a membranous (tissuelike) body.

MATERIALS

Per lab room:

- demonstration specimen of *Ulva*

PROCEDURE

Examine living or preserved specimens of *Ulva*, commonly known as sea lettuce (Figure 17-18). The broad, leaflike body is called a *thallus*, a general term describing a vegetative body with relatively little cell differentiation. The *Ulva* body is only two cells thick!

Figure 17-18 The sea lettuce, *Ulva* (0.5×).

17.7 Phylum Charophyta: Desmids and Stoneworts

A. Spirogyra

A filamentous charophyte common to freshwater ponds is *Spirogyra*. This alga is often called pond scum because it forms a bright green, frothy mass on and just below the surface of the water.

MATERIALS

Per student:

- microscope slide
- coverslip
- dissecting needle
- prepared slide of *Spirogyra*
- compound microscope

Per student pair:

- diluted India ink in dropping bottle

Per lab room:

- living culture of *Spirogyra*

PROCEDURE

1. Observe the *Spirogyra* in the large culture dish. Pick up some of the mass, noting the slimy sensation. This is due to the watery sheath surrounding each filament.
2. Using a dissecting needle, place a few filaments on a slide and add a drop of diluted India ink before adding a coverslip. (Prepared slides may be used if living filaments are not available.)
3. Observe the filaments with the medium-power and high-dry objectives of your compound microscope (Figure 17-19).
4. Identify the **sheath,** which will appear as a bright area off the edge of the cell wall.
5. Note the spiral-shaped **chloroplast** with the numerous **pyrenoids.**
6. Each cell contains a large central **vacuole** and a single **nucleus.** Remember, the chloroplast is located within the cytoplasm, as is the nucleus. The nucleus, however, is suspended in strands of cytoplasm, much as a spider might be found in the center of a web. Look again at Figure 17-19 and locate the nucleus.

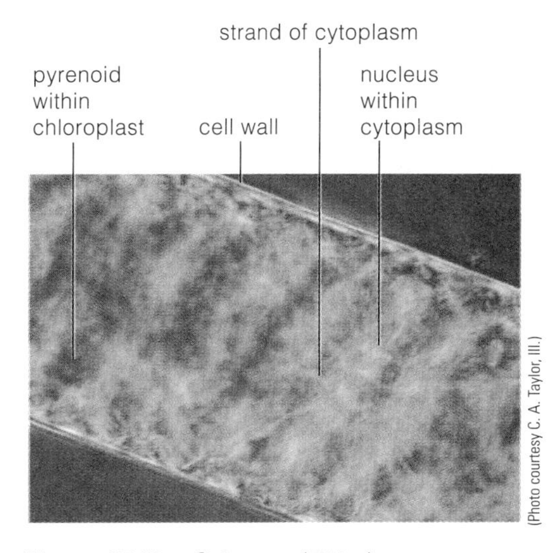

Figure 17-19 *Spirogyra* (550×).

Asexual reproduction occurs when a small portion of the filament simply breaks off and continues to grow. Zoospores are not formed. Sexual reproduction is illustrated in Figure 17-20. Male and gametes, and for that matter sex organs, look identical to one another.

7. Obtain a prepared slide illustrating sexual reproduction in *Spirogyra.* Examine the slide with your medium or high-dry objective.
8. Find two filaments that are joined by cytoplasmic bridges known as **conjugation tubes.** The entire cytoplasmic contents serve as **isogametes** (that is, gametes of similar size) in *Spirogyra,* with one isogamete moving through the conjugation tube into the other cell, where it fuses with the other gamete.
9. Find stages illustrating conjugation as in Figure 17-20.

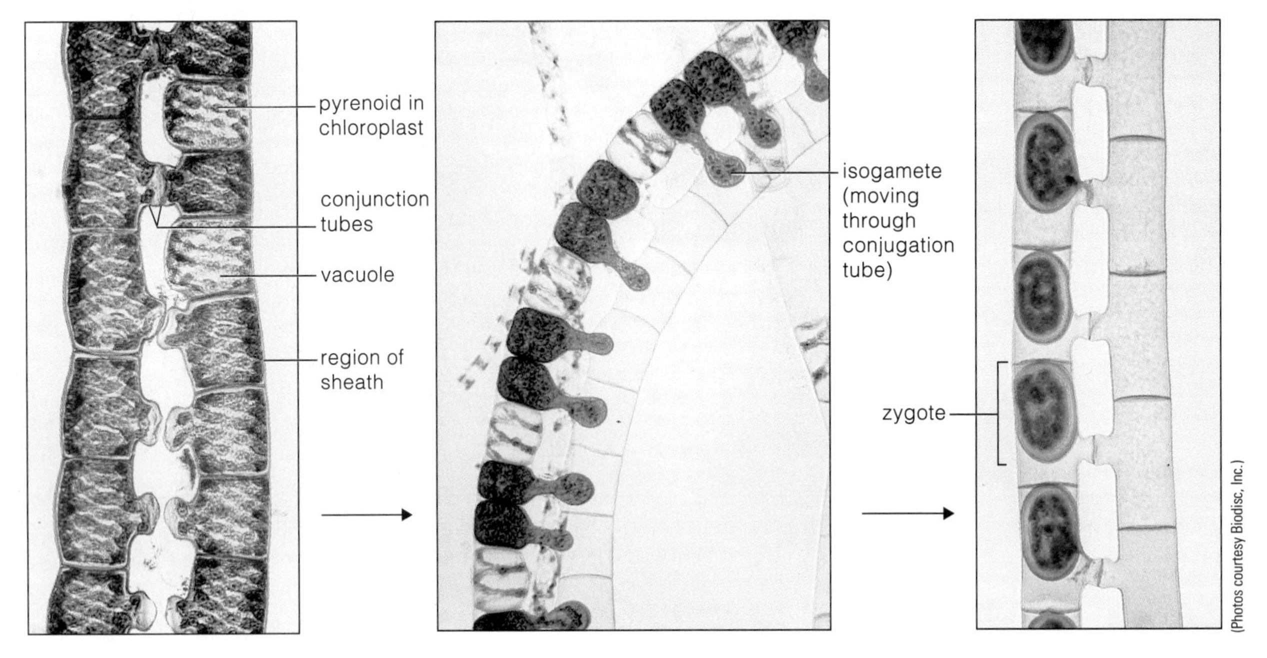

Figure 17-20 *Spirogyra,* stages in sexual reproduction (170×).

B. Chara *or* Nitella–*Stoneworts*

Stoneworts are within a green alga lineage that million of years ago gave rise to the land plants. They have a distinctive body form and sex organs, which are unlike those found in other green algae, and indeed remind one of tiny aquatic plants.

Stoneworts are interesting organisms found "rooted" in brackish and fresh waters, particularly those high in calcium. (Two of your lab manual authors—the Perrys—have ponds that are delightfully full of stoneworts.) They get their common names by virtue of being able to precipitate calcium carbonate over their surfaces, encrusting them and rendering them somewhat stony and brittle. (The suffix *-wort* is from a Greek word meaning "herb.")

MATERIALS

Per student:
- small culture dish
- dissecting microscope

Per lab room:
- living culture (or preserved) *Chara* or *Nitella*
- demonstration slide of *Chara* with sex organs

PROCEDURE

1. From the classroom culture provided, obtain some of the specimen and place it in a small culture dish partially filled with water.
2. Observe the specimen with a dissecting microscope. Note that the stoneworts resemble what we would think of as a plant. They are divided into "stems" and "branches" (Figure 17-21).
3. Search for flask-shaped and spherical structures along the stem. These are the sex organs. If none is present on the specimen, observe the demonstration slide (Figure 17-22) that has been selected to show these structures.
4. The flask-shaped structures are **oogonia,** each of which contains a single, large egg (Figure 17-23). Notice that the oogonium is covered with cells that twist over the surface of the gametangium. Because of the presence of these cells, the oogonium is considered to be a *multicellular* sex organ. Multicellular sex organs are present in all land plants.

Based on your study of previously examined specimens, would you say the egg is motile or nonmotile?

____

5. Now find a spherical **antheridium** (Figure 17-23). Like the oogonium, the antheridium is covered by cells and is also considered to be multicellular. Cells within the interior of the antheridium produce numerous flagellated sperm cells.

Fertilization of eggs by sperm produces a *zygote* within the oogonium. The zygote-containing oogonium eventually falls off from the parent plant. The zygote can remain dormant for some time before the nucleus undergoes meiosis in preparation for germination. Apparently, three of the four nuclei produced during meiosis disintegrate.

Figure 17-21 The stonewort, *Chara,* with sex organs (2×).

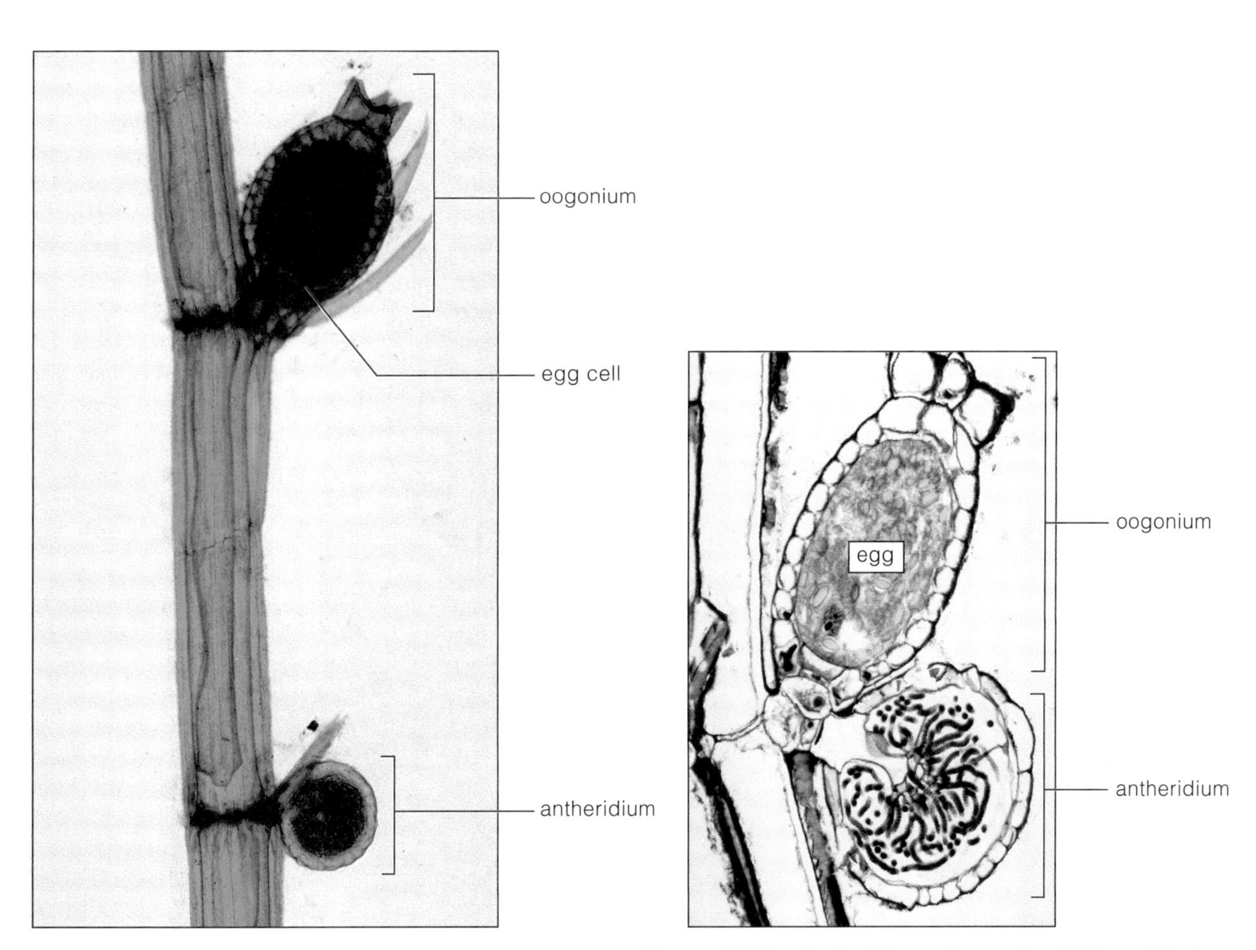

Figure 17-22 *Chara,* showing sex organs (30×). (Photo by J. W. Perry.)

Figure 17-23 Antheridia and oogonia of *Chara* (80×). (Photo by J. W. Perry.)

PRE-LAB QUESTIONS

_____ **1.** Which of these organisms (or parts of the organisms) might you find as an ingredient in toothpaste?
(a) euglenoids
(b) diatoms
(c) dinoflagellates
(d) stoneworts

_____ **2.** Red tides are caused by
(a) dinoflagellates
(b) red algae
(c) brown algae
(d) diatoms

_____ **3.** Agar is derived from
(a) red algae
(b) brown algae
(c) green algae
(d) all of the above

_____ **4.** Which is the correct plural form of the word for the organisms studied in this exercise?
(a) alga
(b) algae
(c) algas
(d) algaes

_____ **5.** Phycobilins are
(a) photosynthetic pigments
(b) found in the red algae
(c) blue and red pigments
(d) all of the above

_____ **6.** The cell wall component algin is
(a) found in the brown algae
(b) used in the production of ice cream
(c) used as a medium on which microorganisms are grown
(d) both a and b

_____ **7.** Specifically, female sex organs are known as
(a) oogonia
(b) gametangia
(c) antheridia
(d) zygotes

_____ **8.** A reagent that stains the stored food of a green alga black is
(a) India ink
(b) I_2KI
(c) methylene blue
(d) both a and b

_____ **9.** The starch production center within many algal cells is the
(a) nucleus
(b) cytoplasm
(c) stipe
(d) pyrenoid

_____ **10.** The phylum of organisms *most* closely linked to the evolution of land plants is the
(a) Chlorophyta
(b) Charophyta
(c) Phaeophyta
(d) Euglenozoa

Name ____________________ Section Number ____________________

EXERCISE 17

Protists II

POST-LAB QUESTIONS

17.1 Phylum Euglenozoa: Euglenoids

1. The protistan *Euglena* is often studied in plant-related courses because it is photosynthetic. What characteristic of the pellicle makes *Euglena* different from true plants?

17.2 Phylum Chrysophyta: Diatoms

2. On a field trip to a stream, you collect a leaf that has fallen into the water and scrape some of the material from its surface, and prepare a wet mount. You examine your preparation with the high-dry objective of your compound microscope, finding the organism pictured at the right. What substance makes up a significant portion of the cell wall?

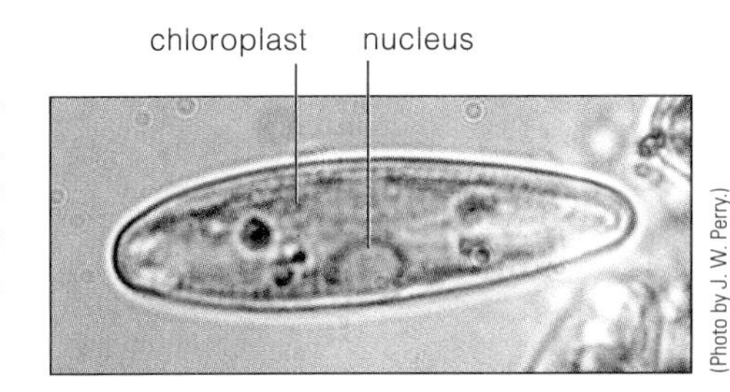

(235×).

(Photo by J. W. Perry.)

17.4 Phylum Rhodophyta: Red Algae

3. While wading in the warm salt water off the beaches of the Florida Keys on spring break, you stoop down to look at the feathery alga shown here.
   a. What pigment gives this organism its coloration?
   b. What commercial products are derived from this phylum?

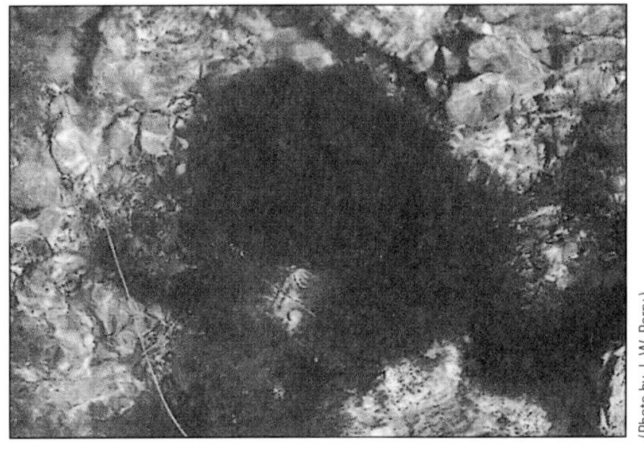

(0.25×).

(Photo by J. W. Perry.)

17.5 Phylum Phaeophyta: Brown Algae

4. While walking along the beach at Point Lobos, California, the fellow pictured walks up to you with alga in hand. Figuring you to be a college student who has probably had a good introductory biology course, he asks if you know what it is.
   a. What is the cell wall component of the organism that has commercial value?
   b. Name three uses for that cell wall component.

(Photo by J. W. Perry.)

17.6 Phylum Charophyta: Desmids and Stoneworts

5. Your class takes a field trip to a freshwater stream, where you collect the organism shown microscopically here.

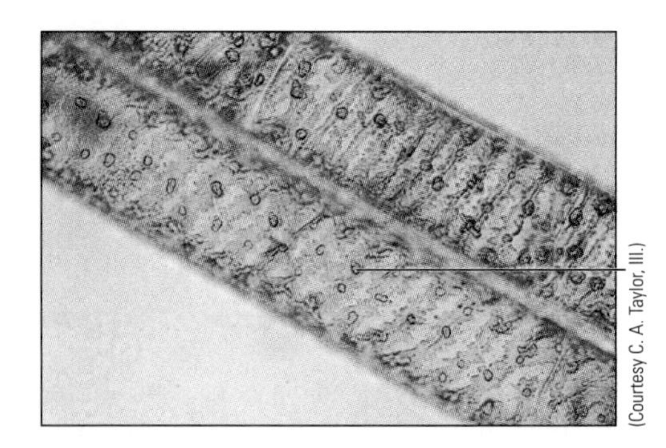

(160×).

a. What is the genus name of this organism?

b. Identify and give the function of the structure within the chloroplast at the end of the leader (line).

Food for Thought

6. Some botanists consider the stoneworts to be a link between the higher plants and the algae. As you will learn in future exercises, higher plants, such as the mosses, have both haploid and diploid stages that are *multicellular*.

a. Describe the multicellular organism in the charophytes.

b. Is this organism haploid or diploid?

c. Is the zygote haploid or diploid?

d. Is the zygote unicellular or multicellular?

7. a. What color are the marker lights at the edge of an airport taxiway?

____

b. Are the wavelengths of this color long or short, relative to the other visible wavelengths?

____

c. Which wavelengths penetrate deepest into water, long or short?

____

d. Make a statement regarding why phycobilin pigments are present in deep-growing red algae.

e. What benefit is there to the color of airport taxiway lights for a pilot attempting to taxi during foggy weather?

8. a. How is an algal holdfast similar to a root?

b. How is it different?

9. Why do you suppose the Swedish automobile manufacturer Volvo chose this company name?

10. List three reasons why algae are useful and important to life.

a.

b.

c.

Fungi

OBJECTIVES

After completing this exercise, you will be able to

1. define *parasite, saprobe, mutualist, gametangium, hypha, mycelium, multinucleate, sporangium, rhizoid, zygosporangium, ascus, conidium, ascospore, ascocarp, basidium, basidiospore, basidiocarp, lichen, mycorrhiza;*
2. recognize representatives of the major phyla of fungi;
3. distinguish structures that are used to place various representatives of the fungi in their proper phyla;
4. list reasons why fungi are important;
5. distinguish between the structures associated with asexual and sexual reproduction described in this exercise;
6. identify the structures (in **boldface**) of the fungi examined;
7. determine the effect of light on certain species of fungi.

INTRODUCTION

As you walk in the woods following a warm rain, you are likely to be met by a vast assemblage of colorful fungi. Some of them grow on dead or diseased trees, some on the surface of the soil, others in pools of water. Some are edible, some deadly poisonous.

Fungi (kingdom Fungi) are *heterotrophic* organisms; that is, they are incapable of producing their own food material. They secrete enzymes from their bodies that digest their food externally. The digested materials are then absorbed into the body.

Depending on the relationship between the fungus and its food source, fungi can be characterized in one of three ways:

1. Parasitic fungi (**parasites**) obtain their nutrients from the organic material of another living organism, and in doing so adversely affect the food source, often causing death.
2. Saprotrophic fungi (**saprobes**) grow on nonliving organic (carbon-containing) matter. Some are even **mutualists.**
3. Mutualistic fungi (**mutualists**) form a partnership beneficial to both the fungus and its host.

Fungi, along with the bacteria, are essential components of the ecosystem as decomposers. These organisms recycle the products of life, making the products of death available so that life may continue. Without them we would be hopelessly lost in our own refuse. Fungi and fungal metabolism are responsible for some of the food products that enrich our lives—the mushrooms of the field, the blue cheese of the dairy case, even the citric acid used in making soft drinks.

The kingdom Fungi is divided into four separate phyla, based on structures formed during sexual reproduction (Table 18.1). Certain fungi (the Imperfect Fungi) do not reproduce sexually; hence they are considered an informal "group."

TABLE 18-1 Classification of the Fungi

Phylum	Common Name
Chytridiomycota	Chytrids
Zygomycota	Zygosporangium-forming fungi
Ascomycota	Sac fungi
Basidiomycota	Club fungi
"Imperfect Fungi"	Fungi without sexual reproduction

18.1 Phylum Chytridiomycota *(About 5 min.)*

Commonly known as chytrids, these fungi are among the most simple in body form. While most live on dead organic matter, some are parasitic on economically important plants, resulting in damage and/or death of the host plant. Parasitic organisms that cause disease are called **pathogens.**

MATERIALS

Per lab room:

- preserved specimen of potato tuber with black wart disease

PROCEDURE

1. Examine the preserved specimen (if available) and Figure 18-1 of the potato tuber that exhibits the disease known as black wart. Notice the warty eruptions on the surface of the tuber. The warts are caused by the presence of numerous cells of a chytrid infecting the tuber.
2. Now look at Figure 18-2, a photomicrograph of a chytrid similar to that which causes black wart disease.

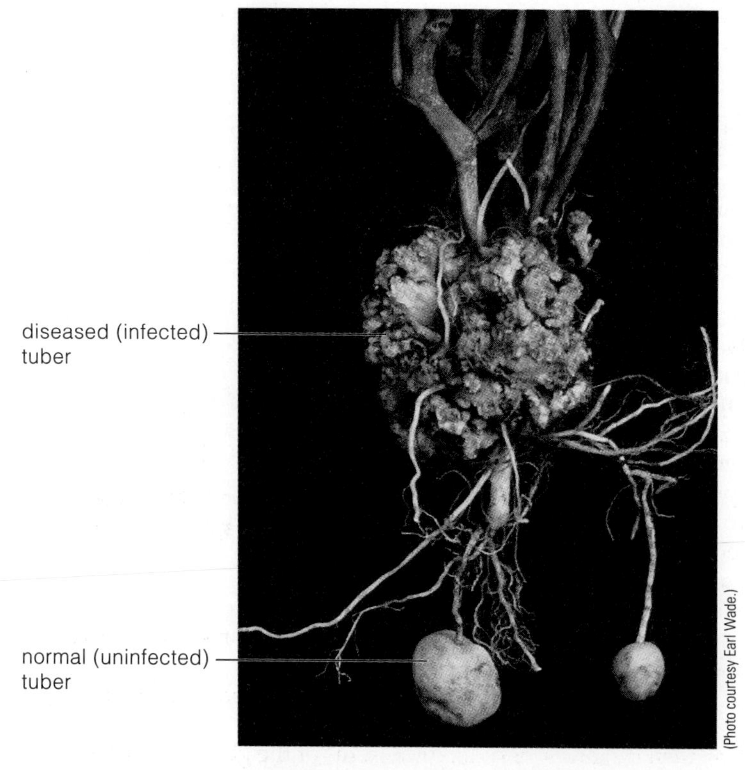

Figure 18-1 Potato tuber infected by black wart pathogen (0.5×).

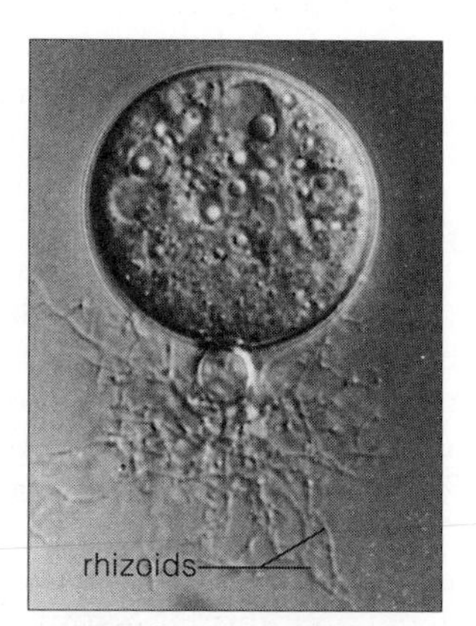

Figure 18-2 Chytrid. (71×). (Photo by M. S. Fuller, *Zoosporic Fungi in Teaching and Research,* Fuller and A. Jaworski (eds.), 1987, Southeastern Publishing Company, Athens, GA.)

18.2 Phylum Zygomycota: Zygosporangium-Forming Fungi *(About 30 min.)*

Commonly called "zygomycetes," all members of this phylum produce a thick-walled zygote called a **zygosporangium.** Most zygomycetes are saprobes. The common black bread mold, *Rhizopus,* is a representative zygomycete. Before the introduction of chemical preservatives into bread, *Rhizopus* was an almost certain invader, especially in high humidity.

MATERIALS

Per student:

- culture of *Rhizopus*
- prepared slide of *Rhizopus*
- dissecting needle
- glass microscope slide
- coverslip
- compound microscope

Per student pair:

- dH_2O in dropping bottle

Per lab room:

- demonstration culture of *Rhizopus* zygosporangia, on dissecting microscope

PROCEDURE

Examine Figure 18-7, the life cycle of *Rhizopus,* as you study this organism.

1. Obtain a petri dish culture of *Rhizopus.*
2. Observe the culture, noting that the body of the organism consists of many fine strands. These are called **hyphae** (Figure 18-3). All of the hyphae together are called the **mycelium.** Biologists use the term *mycelium* instead of saying "the fungal body."
3. Now note that the culture contains numerous black "dots." These are the **sporangia** (singular, *sporangium;* Figures 18-7a and b). Sporangia contain **spores** (Figures 18-7c–d) by which *Rhizopus* reproduces asexually.
4. Using a dissecting needle, remove a small portion of the mycelium and prepare a wet mount. Examine your preparation with the high-dry objective of your compound microscope.
5. It's likely that when you added the coverslip, you crushed the sporangia, liberating the spores. Are there many or few spores within a single sporangium? ______________
6. Identify the **rhizoids** (Figure 18-7a) at the base of a sporangium-bearing hypha. Rhizoids anchor the mycelium to the substrate.
7. Now observe the demonstration culture of sexual reproduction in *Rhizopus* (Figure 18-4). Two different mycelia must grow in close proximity before sexual reproduction can occur. (The difference in the mycelia is genetic rather than structural. Because they are impossible to distinguish, the mycelia are simply referred to as + and – mating types, as indicated in Figure 18-7.)
8. Note the black "line" running down the center of the culture plate. These are the numerous **zygosporangia,** the products of sexual reproduction. Here's how the zygosporangia are formed:
   - **(a)** As the hyphae from each mating type grow close together, chemical messengers produced within the hyphae signal them to produce protuberances (Figure 18-7f).
   - **(b)** When the protuberances make contact, gametangia (Figures 18-5, 18-7g) are produced at their tips. Each gametangium contains many haploid nuclei of a single mating type.
   - **(c)** The wall between the two gametangia then dissolves, and the cytoplasms of the gametangia mix.
   - **(d)** Eventually the many haploid nuclei from each gametangium fuse (Figure 18-7h). The resulting cell contains many diploid nuclei resulting from the fusion of gamete nuclei of opposite mating types. Each diploid nucleus is considered a *zygote.* This multinucleate cell is called a zygospore.

(Photo by J. W. Perry.)

Figure 18-3 Bread mold culture (0.5×).

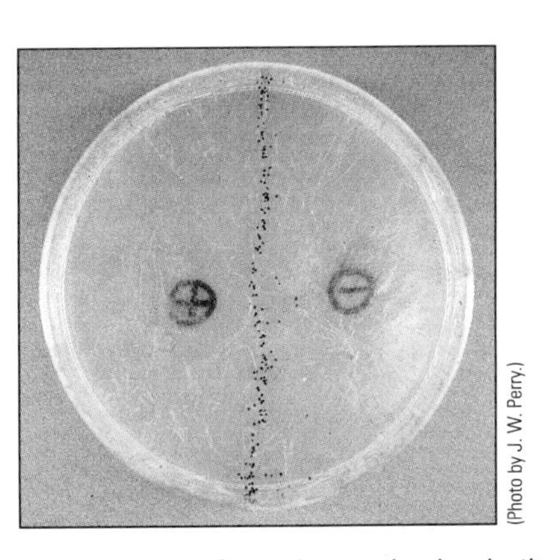

(Photo by J. W. Perry.)

Figure 18-4 Sexual reproduction in the bread mold. Black line consists of zygosporangia (0.5×).

(e) Eventually a thick, bumpy wall forms around this zygospore. Because germination of the zygospore results in the production of a sporangium, the thick-walled zygospore-containing structure is called a zygosporangium. Meiosis takes place inside the zygopsporangium so that the spores formed are haploid, some of one mating type, some of the opposite type (Figures 18-6, 18-7i).

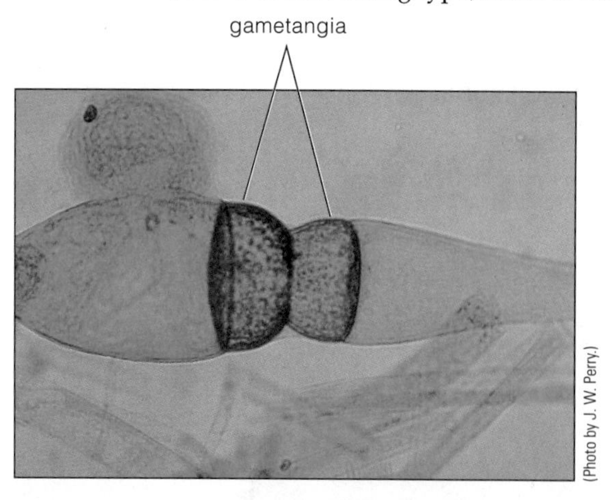

Figure 18-5 Gametangia of a bread mold (230×).

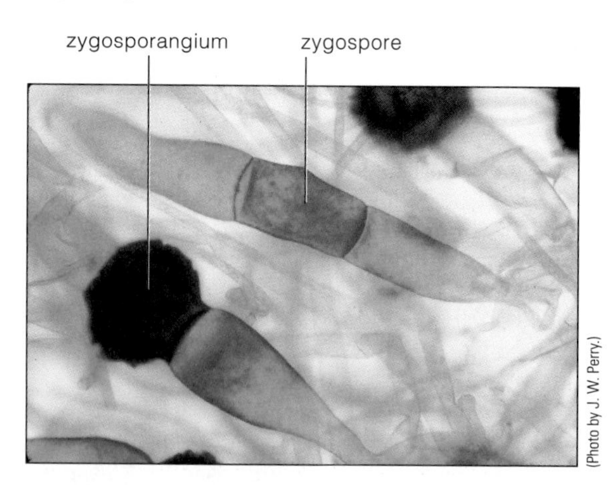

Figure 18-6 Zygospore and zygosporangium of the bread mold (230×).

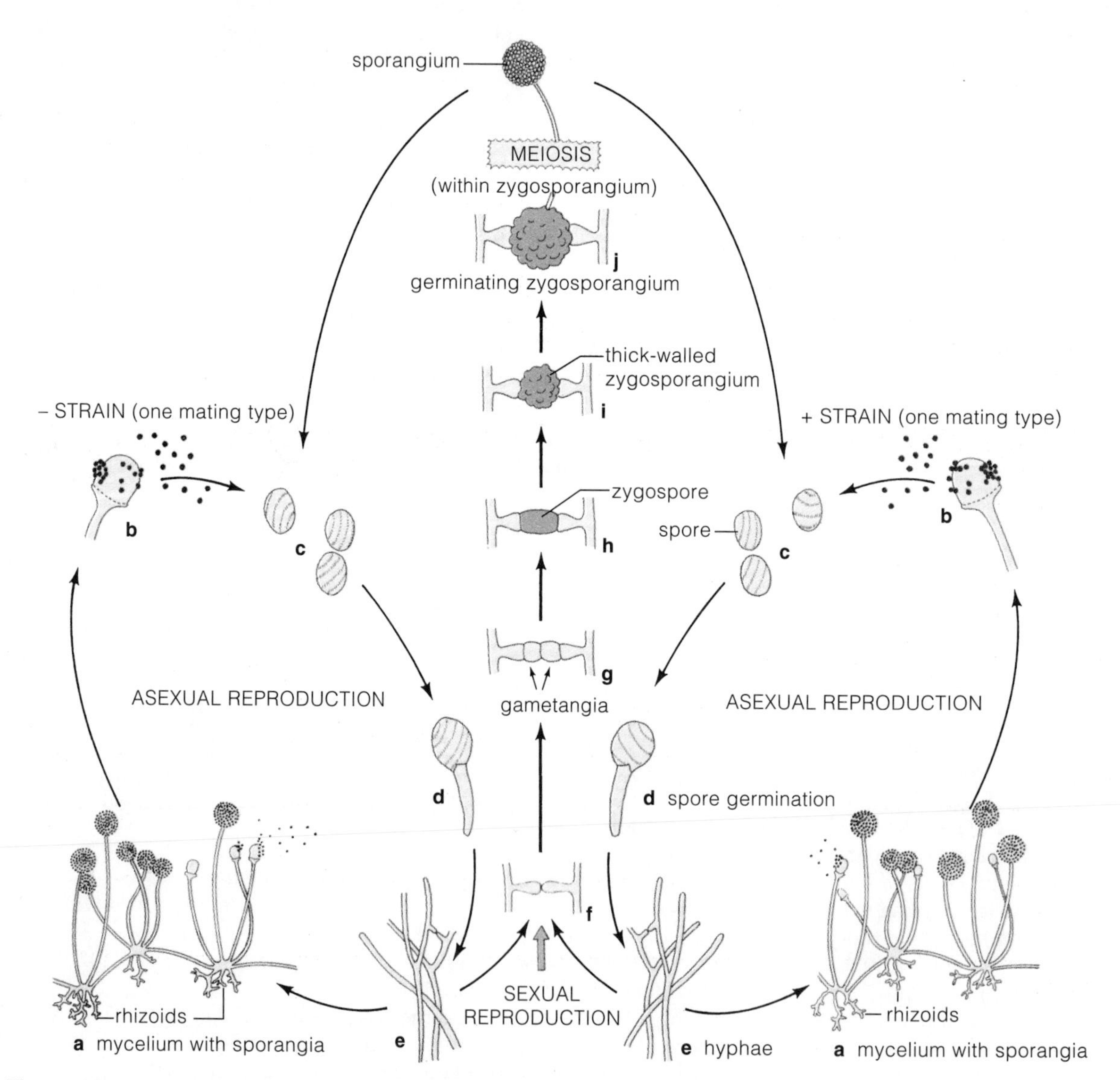

Figure 18-7 Life cycle of *Rhizopus*, black bread mold. (Green structures are 2n.)

9. Obtain a prepared slide of *Rhizopus*. Examine it with the medium- and high-power objective of your compound microscope.
10. Find the stages of sexual reproduction in *Rhizopus,* including gametangia, zygospores, and zygosporangia.

18.3 Experiment: Bread Mold and Food Preservatives *(Setup: about 10 min.)*

Knowing that fungal spores are all around us, why don't we have fungi covering everything we own? Well, given the proper environmental conditions just such a situation can occur. This is particularly true of food products. That's one reason we refrigerate many of our foodstuffs. Manufacturers often incorporate chemicals into some of our foods to retard spoilage, which includes preventing the growth of fungi.

This experiment addresses the hypothesis that *the chemicals placed in bread retard the growth of Rhizopus.* You may wish to review the discussion concerning experimental design in Exercise 1, "The Scientific Method."

MATERIALS

Per student group:

- one slice of bread without preservatives
- one slice of bread with preservatives
- suspension culture of bread mold fungus
- sterile dH_2O
- 2 large culture bowls
- plastic film
- pipet and bulb
- china marker
- metric ruler

PROCEDURE

1. With the china marker, label the side of one culture bowl "NO PRESERVATIVES," the other "WITH PRESERVATIVES." Place the appropriate slice of bread in its respective bowl.
2. Add a small amount of dH_2O to the bottom of each culture bowl. This provides humidity during incubation.
3. Add 5 drops of the spore suspension of the bread mold fungus to the center of each slice of bread.
4. Cover the bowls with plastic film and place them in a warm place designated by your instructor.
5. In Table 18-2, list the ingredients of both types of bread.
6. Make a prediction of what you think will be the outcome of the experiment, writing it in Table 18-2.
7. Check your cultures over the next three days, looking for evidence of fungal growth. Measure the diameter of the mycelium present in each culture, using a metric ruler.
8. Record your measurements and conclusion in Table 18-2, accepting or rejecting the hypothesis.

TABLE 18-2 Growth of Bread Mold

Ingredients			
Bread without preservatives:			
Bread with preservatives:			
Prediction:			
	Mycelial Diameter (cm)		
Culture	Day 1	Day 2	Day 3
With preservatives			
No preservatives			
Conclusion:			

Design an experiment that would allow you to identify the substance that has the effect you observed.

____

____

18.4 Phylum Ascomycota: Sac Fungi *(About 30 min.)*

Members of the ascomycetes produce spores in a sac, the **ascus** (plural, *asci*), which develops as a result of sexual reproduction. Asexual reproduction takes place when the fungus produces asexual spores called **conidia** (singular, *conidium*). The phylum includes organisms of considerable importance, such as the yeasts crucial to the baking and brewing industries, as well as numerous plant pathogens. A few are highly prized as food, including morels and truffles. Truffles cost in excess of $400 per pound!

MATERIALS

Per student:

- glass microscope slide
- coverslip
- dissecting needle
- prepared slide of *Peziza*
- compound microscope

Per student pair:

- culture of *Eurotium*
- dH_2O in dropping bottle
- large preserved specimen of *Peziza* or another cup fungus

A. Eurotium: *A Blue Mold*

The blue mold *Eurotium* gets its common name from the production of blue-walled asexual conidia.

PROCEDURE

1. Use a dissecting needle to scrape some **conidia** from the agar surface of the culture provided, then prepare a wet mount. (Try to avoid the yellow bodies—more about them in a bit.)
2. Refer to Figure 18-8 as you observe your preparation with the high-dry objective of your compound microscope.
3. Note that the conidia are produced at the end of a specialized hypha that has a swollen tip.

This arrangement has been named *Aspergillus.* This structure, the *conidiophore,* somewhat resembles an aspergillum used in the Roman Catholic Church to sprinkle holy water, from which its name is derived.

These tiny conidia are carried by air currents to new environments, where they germinate to form new mycelia.

(You may be confused about why we italicize *Aspergillus.* The reason is that this is a scientific name. Fungi other than *Eurotium* produce the same type of asexual structure; hence the asexual structure itself is given a scientific name.)

4. Note the yellow bodies on the culture medium. These are the *fruiting bodies,* known as **ascocarps,** which are the products of sexual reproduction (Figure 18-9).
5. With your dissecting needle, remove an ascocarp from the culture and prepare a wet mount.

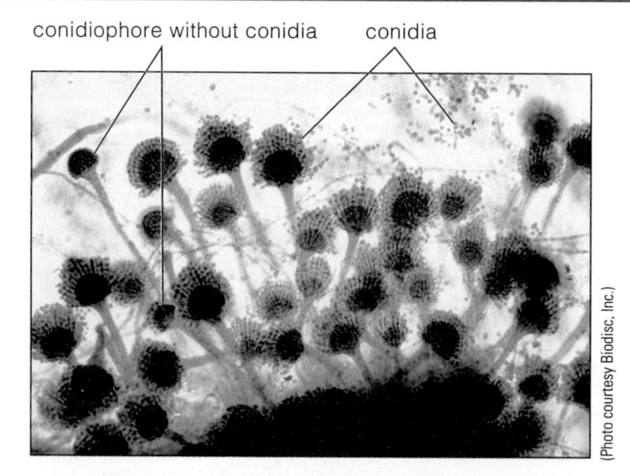

Figure 18-8 Conidiophore and conidia of the blue mold, *Eurotium* (138×).

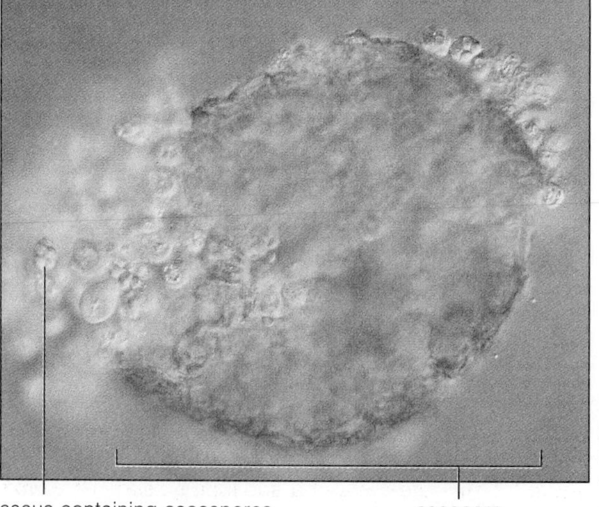

Figure 18-9 Crushed asocarp of *Eurotium* (287×).

6. Using your thumb, carefully press down on the coverslip to rupture the ascocarp.
7. Observe your preparation with the medium-power and high-dry objectives of your compound microscope. Identify the **asci,** which contain dark-colored, spherical **ascospores** (Figure 18-9).

The sexual cycle of the sac fungi is somewhat complex and is summarized in Figure 18-10.

The female gametangium, the **ascogonium** (Figure 18-10a; plural, *ascogonia*) is fertilized by male nuclei from antheridia (Figure 18-10a).

The male nuclei (darkened circles in Figure 18-10a) pair with the female nuclei (open circles) but do not fuse immediately.

Papillae grow from the ascogonium, and the paired nuclei flow into these papillae (Figure 18-10b).

Now cell walls form between each pair of sexually compatible nuclei (Figure 18-10c).

Subsequently, the two nuclei fuse; the resultant cell is the diploid ascus (Figure 18-10d). (Consequently, the ascus is a zygote.)

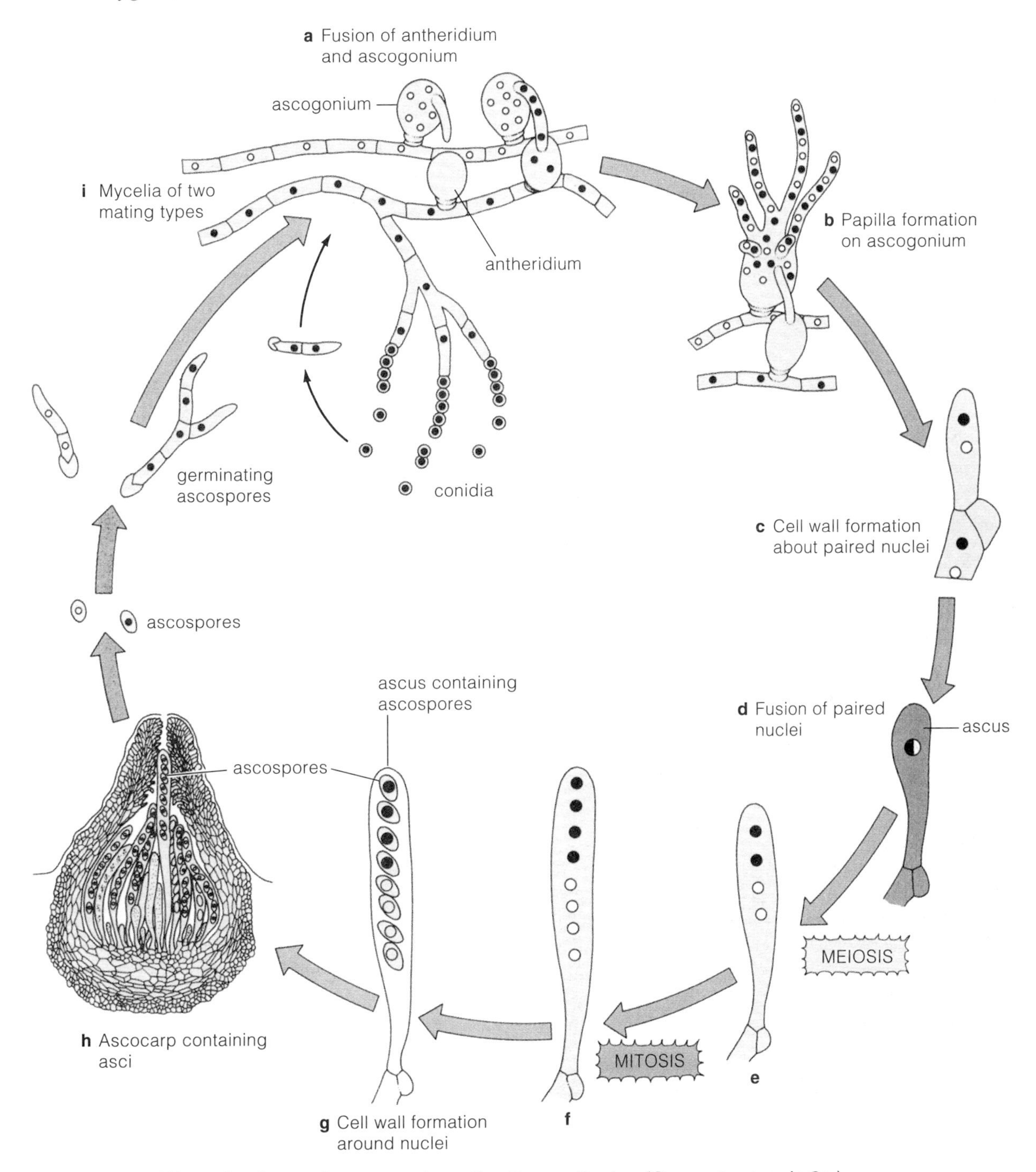

Figure 18-10 Life cycle of a sac fungus, such as *Eurotium* or *Peziza.* (Green structure is 2n.)

The nucleus of the ascus undergoes meiosis to form four nuclei (Figure 18-10e).

Mitosis then produces eight nuclei from these four (Figure 18-10f). (Notice that cytokinesis—cytoplasmic division—does not follow meiosis.)

Next, a cell wall forms about each nucleus, some of the cytoplasm of the ascus being included within the cell wall (Figure 18-10g).

Thus, eight haploid uninucleate ascospores (Figure 18-10g) have been formed.

While asci and ascospores are forming, surrounding hyphae proliferate to form the ascocarp (Figure 18-10h) around the asci.

As you see, two different types of ascospores are produced during meiosis. Each ascospore gives rise to a mycelium having only one mating type (Figure 18-10i).

B. Peziza: A Cup Fungus

The cup fungi are commonly found on soil during cool early spring and fall weather. Their sexual spore-containing structures are cup shaped, hence their common name.

PROCEDURE

1. Observe the preserved specimen of a cup fungus (Figure 18-11).

Actually, the structure we identify as a cup fungus is the fruiting body, produced as a result of sexual reproduction by the fungus. Most of the organism is present within the soil as an extensive mycelium. Specifically, the fruiting body is called an **ascocarp.**

2. Obtain a prepared slide of the ascocarp of *Peziza* or a related cup fungus. Examine the slide with the medium-power and high-dry objectives of your compound microscope.
3. Identify the elongate fingerlike **asci**, which contain dark-colored, spherical **ascospores** (Figure 18-12).

Figure 18-11 Cup fungi (0.25×). (Photo by J. W. Perry.)

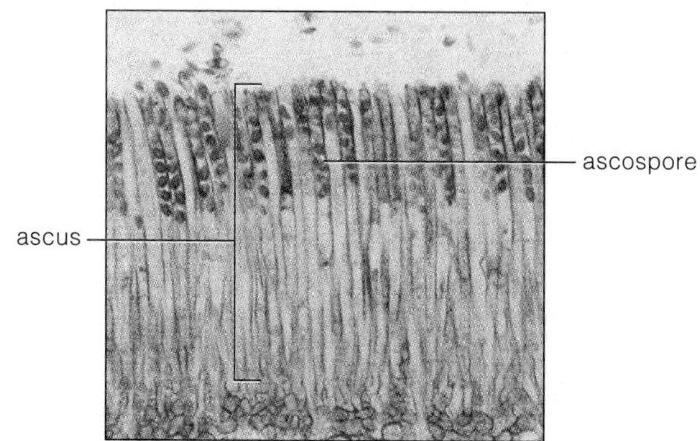

Figure 18-12 Cross section of an ascocarp from a cup fungus (186×).

18.5 Phylum Basidiomycota: Club Fungi *(About 30 min.)*

Members of this group of fungi are probably what first come to mind when we think of fungi, because this phylum contains those organisms called mushrooms. Actually the mushroom is only a portion of the fungus—it's the fruiting body, specifically a **basidiocarp,** containing the sexually produced haploid **basidiospores.** These basidiospores are produced by a club-shaped **basidium** for which the group is named. Much (if not most) of the fungal mycelium grows out of sight, within the substrate on which the basidiocarp is found.

MATERIALS

Per student:

- commercial mushroom
- prepared slide of mushroom pileus (cap), c.s. (*Coprinus*)
- compound microscope

Per lab room:

- demonstration specimens of various club fungi

Mushrooms called gill fungi have their sexual spores produced on sheets of hyphae that look like the gills of fish.

PROCEDURE

1. Obtain a fresh fruiting body, more properly called a **basidiocarp** (Figure 18-13). Identify the **stalk** and **cap.**
2. Look at the bottom surface of the cap, noting the numerous gills. It is on the surface of these gills that the haploid basidiospores are produced. Remember that all the structures you are examining are composed of aggregations of fungal hyphae.
3. Obtain a prepared slide of a cross section of the cap of a mushroom (Figure 18-14). Observe the slide first with the low-power objective of your compound microscope.
4. In the center of the cap, identify the **stalk.** The **gills** radiate from the stalk to the edge of the cap, much as spokes of a bicycle wheel radiate from the hub to the rim.
5. Switch to the high-dry objective to study a single gill (Figure 18-15). Note that the component hyphae produce club-shaped structures at the edge. These are the **basidia** (singular, *basidium*).
6. Each basidium produces four haploid **basidiospores.** Find them. (All four may not be in the same plane of section.)

Each basidiospore is attached to the basidium by a tiny hornlike projection. As the basidiospore matures, it is shot off the projection due to a buildup of turgor pressure within the basidium.

The life cycle of a typical mushroom is shown in Figure 18-16.

Because of genetic differences, basidiospores are of two different types. Mycelia produced by the two types of basidiospores are of two mating strains. Figure 18-16 shows these two different nuclei as open and closed (darkened) circles.

When a haploid basidiospore (Figure 18-16a) germinates, it produces a haploid *primary mycelium* (Figure 18-16b). The primary mycelium is incapable of producing a fruiting body.

Fusion between two sexually compatible mycelia (Figure 18-16b) must occur to continue the life cycle.

Surprisingly, the nuclei of the two mycelia don't fuse immediately; thus, each cell of this so-called *secondary mycelium* (Figure 18-16c) contains two genetically *different* nuclei. This condition is called "dikaryotic."

The secondary mycelium forms an extensive network within the substrate. An environmental or genetic trigger eventually stimulates the formation of the aerial basidiocarp (Figure 18-16d).

Each cell of the basidiocarp has two genetically different nuclei, including the basidia (Figure 18-16e) on the gills.

Within the basidia, the two nuclei fuse; the basidia are now diploid (Figure 18-16f).

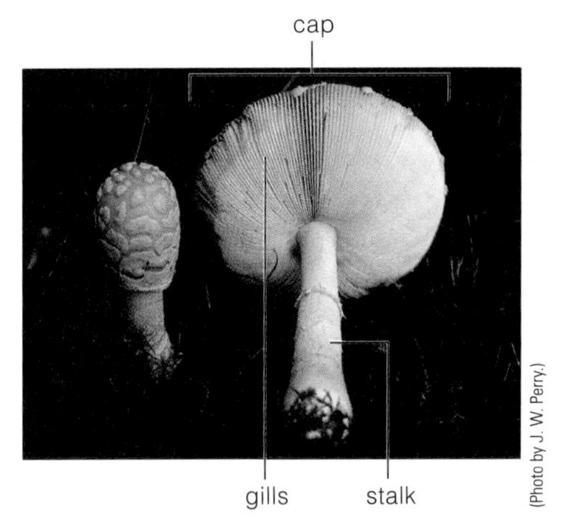

Figure 18-13 Mushroom basidiocarps. The one on the left is younger than that on the right (0.25×).

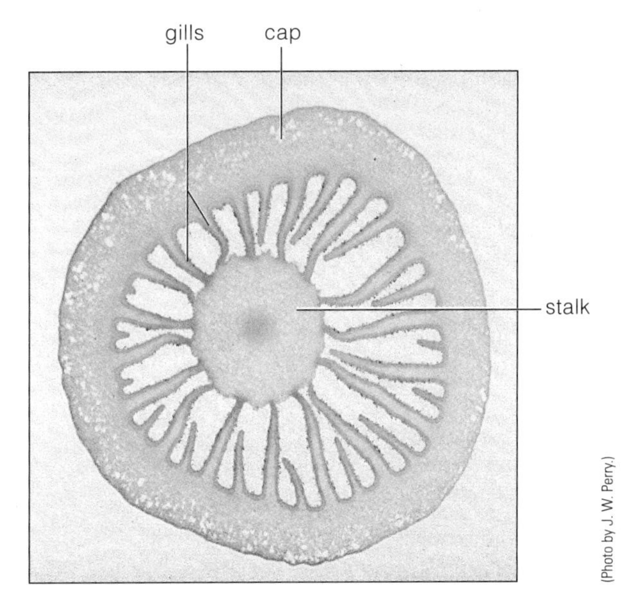

Figure 18-14 Cross section of mushroom cap (23×).

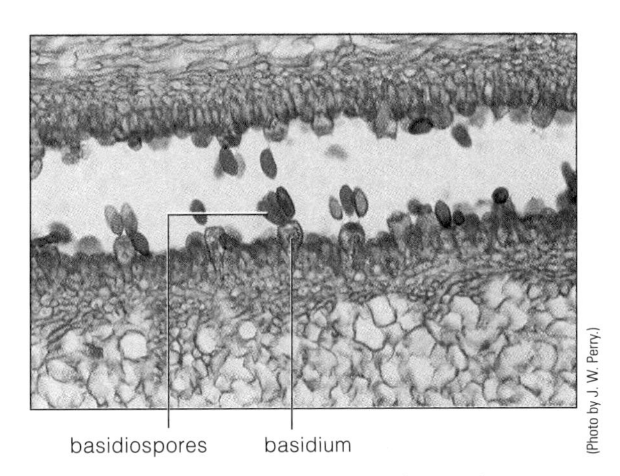

Figure 18-15 High magnification of a mushroom gill (287×).

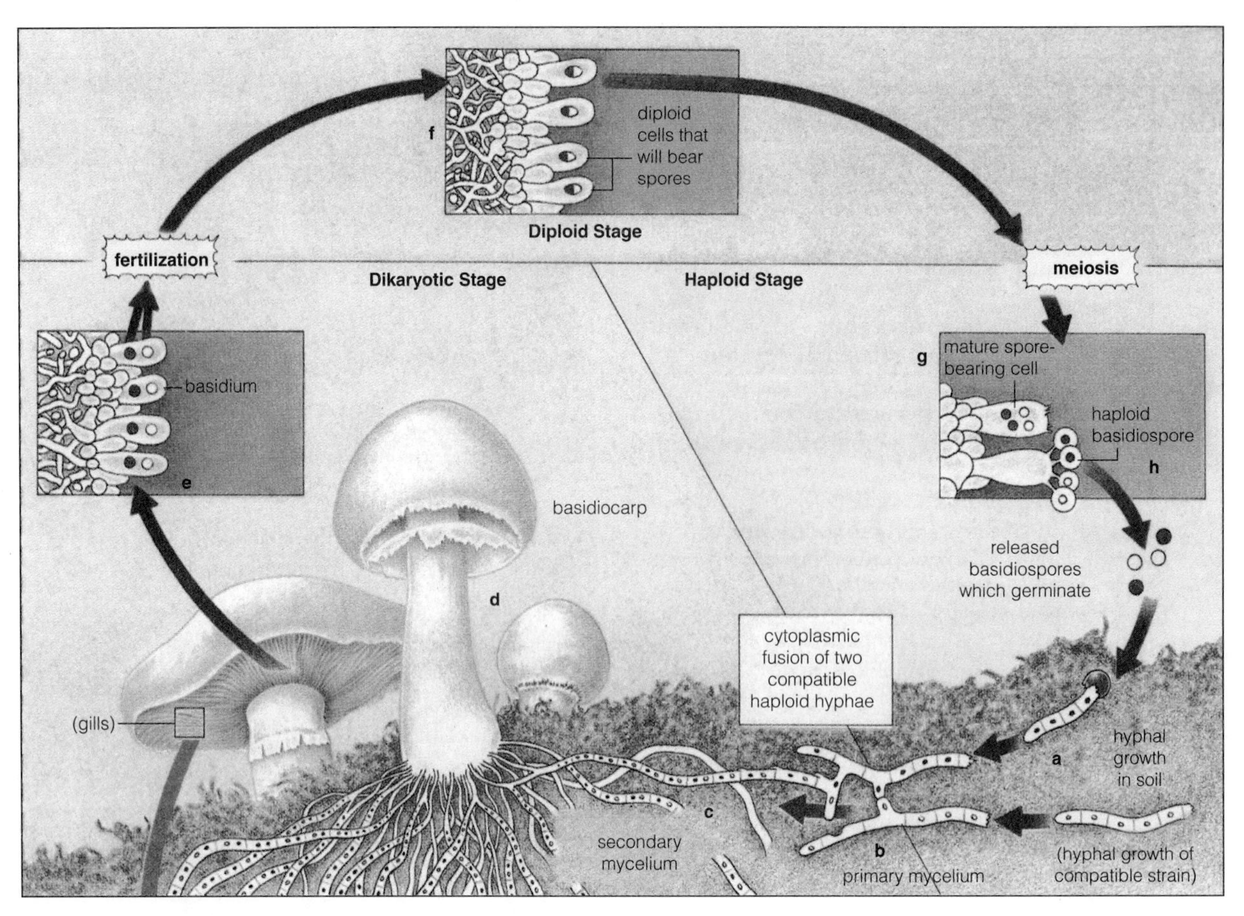

Figure 18-16 Life cycle of a mushroom.

Subsequently, these diploid nuclei undergo meiosis, forming genetically distinct nuclei (Figure 18-16g).

Each nucleus flows with a small amount of cytoplasm through the hornlike projections at the tip of the basidium to form a haploid basidiospore (Figure 18-16h).

Note that the gilled mushrooms do *not* reproduce by means of asexual conidia.

B. Other Club Fungi

A wide variety of basidiomycetes are not mushrooms. Let's examine a few representatives that you're likely to see in the field.

PROCEDURE

1. Examine the representatives of fruiting bodies of other members of the club fungi available in the laboratory.
2. Observe "puffballs" (Figure 18-17).The basidiospores of puffballs are contained within a spherical basidiocarp that develops a pore at the apex. Basidiospores are released when the puffball is crushed or hit by driving rain.
3. Examine "shelf fungi" (Figure 18-18). What you are looking at is actually the basidiocarp of the fungus that has been removed from the surface of a tree.

The presence of a basidiocarp of the familiar shelf fungi indicates that an extensive network of fungal hyphae is growing within a tree, digesting the cells of the wood. As a forester assesses a woodlot to determine the potential yield of usable wood, one of the things noted is the presence of shelf fungi, which indicates low-value (diseased) trees.

The most common shelf fungi are called polypores because of the numerous holes ("pores") on the lower surface of the fruiting body. These pores are lined with basidia bearing basidiospores.

4. Observe the undersurface of the basidiocarp that is on demonstration at a dissecting microscope and note the pores.

Figure 18-17 Puffballs. Note pore for spore escape (0.31×).

Figure 18-18 Shelf fungus (0.12×).

18.6 "Imperfect Fungi" *(About 15 min.)*

This group consists of fungi for which no sexual stage is known. Someone once decided that in order for a life to be complete—to be "perfect"—sex was necessary. Thus, these fungi are "imperfect." Reproduction takes place primarily by means of asexual conidia.

These fungi are among the most economically important, producing antibiotics (for example, one species of *Penicillium* produces penicillin). Others produce the citric acid used in the soft-drink industry; still others are used in the manufacture of cheese. Some are important pathogens of both plants and animals.

MATERIALS

Per student:

- dissecting needle
- glass microscope slide
- coverslip
- prepared slide of *Penicillium* conidia (optional)
- compound microscope

Per student pair:

- dH_2O in dropping bottle
- culture of *Alternaria*

Per lab room:

- demonstration of *Penicillium*-covered food *and/or* plate cultures

A. Penicillium*

The genus *Penicillium* has numerous species. The common feature of all species in this genus is the distinctive shape of the hypha that produce the conidia (Figure 18-19).

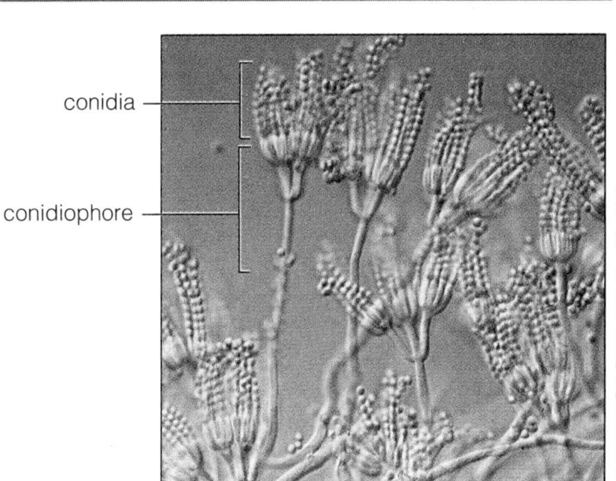

Figure 18-19 *Penicillium* (300×).

PROCEDURE

1. Examine demonstration specimens of moldy oranges or other foods. The blue color is attributable to a pigment in the numerous conidia produced by this fungus, *Penicillium*.
2. With a dissecting needle, scrape some of the conidia from the surface of the moldy specimen (or from a petri dish culture plate containing

*Some species of *Penicillium* reproduce sexually, forming ascocarps. These species are classified in the phylum Ascocarpa. However, not *all* fungi producing the conidiophore form called *Penicillium* reproduce sexually. Those that reproduce only by asexual means are considered "imperfect fungi."

Penicillium) and prepare a wet mount. Observe your preparation using the high-dry objective. (Prepared slides may also be available.)
3. Identify the **conidiophore** and the numerous tiny, spherical **conidia** (Figure 18-19). The name *Penicillium* comes from the Latin word *penicillus,* meaning "a brush." (Appropriate, isn't it?)

B. Alternaria

Perhaps no other fungus causes more widespread human irritation than *Alternaria,* an allergy-causing organism. During the summer, many weather programs announce the daily pollen (from flowering plants) and *Alternaria* spore counts as an index of air quality for allergy sufferers.

PROCEDURE

1. From the petri dish culture plate provided, remove a small portion of the mycelium and prepare a wet mount.
2. Examine with the high-dry objective of your compound microscope to find the **conidia** (Figure 18-20). Produced in chains, *Alternaria* conidia are multicellular, unlike those of *Penicillium* or *Aspergillus.* This makes identification very easy, since the conidia are quite distinct from all others produced by the fungi.

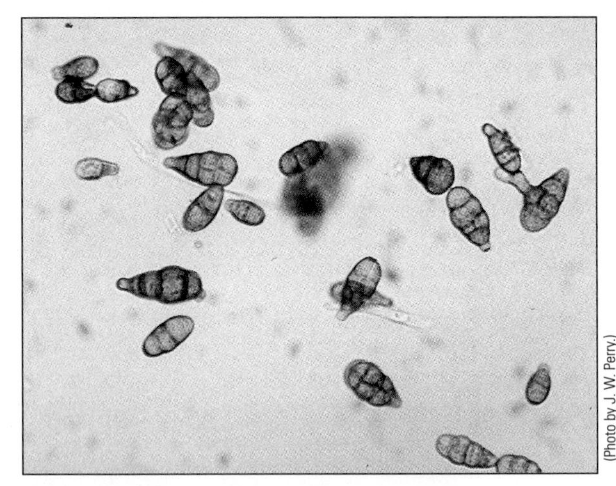
(Photo by J. W. Perry.)

Figure 18-20 Conidia of *Alternaria* (332×).

18.7 Mutualistic Fungi *(About 20 min.)*

Fungi team up with plants or members of the bacterial or protist kingdoms to produce some remarkable mutualistic relationships. The best way to envision such relationships is to think of them as good interpersonal relationships, in which both members benefit and are made richer than would be possible for each individual on its own.

MATERIALS

Per lab room:

- demonstration specimens of crustose, foliose, and fruticose lichens
- demonstration slide of lichen, c.s. on compound microscope
- demonstration slide of mycorrhizal root, c.s. on compound microscope

A. Lichens

Lichens are organisms made up of fungi and either green algae (kingdom Protista) or cyanobacteria (kingdom Eubacteria). The algal or cyanobacterial cells photosynthesize, and the fungus absorbs a portion of the carbohydrates produced. The fungal mycelium provides a protective, moist shelter for the photosynthesizing cells. Hence, both partners benefit from the relationship.

PROCEDURE

1. Examine the demonstration specimen of a **crustose lichen** (Figure 18-21).
2. What is it growing on?

____

____

3. Write a sentence describing the crustose lichen.

____

____

(Photo by J. W. Perry.)

Figure 18-21 Crustose lichen (0.5×).

4. Observe the demonstration specimen of a **foliose lichen** (Figure 18-22).
5. What is it growing on? ______________________________
6. Write a sentence describing the foliose lichen.

7. Examine the demonstration specimen of a **fruticose lichen** (Figure 18-23).
8. What is it growing on?

9. Write a sentence describing the fruticose lichen.

(Photo by J. W. Perry.)

Figure 18-22 Foliose lichen (0.5×).

(Photo by J. W. Perry.)

Figure 18-23 Fruticose lichen (1×).

10. Observe the cross section of a lichen at the demonstration microscope. This particular lichen is composed of fungal and green algae cells (Figure 18-24).
11. Locate the **fungal mycelium** and then the **algae cells.**
12. It's likely that this slide has a cup-shaped fruiting body. To which fungal phylum does this fungus belong?

____

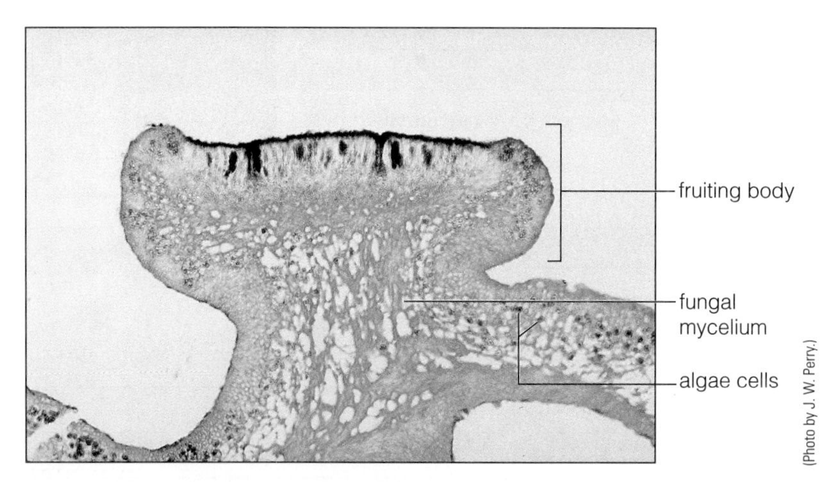

Figure 18-24 Section of lichen (84×).

B. Mycorrhizae

Mycorrhiza literally means "fungus root." A mycorrhiza is a mutualistic association between plant roots and certain species of fungi. In fact, many of the mushroom species (see Section 18-5A) you find in forests (especially those that appear to grow right from the ground) are part of a mycorrhizal association.

The fungus absorbs carbohydrates from the plant roots. The fungal hyphae penetrate much farther into the soil than plant root hairs can reach, absorbing water and dissolved mineral ions that are then released to the plant. Again, both partners benefit. In fact, many plants don't grow well in the absence of mycorrhizal associations.

There are two different types of mycorrhizal roots, **ectomycorrhizae** and **endomycorrhizae.** "Ectos" produce a fungal sheath around the root, and the hyphae penetrate between the cell walls of the root's cortex, but not into the root cells themselves. By contrast, "endos" do not form a covering sheath, and the fungi are found within the cells. We'll look at both types.

MATERIALS

Per student:

- ectomyccorhizal pine root, prep. slide, c.s.
- endomyccorhizal root, prep. slide, c.s.
- compound microscope

PROCEDURE

1. Obtain a prepared slide of an ectomycorrhizal pine root. Examine it with the medium- and high-power objective of your compound microscope.
2. Identify the sheath of fungal hyphae that surround the root.
3. Now look for hyphae between the cell walls within the cortex of the root.
4. Draw what you see in Figure 18-25.
5. Obtain a prepared slide of an endomycorrhizal root (Figure 18-26) and examine it as you did above.
6. Identify the hyphae within the plant root cells.

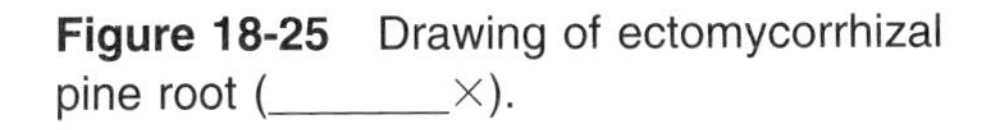

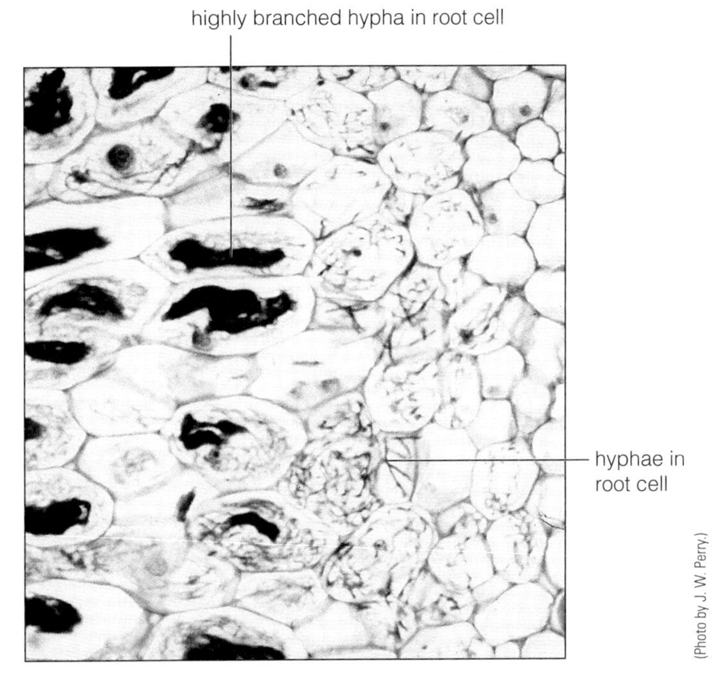

Figure 18-25 Drawing of ectomycorrhizal pine root (________×).

Figure 18-26 Cross section of endomycorrhizal root (110×).

18.8 Experiment: Environmental Factors and Fungal Growth

(About 45 min. to set up)

Like all living organisms, environmental cues are instrumental in causing growth responses in fungi. Intuitively, we know that light is important for plant growth. Light also affects animals in a myriad of ways, from triggering reproductive events in deer to affecting the complex psyche of humans. What effect might light have on fungi? These experiments enable you to determine whether this environmental factor has any effect on fungal growth.

A. Light and Darkness

This experiment addresses the hypothesis that *light triggers spore production of certain fungi.*

MATERIALS

Per experimental group (student pair):

- transfer loop
- bunsen burner (or alcohol lamp)
- matches or striker
- 3 petri plates containing potato dextrose agar (PDA)
- grease pencil or other marker

Per lab bench:

- test tubes with spore suspensions of *Trichoderma viride*, *Penicillium claviforme*, and *Aspergillus ornatus*

PROCEDURE

Work in pairs, and refer to Figure 18-27, which shows how to inoculate the cultures.

1. Receive instructions from your lab instructor as to which fungal culture you should use.

   Note: **Do not open the petri plates until step 3.**

2. Use a grease pencil or other lab marker to label each cover with your name and the name of the fungus you have been assigned. Write "CONTINUOUS LIGHT" on one petri plate, "CONTINUOUS DARKNESS" on a second plate, and "ALTERNATING LIGHT AND DARKNESS" on the third one.
3. Following the directions in Figure 18-27, inoculate each plate by touching the loop gently on the agar surface approximately in the middle of the plate.

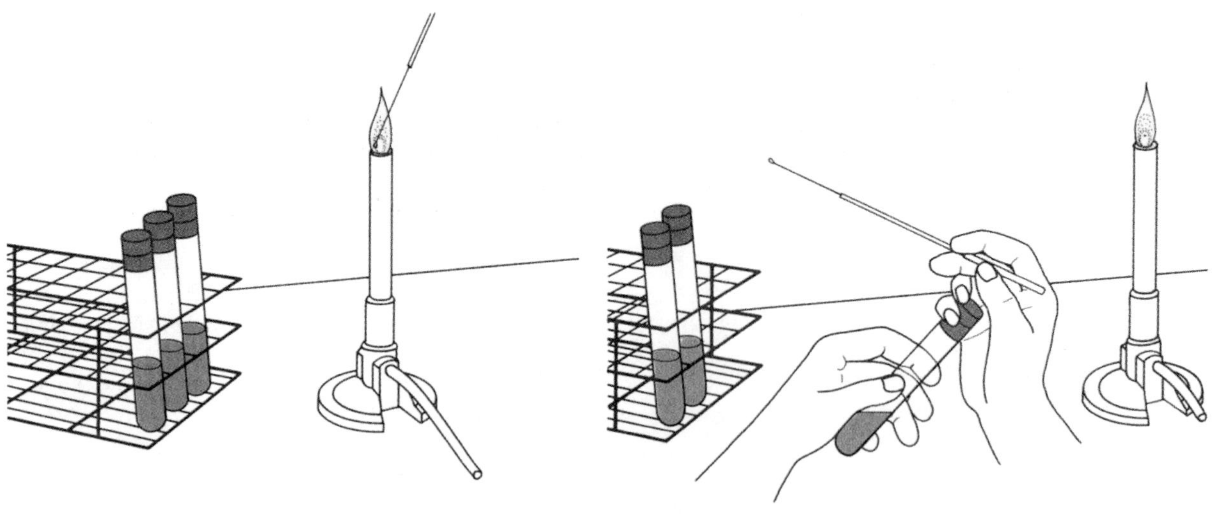

a Sterilize the loop by holding the wire in a flame until it is red hot. Allow it to cool before proceeding.

b While holding the sterile loop and the bacterial culture, remove the cap as shown.

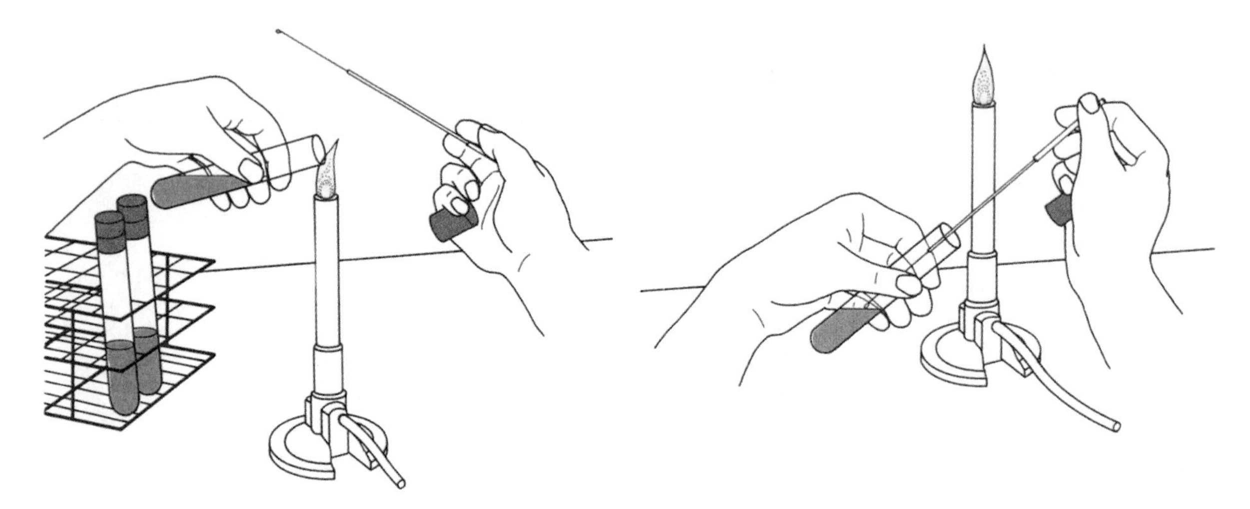

c Briefly heat the mouth of the tube in a burner flame before inserting the loop for an inoculum.

d Get a loopful of culture, withdraw the loop, heat the mouth of the tube, and replace the cap.

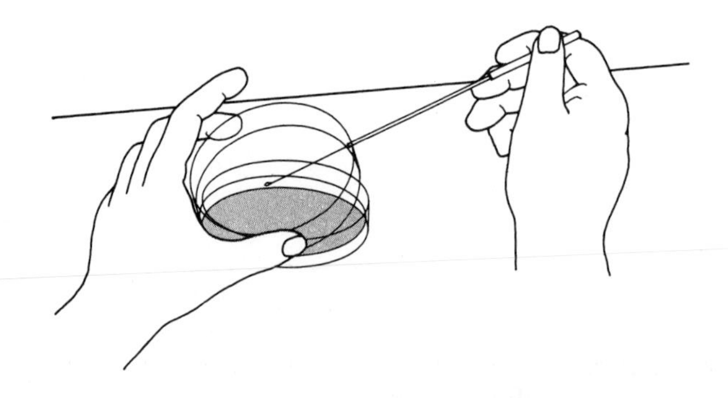

e To inoculate a solid medium in a petri plate, place the plate on a table and lift one edge of the cover.

Figure 18-27 Procedure for inoculating a culture plate. (After Case and Johnson, 1984.)

4. Place one plate in each condition:
   - continuous light
   - continuous darkness
   - alternating light and dark
5. In Table 18-3, write your prediction of what will occur in the different environmental conditions.
6. Check the results after 5–7 days and make conclusions concerning the effect of light on the growth of the cultures. Record your results in Table 18-3. Draw what you observe in Figure 18-28.
7. Compare the results you obtained with the results of other experimenters who used different species. Record your conclusions in Table 18-3.

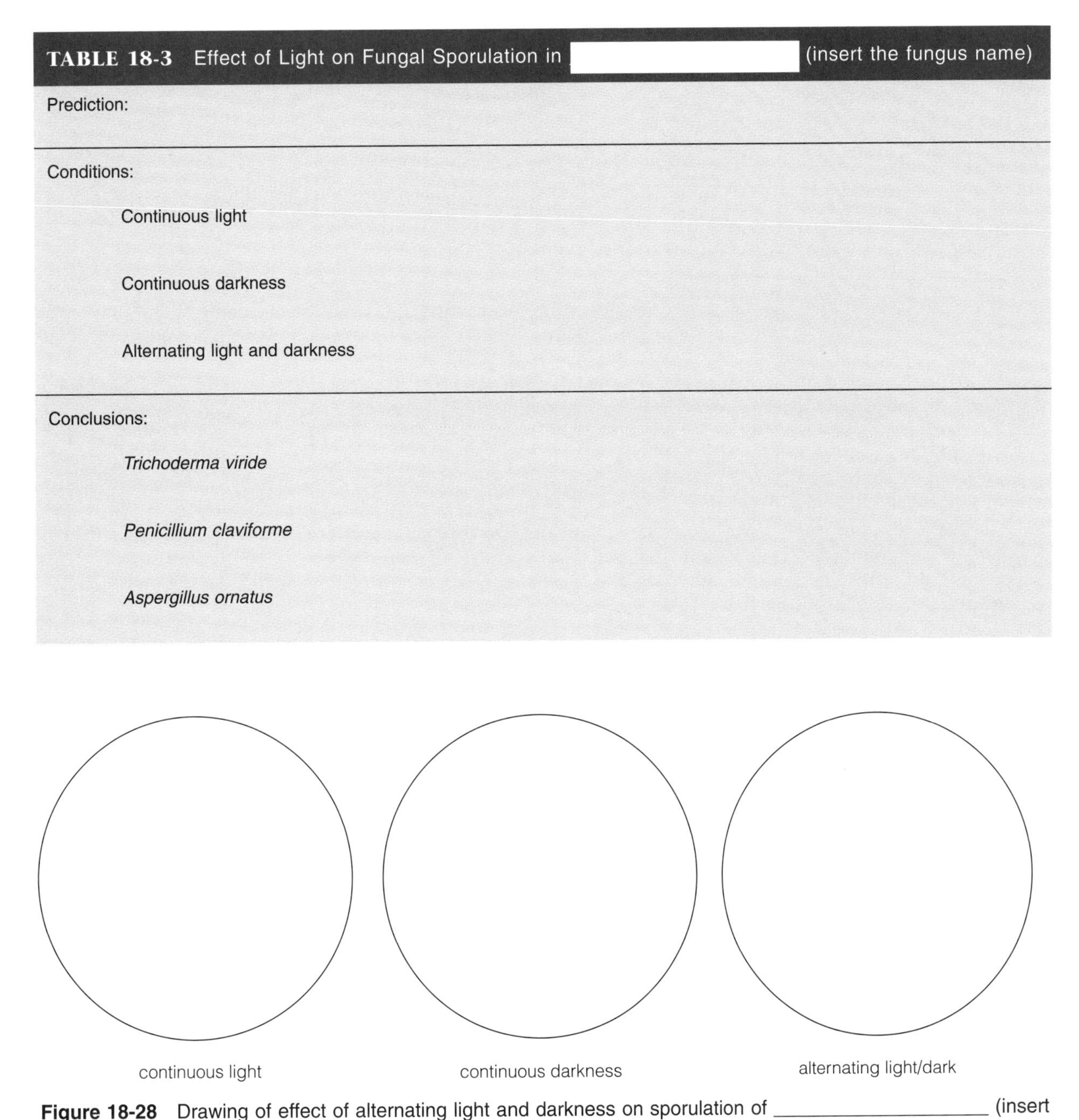

TABLE 18-3 Effect of Light on Fungal Sporulation in ______ (insert the fungus name)

Prediction:	
Conditions:	
Continuous light	
Continuous darkness	
Alternating light and darkness	
Conclusions:	
Trichoderma viride	
Penicillium claviforme	
Aspergillus ornatus	

Figure 18-28 Drawing of effect of alternating light and darkness on sporulation of ______ (insert name of fungus).

B. Directional Illumination

This experiment addresses the hypothesis that *light from one direction causes the spore-bearing hyphae of certain fungi to grow in the direction of the light.*

MATERIALS

Per experimental group (student pair):

- transfer loop
- bunsen burner (or alcohol lamp)
- matches or striker
- 2 petri plates containing either potato dextrose agar (PDA) or cornmeal (CM) agar
- grease pencil or other marker

Per lab bench:

- test tubes with spore suspensions of
  Phycomyces blakesleeanus (inoculate onto PDA)
  Penicillium isariforme (inoculate onto PDA)
  Aspergillus giganteus (inoculate onto CM)

PROCEDURE

Work in pairs and refer to Figure 18-27, which shows how to inoculate cultures.

1. Receive instructions from your lab instructor as to which fungal culture you should use.

   Note: **Do not open the petri plates until step 3.**

2. Use a grease pencil or other lab marker to label each cover with your name and the name of the fungus you have been assigned. Without removing the lid, turn the petri plate upside down and place a line across its center from edge to edge.
3. Following the directions in Figure 18-27, inoculate each petri plate by dragging a loop full of spore suspension over the agar surface above the line you marked on the bottom of the plate.
4. Place one plate in each condition:
   - continuous unilateral illumination
   - continuous illumination from above the plate
5. In Table 18-4, write your prediction of what will occur in the two environmental conditions.
6. Check the results after 5–7 days and make conclusions concerning the effect of light on the growth of the cultures. Record your results in Table 18-4. Draw what you see in Figure 18-29.
7. Compare the results you obtained with the results of other experimenters who used different species. Record your conclusions in Table 18-4.

TABLE 18-4 Effect of Directional Illumination on Growth of the Fungus ______ (insert fungus name)

Prediction:	
Conditions:	
Unilateral illumination	
Illumination from above	
Conclusions:	
Phycomyces blakesleeanus	
Penicillium isariforme	
Aspergillus giganteus	

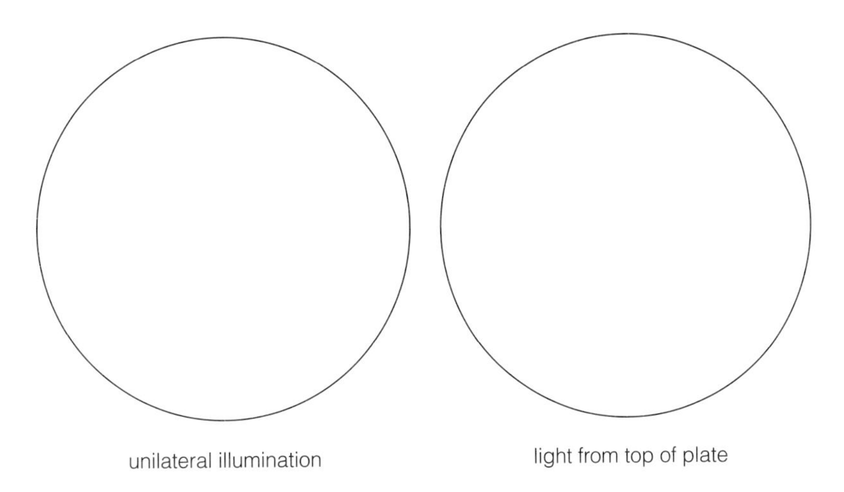

Figure 18-29 Drawing of effect of unilateral illumination on growth of ______________________ (insert name of fungus).

PRE-LAB QUESTIONS

_____ **1.** An organism that grows specifically on nonliving organic material is called
(a) an autotroph
(b) a heterotroph
(c) a parasite
(d) a saprophyte

_____ **2.** Taxonomic separation into fungal phyla is based on
(a) sexual reproduction, or lack thereof
(b) whether the fungus is a parasite or saprophyte
(c) the production of certain metabolites, like citric acid
(d) the edibility of the fungus

_____ **3.** Fungi is to fungus as __________ is to __________.
(a) mycelium, mycelia
(b) hypha, hyphae
(c) sporangia, sporangium
(d) ascus, asci

_____ **4.** Which statement is not true of the zygospore-forming fungi?
(a) They are in the phylum Zygomycota
(b) Ascospores are found in an ascus
(c) *Rhizopus* is a representative genus
(d) A zygospore is formed after fertilization

_____ **5.** Which structures would you find in a sac fungus?
(a) ascogonium, antheridium, zygospores
(b) ascospores, oogonia, asci, ascocarps
(c) basidia, basidiospores, basidiocarps
(d) ascogonia, asci, ascocarps, ascospores

_____ **6.** The club fungi are placed in the phylum Basidiomycota
(a) because of their social nature
(b) because they form basidia
(c) because of the presence of an ascocarp
(d) because they are dikaryotic

_____ **7.** Which statement is *not* true of the imperfect fungi?
(a) They reproduce sexually by means of conidia
(b) They form an ascocarp
(c) Sex organs are present in the form of oogonia and antheridia
(d) All of the above are false

_____ **8.** A relationship between two organisms in which both members benefit is said to be
(a) parasitic
(b) saprotrophic
(c) mutualistic
(d) heterotrophic

_____ **9.** An organism that is made up of a fungus and an associated green alga or cyanobacterium is known as a
(a) sac fungus
(b) lichen
(c) mycorrhizal root
(d) club fungus

_____ **10.** A mutualistic association between a plant root and a fungus is known as a
(a) sac fungus
(b) lichen
(c) mycorrhizal root
(d) club fungus

Name ______________________ Section Number ______________________

EXERCISE 18

Fungi

POST-LAB QUESTIONS

18.1 Phylum Zygomycota: Zygosporangium-Forming Fungi

1. Observe this photo at the right of a portion of a fungus you examined in lab.
   a. What is structure a?
   b. Are the contents of structure a haploid or diploid?

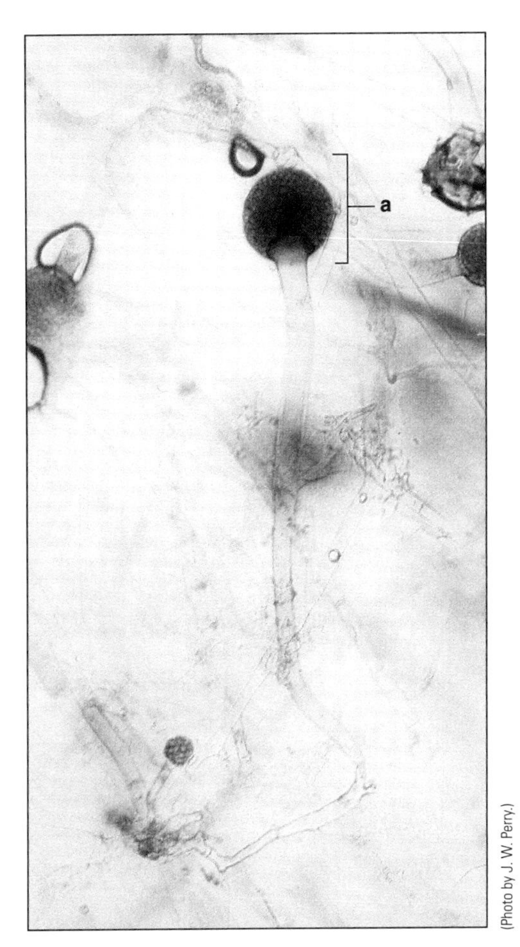

(130×).

2. Distinguish between a *hypha* and a *mycelium*.

3. Examine the below photomicrographs of a fungal structure you studied in lab. Is this structure labeled **b** in the photomicrograph below the product of sexual or asexual reproduction?

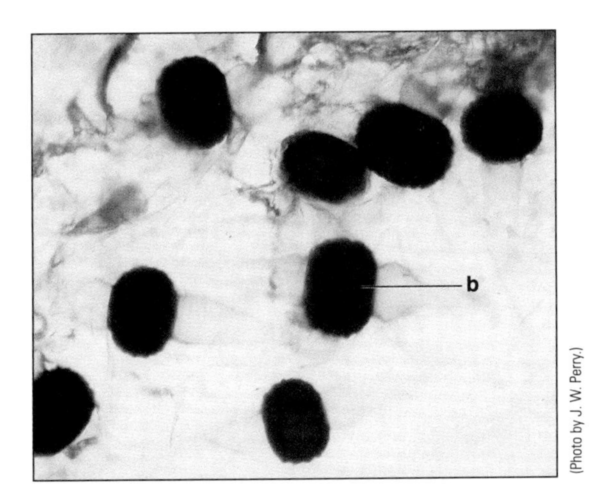

(120×).

18.3 *Phylum Ascomycota: Sac Fungi*

4. Distinguish among an *ascus,* an *ascospore,* and an *ascocarp.*

5. Walking in the woods, you find a cup-shaped fungus. Back in the lab, you remove a small portion from what appears to be its fertile surface and crush it on a microscope slide, preparing the wet mount that appears at the right. Identify the fingerlike structures present on the slide.

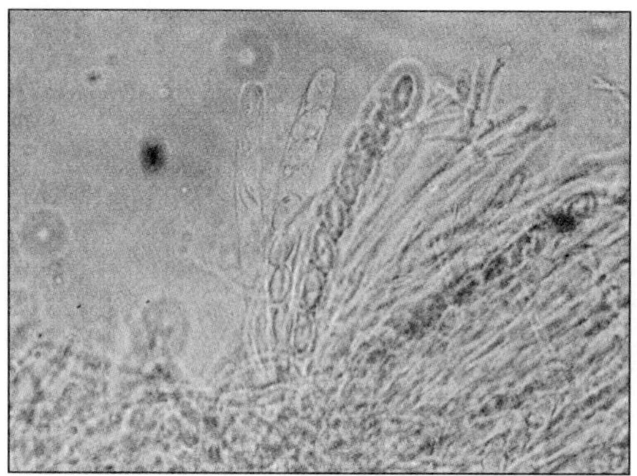
(Photo by J. W. Perry.)

(287×).

18.4 *Phylum Basidiomycota: Club Fungi*

6. What type of spores are produced by the fungus pictured at the right?

(Photo by J. W. Perry.)

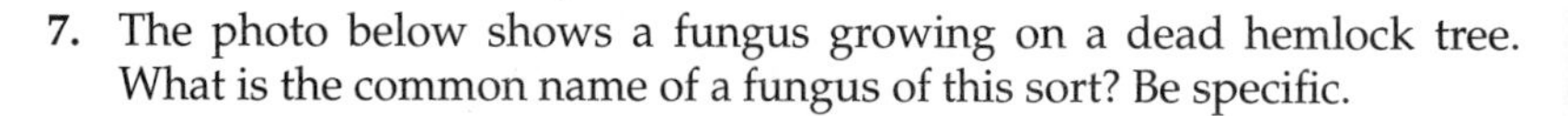

7. The photo below shows a fungus growing on a dead hemlock tree. What is the common name of a fungus of this sort? Be specific.

(Photo by J. W. Perry.)

18.5 *"Imperfect Fungi"*

8. Explain the name "imperfect fungi."

Food for Thought

9. Give the correct singular or plural form of the following words in the blanks provided.

	Singular	Plural
a.	hypha	______
b.	______	mycelia
c.	zygospore	______
d.	______	asci
e.	basidium	______
f.	______	conidia

10. Lichens are frequently the first colonizers of hostile growing sites, including sunbaked or frozen rock, recently hardened lava, and even gravestones. How can lichens survive in habitats so seemingly devoid of nutrients and under such harsh physical conditions?

Seedless Vascular Plants: Club Mosses and Ferns

OBJECTIVES

After completing this exercise, you will be able to

1. define *tracheophyte, rhizome, mutualism, sporophyte, sporangium, gametophyte, gametangium, antheridium, archegonium, epiphyte, strobilus, node, internode, frond, sorus, annulus, hygroscopic, chemotaxis;*
2. recognize whisk ferns, club mosses, horsetails, and ferns when you see them, placing them in the proper taxonomic phylum;
3. identify the structures of the fern allies and ferns in **boldface;**
4. describe the life cycle of ferns;
5. explain the mechanism by which spore dispersal occurs from a fern sporangium;
6. describe the significant differences between the life cycles of the bryophytes, and the club mosses and ferns.

INTRODUCTION

If you did the exercise on liverworts and mosses, you learned about alternation of generations. (If you did not do this exercise, you now need to read the Introduction for Exercise 22, page 323.) Indeed, this theme is present in all plants. However, a major distinction exists between the bryophytes (liverworts and mosses) and the fern allies, ferns, gymnosperms, and flowering plants: Whereas the dominant and conspicuous portion of the life cycle in the bryophytes is the **gametophyte** (the gamete-producing part of the life cycle), in all other plants that you will examine in this and subsequent exercises, it is the **sporophyte** (that portion of the life cycle producing spores).

A second, and perhaps more important, distinction also exists between bryophytes and the fern allies, ferns, gymnosperms, and flowering plants: The latter group contains *vascular tissue.* Vascular tissues include *phloem,* the tissue that conducts the products of photosynthesis, and *xylem,* the tissue that conducts water and minerals. As a result of the presence of xylem and phloem, fern allies, ferns, gymnosperms, and flowering plants are sometimes called **tracheophytes.**

Push gently on the region of your throat at the base of your larynx (voicebox or Adam's apple). If you move your fingers up and down, you should be able to feel the cartilage rings in your *trachea* (windpipe). To visualize its structure, imagine that your trachea is a pipe with donuts inside of it (Figure 19-1). This same arrangement exists in some cell types within the xylem of plants. Thus, early botanical microscopists called plants having such an arrangement *tracheophytes.*

In this exercise, you will study the tracheophytes that lack seeds. Two phyla of plants make up the seedless vascular plants:

Phylum	Common Names
Lycophyta	Club mosses
Moniliophyta	Whisk ferns, horsetails, and ferns

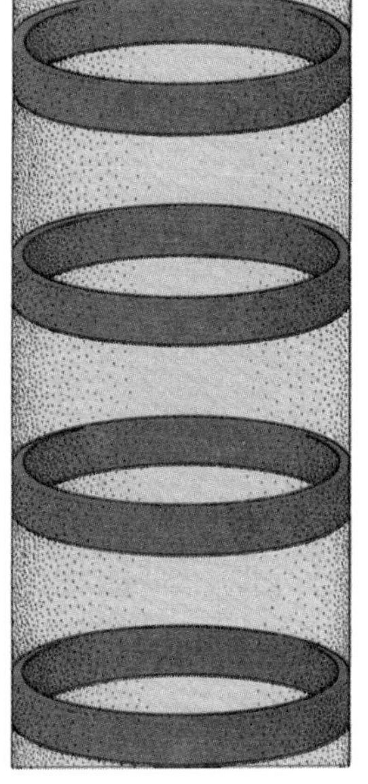

Figure 19-1 Three-dimensional representation of the cartilage in your trachea and some xylem cells of vascular plants.

Whisk ferns and horsetails are very unlike what we often think of as ferns, but molecular evidence confirms their relationship to more typical ferns. We will study most closely the life cycle of the ferns, since they are common in our environment and have gametophytes and sporophytes that beautifully illustrate the concept of alternation of generations.

19.1 Phylum Lycophyta: Club Mosses *(About 20 min.)*

Lycopodium and *Selaginella* are the most commonly found genera in the phylum Lycophyta, the club mosses. The common name of this phylum comes from the presence of the so-called **strobilus** (plural, *strobili*), a region of the stem specialized for the production of spores. The strobilus looks like a very small club. Strobili are sometimes called cones.

MATERIALS

Per lab room:

- *Lycopodium,* living, preserved, or herbarium specimens with strobili
- *Lycopodium* gametophyte, preserved (optional)
- *Selaginella,* living, preserved, or herbarium specimens
- *Selaginella lepidophylla,* resurrection plant, 2 dried specimens
- culture bowl containing water

A. Lycopodium

The club moss *Lycopodium* is a small, forest-dwelling plant. These plants are often called ground pines or trailing evergreens. The scientific name comes from the appearance of the rhizome—it looked like a wolf's paw to early botanists. *Lycos* is Greek for "wolf," and *podium* means "foot."

Figure 19-2 Sporophyte of *Lycopodium* (0.6×).

PROCEDURE

1. Observe the living, preserved, or herbarium specimens of *Lycopodium* (Figure 19-2).
2. On your specimen, identify the true **roots, stems,** and **leaves.**
3. The adjective *true* is used to indicate that the organs contain vascular tissue—xylem and phloem.
4. Identify the **rhizome** to which the upright stems are connected. In the case of *Lycopodium,* the rhizome may be either beneath or on the soil surface, depending on the species. Notice that leaves cover the rhizome, as are the upright stems.
5. At the tip of an upright stem, find the **strobilus** (Figure 19-2). Look closely at the strobilus. As you see, it too is made up of leaves, but these leaves are much more tightly aggregated than the sterile (nonreproductive) leaves on the rest of the stem.

The leaves of the strobilus produce *spores* within a *sporangium,* the spore container. As was pointed out in the introduction, the plant you are looking at is a diploid sporophyte.

Is the sporophyte haploid (n) or diploid (2n)? ______________________

Since the spores are haploid, what process must have taken place within the sporangium?

____

As maturation occurs, the internodes between the sporangium-bearing leaves elongate slightly, the sporangium opens, and the spores are carried away from the parent plant by wind. If they land in a suitable habitat, the spores germinate to produce small and inconspicuous subterranean gametophytes, which bear sex organs—antheridia and archegonia. For fertilization and the development of a new sporophyte to take place, free water must be available, since the sperm are flagellated structures that must swim to the archegonium.

6. If available, examine a preserved *Lycopodium* gametophyte (Figure 19-3).

B. Selaginella

Selaginella is another genus in the phylum Lycophyta. These plants are usually called spike mosses. (This points out the problem with common names—the common name for the phylum is club mosses, but individual genera have their own common names.) Some species are grown ornamentally for use in terraria.

PROCEDURE

1. Examine living representatives of *Selaginella* (Figure 19-4).
2. Look closely at the tips of the branches and identify the rather inconspicuous **strobili.**
3. Examine the specimens of the resurrection plant (yet another common name!), a species of *Selaginella* sold as a novelty, often in grocery stores. A native of the southwestern United States, this plant grows in environments that are subjected to long periods without moisture. It becomes dormant during these periods.
4. Describe the color and appearance of the dried specimen.

____

____

5. Now place a dried specimen in a culture bowl containing water. Observe what happens in the next hour or so, and describe the change in appearance of the plant.

____

____

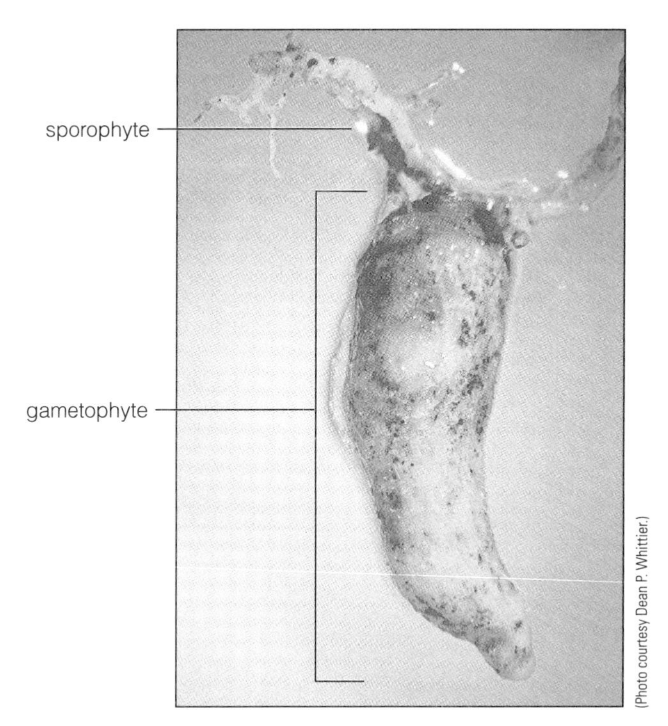

Figure 19-3 Gametophyte with young sporophyte of *Lycopodium* (2×).

a b

Figure 19-4 (**a**) Sporophyte of *Selaginella* (0.4×). (**b**) Higher magnification showing strobili at branch tips (0.9×).

19.2 Phylum Moniliophyta, Subphylum Psilophyta: Whisk Ferns *(About 15 min.)*

The whisk ferns consist of only two genera of plants, *Psilotum* (the *P* is silent) and *Tmesipteris* (the *T* is silent). Neither has any economic importance, but they have intrigued botanists for a long time, especially because *Psilotum* resembles the first vascular plants that colonized the earth. Current evidence indicates that the psilophytes are not directly related to those very early land plants, but instead are ferns. We'll examine only *Psilotum.*

MATERIALS

Per lab room:

- *Psilotum,* living plant
- *Psilotum,* herbarium specimen showing sporangia and rhizome
- *Psilotum* gametophyte
- dissecting microscope

PROCEDURE

1. Observe first the **sporophyte** in a potted *Psilotum* (Figure 19-5). Within the natural landscapes of the United States, this plant grows abundantly in parts of Florida and Hawaii.
2. If the pot contains a number of stems, you can see how it got its common name, the whisk fern, because it looks a bit like a whisk broom. Closely examine a single aerial stem.

What color is it? ____

3. From this observation, make a conclusion regarding one function of this stem.

____

4. Observe the herbarium specimen (mounted plant) of *Psilotum.* Identify the nongreen underground stem, called a **rhizome.**

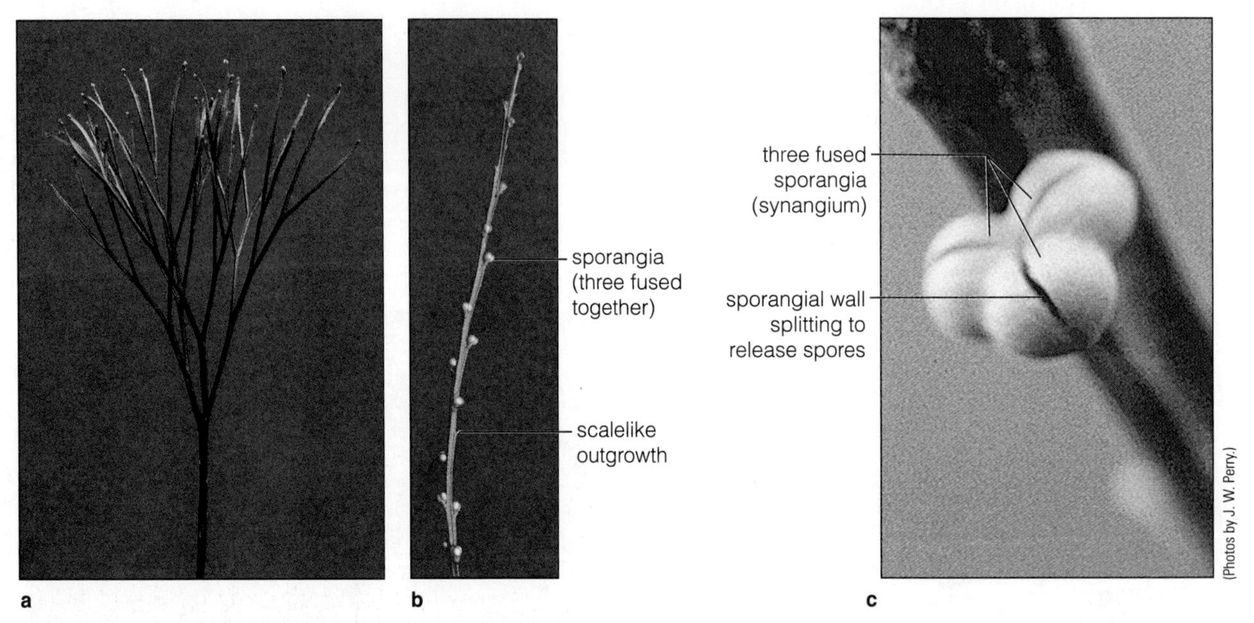

Figure 19-5 Sporophytes of *Psilotum.* (**a**) Single stem without sporangia. (**b**) Portion of stem with sporangia (0.3×). (**c**) Sporangia (10×).

5. *Psilotum* is unique among vascular plants in that it lacks roots. Absorption of water and minerals takes place through small rhizoids attached to the rhizome. Additionally, a fungus surrounds and penetrates into the outer cell layers of the rhizome. The fungus absorbs water and minerals from the soil and transfers them to the rhizome. This beneficial association is similar to the mycorrhizal association between plant roots and fungi described in the fungi exercise (page 267). A beneficial relationship like that between the fungus and *Psilotum* is called a **mutualistic symbiosis** or, more simply, **mutualism.**
6. On either the herbarium specimen or the living plant, identify the tiny scalelike outgrowths that are found on the aerial stems. Because these lack vascular tissue, they are not considered true leaves. Their size precludes any major role in photosynthesis.
7. The plant you are observing is a sporophyte. Thus, it must produce spores. Find the three-lobed structures on the stem (Figure 19-5b). Each lobe is a single **sporangium** containing spores.
8. If these spores germinate after being shed from the sporangium, they produce a small and infrequently found gametophyte that grows beneath the soil surface. The gametophyte survives beneath the soil thanks to a symbiotic relationship similar to that for the sporophyte's rhizome.
9. If a gametophyte is available, observe it with the aid of a dissecting microscope (Figure 19-6). Note the numerous **rhizoids.** If you look very carefully, you may be able to distinguish **gametangia** (sex organs; singular, *gametangium*).

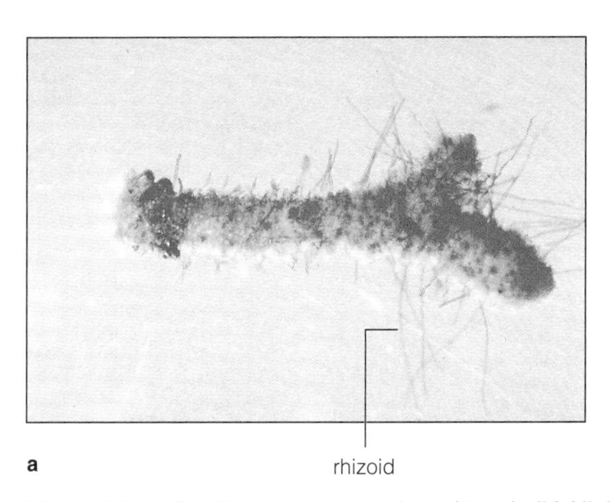

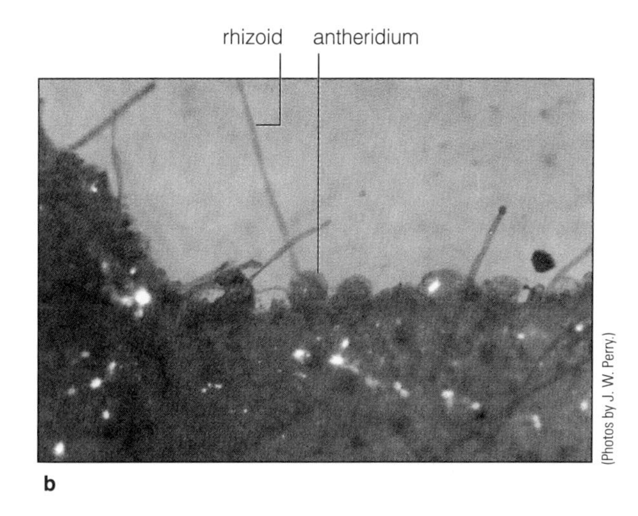

Figure 19-6 (**a**) *Psilotum* gametophyte (7.5×). (**b**) Higher magnification showing antheridia (23×).

10. The male sex organs are **antheridia** (Figure 19-6b; singular, *antheridium*); the female sex organs are **archegonia** (singular, *archegonium*). Both antheridia and archegonia are on the same gametophyte. Is the gametophyte dioecious or monoecious? ______________________
11. Fertilization of an egg within an archegonium results in the production of a new sporophyte.

19.3 Phylum Moniliophyta, Subphylum Sphenophyta: Horsetails *(About 15 min.)*

During the age of the dinosaurs, tree-sized representatives of this phylum flourished. But like the dinosaurs, they too became extinct. Only one genus remains. However, the remnants of these plants, along with others that were growing at that time, remain in the form of massive coal deposits. *Speno-*, a Greek prefix meaning "wedge," gives the phylum its name, presumably because the leaves of these plants are wedge shaped.

A single genus, *Equisetum,* is the only living representative of this subphylum. Different species of *Equisetum* are common throughout North America. Many are highly branched, giving the appearance of a horse's tail, and hence the common name. (*Equus* is Latin for "horse"; *saeta* means "bristle.")

Figure 19-7 Sporophyte of the horsetail, *Equisetum* (0.2×).

MATERIALS

Per lab room:

- *Equisetum,* living, preserved, or herbarium specimens with strobili
- *Equisetum* gametophytes, living or preserved

PROCEDURE

1. Examine the available specimens of *Equisetum* (Figure 19-7). Depending on the species, it will be more or less branched.
2. Note that the plant is divided into **nodes** (places on the stem where the leaves arise) and **internodes** (regions on the stem between nodes). Identify the small leaves at the node.

If the specimen you are examining is a highly branched species, don't confuse the branches with leaves. The leaves are small, scalelike structures, often brown. (They *do* have vascular tissue, so they are true leaves.) Distinguish the leaves.

3. Look at the herbarium mount and identify the underground **rhizome** that bears **roots.**
4. Examine both the aerial stem and rhizome closely.

Do both have nodes? ______________________________

Do both have leaves? ______________________________

Which portion of *Equisetum* is primarily concerned with photosynthesis? ______________________________

5. Find the **strobilus** (Figure 19-7). Where on the plant is it located? ______________________________

Based on the knowledge you've acquired in your study of *Lycopodium,* what would you expect to find within the strobilus? ______________________________

When spores fall to the ground, what would you expect them to grow into after germination? ______________________________

6. Now observe the horsetail gametophytes (Figure 19-8).

What color are they? ______________________________

Would you expect them to be found on or below the soil surface? Why?

____

____

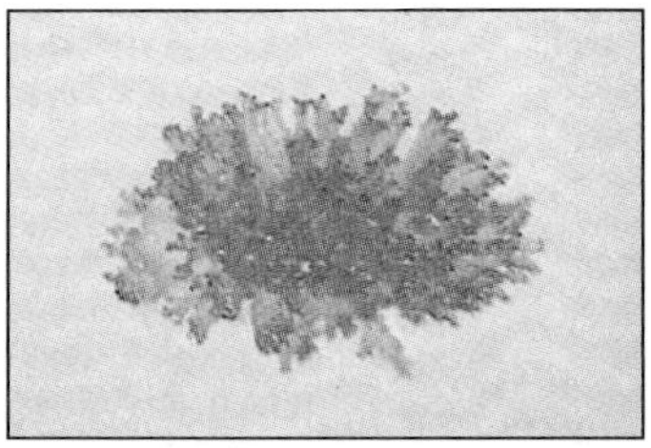

(Photo courtesy Dean Whittier.)

Figure 19-8 Horsetail gametophyte (5×).

19.4 Phylum Moniliophyta, Subphylum Pterophyta: Ferns *(About 60 min.)*

The variation in form of ferns is enormous, as is their distribution and ecology. We typically think of ferns as inhabitants of moist environments, but some species are also found in very dry locations. Tropical species are often grown as houseplants.

This subphylum gets its name from the Greek word *pteri,* meaning "fern."

MATERIALS

Per student:

- fern sporophytes, fresh, preserved, or herbarium specimens
- prepared slide of fern rhizome, c.s.
- fern gametophytes, living, preserved, or whole mount prepared slides
- fern gametophyte with young sporophyte, living or preserved
- microscope slide
- compound microscope
- dissecting microscope

Per lab room:

- squares of fern sori, in moist chamber (*Polypodium aureum* recommended)
- demonstration slide of fern archegonium, median l.s.
- other fern sporophytes, as available

PROCEDURE

1. Obtain a fresh or herbarium specimen of a typical fern **sporophyte.** As you examine the structures described in this experiment, refer to Figure 19-16, a diagram of the life cycle of a fern.

The sporophyte of many ferns (Figures 19-9, 19-16a) consists of *true* roots, stems, and leaves; that is, they contain vascular tissue.

2. Identify the horizontal stem, the **rhizome** (which produces true **roots**), and upright leaves. The leaves of ferns are called **fronds** and are often highly divided.

Ferns, unlike bryophytes, are vascular plants, their sporophytes containing xylem and phloem. Let's examine this vascular tissue.

3. Obtain a prepared slide of a cross section of a fern stem (rhizome). Study it using the low-power objective of your compound microscope. Use Figure 19-10 as a reference.
   (a) Find the **epidermis, cortex,** and **vascular tissue.**
   (b) Within the vascular tissue, distinguish between the **phloem** (of which there are outer and inner layers) and the thick-walled **xylem** sandwiched between the phloem layers.
4. Now examine the undersurface of the frond. Locate the dotlike **sori** (singular, *sorus;* Figure 19-9).

Figure 19-9 (**a**) Morphology of a typical fern (0.25×). (**b**) Enlargement showing the sori (0.5×).

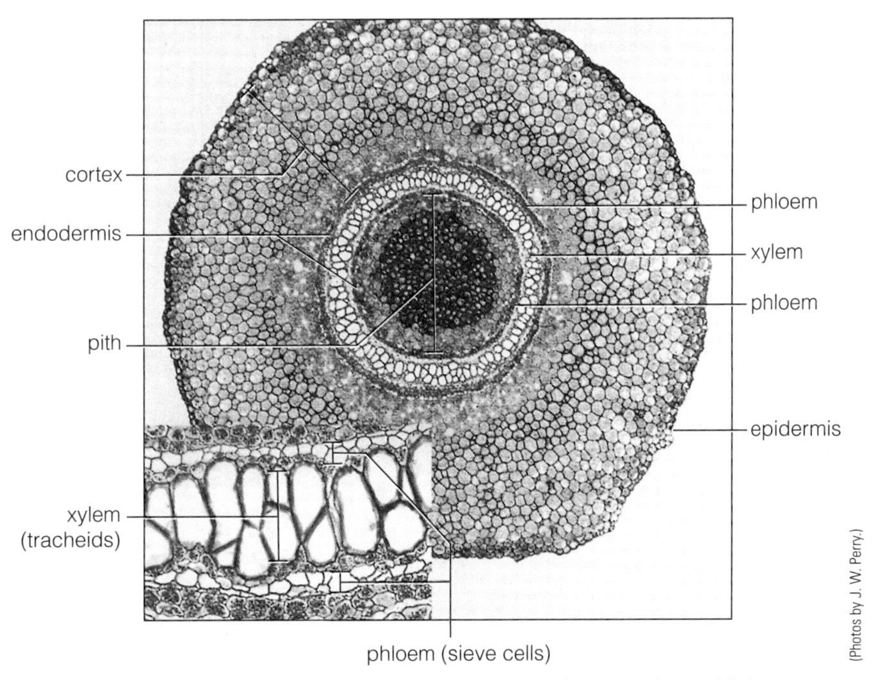

Figure 19-10 Cross section of a fern rhizome (29×). Inset shows higher magnification of vascular tissue (80×).

5. Each sorus is a cluster of **sporangia.** Using a dissecting microscope, study an individual sorus. Identify the sporangia (Figures 19-11, 19-16b).

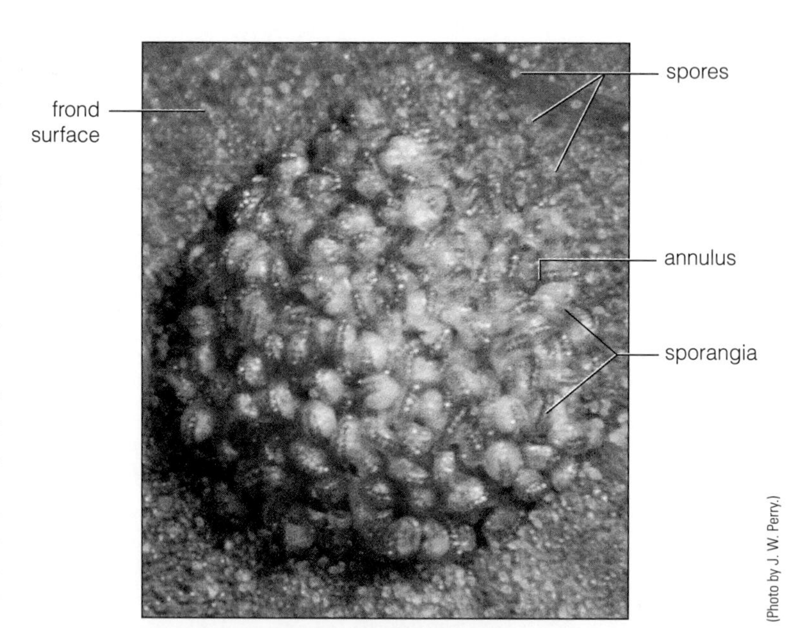

Figure 19-11 Sorus containing sporangia (30×).

Each sporangium contains *spores* (Figure 19-16c). Although the sporangium is part of the diploid (sporophytic) generation, spores are the first cells of the haploid (gametophytic) generation.

What process occurred within the sporangium to produce the haploid spores? ______________________

6. Obtain a single sorus-containing square of the hare's foot fern (*Polypodium aureum*).
   - **(a)** Place it sorus-side up on a glass slide (DON'T ADD A COVERSLIP), and examine it with the low-power objective of your compound microscope.
   - **(b)** Note the row of brown, thick-walled cells running over the top of the sporangium, the **annulus** (Figures 19-11, 19-16b).

The annulus is **hygroscopic.** Changes in moisture content within the cells of the annulus cause the sporangium to crack open. Watch what happens as the sporangium dries out.

As the water evaporates from the cells of the annulus, a tension develops that pulls the sporangium apart. Separation of the halves of the sporangium begins at the thin-walled *lip cells.* As the water continues to evaporate, the annulus pulls back the top half of the sporangium, exposing the spores. The sporangium continues to be pulled back until the tension on the water molecules within the annulus exceeds the strength of the hydrogen bonds holding the water molecules together. When this happens, the top half of the sporangium flies forward, throwing the spores out. The fern sporangium is a biological catapult!

If spores land in a suitable environment, one that is generally moist and shaded, they germinate (Figures 19-16d and e), eventually growing into the heart-shaped adult **gametophyte** (19-16f).

7. Using your dissecting microscope, examine a living, preserved, or prepared slide whole mount of the gametophyte (Figures 19-12, 19-16f).

What color is the gametophyte? ____

What does the color indicate relative to the ability of the gametophyte to produce its own carbohydrates?

____

   - **(a)** Examine the undersurface of the gametophyte. Find the **rhizoids,** which anchor the gametophyte and perhaps absorb water.
   - **(b)** Locate the gametangia (sex organs) clustered among the rhizoids (Figure 19-16f). There are two types of gametangia: **antheridia** (Figure 19-16g), which produce the flagellated *sperm,* and **archegonia** (Figures 19-13, 19-16h), which produce *egg cells.*

8. Study the demonstration slide of an archegonium (Figure 19-14). Identify the **egg** within the swollen basal portion of the archegonium. Note that the *neck* of the archegonium protrudes from the surface of the gametophyte.

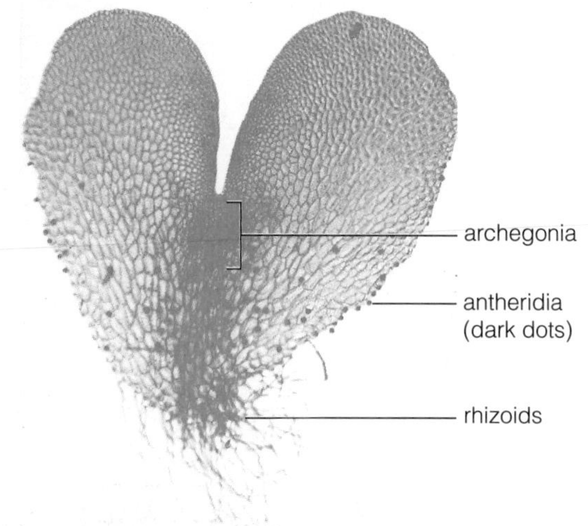

Figure 19-12 Whole mount of fern gametophyte, undersurface (10×).

The archegonia secrete chemicals that attract the flagellated sperm, which swim in a water film down a canal within the neck of the archegonium. One sperm fuses with the egg to produce the first cell of the sporophyte generation, the *zygote* (Figure 19-16i). With subsequent cell divisions, the zygote develops into an embryo (embryo sporophyte). As the embryo grows, it pushes out of the gametophyte and develops into a young sporophyte (Figure 19-16j).

9. Obtain a specimen of a young sporophyte that is attached to the gametophyte (Figures 19-15, 19-16j). Identify the **gametophyte** and then the *primary leaf* and *primary root* of the young sporophyte. As the sporophyte continues to develop, the gametophyte withers away.
10. Examine any other specimens of ferns that may be on demonstration, noting the incredible diversity in form. Look for sori on each specimen.

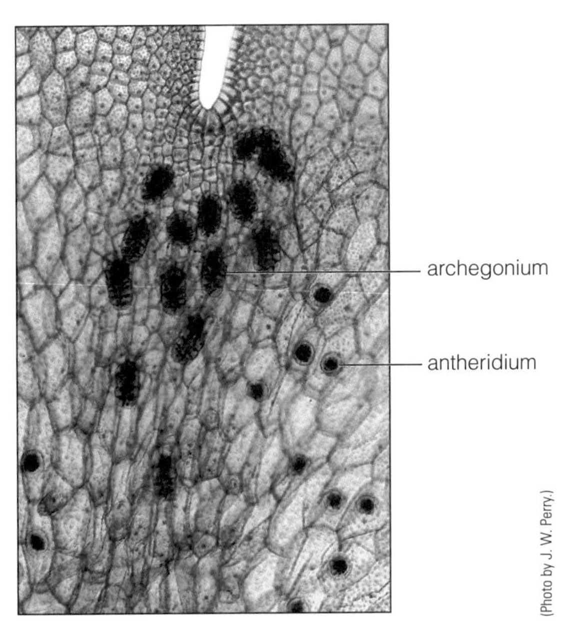

Figure 19-13 Gametangia (antheridia and archegonia) on undersurface of fern gametophyte (57×).

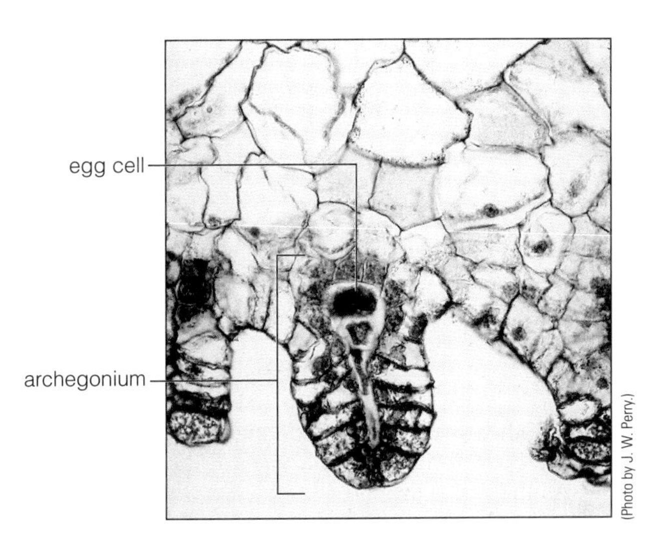

Figure 19-14 Longitudinal section of a fern archegonium (218×).

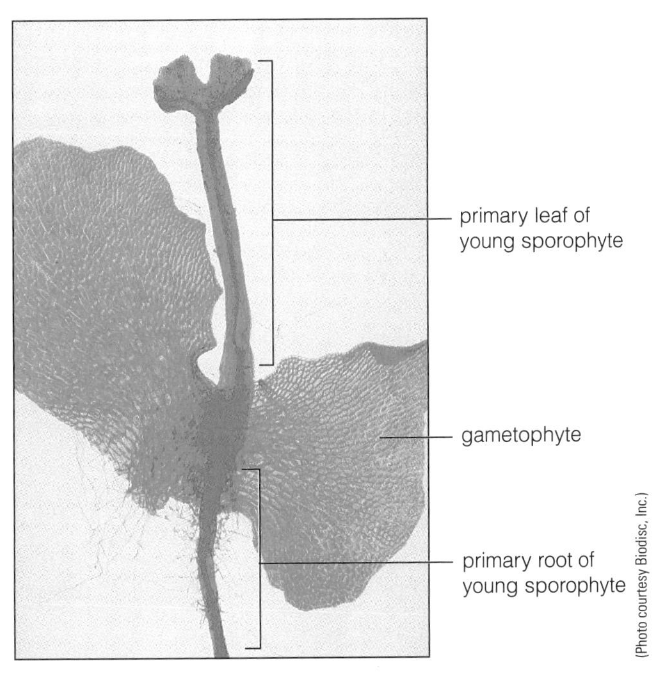

Figure 19-15 Fern gametophyte with attached sporophyte (12×).

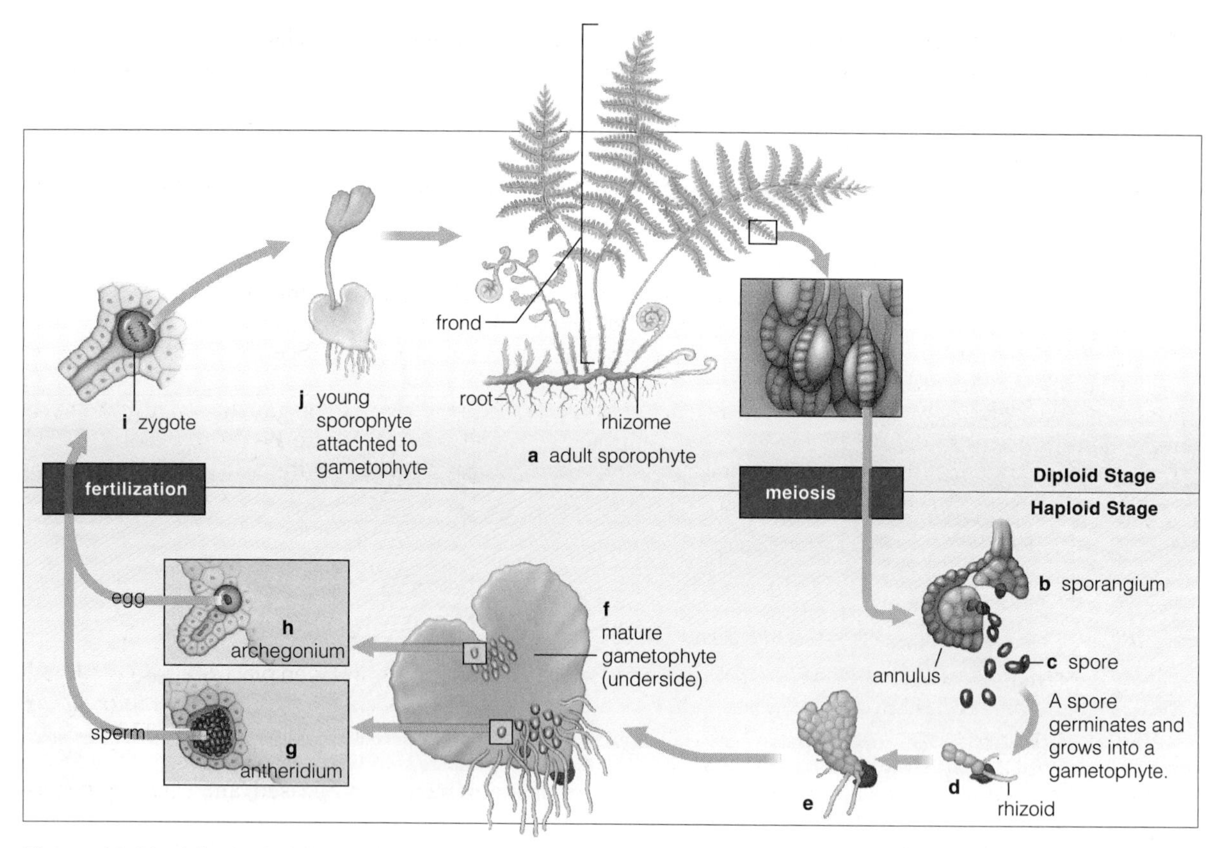

Figure 19-16 Life cycle of a typical fern.

19.5 Experiment: Fern Sperm Chemotaxis *(About 60 min.)*

Chemotaxis is a movement in response to a chemical gradient. This is an important biological process as it allows organisms to find their way or locate things. A dog following a scent is a common example of chemotaxis. How does a sperm cell shed from an antheridium find an archegonium-containing egg cell? The archegonium secretes a sperm-attracting chemical. The farther away the sperm from the archegonium, the more dilute the chemical. The sperm swim up the chemical concentration gradient, right to the egg cell.

In this experiment, you will determine whether certain chemical substances evoke a chemotactic response by fern sperm. This experiment addresses the hypothesis that *some chemicals will attract sperm more strongly and for greater duration than others.*

MATERIALS

Per experimental group (pair of students):

- 6 sharpened toothpicks
- 5 test solutions
- dissecting needle
- depression slide
- dissecting microscope
- petri dish culture containing *C-Fern*™ gametophytes

PROCEDURE

Work in pairs.

1. Obtain six wooden toothpicks whose tips have been sharpened to a fine point.

Caution

Do not touch the sharpened end with your fingers!

2. To identify the toothpicks, place one to five small dots along the side of the toothpicks near the unsharpened end so that each toothpick has a unique marking (one dot, two dots, and so on). One toothpick remains unmarked.
3. Dip the sharpened end of the toothpick with one dot into test solution 1. Insert it with the sharpened end up in the foam block toothpick holder.
4. Repeat step 3 with the four remaining toothpicks and four other test solutions. Leave the unmarked toothpick dry.

Which toothpick serves as the control? ______________________________

5. Obtain a depression slide and pipet one drop of Sperm Release Buffer (SRB) into the depression.
6. Obtain a 12- to 18-day-old culture of *C-Fern*™ gametophytes. Remove the cover, place it on the stage of a dissecting microscope, and observe the gametophytes using transmitted light. (See Exercise 3, page 36, if you need to learn how to choose transmitted light.)
7. Identify the two types of gametophytes present—the smaller thumb-shaped male gametophytes that have many bumps on them and the larger, heart-shaped bisexual gametophytes (Figure 19-17).
8. Using a dissecting needle, pick up 7–10 male gametophytes and place them into the SRB in the depression slide well. They don't need to be completely submerged.

Caution

If you wound or damage a gametophyte, discard it.

9. Place the petri dish cover on the stage of the dissecting microscope, edges up, and then place the gametophyte-containing depression slide on the cover. (This keeps the gametophytes cooler, particularly important if your dissecting microscope has an incandescent light.) Adjust the magnification so it is at least 12×.
10. Within 5 minutes, you should be able to observe sperm being released from the antheridia.
11. Wait another 5 minutes to ensure that a sufficient number of sperm have been released, and then begin testing for chemotactic response by following steps 12–15.
12. While using 12–20× magnification, focus on the TOP surface of the sperm suspension droplet in an area devoid of male gametophytes.

Caution

As you make the tests, do not insert the toothpick fully into the drop—just touch the surface briefly (Figure 19-18).

13. Remove one test toothpick from the block and, while looking through the oculars of the dissecting microscope, gently and briefly touch the sharpened end of the toothpick to the surface of the drop (Figure 19-18).
14. Observe whether any chemotactic response takes place. Record your observation in Table 19-1.
15. Repeat the procedure with the remaining toothpicks, stirring the sperm suspension (if necessary) with a dissecting needle after each test to redistribute the sperm.
16. Place your data on the lab chalk or marker board. Your instructor will pool the data for all experimental groups.

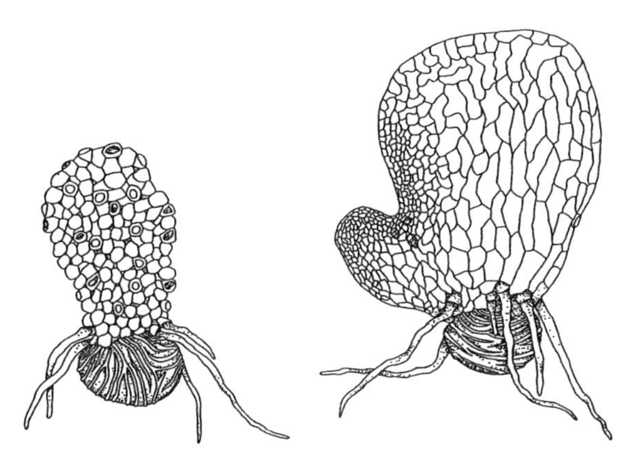

Figure 19-17 Mature male (left) and bisexual (right) *C-Fern*™ gametophytes.

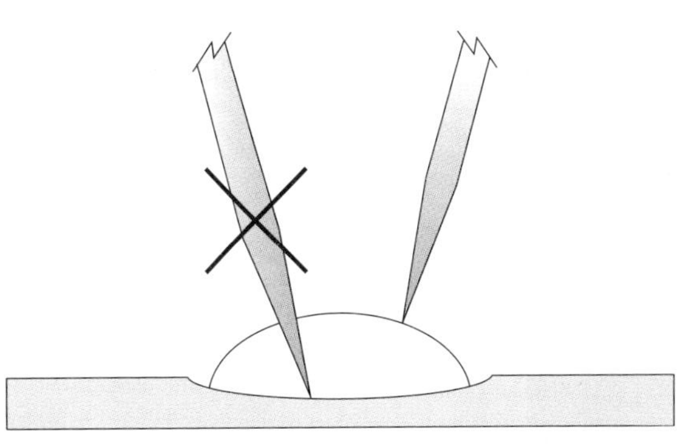

Figure 19-18 Method for applying test substances. Proper technique is important. Gently and briefly touch the end of the toothpick to the surface of the sperm suspension. *Note:* The size of the drop of suspension is exaggerated to show detail.

17. Make a conclusion about the chemotactic properties of the various chemical test substances, and note it in Table 19-1.

Examine the chemical structures of each test solution (Figure 19-19). Relate the biological (chemotactic) response you observed to any of the chemical structural differences of the test substances. ____________________

____

____

____

What is the advantage to using the pooled data to draw your conclusion? ____________________

____

____

____

What do you hypothesize the naturally occurring chemical produced by mature archegonia to be? ____________________

____

____

____

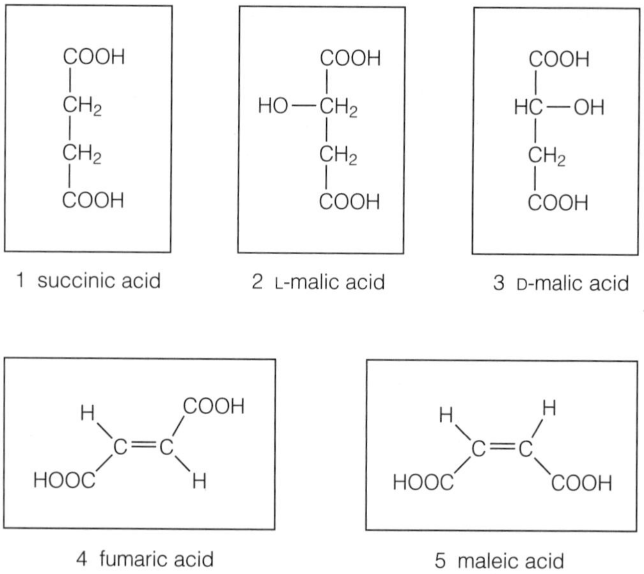

Figure 19-19 Chemical structures of test substance.

TABLE 19-1 My Data: Chemotactic Response of *C-Fern*™ Sperm

Prediction:

	Swarming Response[b]			
	Intensity (low, medium, high)		Duration (short, medium, long)	
Test Substance[a]	**My Data**	**Pooled Data**	**My Data**	**Pooled Data**
0				
1				
2				
3				
4				
5				

Conclusion:

[a] Identified by number of dots on toothpick.

[b] Quantify the response by using the symbols (0) for no response, (–) if the sperm move away from the test site, and (+, ++, +++) for varying degrees of attraction to the test site.

PRE-LAB QUESTIONS

_____ **1.** The sporophyte is the dominant and conspicuous generation in
(a) the fern allies and ferns
(b) gymnosperms
(c) flowering plants
(d) all of the above

_____ **2.** A tracheophyte is a plant that has
(a) xylem and phloem
(b) a windpipe
(c) a trachea
(d) the gametophyte as its dominant generation

_____ **3.** *Psilotum* lacks
(a) roots
(b) a mechanism to take up water and minerals
(c) vascular tissue
(d) alternation of generations

_____ **4.** Spore germination followed by cell divisions results in the production of
(a) a sporophyte
(b) an antheridium
(c) a zygote
(d) a gametophyte

_____ **5.** Which phrase *best* describes a plant that is an epiphyte?
(a) a plant that grows upon another plant
(b) a parasite
(c) a plant with the gametophyte generation dominant and conspicuous
(d) a plant that has a mutually beneficial relationship with another plant

_____ **6.** Club mosses
(a) are placed in the subphylum Pterophyta
(b) are so called because of the social nature of the plants
(c) do not produce gametophytes
(d) are so called because most produce strobili

_____ **7.** The resurrection plant
(a) is a species of *Selaginella*
(b) grows in the desert Southwest of the United States
(c) is a member of the phylum Lycophyta
(d) all of the above

_____ **8.** Which statement is *not* true?
(a) Nodes are present on horsetails
(b) The rhizome on horsetails bears roots
(c) The internode of a horsetail is the region where the leaves are attached
(d) Horsetails are members of the subphylum Sphenophyta

_____ **9.** In ferns
(a) xylem and phloem are present in the sporophyte
(b) the sporophyte is the dominant generation
(c) the leaf is called a frond
(d) all of the above are true

_____ **10.** The spores of a fern are
(a) produced by mitosis within the sporangium
(b) diploid cells
(c) the first cells of the gametophyte generation
(d) both a and b

Name ______________________________ Section Number ______________________

EXERCISE 19

Seedless Vascular Plants: Club Mosses and Ferns

POST-LAB QUESTIONS

Introduction

1. List two features that distinguish the seedless vascular plants from the bryophytes.
   a.

   b.

2. After consulting a biological or scientific dictionary, explain the derivation from the Greek of the word *symbiosis*.

19.1 Phylum Lycophyta: Club Mosses

3. Both *Lycopodium* and *Equisetum* have strobili, roots, and rhizomes. What did you learn in this exercise that enables you to distinguish these two plants?

4. Examine the photo at the right.
   a. Give the genus of this plant.
   ____
   b. Label the structure indicated.

(0.3×).

5. Some species of *Lycopodium* produce gametophytes that grow beneath the surface of the soil, while others grow on the soil surface. Basing your answer on what you've learned from other plants in this exercise, predict how each respective type of *Lycopodium* gametophyte might obtain its nutritional needs.

19.2 Phylum Moniliophyta, Subphylum Psilophyta: Whisk Ferns

6. Using a biological or scientific dictionary, or a reference in your textbook, determine the meaning of the root word *psilo,* and relate it to the appearance of *Psilotum.*

19.3 Phylum Moniliophyta, Subphylum Sphenophyta: Horsetails

7. Examine the photo on the right. You studied this genus in lab, but this is a different species. This species has two separate stems produced at different times in the growing season. The stem on the left is a reproductive branch, while that on the right is strictly vegetative. Based on the characteristics obvious in this photo, identify the plant, give its scientific name, and then identify the labeled structures.

   Scientific name

   ___

   A

   ___

   B

   ___

   C

   ___

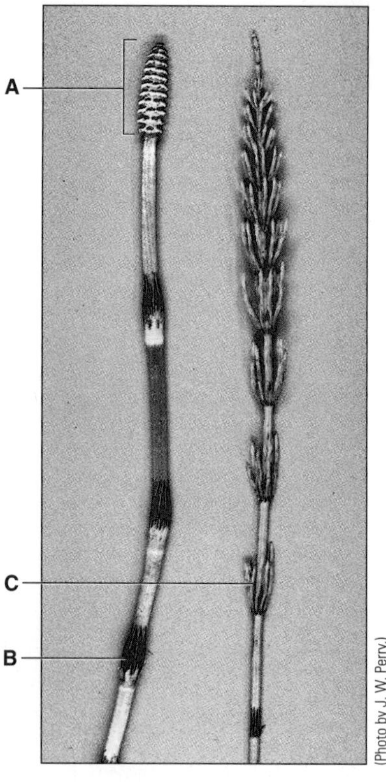

(0.75×).

8. Explain the distinction between a *node* and an *internode*.

19.4 Phylum Moniliophyta, Subphylum Pterophyta: Ferns

9. While rock climbing, you encounter the plant shown in the photo on the right growing out of a crevice.

   a. Is this the sporophyte or the gametophyte?

   b. What special name is given to the leaf of this plant?

(0.5×).

10. The environments in which ferns grow range from standing water to very dry areas. Nonetheless, all ferns are dependent upon free water in order to complete their life cycles. Explain why this is the case.

EXERCISE 20

Seed Plants I: Gymnosperms

OBJECTIVES

After completing this exercise, you will be able to

1. define *gymnosperm, pulp, heterosporous, homosporous, pollination, fertilization, dioecious, monoecious;*
2. describe the characteristics that distinguish seed plants from other vascular plants;
3. produce a cycle diagram of heterosporous alternation of generations;
4. list uses for conifers;
5. recognize the structures in **boldface** and describe the life cycle of a pine;
6. distinguish between a male and a female pine cone;
7. describe the method by which pollination occurs in pines;
8. describe the process of fertilization in pines;
9. recognize members of the four phyla: Coniferophyta, Cycadophyta, Ginkgophyta, and Gnetophyta.

INTRODUCTION

The development of the seed was a significant event in the evolution of vascular plants. Seeds have remarkable ability to survive under adverse conditions. This is one reason for the dominance of seed plants today.

Let's examine the characteristics of seeds and seed plants.

1. All seed plants produce **pollen grains.** Pollen grains serve as carriers for sperm. This characteristic is one factor accounting for the widespread distribution of seed plants. As a benefit of pollen production, the sperm of seed plants do *not* need free water to swim to the egg. Thus, seed plants can reproduce in harsh climates where nonseed plants are much less successful.
2. Almost all seeds have some type of **stored food** for the embryo to use as it emerges from the seed during germination. (The sole exception is orchid seeds, which rely on symbiotic association with a fungus to obtain nutrients.)
3. All seeds have a **seed coat,** a protective covering enclosing the embryo and its stored food.

A seed coat and stored food are particularly important for survival. An embryo within a seed is protected from an inhospitable environment. Consider a seed produced during a severe drought. Water is necessary for the embryo to grow. If none is available, the seed can remain dormant until growing conditions are favorable. When germination occurs, a ready food source is present to get things underway, providing nutrients until the developing plant can produce its own carbohydrates by photosynthesis.

4. Like the bryophytes, fern allies, and ferns, seed plants exhibit **alternation of generations.**
5. All seed plants are **heterosporous;** that is, they produce *two* types of spores. Bryophytes and most fern allies and ferns are **homosporous,** producing only *one* spore type.

Examine Figure 20-1, a diagram of the heterosporous alternation of generations.

Gymnosperms are one of two groups of seed plants. *Gymnosperm* translates literally as "naked seed," referring to the production of seeds on the *surface* of reproductive structures. This contrasts with the situation in the angiosperms, whose seeds are contained within a fruit.

The general assemblage of plants known as gymnosperms contains plants in four separate phyla:

Phylum	Common Name
Coniferophyta	Conifers
Cycadophyta	Cycads
Ginkgophyta	Ginkgos
Gnetophyta	Gnetophytes or vessel-containing gymnosperms

20.1 Phylum Coniferophyta: Conifers *(About 90 min.)*

By far the most commonly recognized gymnosperms are the conifers. Among the conifers, perhaps the most common is the pine (*Pinus*). Many people think of conifers and pines as one and the same. However, while all members of the genus *Pinus* are conifers, not all conifers are pines. Examples of other conifers include spruce (*Picea*), fir (*Abies*), and even some trees that lose their leaves in the fall such as larch.

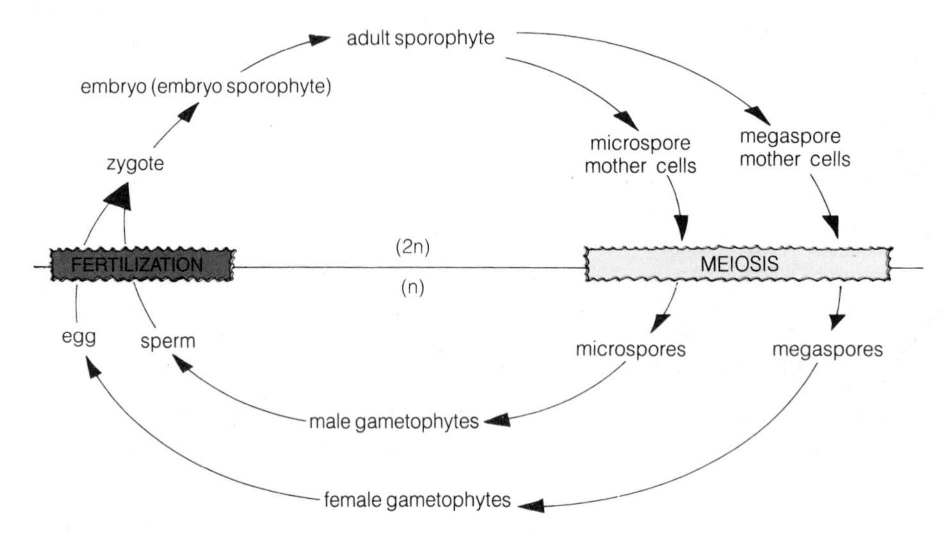

Figure 20-1 Heterosporous alternation of generations.

The conifers are among the most important plants economically, because their wood is used in building construction. Millions of hectares (1 hectare = 2.47 acres) are devoted to growing conifers for this purpose, not to mention the numerous plantations that grow Christmas trees. In many areas conifers are used for **pulp,** the moistened cell wall–derived cellulose used to manufacture paper.

The structures and events associated with reproduction in pine (*Pinus*) will be studied here as a representative conifer.

MATERIALS

Per student:

- cluster of male cones
- prepared slide of male strobilus, l.s., with microspores
- young female cone
- prepared slide of female strobilus, l.s., with megaspore mother cell
- prepared slide of pine seed, l.s.
- compound microscope
- dissecting microscope
- single-edged razor blade

Per student group (table):

- young sporophyte, living or herbarium specimen

Per lab room:

- demonstration slide of female strobilus with archegonium
- demonstration slide of fertilization
- pine seeds, soaking in water
- pine seedlings, 12 weeks old
- pine seedlings, 36 weeks old

PROCEDURE

As you do this activity, examine Figure 20-12, representing the life cycle of a pine tree. To refresh your memory, look at a specimen of a small pine tree. This is the adult **sporophyte** (Figure 20-12a). Identify the stem and leaves. You probably know the main stem of a woody plant as the trunk. The leaves of conifers are often called needles because most are shaped like needles.

A. Male Reproductive Structures and Events

1. Obtain a cluster of **male cones** (Figures 20-2, 20-12b). The function of male cones is to produce **pollen;** consequently, male cones are typically produced at the ends of branches, where the wind currents can catch the pollen as it is being shed.

Figure 20-2 Cluster of male cones shedding pollen (0.75×).

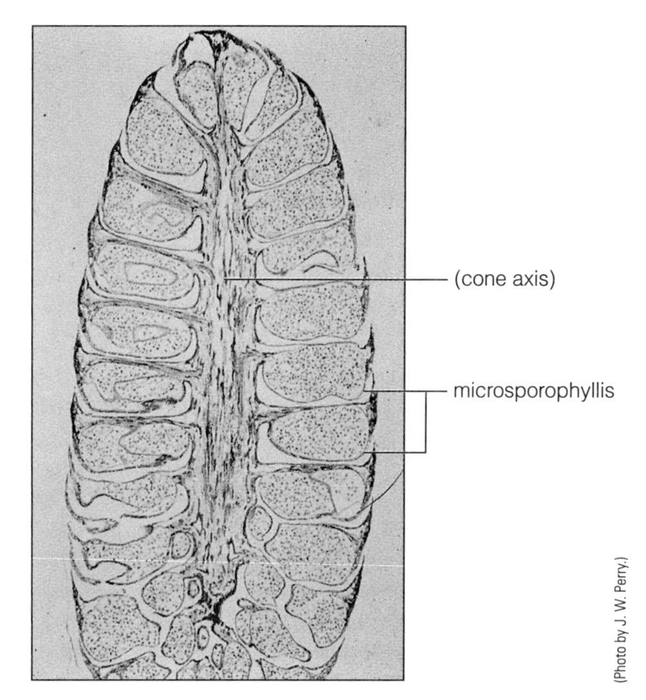

Figure 20-3 Male pine cone, l.s. (12×).

2. Note all the tiny scalelike structures that make up the male cones. These are *microsporophylls.*

Translated literally, a sporophyll would be a "spore-bearing leaf." The prefix *micro-* refers to "small." But the literal interpretation "small, spore-bearing leaf" does not convey the full definition; biologists use *microsporophyll* to mean a leaf that produces *male* spores, called *microspores.* These develop into winged, immature male *gametophytes* called **pollen grains.** (Why they are immature is a logical question. The male gametophyte is not mature until it produces sperm.)

3. Remove a single microsporophyll and examine its lower surface with a dissecting microscope. Identify the two **microsporangia,** also called **pollen sacs** (Figure 20-12c).
4. Study a prepared slide of a longitudinal section of a male cone (also called a male *strobilus*), first with a dissecting microscope to gain an impression of the cone's overall organization and then with the low-power objective of your compound microscope. Identify the *cone axis* bearing numerous **microsporophylls** (Figures 20-3).
5. Switch to the medium-power objective to observe a single microsporophyll more closely. Note that it contains a cavity; this is the **microsporangium** (also called a *pollen sac*), which contains numerous *pollen grains.*

As the male cone grows, several events occur in the microsporophylls that lead to the production of pollen grains. Microspore mother cells within the microsporangia undergo meiosis to form *microspores.* Cell division within the microspore wall and subsequent differentiation result in the formation of the pollen grain.

6. Examine a single **pollen grain** with the high-dry objective (Figures 20-4, 20-12d). Identify the earlike *wings* on either side of the body.
7. Identify the four cells that make up the body: The two most obvious are the **tube cell** and smaller **generative cell.** (The nucleus of the tube cell is almost as large as the entire generative cell.) Note that this male gametophyte is made up of only four cells!

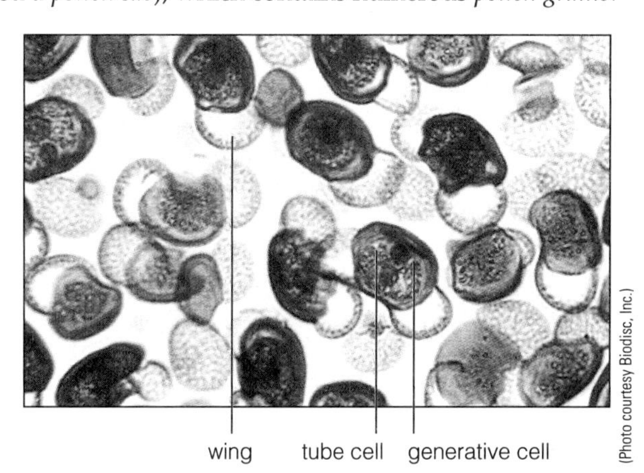

Figure 20-4 Pine pollen grains (400×).

In conifers, **pollination,** the transfer of pollen from the male cone to the female cone, is accomplished by wind and occurs during spring. Pollen grains are caught in a sticky *pollination droplet* produced by the female cone.

B. Female Reproductive Structures and Events

Development and maturation of the female cone take 2–3 years; the exact time depends on the species. Female cones are typically produced on higher branches of the tree. Because the individual tree's pollen is generally shed downward, this arrangement favors crossing between *different* individuals.

Figure 20-5 Female cones of pine (1×).

1. Obtain a young **female cone** (Figures 20-5, 20-12e), and note the arrangement of the cone scales. Unlike the male cone, the female cone is a complex structure, each scale consisting of an **ovuliferous** scale fused atop a *sterile bract.*
2. Remove a single scale-bract complex (Figure 20-12f). On the top surface of the complex find the two **ovules,** the structures that eventually will develop into the **seeds.**
3. Examine a prepared slide of a longitudinal section of a female cone (Figures 20-6, 20-12g) first with the dissecting microscope. Note the spiral arrangement of the scales on the cone axis. Distinguish the smaller *sterile bract* from the *ovuliferous scale.*

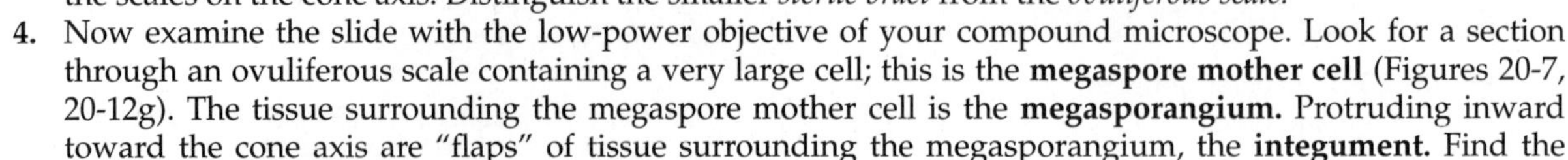

4. Now examine the slide with the low-power objective of your compound microscope. Look for a section through an ovuliferous scale containing a very large cell; this is the **megaspore mother cell** (Figures 20-7, 20-12g). The tissue surrounding the megaspore mother cell is the **megasporangium.** Protruding inward toward the cone axis are "flaps" of tissue surrounding the megasporangium, the **integument.** Find the integuments and the opening between them, the **micropyle.**

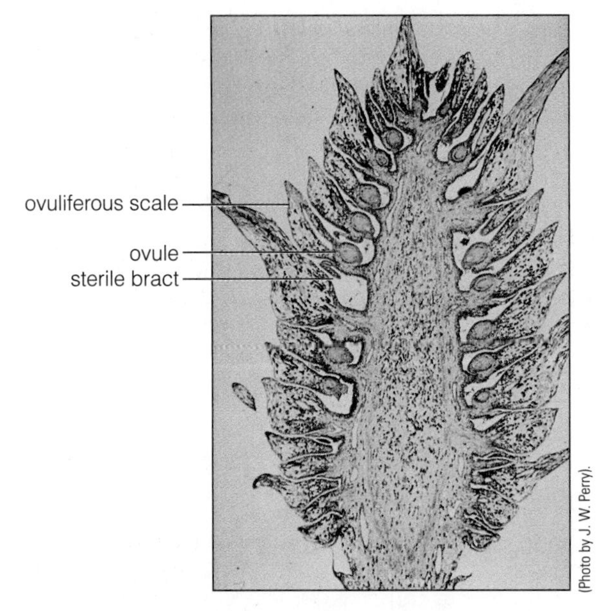

Figure 20-6 Female pine cone, l.s. (5×).

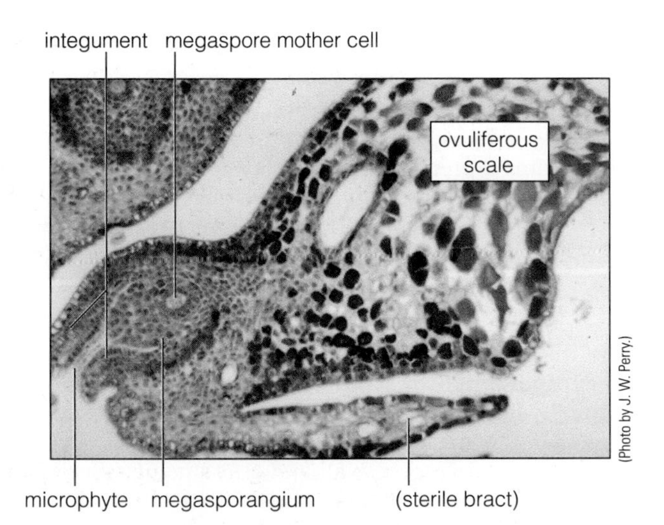

Figure 20-7 Portion of female cone showing megaspore mother cell (75×).

Think for a moment about the three-dimensional nature of the ovule: It's much like a short vase lying on its side on the ovuliferous scale. The neck of the vase is the integument, the opening the micropyle. The integument extends around the base of the vase. If you poured liquid rubber inside the base of a vase, suspended a marble in the middle, and allowed the rubber to harden, you'd have a model of the megasporangium and the megaspore mother cell. Figure 20-8 gives you an idea of the three-dimensional structure.

The megaspore mother cell undergoes meiosis to produce four haploid *megaspores* (Figure 20-12h), but only one survives; the other three degenerate. The functional megaspore repeatedly divides mitotically to produce the multicellular **female gametophyte** (Figure 20-12i). At the same time, the female cone is continually increasing in size to accommodate the developing female gametophytes. (Remember, there are numerous ovuliferous scale/sterile bract complexes on each cone.)

The female gametophyte of pine is produced _______________ (within *or* outside of) the megasporangium.

5. Examine the demonstration slide of **archegonia** (Figures 20-9, 20-12i) that have developed within the female gametophyte. Identify the single large **egg cell** that fills the entire archegonium. (The nucleus of the egg cell may be visible as well. The other generally spherical structures are protein bodies within the egg.)

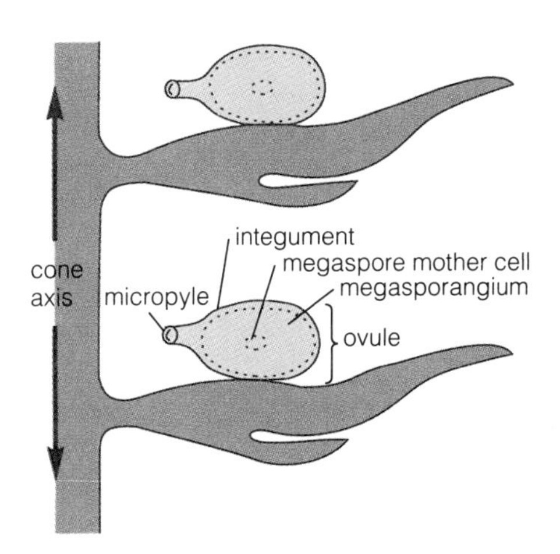

Figure 20-8 Ovule and ovuliferous scale/sterile bract complex.

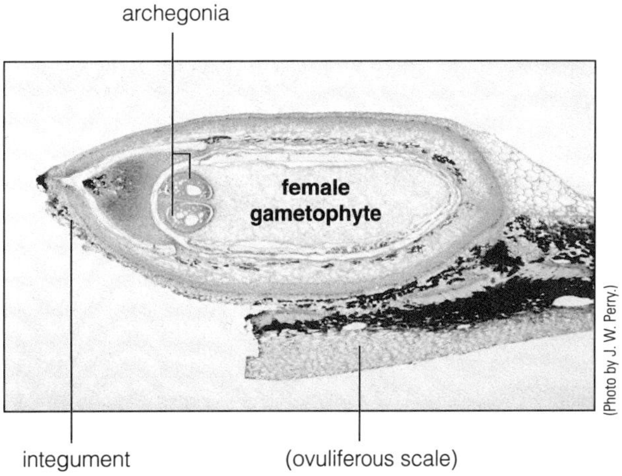

archegonia
female gametophyte
integument
(ovuliferous scale)
(Photo by J. W. Perry.)

Figure 20-9 Pine ovule with female gametophyte and archegonia, l.s. (12×).

Recall that the pollen grains produced within male cones are caught in a sticky pollination droplet produced by the female cone. As the pollination droplet dries, the pollen grain is drawn through the micropyle and into a cavity called the *pollen chamber.*

Fertilization—the fusion of egg and sperm—occurs after the *pollen tube,* an outgrowth of the pollen grain's tube cell, penetrates the megasporangium and enters the archegonium. The generative cell of the pollen grain has divided to produce two sperm, one of which fuses with the egg. (The second sperm nucleus degenerates.)

6. Examine the demonstration slide of fertilization in *Pinus.* Identify the **zygote,** the product of fusion of egg and sperm (Figures 20-10, 20-12j).

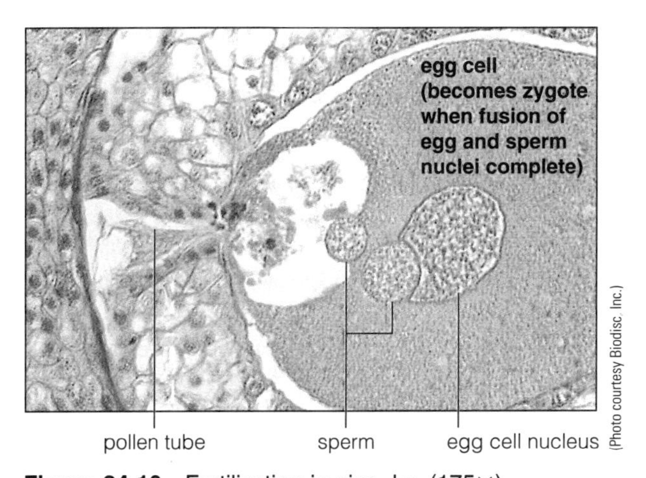

Figure 24-10 Fertilization in pine, l.s. (175×).

After fertilization, numerous mitotic divisions of the zygote take place, eventually producing an **embryo** (*embryo sporophyte*). Fertilization also triggers changes in the integument, causing it to harden and become the seed coat.

7. With the low-power objective of your compound microscope, study a prepared slide of a longitudinal section through a pine seed (Figures 20-11, 20-12k). Starting from the outside, identify the **seed coat, megasporangium** (a very thin, papery remnant), the female gametophyte, and **embryo** (embryo sporophyte).
8. Within the embryo, identify the **hypocotyl-root axis** and numerous **cotyledons,** in the center of which is the **epicotyl.** (*Hypo-* and *epi-* are derived from Greek, meaning "under" and "over," respectively. Thus, these terms refer to orientation with reference to the cotyledons.) The female gametophyte will serve as a food source for the embryo sporophyte when germination takes place.
9. Obtain a pine seed that has been soaked in water to soften the seed coat. Remove it and make a freehand longitudinal section with a sharp razor blade.
10. Identify the papery remnant of the **megasporangium,** the white **female gametophyte,** and **embryo** (embryo sporophyte).

How many cotyledons are present? ______________________________

11. Examine the culture of pine seeds that were planted in sand 12 weeks ago. Note the germinating seeds (Figure 20-12l).
12. Identify the **hypocotyl-root axis, cotyledons, female gametophyte,** and **seed coat.**

The cotyledons serve two functions. One is to absorb the nutrients stored in the female gametophyte during germination. As the cotyledons are exposed to light, they turn green. What then is the second function of the cotyledons?

____

____

____

13. Finally, examine the 36-week-old sporophyte seedlings. Notice that the cotyledons eventually wither away as the epicotyl produces new leaves.

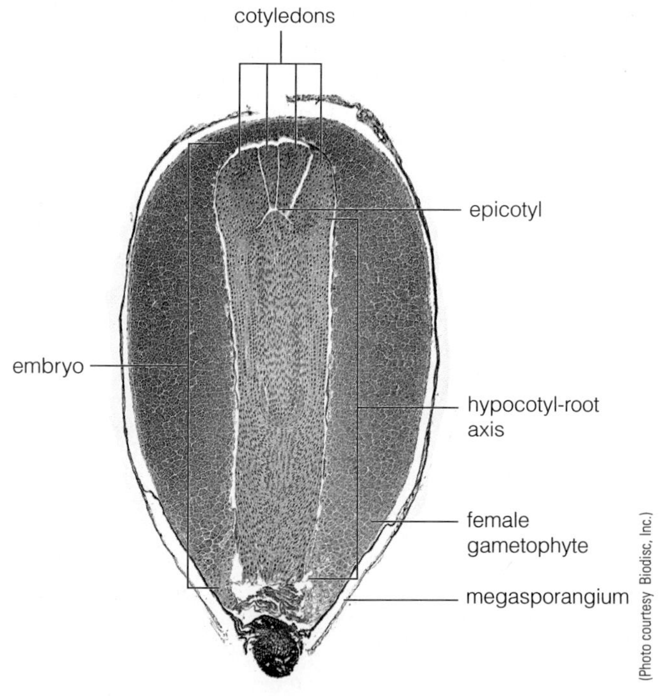

Figure 20-11 Pine seed, l.s. (24×).

20.2 Phylum Cycadophyta: Cycads *(About 10 min.)*

During the age of the dinosaurs (240 million years ago), cycads were extremely abundant in the flora. In fact, botanists often call this "the age of the cycads."

Zamia and *Cycas* are two cycads most readily available in North America. In nature, these two species are limited to the subtropical regions. They're often planted as ornamentals in Florida, Gulf Coast states, and California.

All cycads have separate male and female plants, unlike pine.

The female cones of some other genera become extremely large, weighing as much as 30 kg!

MATERIALS

Per lab room:

- demonstration specimens of *Zamia* and/or *Cycas*

PROCEDURE

1. Examine the demonstration specimen of *Zamia* and/or *Cycas* (Figure 20-13). Both have the common name *cycad*. Do these plants resemble any of the conifers you know? ____

   ____

2. Notice the leaves of the cycads. They more closely resemble the leaves of the ferns than those of the conifers.

20.3 Phylum Ginkgophyta: Ginkgo *(About 10 min.)*

A single species of this phylum is all that remains of what was once a much more diverse assemblage of plants. *Ginkgo biloba* is sometimes called a living fossil because it has changed little in the last 80 million years. In fact, at one time it was believed to be extinct; the Western world knew it from the fossil record before living trees were discovered in the Buddhist temple courtyards of remote China. Today *Ginkgo* is a highly prized ornamental tree that is commonly planted in our urban areas. The tree has a reputation for being resistant to most insect pests and atmospheric pollution. The seeds are ground up and sold as a memory-enhancing dietary supplement, although the supplement's ability to do that is in dispute.

MATERIALS

Per lab room:

- demonstration specimen of *Ginkgo* (living plant or herbarium specimen)

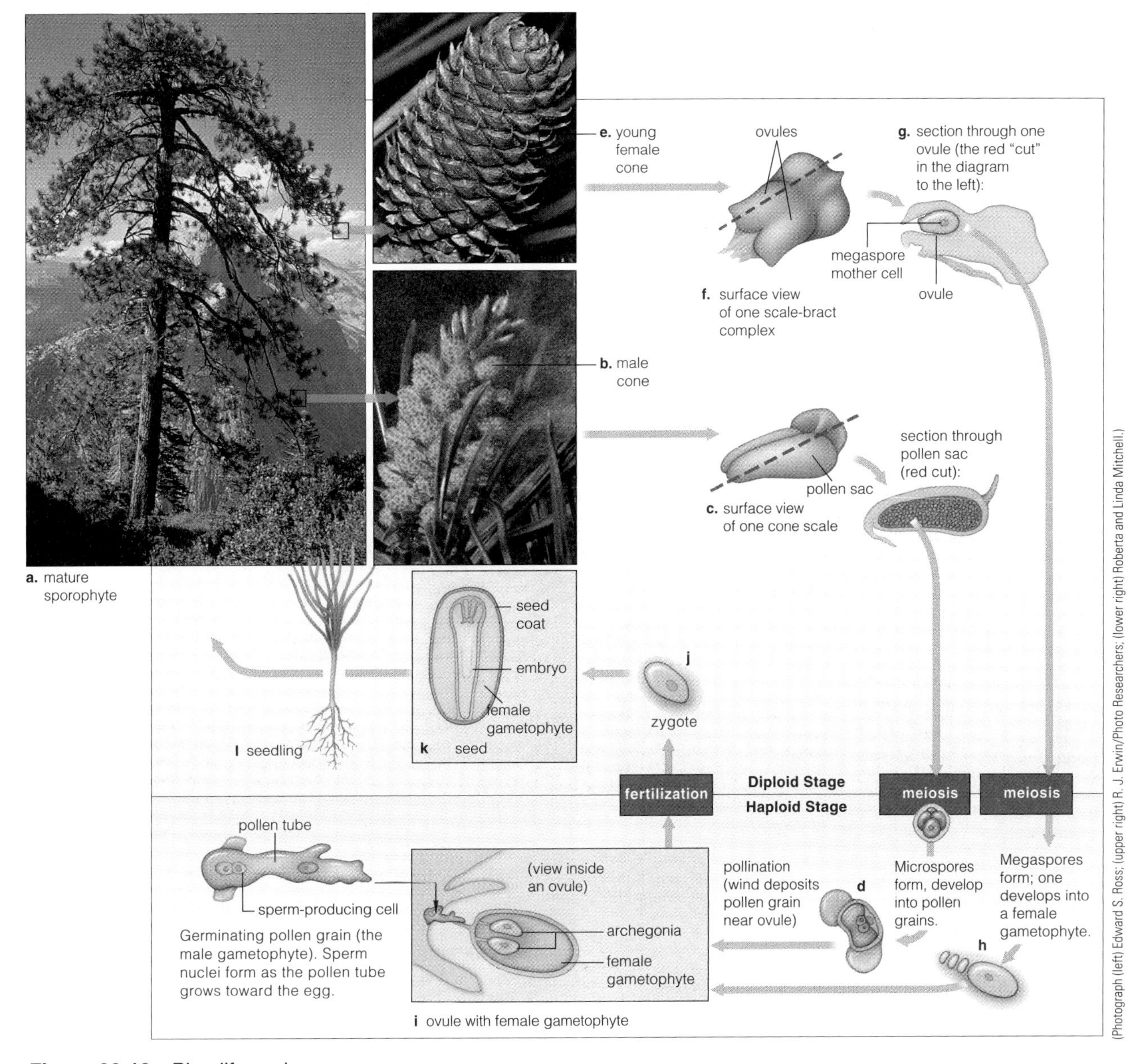

Figure 20-12 Pine life cycle.

Figure 20-13 Cycads. (**a**) *Cycas* (0.01×). (**b**) *Zamia* (0.1×).

PROCEDURE

1. Examine the demonstration specimen of *Ginkgo biloba,* the maidenhair tree (Figure 20-14a).
2. Note the fan-shaped leaves (Figure 20-14b).

a

b

(Photos by J. W. Perry.)

Figure 20-14 *Ginkgo.* (**a**) Tree. (**b**) Branch with leaves (0.5×).

20.4 Phylum Gnetophyta: Gnetophytes (Vessel-Containing Gymnosperms)

(About 10 min.)

The gnetophytes are a small assemblage of plants that have several characteristics found only in the flowering plants (angiosperms). Their reproductive structures look much more like flowers than cones. "Double fertilization", a feature thought to be unique to the flowering plants, has been reported in one gnetophyte species. And the water-conducting xylem tissue contains vessels in all species. Consequently, many scientists now believe these plants are very closely related to the flowering plants. Most species are found in desert or arid regions of the world. In the desert Southwest of the United States, *Ephedra* is a common shrub known as Mormon tea, because its stems were once harvested by Mormon settlers in Utah and used to make a tea.

MATERIALS

Per lab room:

- demonstration specimen of *Ephedra* and/or *Gnetum* (living plant or herbarium specimen)

PROCEDURE

1. Examine the demonstration specimen of *Ephedra* (Figure 20-15).
2. A second representative gnetophyte is the genus *Gnetum* (pronounced "neat-um"). A few North American college and university greenhouses—and some tropical gardens—keep these specimens. *Gnetum* is native to Brazil, tropical west Africa, India, and Southeast Asia. Different species vary in form from vines to trees (Figures 20-16a, b). Examine the living specimen if one is on display, noting particularly the broad, flat leaves (Figure 20-16c).

Figure 20-15 *Ephedra.* (**a**) Several plants (1×). (**b**) Close-up of stems (0.5×).

Figure 20-16 *Gnetum.* (**a**) A species that is a vine (0.02×). (**b**) A species that is a tree (0.02×). (**c**) Leaves (0.38×).

PRE-LAB QUESTIONS

_____ **1.** Which statement is *not* true about conifers?
(a) Conifers are gymnosperms
(b) All conifers belong to the genus *Pinus*
(c) All conifers have naked seeds
(d) Conifers are heterosporous

_____ **2.** Seed plants
(a) have alternation of generations
(b) are heterosporous
(c) develop a seed coat
(d) are all of the above

_____ **3.** A pine tree is
(a) a sporophyte
(b) a gametophyte
(c) diploid
(d) both a and c

_____ **4.** The male pine cone
(a) produces pollen
(b) contains a female gametophyte
(c) bears a megasporangium containing a megaspore mother cell
(d) gives rise to a seed

_____ **5.** The male gametophyte of a pine tree
(a) is produced within a pollen grain
(b) produces sperm
(c) is diploid
(d) is both a and b

_____ **6.** Which of these are produced directly by meiosis in pine?
(a) sperm cells
(b) pollen grains
(c) microspores
(d) microspore mother cells

_____ **7.** An ovule
(a) is the structure that develops into a seed
(b) contains the microsporophyll
(c) is produced on the surface of a male cone
(d) is all of the above

_____ **8.** The process by which pollen is transferred to the ovule is called
(a) transmigration
(b) fertilization
(c) pollination
(d) all of the above

_____ **9.** Which statement is true of the female gametophyte of pine?
(a) It's a product of repeated cell divisions of the functional megaspore
(b) It's haploid
(c) It serves as the stored food to be used by the embryo sporophyte upon germination
(d) All of the above are true

_____ **10.** The seed coat of a pine seed
(a) is derived from the integuments
(b) is produced by the micropyle
(c) surrounds the male gametophyte
(d) is divided into the hypocotyl-root axis and epicotyl

Name ______________________ Section Number ______________________

EXERCISE 20

Seed Plants I: Gymnosperms

POST-LAB QUESTIONS

Introduction

1. What survival advantage does a seed have that has enabled the seed plants to be the most successful of all plants?

2. In the diagram below of a seed, give the ploidy level (n or 2n) of each part listed.

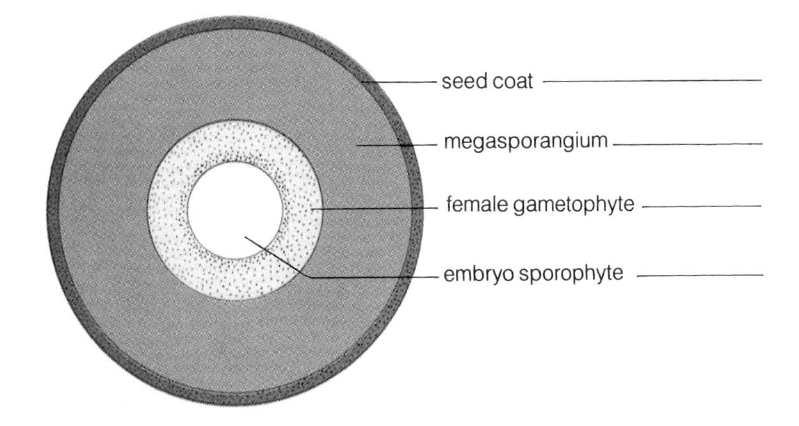

3. Distinguish between a homosporous and a heterosporous type of life cycle.

20.1 Phylum Coniferophyta: Conifers

4. List four uses for conifers.

   a.

   b.

   c.

   d.

5. While snowshoeing through the winter woods, you stop to look at a tree branch pictured at the right. Specifically, what are the brown structures hanging from the branch?

(Photo by J. W. Perry.)

(0.5×).

6. Distinguish between *pollination* and fertilization.

7. Are antheridia present in conifers? ____________
   Are archegonia present? ____________

8. Suppose you saw the seedling pictured at the right while walking in the woods.
   a. To which gymnosperm phylum does the plant belong?
   b. Identify structures A and B.

A ____________
B ____________
(Photo by J. W. Perry.)

(0.5×).

20.2 Phylum Cycadophyta: Cycads

9. On spring break, you are strolling through a tropical garden in Florida and encounter the plant pictured at the right. Would the xylem of this plant contain tracheids, vessels, or both?

(Photo by A. B. Russell.)

(0.01×).

20.3 Phylum Ginkgophyta: Ginkgo

10. A friend of yours picks up a branch like the one pictured at the right. Knowing that you have taken a biology course and studied plants, she asks you what the plant is. Identify this branch, giving your friend the full scientific name for the plant.

(Photo by J. W. Perry.)

(0.75×).

EXERCISE 21

Seed Plants II: Angiosperms

OBJECTIVES

After completing this exercise, you will be able to

1. define *angiosperm, fruit, pollination, double fertilization, endosperm, seed, germination, annual, biennial, perennial;*
2. describe the significance of the flower, fruit, and seed for the success of the angiosperms;
3. identify the structures of the flower;
4. recognize the structures and events (those in **boldface**) that take place in angiosperm reproduction;
5. describe the origin and function of fruit and seed;
6. identify the characteristics that distinguish angiosperms from gymnosperms.

INTRODUCTION

The **angiosperms,** Phylum Anthophyta, are seed plants that produce flowers. *"Antho"* means flower, and *"phyta"* plant. The word *angiosperm* literally means "vessel seed" and refers to the seeds borne within a fruit.

There are more flowering plants in the world today than any other group of plants. Assuming that numbers indicate success, flowering plants are the most successful plants to have evolved thus far.

The most important characteristic that distinguishes the Anthophyta from other seed plants is the presence of flower parts that mature into a **fruit**, a container that protects the seeds and allows them to be dispersed without coming into contact with the rigors of the external environment. In many instances, the fruit also contributes to the dispersal of the seed. For example, some fruits stick to fur (or clothing) of animals and are brushed off some distance from the plant that produced them. Animals eat others. The undigested seeds pass out of the digestive tract and fall into environments often far removed from the seeds' source.

Our lives and diets revolve around flowering plants. Fruits enrich our diet and include such things as apples, oranges, tomatoes, beans, peas, corn, wheat, walnuts, pecans . . . the list goes on and on. Moreover, even when we are not eating fruits, we're eating flowering plant parts. Cauliflower, broccoli, potatoes, celery, and carrots all are parts of flowering plants.

Biologists believe that flower parts originated as leaves modified during the course of evolution to increase the probability for fertilization. For instance, some flower parts are colorful and attract animals that transfer the sperm-producing pollen to the receptive female parts.

Figure 21-15 depicts the life cycle of a typical flowering plant. Refer to it as you study the specimens in this exercise.

***Note:* This exercise provides two alternative paths to accomplish the objectives described above, a traditional approach (21.1–21.2) and an investigative one (21.3). Your instructor will indicate which alternative you will follow.**

21.1 External Structure of the Flower *(About 20 min.)*

The number of different kinds of flowers is so large that it's difficult to pick a single example as representative of the entire division. Nonetheless, there is enough similarity among flowers that, once you've learned the structure of one representative, you'll be able to recognize the parts of most.

MATERIALS

Per student:

- flower for dissection (gladiolus or hybrid lily, for example)
- single-edged razor blade
- dissecting microscope

PROCEDURE

1. Obtain a flower provided for dissection.
2. At the base of the flower, locate the swollen stem tip, the **receptacle,** upon which the whorls of floral parts are arranged.
3. Identify the **calyx,** comprising the outermost whorl. Individual components of the calyx are called **sepals.** The sepals are frequently green (although not always). The calyx surrounds the rest of the flower in the bud stage (see Figure 21-15a).
4. Moving inward, locate the next whorl of the flower, the usually colorful **corolla** made up of **petals.**

It is usually the petals that we appreciate for their color. Remember, however, that the evolution of colorful flower parts was associated with the presence of color-visioned *pollinators,* animals that carry pollen from one flower to another. The colorful flowers attract those animals, enhancing the plants' chance of being pollinated, producing seeds, and perpetuating their species.

Both the calyx and corolla are sterile, meaning they do not produce gametes.

5. The next whorl of flower parts consists of the male, pollen-producing parts, the **stamens** (also called *microsporophylls,* "microspore-bearing leaves"; Figure 21-15b). Examine a single stamen in greater detail. Each stamen consists of a stalklike **filament** and an **anther.** The anther consists of four **microsporangia** (also called *pollen sacs*).
6. Next locate the female portion of the flower, the **pistil** (Figures 21-15a and c). A pistil consists of one or more **carpels,** also called *megasporophylls,* "megaspore-bearing leaves."

If the pistil consists of more than one carpel, they are usually fused, making it difficult to distinguish the individual components. However, you can usually determine the number of carpels by counting the number of lobes of the stigma.

7. How many carpels does your flower contain? __________
8. Identify the different parts of the pistil (Figure 21-15c): at the top, the **stigma,** which serves as a receptive region on which pollen is deposited; a necklike **style;** and a swollen **ovary.** The only members of the plant kingdom to have ovaries are angiosperms!
9. With a sharp razor blade, make a section of the ovary. (Some students should cut the pistil longitudinally; others should cut the ovary crosswise. Then compare the different sections.)
10. Examine the sections with a dissecting microscope, finding the numerous small **ovules** within the ovary. The diagram in Figure 21-15c is oversimplified; many flowers have more than one ovule per ovary.
11. In Figure 21-1, make and label two sketches: one of the cross section and the other of a longitudinal section of the ovary. Notice that the ovules are completely enclosed within the ovary. After fertilization, the ovules will develop into *seeds,* and the ovary will enlarge and mature into the *fruit.*

There are two groups of flowering plants, monocotyledons and dicotyledons. The number of flower parts indicates which group a plant belongs to. Generally, monocots have the flower parts in threes or multiples of three. Dicots have their parts in fours or fives or multiples thereof.

12. Count the number of petals or sepals in the flower you have been examining. Are you studying a monocot or dicot?

____

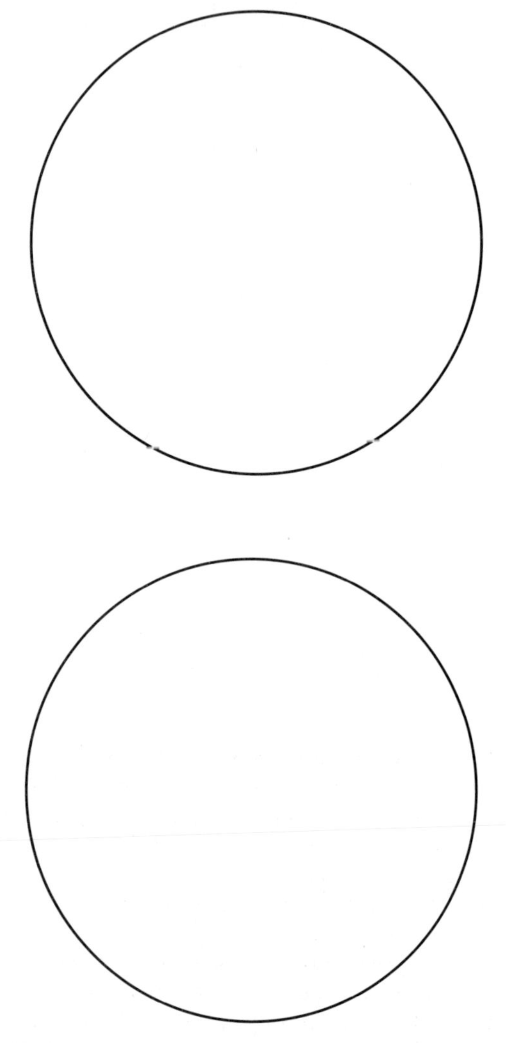

Figure 21-1 Drawings of cross section and longitudinal section of an ovary.
Label: ovules

21.2 The Life Cycle of a Flowering Plant

A. *Male Gametophyte (Pollen Grain) Formation in the Microsporangia* (About 20 min.)

The male gametophyte in flowering plants is the pollen grain; a microscopic sperm-producing structure.

MATERIALS

Per student:

- prepared slide of young lily anther, c.s.
- prepared slide of mature lily anther (pollen grains), c.s.
- *Impatiens* flowers, with mature pollen
- glass microscope slide
- coverslip
- compound microscope

Per student pair:

- 0.5% sucrose, in dropping bottle

PROCEDURE

1. With the low-power objective of your compound microscope, examine a prepared slide of a cross section of an immature anther (Figures 21-2, 21-15d). Find sections of the four **microsporangia** (also called *pollen sacs*), which appear as four clusters of densely stained cells within the anther.

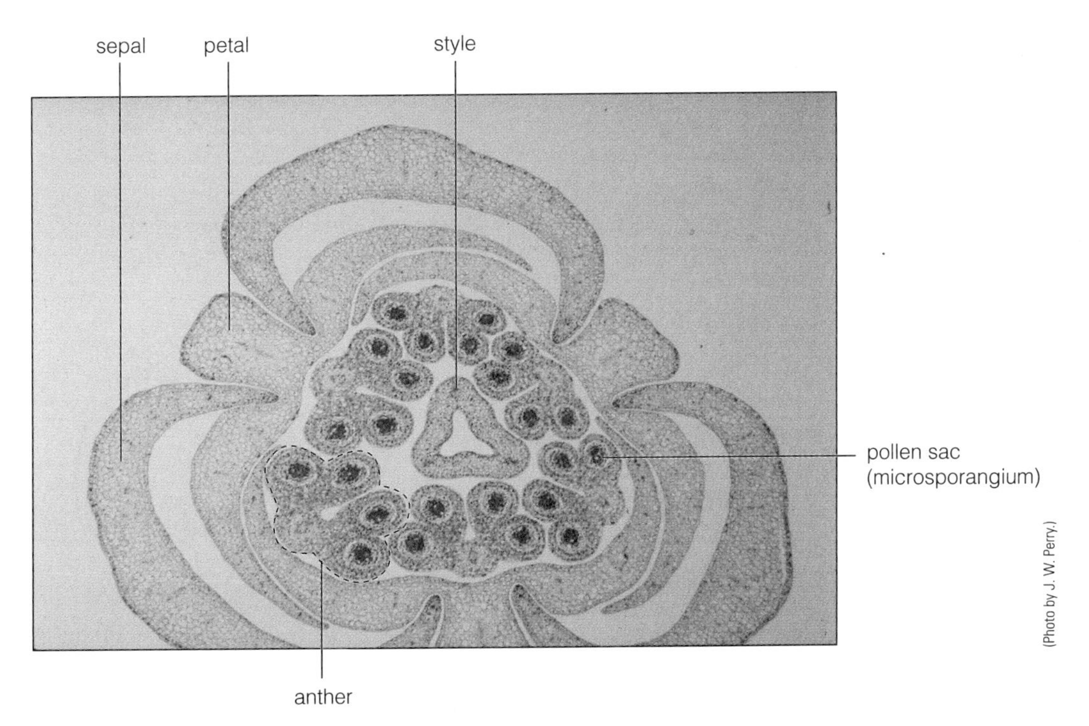

Figure 21-2 Immature anthers within flower bud, c.s. (4×).

2. Study the contents of a single microsporangium. Depending on the stage of development, you will find either diploid *microspore mother cells* (Figure 21-15e) or haploid *microspores* (Figures 21-15f and g).
3. Obtain a prepared slide of a cross section of a mature anther (Figure 21-3). Observe it first with the low-power objective, noting that the walls have split open to allow the pollen grains to be released, as shown in Figure 21-15i.
4. Pollen grains are immature *male gametophytes,* and very small ones at that. Switch to the high-dry objective to study an individual pollen grain more closely (Figure 21-4).
5. The pollen grain consists of only two cells. Identify the large **tube cell** and a smaller, crescent-shaped **generative cell** that floats freely in the cytoplasm of the tube cell (Figure 21-15h).
6. Note the ridged appearance of the outer wall layer of a pollen grain. Within the ridges and valleys of the wall, glycoproteins are present that appear to play a role in recognition between the pollen grain and the stigma.

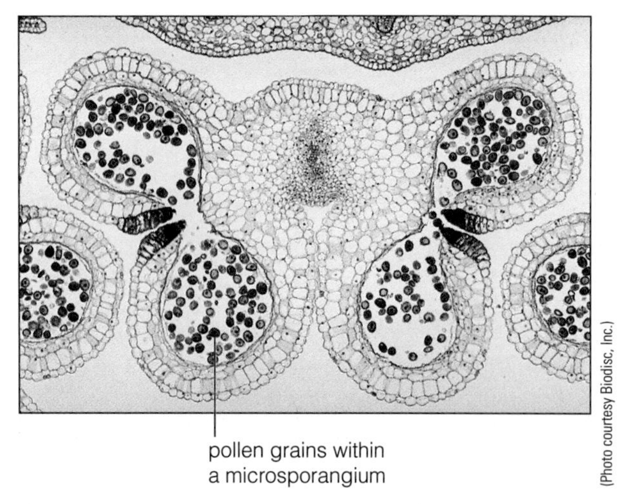

Figure 21-3 Mature anther, c.s. (21×).

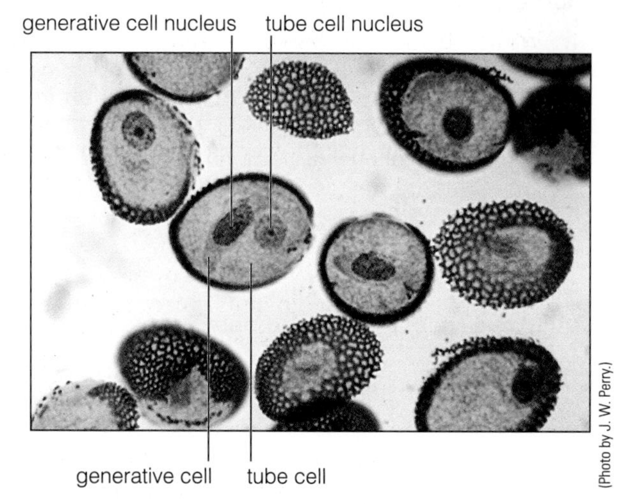

Figure 21-4 Pollen grain, c.s. (145×).

Transfer of pollen from the microsporangia to the stigma, called **pollination,** occurs by various means—wind, insects, and birds being the most common carriers of pollen. When a pollen grain lands on the stigma of a compatible flower, it germinates and produces a **pollen tube** that grows down the style (Figure 21-15j). The generative cell flows into the pollen tube, where it divides to form two *sperm* (Figure 21-15k). Because it bears two gametes, the pollen grain is now considered to be a *mature* male gametophyte.

7. Obtain an *Impatiens* flower and tap some pollen onto a clean microslide. Add a drop of 0.5% sucrose, cover with a coverslip, and observe with the medium-power objective of your compound microscope.
8. Look for the *pollen tube* as it grows from the pollen grain. (You may wish to set the slide aside for a bit and re-examine it 15 minutes later to check the progress of the pollen tubes.)

In Figure 21-5, draw a sequence showing the germination of an *Impatiens* pollen grain.

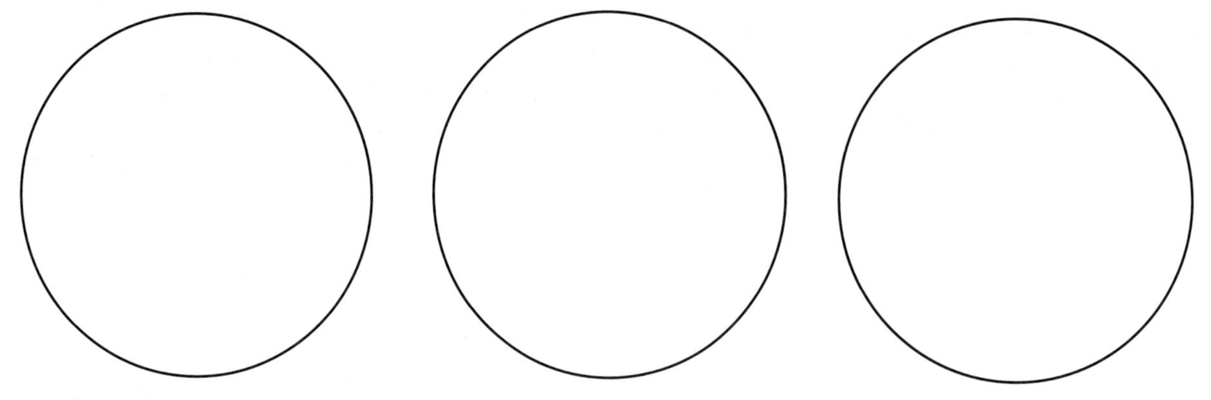

Figure 21-5 Germination of *Impatiens* pollen grain.

B. Female Gametophyte Formation in the Megasporangia (About 20 min.)

The female gametophyte is formed inside each ovule within the ovary of the flower. Like the male gametophyte, the female gametophyte in flowering plants consists of only a few cells.

MATERIALS

Per student:

- prepared slide of lily ovary, c.s., megaspore mother cell
- compound microscope

Per lab room:

- demonstration slide of lily ovary, c.s., 7-celled, 8-nucleate gametophyte
- demonstration slide of lily ovary, c.s., double fertilization

PROCEDURE

1. With the medium-power objective of your compound microscope, examine a prepared slide of a cross section of an ovary (Figures 21-6, 21-15l).
2. Find the several **ovules** that have been sectioned. One ovule will probably be sectioned in a plane so that the very large, diploid **megaspore mother cell** is obvious (Figures 21-7, 21-15m).
3. The megaspore mother cell is contained within the **megasporangium,** the outer cell layers of which form two flaps of tissue called **integuments.** Identify the structures in boldface print. After fertilization, the integuments develop into the *seed coat.*
4. Identify the **placenta,** the region of attachment of the ovule the ovary wall.

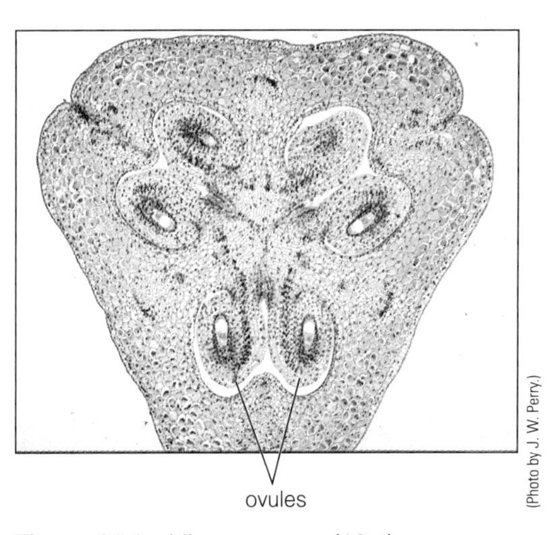

Figure 21-6 Lily ovary, c.s. (19×).

As the megasporangium develops, the integuments grow, enveloping the megasporangium. However, a tiny circular opening remains. This opening is the **micropyle.** (The micropyle is obvious in Figure 21-15n.) Remember, the micropyle is an opening in the ovule. After pollination, the pollen grain germinates on the surface of the stigma, and the pollen tube grows down the style, through the space surrounding the ovule, through the micropyle, and penetrates the megasporangium (Figure 21-15p).

5. Identify the micropyle present in the ovule you are examining.

Considerable variation exists in the next sequence of events; we describe the pattern found in the lily.

The diploid megaspore mother cell undergoes meiosis, producing four haploid *nuclei* (Figure 21-15n). (Note that cytokinesis does *not* follow meiosis, and thus only nuclei—not cells—are formed.) The cell containing the four nuclei (the old megaspore mother cell) is now called the **female gametophyte** (also called the *embryo sac*).

Three of these four nuclei fuse. Thus, the female gametophyte contains one triploid (3n) nucleus and one haploid (n) nucleus (Figure 21-15o). Subsequently, the nuclei undergo two *mitotic* divisions, forming eight nuclei in the female gametophyte. Cell walls form around six of the eight nuclei; the large cell remaining—the **central cell**—contains two nuclei, one of which is triploid, the other haploid. These two nuclei are called the **polar nuclei.** At the micropylar end of the ovule there are three cells, all haploid. One of these is the **egg cell.** Opposite the micropylar end are three 3n cells. This stage of development is often called the *seven-celled, eight-nucleate female gametophyte* (Figures 21-8, 21-15p). Fertilization takes place at this stage.

6. Study the demonstration slide of the seven-celled, eight-nucleate female gametophyte. Identify the **placenta, integuments, micropyle, egg cell, central cell,** and **polar nuclei** (Figures 21-8, 21-15p).

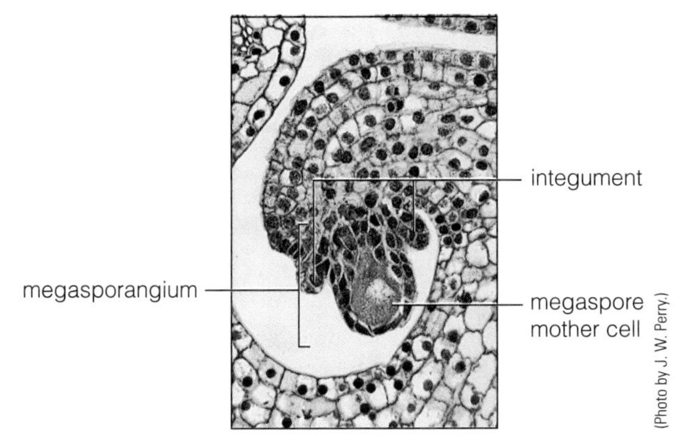

Figure 21-7 Megaspore mother cell within megasporangium (94×).

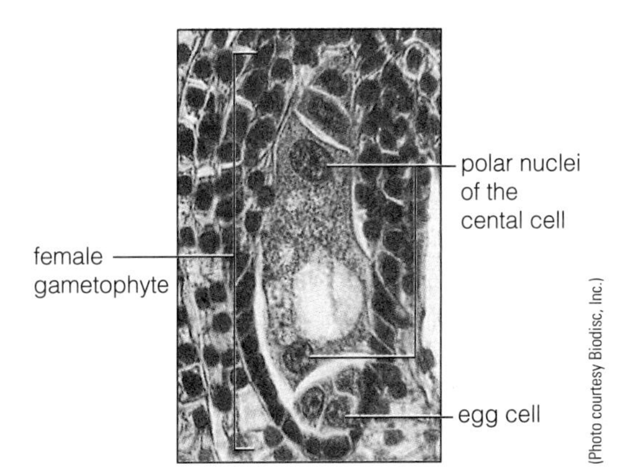

Figure 21-8 Seven-celled, eight-nucleate female gametophyte (170×).

As the pollen tube penetrates the female gametophyte, it discharges the sperm; one of the sperm nuclei fuses with the haploid egg nucleus, forming the *zygote.* Figure 21-15q shows the female gametophyte after fertilization has occurred.

The zygote is a ______________________ (haploid, diploid) cell.

The other sperm nucleus enters the central cell and fuses with the two polar nuclei, forming the primary endosperm nucleus (Figure 21-15q). Thus, the primary endosperm nucleus is ______________________ (haploid, diploid, triploid, tetraploid, pentaploid).

The cell containing the primary endosperm nucleus (the old central cell) is now called the **endosperm mother cell.** Traditionally, the process whereby one sperm nucleus fuses with the egg nucleus and the other sperm nucleus fuses with the two polar nuclei has been called **double fertilization.**

7. Observe the demonstration slide of double fertilization (Figure 21-9), and identify the **zygote, primary endosperm nucleus,** and **central cell** of the female gametophyte.

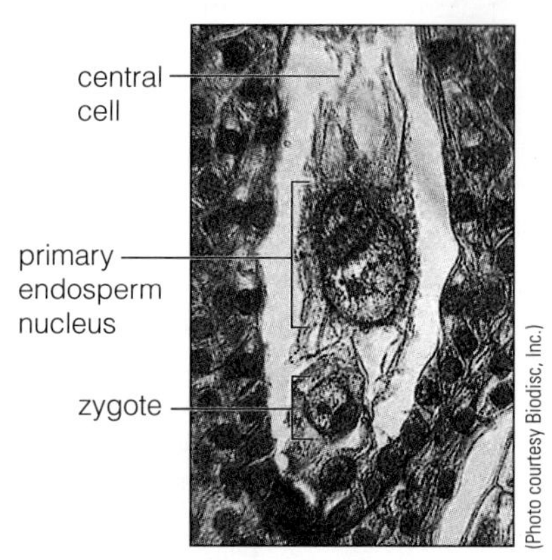

Figure 21-9 Double fertilization (170×).

Numerous mitotic and cytoplasmic divisions of the endosperm mother cell form the **endosperm,** a tissue used to nourish the embryo sporophyte as it develops within the seed.

C. Embryogeny (*About 10 min.*)

The zygote undergoes mitosis and cytokinesis to produce a two-celled **embryo** (also called the *embryo sporophyte*). Numerous subsequent divisions produce an increasingly large and complex embryo.

MATERIALS

Per lab room:

- demonstration slides of *Capsella* embryogeny: globular embryo, emerging cotyledons, torpedo-shaped embryo, mature embryo

PROCEDURE

1. Observe the series of four demonstration slides, which show the stages of embryo development in the female gametophyte of *Capsella* (Figure 21-10).
2. The first slide shows the so-called globular stage (Figure 21-10a), in which a chain of cells (the suspensor) attaches the embryo, a spherical mass of cells, to the wall of the female gametophyte (embryo sac). The very enlarged cell at the base of the suspensor is the *basal cell* and is active in uptake of nutrients that the developing embryo will use. Note the endosperm within the female gametophyte.
3. The second slide shows the heart-shaped stage (Figure 21-10b). Now you can distinguish the emerging **cotyledons** (seed leaves). In many plants, the cotyledons absorb nutrients from the endosperm and thus serve as a food reserve to be used during seed germination.
4. Further development of the embryo has occurred in the third slide, the torpedo stage (Figure 21-10c). Notice that the entire embryo has elongated. Find the *cotyledons.* Between the cotyledons, locate the **epicotyl,** which is the *apical meristem of the shoot.* Beneath the epicotyl and cotyledons find the **hypocotyl-root axis.** At the tip of the hypocotyl-root axis, locate the *apical meristem of the root* and the **root cap** covering it.
5. The final slide (Figure 21-10d) shows a mature embryo, neatly packaged inside the **seed coat.** Identify the seed coat and other regions previously identified in the torpedo stage.

How many cotyledons were there in the slides you examined? ______________________

Thus, is *Capsella* a monocot or dicot? ______________________

D. Fruit and Seed (*About 20 min.*)

Simply stated, a **fruit** is a matured ovary, while a **seed** is a matured ovule. Bear in mind that for each seed produced, a pollen grain had to fertilize the egg cells in the ovules. Fertilization not only causes the integuments of the ovule to develop into a seed coat, it also causes the ovary wall to expand into the fruit.

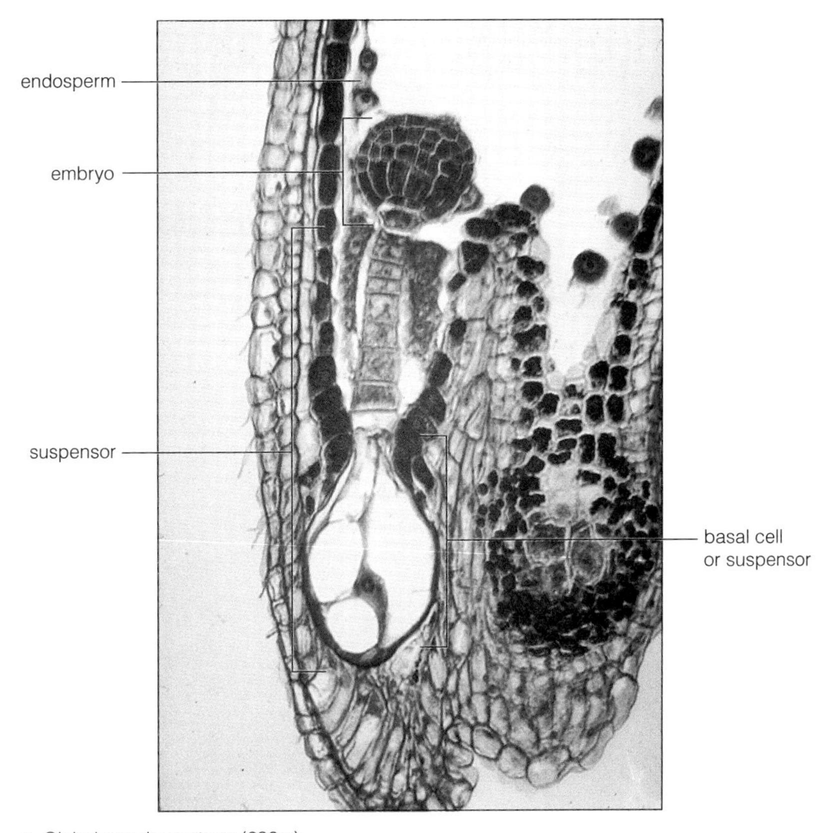

a Globular embryo stage (232×).

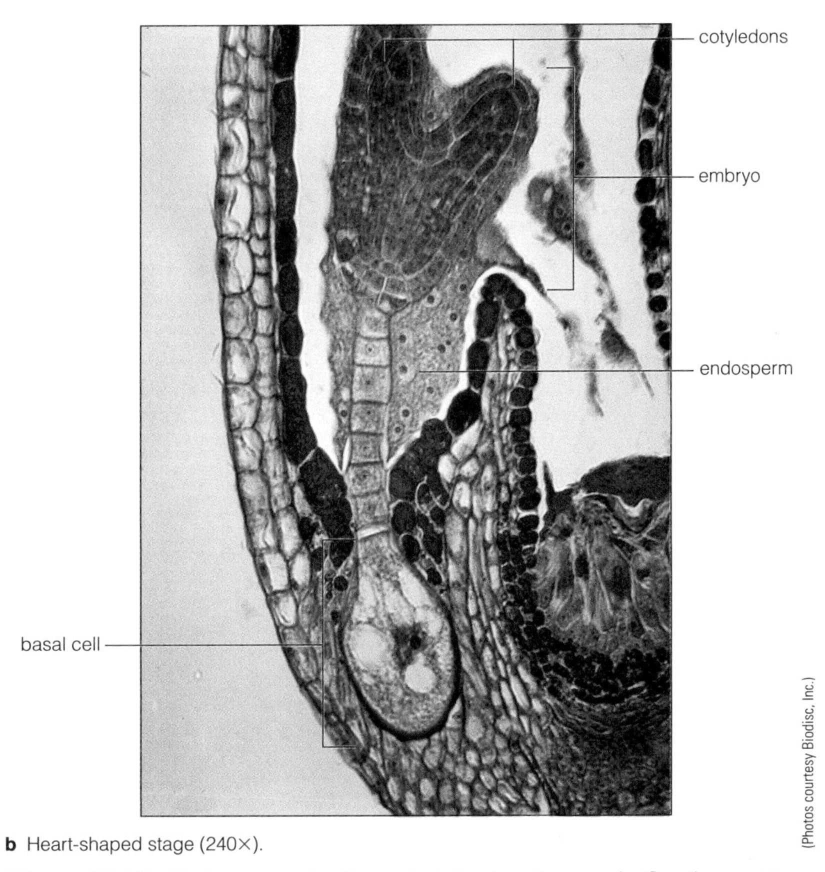

b Heart-shaped stage (240×).

(Photos courtesy Biodisc, Inc.)

Figure 21-10 Embryogeny in *Capsella* (shepherd's purse). *Continues on page 326.*

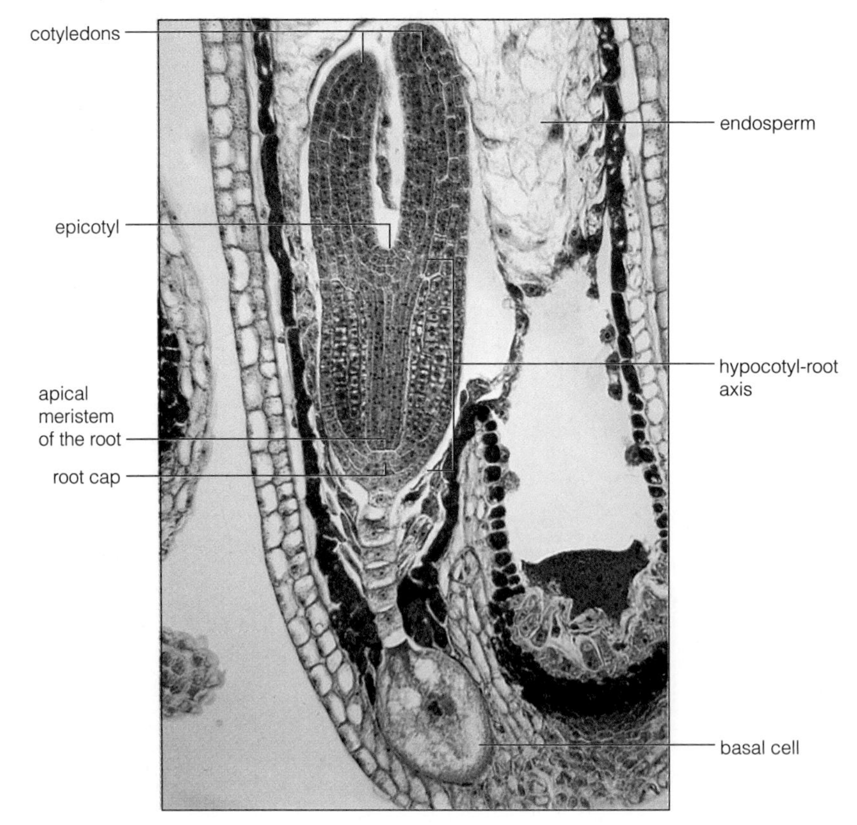

c Torpedo stage (170×).

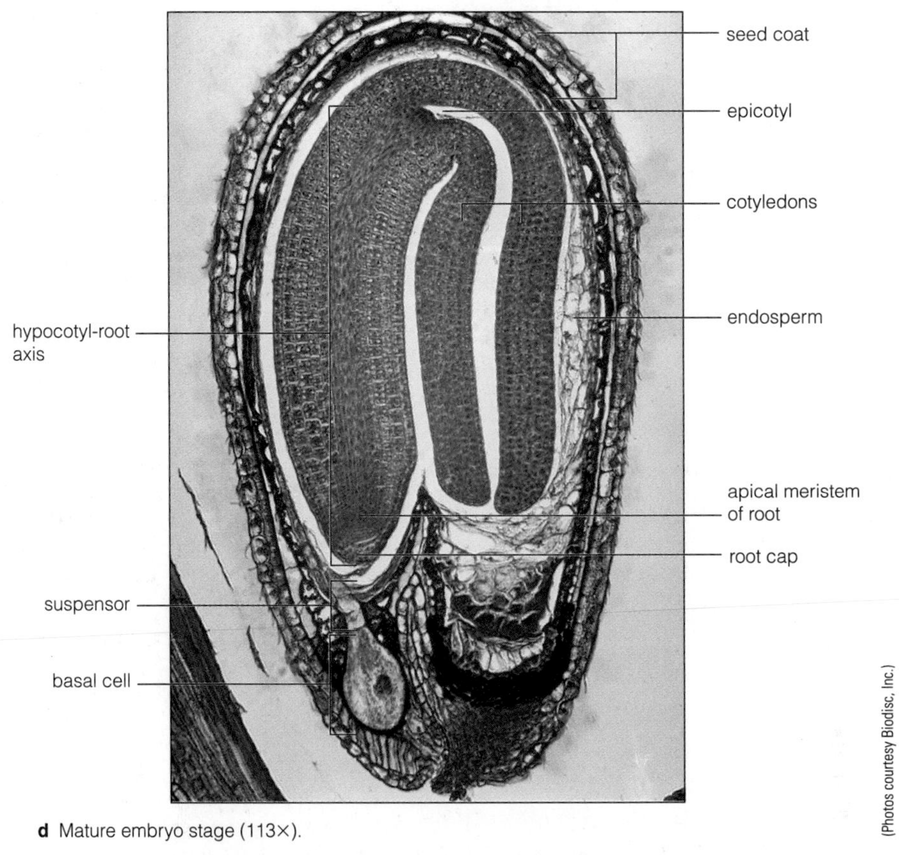

d Mature embryo stage (113×).

Figure 21-10 Embryogeny in *Capsella* (shepherd's purse). *Continued.*

(Photos courtesy Biodisc, Inc.)

MATERIALS

Per student:

- bean fruits
- soaked bean seeds
- iodine solution (I_2KI), in dropping bottle

Per lab room:

- herbarium specimen of *Capsella,* with fruits
- demonstration slide of *Capsella* fruit, c.s.

PROCEDURE

1. Examine the herbarium specimen and demonstration slide of the fruits of *Capsella*. On the herbarium specimen, identify the **fruits,** which are shaped like the bag that shepherds carried at one time (Figure 21-11; the common name of this plant is "shepherd's purse").
2. Now study the demonstration slide of a cross section through a single fruit (Figure 21-12). Note the numerous **seeds** in various stages of embryo development.
3. Obtain a bean pod and carefully split it open along one seam.

   The pod is a matured ovary and thus is a ________________________.

   Find the *sepals* at one end of the pod and the shriveled *style* at the opposite end. The "beans" inside are

   ________________________________.
4. Note the point of attachment of the bean to the pod. This is the placenta, a term shared by the plant and animal kingdom that describes the nutrient bridge between the "unborn" offspring and parent.
5. In Figure 21-13, draw the split-open bean pod, labeling it with the correct scientific terms. Figure 21-15r shows a section of a typical fruit.

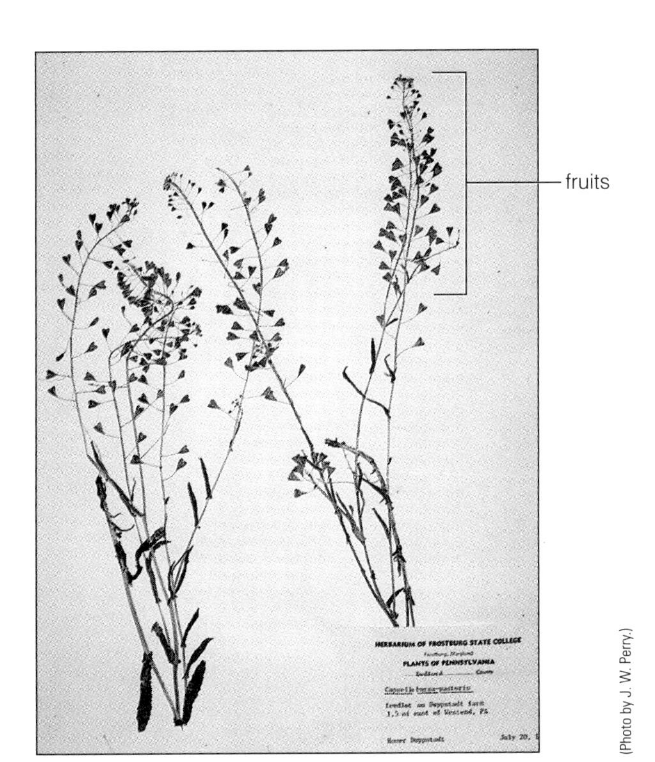

(Photo by J. W. Perry.)

Figure 21-11 Herbarium specimen of *Capsella,* shepherd's purse (0.25×).

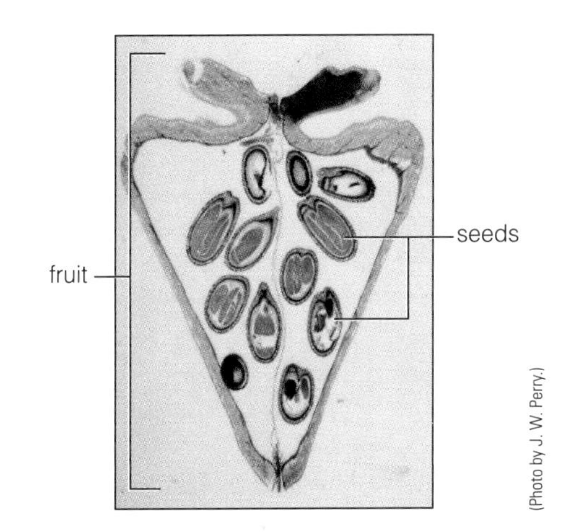

(Photo by J. W. Perry.)

Figure 21-12 Cross section of *Capsella* fruit (8×).

Figure 21-13 Drawing of an open bean pod. **Labels:** fruit, seeds, placenta

6. Closely study one of the beans from within the pod or a bean that has been soaked overnight to soften it. Find the scar left where the seed was attached to the fruit wall.
7. Near the scar, look for the tiny opening left in the seed coat.

What is this tiny opening? ________________ (*Hint:* The pollen tube grew through it.)

8. Remove the seed coat to expose the two large **cotyledons.** Split the cotyledons apart to find the **epicotyl** and **hypocotyl-root axis.** During maturation of the bean embryo, the cotyledons absorb the endosperm. Thus, bean cotyledons are very fleshy because they store carbohydrates that will be used during seed germination. Add a drop of iodine I_2KI to the cotyledon.

What substance is located in the cotyledon? ________________
(*Hint:* Return to Exercise 7 if you've forgotten what is stained by I_2KI.)

E. Seedling *(About 20 min.)*

When environmental conditions are favorable for growth (adequate moisture, oxygen, and proper temperatures), the seed germinates; that is, the seedling (young sporophyte) begins to grow.

MATERIALS

Per student:

- germinating bean seeds
- bean seedlings

Per table:

- dishpan of water

PROCEDURE

1. Obtain a germinating bean seed from the culture provided (Figure 21-14). Wash the root system in the dishpan provided, *not* in the sink.
2. Identify the **primary root** with the smaller **secondary roots** attached to it. Emerging in the other direction will be the **hypocotyl**, the **cotyledons**, the **epicotyl** (above the cotyledons), and the first **true leaves** above the epicotyl. Identify these structures.

When some seeds (like the pea) germinate, the cotyledons remain *below* ground. Others, like the bean, emerge from the ground because of elongation of the hypocotyl-root axis.

3. Now, obtain a bean seedling from the growing medium. Be careful as you pull it up so as not to damage the root system. Wash the root system in the dishpan provided.

Your seedling should be in a stage of development similar to that shown in Figure 21-15s. Much of the growth that has taken place is the result of cellular elongation of the parts present in the seed.

Figure 21-14 Germinating bean seed (2×).

4. On the root system, identify the **primary** and **secondary roots.** As the root system merges into the shoot (aboveground) system, find the **hypocotyl.** (The prefix *hypo-* is derived from Greek, meaning "below" or "underneath.")
5. Next, identify the cotyledons. Notice their shriveled appearance. Now that you know the function of the cotyledons from your previous study, why do you suppose the cotyledons are shriveled?

____

6. Above the cotyledons, find that portion of the stem called the **epicotyl.**

Knowing what you do about the prefix *hypo-*, speculate on what the prefix *epi-* means.

____

As noted earlier, the seedling has *true leaves.* Contrast the function of the cotyledons ("seed leaves") with the true leaves. ________________

____

Depending on the stage of development, your seedling may have even more leaves that have been produced by the shoot apex.

The amount of time between seed germination and flowering largely depends on the particular plant. Some plants produce flowers and seeds during their first growing season, completing their life cycle in that growing season. These plants are called **annuals.** Marigolds are an example of an annual. Others, known as **biennials,** grow vegetatively during the first growing season and do not produce flowers and seeds until the second growing season (carrots, for example). Both annuals and biennials die after seeds are produced.

Perennials are plants that live several to many years. The time between seed germination and flowering (seed production) varies, some requiring many years. Moreover, perennials do not usually die after producing seed, but flower and produce seeds many times during their lifetime.

21.3 Experiment: An Investigative Study of the Life Cycle of Flowering Plants

Recent developments with a fast-growing and fast-reproducing plant in the mustard family allow you to study the life cycle of a flowering plant over the course of only 35 days. The plants were bred by a plant scientist at the University of Wisconsin–Madison and are sometimes called Wisconsin Fast Plants™. You will grow these rapid-cycling *Brassica rapa* (abbreviated RCBr for *R*apid *C*ycling *B*rassica *r*apa) plants from seeds, following the growth cycle and learning about plant structure, the life cycle of a flowering plant, and adaptive mechanisms for pollination.

Because this investigation occurs over several weeks, all Materials and Procedures are listed by the day the activity begins, starting from day 0. Record your measurements and observations in tables at the end of this exercise.

A. Germination

Day 0 (About 30 min.)

MATERIALS

Per student:

- one RCBr seed pod
- metric ruler
- 3 × 5 in. index card
- transparent adhesive tape
- petri dish
- paper toweling or filter paper disk to fit petri dish
- growing quad ***or*** two 35-mm film canisters
- 12 N-P-K fertilizer pellets (24 if using canister method)
- 4 paper wicks (growing quad) ***or*** 2 cotton string wicks (canister method)
- disposable pipet with bulb
- pot label (2 if using canister method)
- marker
- forceps
- model seed
- 500-mL beaker
- 2 wide-mouth bottles (film canister method)

Per student group (4):

- 2-L soda bottle bottom or tray
- moistened soil mix
- watertight tray
- water mat

Per lab room:

- light bank

PROCEDURE

The life cycle of a seed plant has no beginning or end; it's a true cycle. For convenience, we'll start our examination of this cycle with germination of nature's perfect time capsule, the seed. **Germination** is the process of switching from dormancy to growth.

1. Obtain an RCBr seed pod that has been sandwiched between two strips of transparent adhesive tape. This pod is the **fruit** of the plant. A fruit is a matured ovary. Measure and record the length of the pod in Table 21-1.
2. Harvest the seeds by scrunching up the pod until the seeds come out. Then carefully separate the halves of the pod to expose the seeds.
3. Roll a piece of adhesive tape into a circle with the sticky side out. Attach the tape to an index card and then place the seeds on the tape. (This keeps the small seeds from disappearing.) You may wish to handle the seeds with forceps. Count the number of seeds and record the number in Table 21-1 on p. 340.

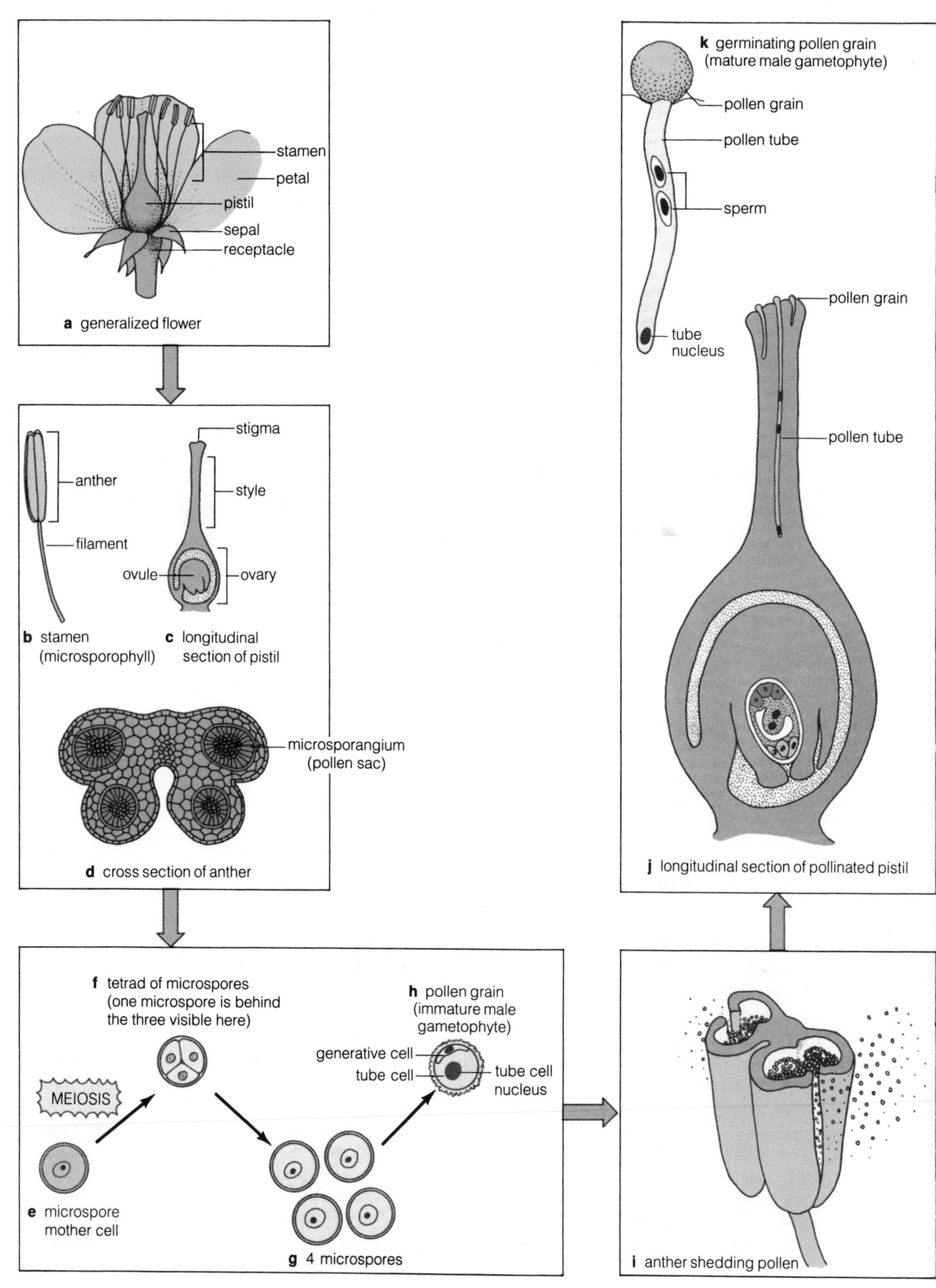

Figure 21-15 Life cycle of an angiosperm. [Color scheme: Haploid (n) structures are yellow, diploid (2n) are green, gold, or red; triploid (3n) are purple.]

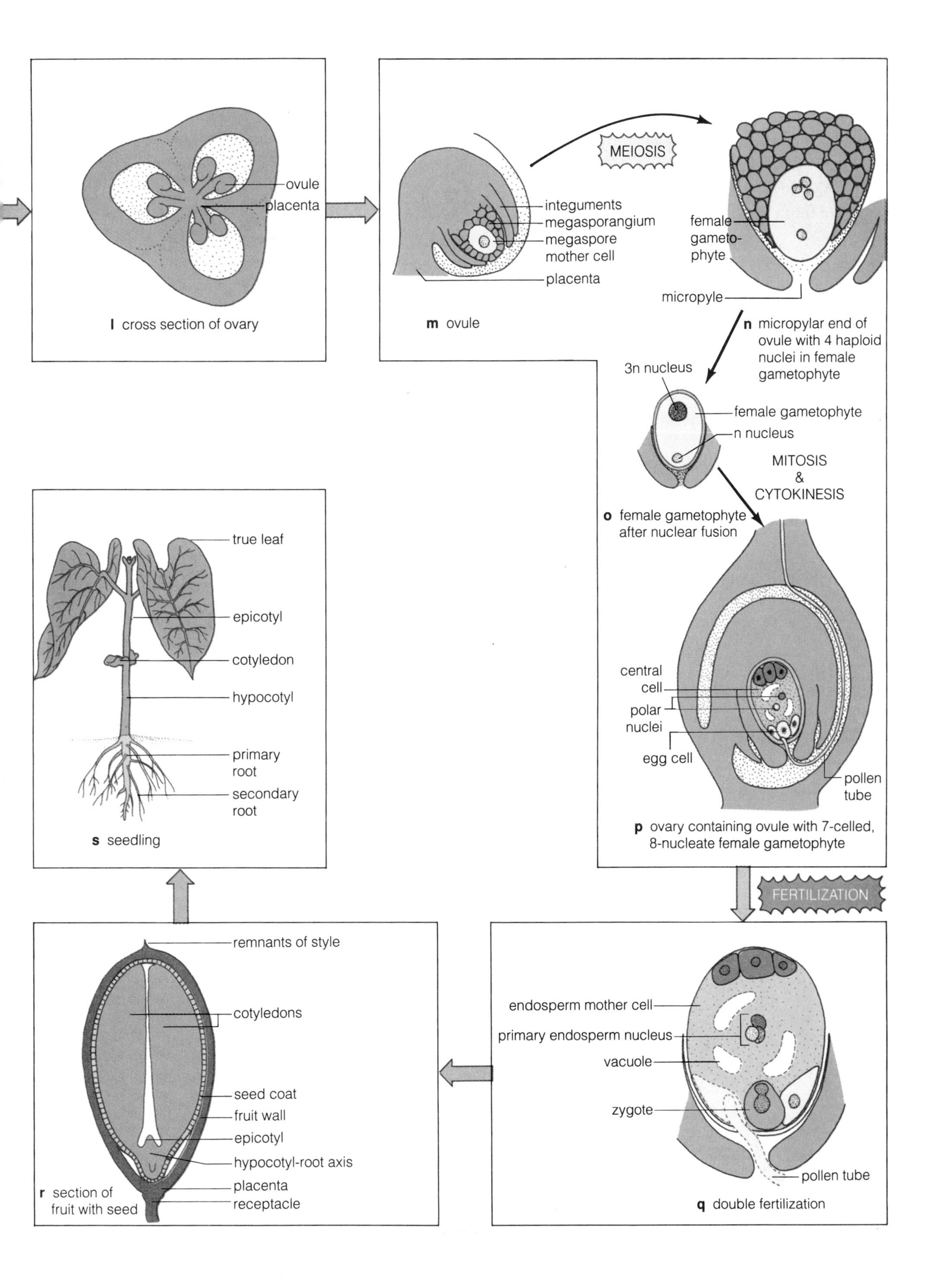
ovule
placenta
l cross section of ovary
MEIOSIS
integuments
megasporangium
megaspore
mother cell
placenta
m ovule
female
gameto-
phyte
micropyle
n micropylar end of ovule with 4 haploid nuclei in female gametophyte
3n nucleus
female gametophyte
n nucleus
MITOSIS
&
CYTOKINESIS
o female gametophyte after nuclear fusion
central
cell
polar
nuclei
egg cell
pollen
tube
p ovary containing ovule with 7-celled, 8-nucleate female gametophyte
true leaf
epicotyl
cotyledon
hypocotyl
primary
root
secondary
root
s seedling
FERTILIZATION
remnants of style
cotyledons
seed coat
fruit wall
epicotyl
hypocotyl-root axis
placenta
receptacle
r section of
fruit with seed
endosperm mother cell
primary endosperm nucleus
vacuole
zygote
pollen tube
q double fertilization

4. Obtain a petri dish to use as a germination chamber. Obtain a filter paper disk or cut a disk from paper toweling and insert it into the larger (top) part of the petri dish. With a *pencil* (**DO NOT** use ink), label the bottom of the circle with your name, the date, and current time of day.
5. Moisten the toweling in the petri dish with tap water. Pour out any excess water.
6. Place five seeds on the top half of the toweling and cover with the bottom (smaller half) of the petri dish.
7. Place the germination chamber at a steep angle in shallow water in the base of a 2-L soda bottle or tray so that about 1 cm of the lower part of the towel is below the water's surface. This will allow water to be evenly "wicked" up to keep the seeds uniformly moist.
8. Set your experiment in a warm location (20–25°C is ideal). In Table 21-1, record the temperature and number of seeds placed in the germination plate.
9. To envision the germination process, obtain a model seed (Figure 21-16). A **seed** consists of a **seed coat, stored food,** and an **embryo.**
10. Germination in the RCBr seed starts with the imbibition of water. Toss your model seed into a beaker of water. Out pops the embryo, consisting of two **cotyledons** and a **hypocotyl-root axis.** (*Hypocotyl* literally means "below the cotyledons.")

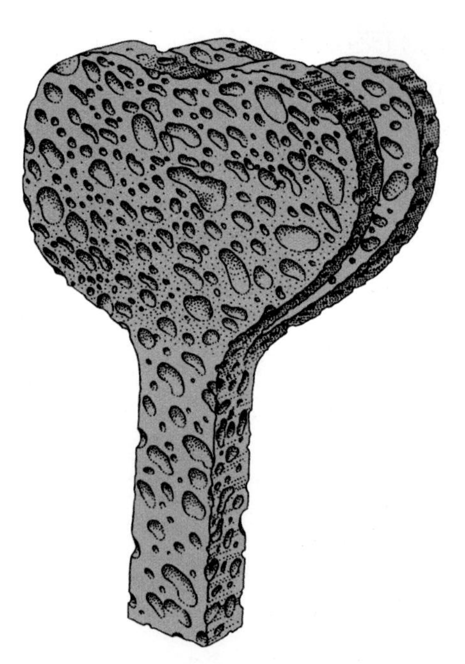

Figure 21-16 Model seed.

You'll now plant seeds for additional studies. If your instructor indicates that you will be using growth quads, proceed to step 11. If you're using the canister method, skip to step 12.

11. *Option A:* Polystyrene Growth Quad Method
    - **(a)** Obtain 4 diamond-shaped paper towel wicks, growth quad, moistened soil mix, 12 fertilizer pellets, and a pot label.
    - **(b)** Drop 1 wick into each cell in the quad. The tip of the wick should protrude about 1 cm from the hole in the bottom of the quad.
    - **(c)** Fill each cell *halfway* with soil.
    - **(d)** Add 3 N-P-K fertilizer pellets to *each* cell.
    - **(e)** Fill each cell to the top with *loose* soil. **DO NOT** compact the soil.
    - **(f)** With your thumb, make a shallow depression in the soil at the top of each cell.
    - **(g)** Drop 3 seeds into each depression. In Table 21-1, record the total number of seeds planted.
    - **(h)** Sprinkle enough soil to cover the seeds in each cell.
    - **(i)** Water very gently with a pipet until water drips from each wick.
    - **(j)** Write your name and date on a pot label and insert the label against the side of one cell of the quad.
    - **(k)** Place the quad on a wet water mat inside a tray under the light bank. The top of the quad should be approximately 5 cm below the lights.
12. *Option B*: Film Canister Method
    - **(a)** Obtain 2 cotton string wicks, 2 film canisters (with holes in bottoms), 24 fertilizer pellets, and 2 pot labels.
    - **(b)** Using tap water, wet the string wicks thoroughly and insert them through the holes in the canisters' bottoms. The wicks should extend about halfway up the length of the canister.
    - **(c)** Fill each canister *halfway* with soil mix.
    - **(d)** Add 12 N-P-K fertilize pellets atop the soil of each canister.
    - **(e)** Fill each canister to the top with soil mix. **DO NOT** compact the soil.
    - **(f)** With your thumb, make a shallow depression in the soil of each canister.
    - **(g)** Drop 6 seeds atop the soil in each canister. Distribute the seeds so they're about equally spaced.
    - **(h)** Sprinkle enough soil to cover the seeds.
    - **(i)** Water the containers very gently with a pipet until water drips from each wick.
    - **(j)** Write your name and date on pot labels and insert them into the soil against the side of each canister.
    - **(k)** Fill 2 wide-mouth bottles that have the same diameter as the canisters with tap water and place each canister in its own water bottle.
    - **(l)** Place the canisters in their water bottles under the light bank. The top of the canister should be approximately 5 cm below the lights.

Day 1 (About 10 min.)

MATERIALS

Per student:

- dissecting microscope
- disposable pipet with bulb
- tap water
- metric ruler

PROCEDURE

1. Water each quad cell or canister from the top using a pipet.
2. Observe your germination experiment in the petri dish chamber with a dissecting microscope. Note the **radicle** (primary root) and **hypocotyl** extending from a split in the brown **seed coat.** What colors are the radicle and the hypocotyl? ____________________
3. What does the color indicate about the source of carbohydrates for the germinating seeds?

   ____
4. Count the number of seeds that have germinated, and record it in Table 21-1. Measure the length of each hypocotyl and attached radicle. Record this in Table 21-1. Replace the cover and return your experiment to its place.

Day 2 (About 15 min.)

MATERIALS

Per student:

- germination experiment started on day 0
- disposable pipet with bulb
- tap water
- metric ruler
- forceps
- dissecting microscope
- 2 additional germination chambers (for gravitropism experiment)
- 2 RCBr seed pods

PROCEDURE

1. Water each quad cell or canister from the top using a pipet.
2. Observe your germination experiment. Count the number of germinated seeds and calculate the percentage of seeds that germinated. Record both numbers in Table 21-1. With a forceps, remove one of the germinating seeds from the toweling, place it on the stage of a dissecting microscope, and observe the numerous **root hairs** attached to the radicle. Root hairs greatly increase the surface area for absorption of water and minerals. Measure the lengths of the hypocotyls and attached radicles and record their average length in Table 21-1.

In what direction (up, down, or horizontal) are the hypocotyl and radicle growing?

hypocotyl: ____________________

radicle: ____________________

This directional growth is called **gravitropism**, a plant growth response to gravity.

Day 3 (About 10 min.)

MATERIALS

Per student:

- germination experiment started on day 0
- metric ruler
- disposable pipet with bulb
- tap water

PROCEDURE

1. Water your plants in the quads or canisters from the top using a pipet. Is any part of the plant visible yet? ______________ (yes or no) If yes, which part? ____________________
2. Observe your germination experiment. By this time, the seed coat should be shed, and the **cotyledons**, the so-called seed leaves that were packed inside the seed coat, should be visible. The cotyledons absorb food stored within the seed, making the food available for initial stages of growth before photosynthesis starts to provide carbohydrates necessary for growth.

What color are the cotyledons? ________________

What function does this color indicate for the cotyledons? ________________

3. Measure the length of the hypocotyls and attached radicles and record their average length in Table 21-1.

B. Growth

Day 4 or 5 (About 10 min.)

MATERIALS

Per student:

- forceps
- metric ruler
- dissecting microscope
- germination experiment started on day 0

PROCEDURE

1. Count the number of seedlings present, and record it in Table 21-1. Use a forceps to thin the plants to one per cell if you are using the quad growing method. If a cell has no plants, transplant one of the extra plants from another cell by inserting the tip of a pencil into the soil to form a hole and then inserting the transplant's root. Gently firm the soil around the transplant. Measure the height of each plant's shoot system—that is, from soil to the apex. Compute the average height and record it in Table 21-1.
2. Observe your germination experiment. Note that the radicle has produced additional roots. These are **lateral roots** (secondary roots). Remove one plant and observe it with the dissecting microscope.

Do lateral roots have root hairs? ________________ (yes or no)

At this time, you may discard your germination experiment. Your instructor will indicate what you should do with the materials.

Day 7 (About 10 min.)

MATERIALS

Per student:

- metric ruler

PROCEDURE

1. By now your plant should have its first **true leaves.** Describe how the true leaves differ from the cotyledons.

____

____

____

2. Measure the length of the true leaves, and record their average length in Table 21-1.
3. Measure the height of your plants from soil line to shoot apex, and record the average height in Table 21-1.

C. The Pollinator and Pollination

Pollen grains are actually sperm conveyors that deliver these male gametes to the egg-containing ovules within the ovary. Transfer of pollen from the anther to the stigma, called **pollination,** occurs by various means in different species of flowers, with wind, birds, and insects being the most common carriers of pollen. In the case of *Brassica,* the main pollinator is the common honeybee.

Pollinators and flowers are probably one of the best-known examples of **coevolution.** Coevolution is the process of two organisms evolving together, their structures and behaviors changing to benefit both organisms. Flowering plants and pollinators work together in nearly perfect harmony: The bee receives the food it needs for its respiratory activities from the flower, and the plant receives the male gametes directly to the female structure—it doesn't have to rely on the presence of water, which is necessary in lower plants, algae, and fungi, to get the gametes together.

Flowering plants have considerable advantage over gymnosperms. While gymnosperms also produce pollen grains, the pollinator in these plants is the wind. No organism carries pollen directly to the female structure. Rather, chance and the production of enormous quantities of pollen (which takes energy that might be used otherwise for growth) are integral in assuring reproduction.

In preparation for pollinating your plants' flowers, let's first study the pollinator, the common honeybee.

Day 10 (About 20 min.)

MATERIALS

Per student:

- metric ruler
- 3 toothpicks
- 3 dried honeybees
- dissecting microscope

Per student group (4):

- tube of glue
- scissors

PROCEDURE

1. Measure the lengths of the second set of true leaves and plant height, and record their averages in Table 21-1.
2. Obtain a dried honeybee and observe it with your dissecting microscope. Use Figure 21-2 to aid you in this study. Note the featherlike hairs that cover its body. As a bee probes for nectar, it brushes against the anthers and stigma. Pollen is entrapped on these hairs.
3. Closely study the foreleg of the bee, noticing the notch. This is the *antenna cleaner.* Quick movements over the antennae remove pollen from these sensory organs.
4. Examine the mid- and hind legs, identifying their *pollen brushes,* which are used to remove the pollen from the forelegs, thorax, and head.
5. On the hind leg, note the *pollen comb* and *pollen press.* The comb rakes the pollen from the midleg, and the press is a storage area for pollen collected.

Pollen is rich in the vitamins, minerals, fats, and proteins needed for insect growth. When the baskets are filled, the bee returns to the hive to feed the colony. Partially digested pollen is fed to larvae and to young, emerging bees.

As you can imagine, during pollen- and nectar-gathering at numerous flowers, cross-pollination occurs as pollen grains are rubbed off onto stigmas.

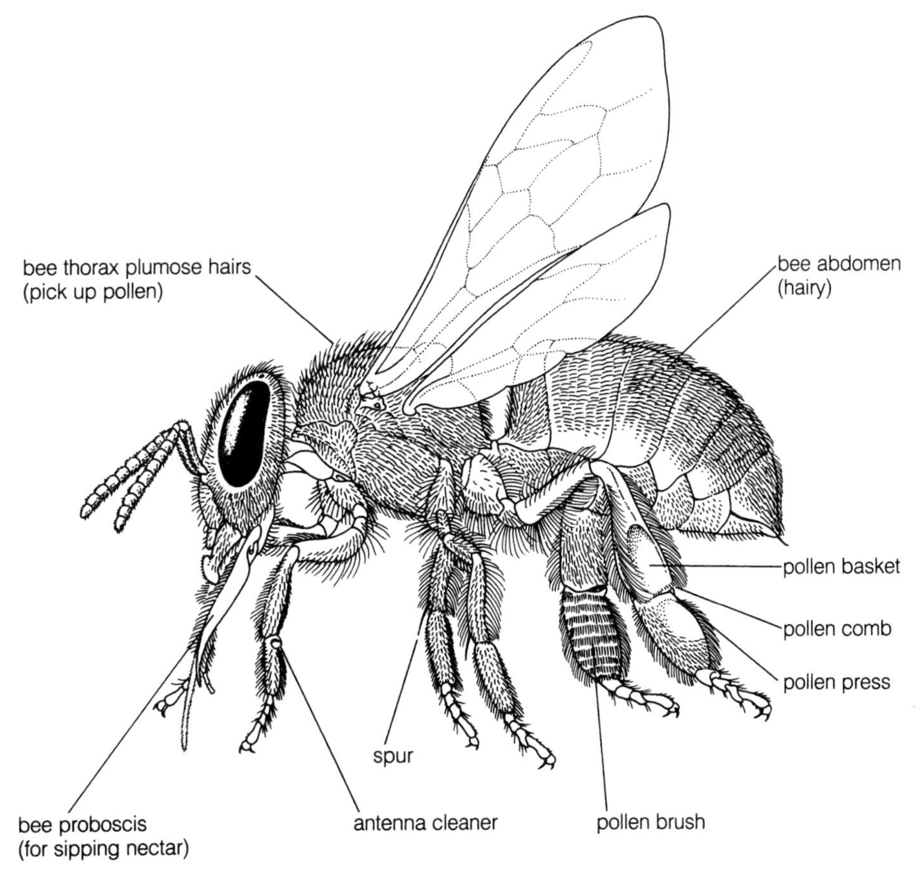

Figure 21-17 Structure of the honeybee.

On day 12, you will start pollinating your flowers. What better means to do it than to use a bee? (In this case, a dead bee that is attached to a stick.) Let's make some beesticks.

6. Add a drop of glue to one end of three round toothpicks. Insert the glue-bearing end of one stick into the top (dorsal) side of the thorax of a bee. Repeat with two more dried bees. Set your beesticks aside to dry.

D. Pollination

It's time to do a pollination study. Some species of flowers are able to **self-pollinate,** meaning that the pollen from one flower of a plant will grow on stigmas of other flowers of that same plant. Hence, the sperm produced by a single plant will fertilize its own flowers. Other species are **self-incompatible.** For fertilization to occur in the latter, pollen must travel from one plant to another plant of the same species. This is **cross-pollination.** In this portion of the exercise, you will determine whether RCBr is self-incompatible or is able to self-pollinate.

Day 12 (About 30 min.)

MATERIALS

Per student:

- beestick prepared on day 10
- metric ruler
- forceps
- index card divider
- pot label
- marker

PROCEDURE

1. Measure the height of each plant, and record the average height in Table 21-1.
2. Separate the cells of your quad or canister with a divider constructed with index cards as shown in Figure 21-18. This keeps the plants from brushing against each other.

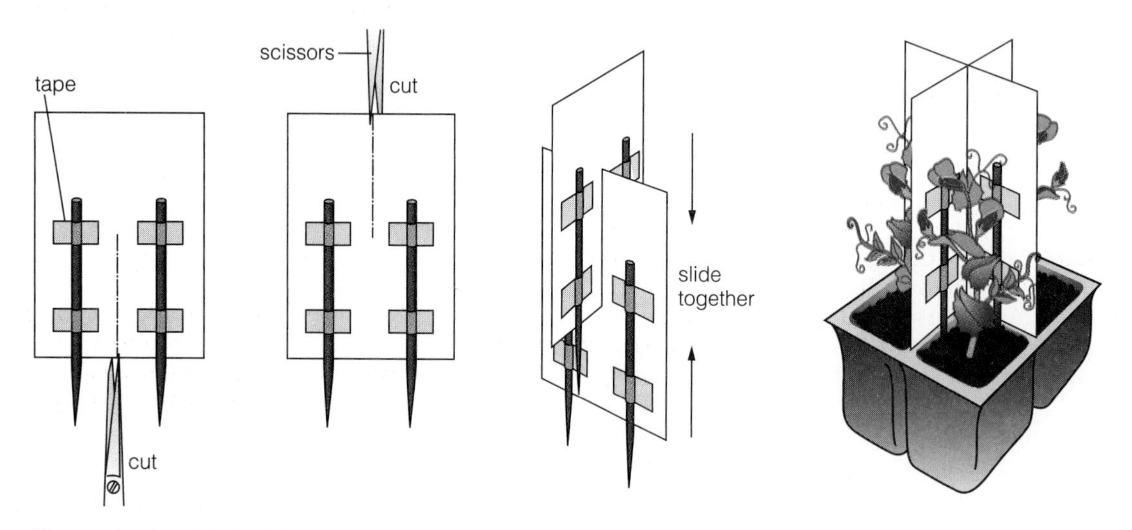

Figure 21-18 Method for separating flowers for pollination study.

3. In one cell of the quad, or in one canister, insert a pot label on which you have written the word ("SELF"), to indicate that this plant (plants if using canister) will be self-pollinated. The other three plants (quad method) or those in the other canister (canister method) will be cross-pollinated.
4. Now use one of your beesticks to pick up pollen from the flowers on the plant in the "SELF" cell. Rub the bee's body against the pollen-containing anthers. Examine your beestick using the dissecting microscope, and note the pollen adhering to the hairs of the bee's body.
5. Brush some of the pollen on the stigmas of the "SELF" plant's flowers. Next, cross-pollinate the other three plants in the quad (or canister) with the *same* beestick.
6. Return your plants to the light-bank growing area and discard the beestick.

Caution

It is important that you DO NOT retouch any of the self-pollinated flowers with the beestick once you have touched the flowers of any of the other plants.

Day 14 (About 15 min.)

MATERIALS

Per student:

- metric ruler
- beestick

PROCEDURE

1. Measure the height of each plant, and record their average height in Table 21-1.
2. Pollinate the newly opened flowers using the same techniques as described for day 12.
3. Return your plants to the light-bank growing area.

Caution

Use a new beestick and be sure not to transfer pollen from any of the cross-pollinated plants to the self-pollinated plant.

Day 17 (About 15 min.)

MATERIALS

Per student:

- metric ruler
- beestick

PROCEDURE

1. Measure the height of each plant, and record their average height in Table 21-1.
2. Count the number of flowers on each plant (opened and unopened), calculate the average number per plant, and record it in Table 21-1. Pinch off the shoot apex of each plant to remove all unopened buds and pinch off any side shoots the plant has produced.
3. Do a final pollination of your flowers, using a fresh beestick and taking the same precautions as on days 12 and 14.
4. Return your plants to the light-bank growing area.

E. Fertilization and Seed Development

Within 24 hours of pollination, a pollen grain that has landed on a compatible stigma will produce a pollen tube that grows down the style and through the micropyle of the ovule.

Within the pollen tube are two **sperm,** the male gametes. As the pollen tube enters the ovule, the two sperm are discharged. One fertilizes the haploid egg nucleus, forming a diploid **zygote.** The second sperm unites with two other nuclei within the ovule, forming a triploid nucleus known as the **primary endosperm nucleus.**

The fusion of one sperm nucleus to form the zygote and the other to form the primary endosperm nucleus is called **double fertilization.**

1. Re-examine Figure 21-9, which illustrates double fertilization.

What is the fate of cells formed by double fertilization? The zygote undergoes numerous cell divisions during a process called **embryogeny** (the suffix *-geny* is derived from Greek, meaning "production") to form the diploid embryo. The triploid primary endosperm nucleus divides numerous times without cytokinesis. The result is the formation of a multinucleate milky **endosperm** (each nucleus is 3n), which serves as the food source for the embryo as it grows. At the same time, changes are taking place in the cell layers of the ovule—they are becoming hardened and will eventually form the seed coat. Also, the ovary is elongating as it matures into the **fruit**—in this case, a pod.

2. Examine your plants, looking for small, elongated pods.
3. At this time, you should be able to tell whether RCBr can self-pollinate or is self-incompatible, because the fruit—a pod—will develop only if fertilization has taken place. Do the cross-pollinated plants have pods? (yes or no) ________
4. Does the self-pollinated plant have pods? (yes or no) _______
5. Make a conclusion regarding the need for cross-pollination in RCBr. ______________________________

____

The young embryo goes through a continuous growth process, starting as a ball of cells (the globular stage) and culminating in the formation of a mature embryo. The early stages are somewhat difficult to observe, but the later stages can be seen quite easily with a microscope. By day 24, many of the ovules will be at the globular stage, similar to that seen in Figure 21-10a.

Day 26 (About 30 min.)

MATERIALS

Per student:

- dissecting microscope
- 2 dissecting needles
- dH_2O in dropping bottle
- glass microscope slide
- iodine solution (I_2KI) in dropping bottle

PROCEDURE

1. Remove a pod from one of the plants, place it on the stage of your dissecting microscope, and tease it open using dissecting needles. RCBr flowers have pistils consisting of two carpels; each carpel makes up one longitudinal half of the pod, and the "seam" represents where the carpels fuse.
2. Carefully remove the ovules (immature seeds) inside the pod, placing them in a drop of water on a clean microscope slide.
3. Rupture the ovule by mashing it with dissecting needles and place the slide on the stage of a dissecting microscope.
4. Attempt to locate the embryo, which should be at the heart-shaped stage, similar to Figure 21-10b, and which may be turning green. A portion of the endosperm has become cellular and may be green, too, so be careful with your observation. The "lobes" of the heart-shaped embryo are the young cotyledons. Identify these. Find the trailing **suspensor,** a strand of eight cells that serves as a sort of "umbilical cord" to transport nutrients from the endosperm to the embryo.
5. Add a drop of I_2KI to stain the endosperm. What color does the endosperm stain? ____________________
6. What can you conclude about the composition of the endosperm? (*Hint:* If you've forgotten, consult Exercise 7.)

   ____

Day 29 (About 20 min.)

MATERIALS

Per student:

- dissecting microscope
- 2 dissecting needles
- dH_2O in dropping bottle
- glass microscope slide

PROCEDURE

1. Remove another pod from a plant and examine it as you did on day 26.
2. Attempt to find the torpedo stage of embryo development, similar to that illustrated in Figure 21-10c.
3. Describe the changes that have taken place in the transition from the heart-shaped to the torpedo stage.

   ____

   ____

Day 31 (About 20 min.)

MATERIALS

Per student:

- dissecting microscope
- 2 dissecting needles
- dH_2O in dropping bottle
- glass microscope slide

PROCEDURE

1. Again, remove and dissect a pod and ovules, and look for the embryo whose cotyledons have now curved, forming the so-called walking-stick stage.
2. Identify the **cotyledons** and the **hypocotyl-root axis** that has curved within the ovules' integuments.

Days 35–37 (About 20 min.)

MATERIALS

Per student:

- dissecting microscope
- 2 dissecting needles

PROCEDURE

The pods of your plants should be drying and have much the same appearance as the pod you started with about a month ago.

1. Remove seeds from mature pods and examine the seeds with your dissecting microscope. Note the bumpy **seed coat,** derived from the integument of the ovule. Look for a tiny hole in the seed coat. This is the **micropyle!**
2. Crack the seed coat and remove the embryo, identifying the cotyledons and hypocotyl-root axis. Separate the cotyledons and look between them, identifying the tiny **epicotyl,** which is the *apical meristem of the shoot.*
3. Harvest the remaining pods and count the number of seeds in each pod. Calculate the average number of seeds per pod and number of seeds per plant. Record this information in Table 21-1.

Congratulations! You have completed the life cycle study of a rapid cycling *Brassica rapa.* Virtually all of the 200,000 species of flowering plants have stages identical to that which you have examined over the course of the past 35–40 days. Most just take longer—from a few days to nearly a century longer!

Conclude your study by preparing the required graphs and answering the questions following Table 21-1.

4. In Figure 21-19, graph the increase in average length of the hypocotyl-root axis produced in the germination chamber.

Does length increase in a constant fashion (linearly), or does extension occur more quickly during a particular time period (logarithmically)? ____________________

Using the data from Table 21-1, graph in Figure 21-20 the rate of growth of your plants' shoot system as a function of time.

How does the number of seeds produced per pod compare with the number in the original pod from which you harvested your seed at the beginning of the study? ____________________

What is the *reproductive potential* (the number of possible plants coming from one parent plant) of your RCBr? ____________________

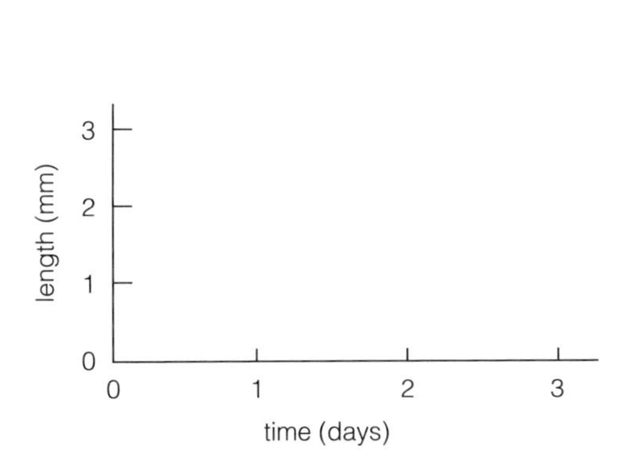

Figure 21-19 Growth of hypocotyl-root axis.

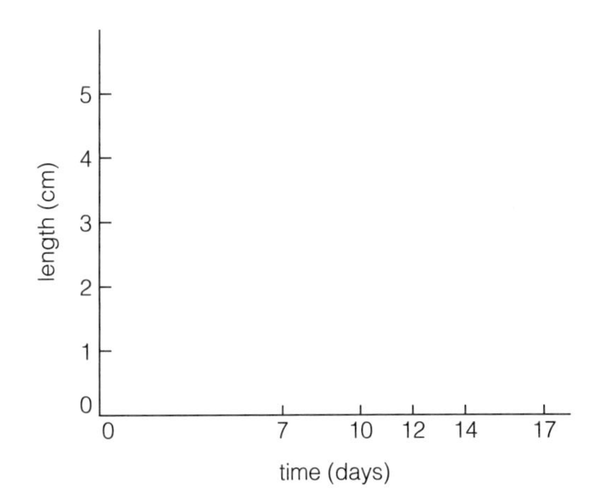

Figure 21-20 Growth of shoot system.

TABLE 21-1 Data for RCBr Life Cycle

Day	Parameter	Measurement
0	Pod length	mm
	Seed number (single pod)	
	Number of seeds sown in germination chamber	
	Temperature, °C	
	Number of seeds sown in quad (total)	
1	Germinated seeds (germination chamber)	
	Average length of hypocotyl-root axis	mm
2	Germinated seeds (germination chamber)	
	Percentage seeds germinated	
	Average length of hypocotyl-root axis	mm
3	Average length of hypocotyl-root axis	mm
4 or 5	Number of seedlings (quad)	
	Average height of shoot system	mm
7	Average length of first set of true leaves	mm
	Average height of shoot system	mm
10	Average length of second set of true leaves	mm
	Average height of shoot system	mm
12	Average number of leaves per plant	
	Average height of shoot system	mm
14	Average height of shoot system	mm
17	Average height of shoot system	mm
	Average number of flowers per plant	
35–37	Average total number seeds produced per plant	
	Average number of seeds per pod	

PRE-LAB QUESTIONS

_____ **1.** Plants that produce flowers are
(a) members of the Anthophyta
(b) angiosperms
(c) seed producers
(d) all of the above

_____ **2.** All of the petals of a flower are collectively called the
(a) corolla
(b) stamens
(c) receptacles
(d) calyx

_____ **3.** Which group of terms refers to the microsporophyll, the male portion of a flower?
(a) ovary, stamens, pistil
(b) stigma, style, ovary
(c) anther, stamen, filament
(d) megasporangium, microsporangium, ovule

_____ **4.** A carpel is the
(a) same as a megasporophyll
(b) structure producing pollen grains
(c) component making up the anther
(d) synonym for microsporophyll

_____ **5.** The portion of the flower containing pollen grains is
(a) the pollen sac
(b) the microsporangium
(c) the anther
(d) all of the above

_____ **6.** Which group of terms is in the correct developmental sequence?
(a) microspore mother cell, meiosis, megaspore, female gametophyte
(b) microspore mother cell, meiosis, microspore, pollen grain
(c) megaspore, mitosis, female gametophyte, meiosis, endosperm mother cell
(d) all of the above

_____ **7.** Where does germination of a pollen grain occur in a flowering plant?
(a) in the anther
(b) in the micropyle
(c) on the surface of the corolla
(d) on the stigma

_____ **8.** Double fertilization refers to
(a) fusion of two sperm nuclei and two egg cells
(b) fusion of one sperm nucleus with two polar nuclei and fusion of another with the egg cell nucleus
(c) maturation of the ovary into a fruit
(d) none of the above

_____ **9.** Ovules mature into _____________, while ovaries mature into _____________.
(a) seeds/fruits
(b) stamens/seeds
(c) seeds/carpels
(d) fruits/seeds

_____ **10.** A bean pod is
(a) a seed container
(b) a fruit
(c) a part of the stamen
(d) both a and b

Name ______________________________ Section Number ____________________

EXERCISE 21

Seed Plants II: Angiosperms

POST-LAB QUESTIONS

Introduction

1. There are two major groups of seed plants, gymnosperms and angiosperms. Compare these two groups of seed plants with respect to the following features:

Feature	Gymnosperms	Angiosperms
a. type of reproductive structure	________	________
b. source of nutrition for developing embryo	________	________
c. enclosure of mature seed	________	________

21.1 External Structure of the Flower

2. What *event*, critical to the production of seeds, is shown at the right?

(Photo by J. W. Perry.)

(0.84×).

3. Distinguish between *pollination* and *fertilization*.

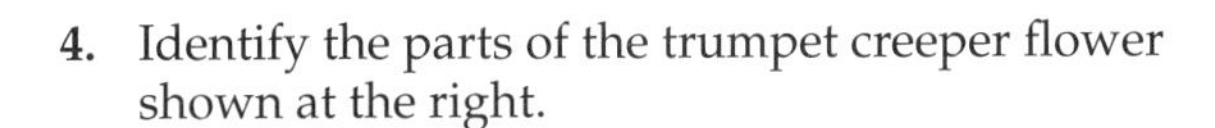

4. Identify the parts of the trumpet creeper flower shown at the right.

   a.

   b.

   c.

   d.

   e.

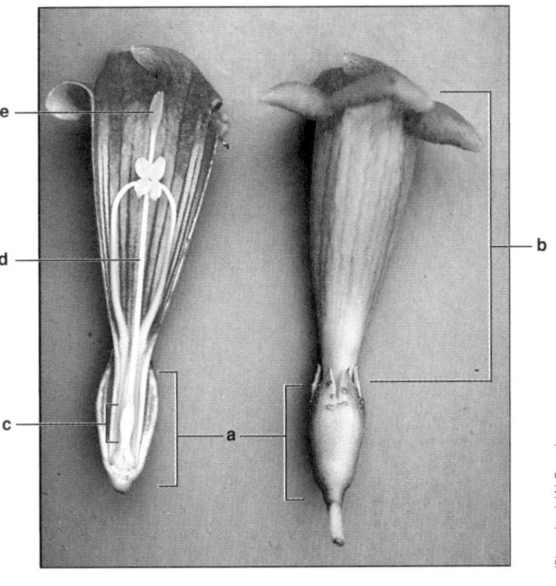

(Photo by J. W. Perry.)

(1×).

5. Examine the photo of the *Trillium* flower pictured at the right.

   a. Is *Trillium* a monocotyledon or dicotyledon?

   b. Justify your answer.

(Photo by J. W. Perry.)

(0.75×).

6. Based on your observation of the stigma of the daylily flower in the photo at the right, how many carpels would you expect to comprise the ovary?

(Photo by J. W. Perry.)

(1×).

21.2 The Life Cycle of a Flowering Plant

7. The photo is a cross section of a (an) ____________
   ____.

   The numerous circles within the four cavities are
   ____.

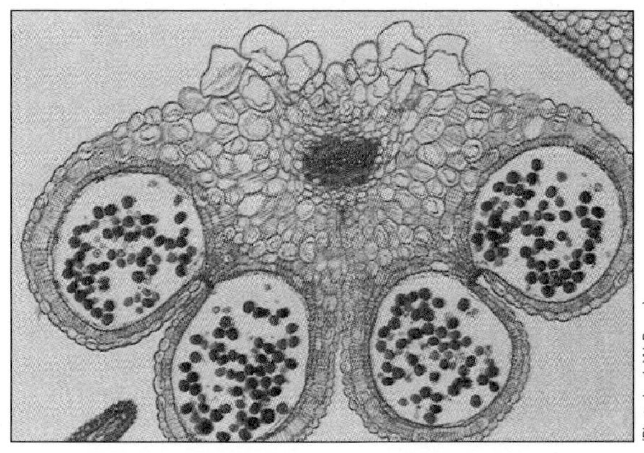

(Photo by J. W. Perry.)

(74×).

8. The photo shows a flower of the pomegranate some time after fertilization. Identify the parts shown.

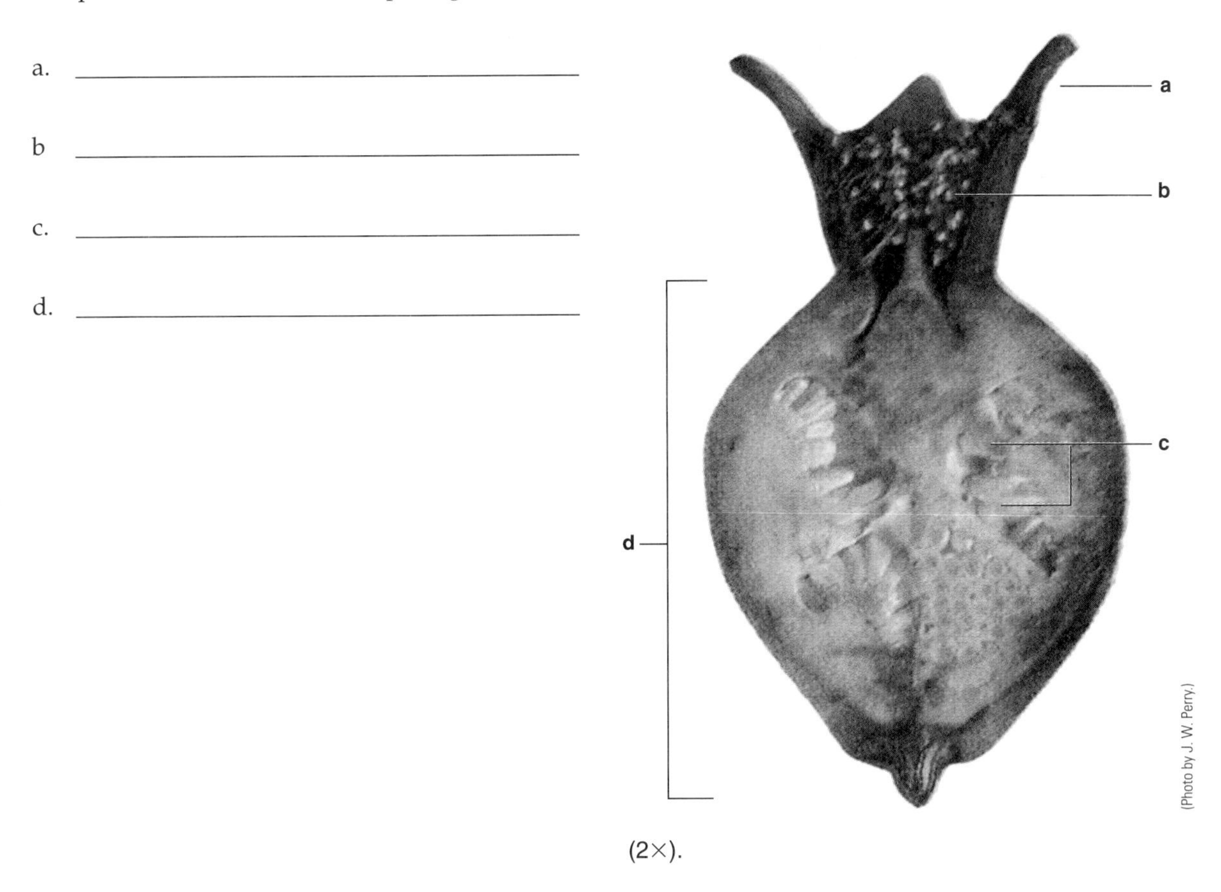

(2×).

9. Some biologists contend that the term *double fertilization* is a misnomer and that the process should be called *fertilization* and *triple fusion.* Why do they argue that the fusion of the one sperm nucleus and the two polar nuclei is *not* fertilization?

Food for Thought

10. Your roommate says that you need vegetables and asks you to pick up tomatoes at the store. To your roommate's surprise, you say a tomato is not a vegetable, but a fruit. Explain.

EXERCISE 22

Vertebrates

OBJECTIVES

After completing this exercise, you will be able to

1. define *vertebrate, vertebral column, cranium, vertebrae, cloaca, lateral line, placoid scale, operculum, atrium, ventricle, artery, vein, ectothermic, endothermic, viviparous;*
2. describe the basic characteristics of members of the subphylum Vertebrata;
3. identify representatives of the vertebrate classes, Cephalaspidomorphi, Chondrichthyes, Osteichthyes, Amphibia, Reptilia, Aves, and Mammalia;
4. identify structures (and indicate associated functions) of representatives of the vertebrate classes;
5. construct a dichotomous key to the animals (optional).

INTRODUCTION

In terms of the numbers of individuals and species, chordates do not constitute a large phylum; however, chordates known as vertebrates have had a disproportionately large ecological impact. The **vertebrates** have the *four basic characteristics of chordates* listed in the previous exercise, plus a cranium and a **vertebral column.** The dorsal hollow nerve cord has differentiated into a *brain* and a *spinal cord.* Bones protect both of these structures, the brain by the bones of the **cranium** (braincase) and the spinal cord by the **vertebrae,** the bones that make up the vertebral column.

Like echinoderm larvae and sea squirt larvae and lancelets, vertebrates are *cephalized, bilaterally symmetrical,* and *segmented.* In adults, segmentation is most easily seen in the musculature, vertebrae, and ribs. The body is typically divided into a *head, neck, trunk,* and *tail.* If appendages are present, they are paired, lateral *thoracic appendages* (pectoral fins, forelimbs, wings, and arms) and *pelvic appendages* (pelvic fins, hindlimbs, and legs), which are linked to the vertebral column and function to support and to help move the body. Seven classes of vertebrates have representatives alive today (lampreys, cartilaginous fishes, bony fishes, amphibians, reptiles, birds, and mammals), and one class is completely extinct (armored fishes).

Caution

Preserved specimens are kept in preservative solutions. Use latex gloves whenever you handle a specimen. Wash any part of your body exposed to this solution with copious amounts of water. If preservative solution is splashed into your eyes, wash them with the safety eyewash bottle for 15 minutes. If you wear contact lenses during a dissection, wear safety goggles.

22.1 Lampreys (Class Cephalaspidomorphi) *(12 min.)*

The ancestors of lampreys were the first vertebrates to evolve. Lampreys have a cartilaginous (made of cartilage), primitive skeleton.

Lampreys are represented by both marine and freshwater species and by some species that use seawater and fresh water at some point in their life span. Most feed on the blood and tissue of fishes by rasping wounds in their sides. A landlocked population of the sea lamprey is infamous for having nearly decimated the commercial fish industry in the Great Lakes. Only a vigorous control program that targeted lamprey larvae restored this industry.

MATERIALS

Per student:

- scalpel
- blunt probe or dissecting needle

Per student pair:

- dissection microscope
- dissection pan

Per student group:

- prepared slide of a whole mount of a lamprey larva or ammocoete
- preserved specimen of sea lamprey

Per lab room:

- boxes of different sizes of vinyl or latex gloves
- box of safety goggles

PROCEDURE

1. With a dissection microscope, observe a prepared slide of a whole mount of a lamprey larva or ammocoete. Sea lampreys spawn in freshwater streams. Ammocoetes hatch from their eggs and, after a period of development, burrow into the sand and mud. They are long-lived, metamorphosing into adults after as long as 7 years. Because of their longevity and dissimilar appearance, adult lampreys and ammocoetes were long thought to be separate species.

The ammocoete is considered by many to be the closest living form to ancestral chordates. Unlike a lancelet, it has an eye (actually two median eyes), internal ears (otic vesicles), and a heart, as well as several other typical vertebrate organs. Identify the structures labeled in Figure 22-1.

2. Examine a preserved sea lamprey (Figure 22-2). Note its slender, rounded body. The skin of the lamprey is soft and lacks scales. Look at the round, suckerlike **mouth,** inside of which are circular rows of horny, rasping **teeth** and a deep, rasping **tongue.** Although there are no lateral, paired appendages, there are two **dorsal fins** and a **caudal fin** (tail fin). Count and record below the number of pairs of external **gill slits.**

____

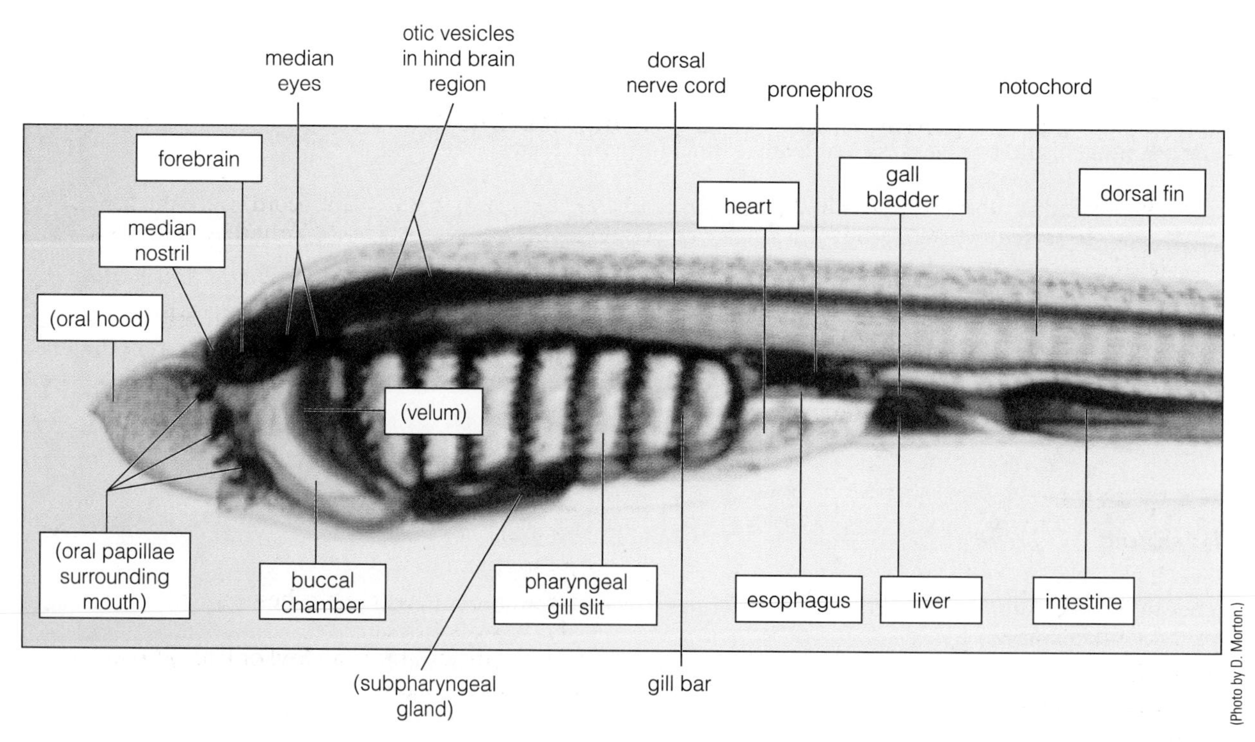

Figure 22-1 Whole mount of ammocoete (40×).

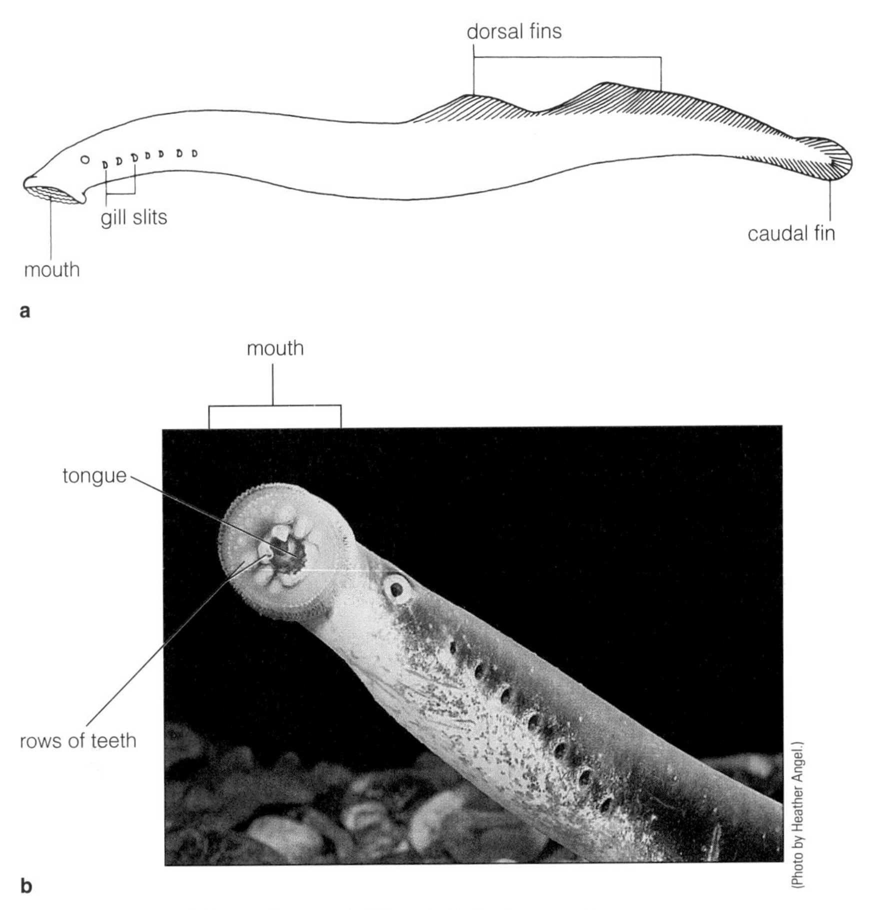

Figure 22-2 (**a**) Side view and (**b**) oral disk of a sea lamprey.

22.2 Cartilaginous Fishes (Class Chondrichthyes) *(12 min.)*

The first vertebrates to evolve jaws and paired appendages, according to the fossil record, were the heavily armored *placoderms* (class Placodermi). These fishes arose about 425 million years ago. Current hypotheses suggest that the ancestors of placoderms relatively quickly gave rise to the cartilaginous fishes.

Members of the class Chondrichthyes are *jawed fishes* with *cartilaginous skeletons* and *paired appendages.* The evolution of jaws and paired appendages and the evolution of more efficient respiration and a better developed nervous system, including sensory structures, were critical events in the history of vertebrate evolution. Together these adaptations allowed cartilaginous fishes to chase, catch, and eat larger and more active prey. This mostly marine group includes the carnivorous skates, rays, and sharks.

MATERIALS

Per student:

- scalpel
- forceps
- dissecting scissors

Per student pair:

- dissection microscope
- dissection pan

Per student group:

- preserved specimen of dogfish (shark)

Per lab room:

- collection of
  - preserved cartilaginous fishes
  - shark jaw with teeth
- boxes of different sizes of vinyl or latex gloves
- box of safety goggles

PROCEDURE

1. Look at the assortment of cartilaginous fishes on display.
2. Refer to Figure 22-3a as you examine the external anatomy of a preserved dogfish (shark).
   (a) How many dorsal fins are there? _______________
   (b) Notice the front pair of lateral appendages, the **pectoral fins,** and the back pair, the **pelvic fins.** The **tail fin** has a *dorsal lobe* larger than the *ventral* one. In the male, the pelvic fins bear **claspers,** thin processes for transferring sperm to the oviducts of the female. Examine the pelvic fins of a female and a male.
   (c) Identify the opening of the **cloaca** just in front of the pelvic fins. The cloaca is the terminal organ that receives the products of the digestive, excretory, and reproductive systems.
   (d) There are five to seven pairs of gill slits in cartilaginous fishes, six in the dogfish. Observe the most anterior gill slit, which is called the *spiracle,* located just behind the eye. Trace the flow of seawater in nature through the mouth, over the pharyngeal gills, and out of the gill slits.

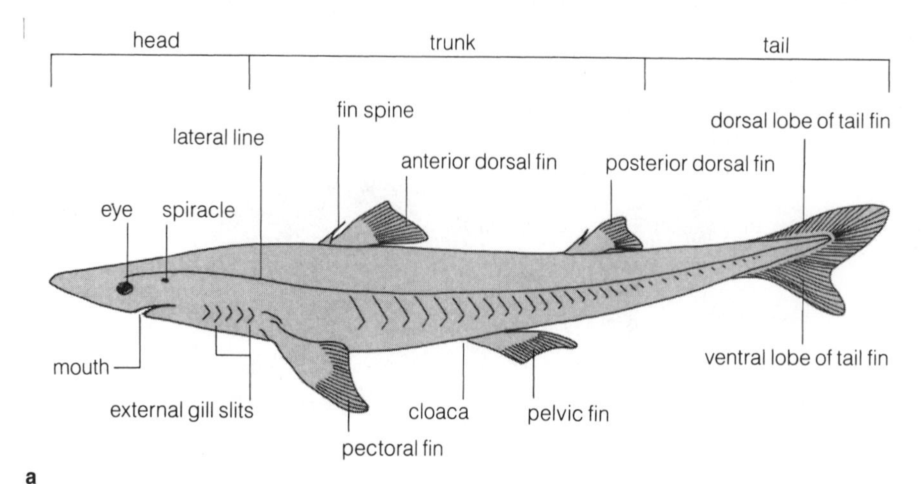

Figure 22-3 (**a**) External appearance of the dogfish (shark), *Squalus acanthias.* (After S. Wischnitzer, *Atlas and Dissection Guide for Comparative Anatomy*, 3rd ed., © 1967, 1972, 1979, W. H. Freeman and Company. Used by permission.) (**b**) Shark jaw.

(e) Locate the **nostrils,** which open into blind olfactory sacs; they don't connect with the pharynx, as your nostrils do. They function solely in olfaction (smell) in the cartilaginous fishes. The **eyes** are effective visual organs at short range and in dim light. There are no eyelids.

(f) Find the dashed line that runs along each side of the body. This is called the **lateral line** and functions to detect vibrations in the water. It consists of a series of minute canals perpendicular to the surface that contain sensory hair cells. When the hairs are disturbed, nerve impulses are initiated; their frequency enables the fish to locate the disturbance.

(g) Placoid scales, toothlike outgrowths of the skin, cover the body. Run your hand from head to tail along the length of the animal. How does it feel?

____

Now run your hand in the opposite direction along the animal. How does it feel this time?

____

Cut out a small piece of skin with your scissors, pick it up with your forceps, and examine its surface with the dissection microscope. Draw what you see in Figure 22-4.

Figure 22-4 Drawing of shark skin (_____×).

3. Examine the demonstration of a shark jaw with its multiple rows of teeth (Figure 22-3b).

22.3 Bony Fishes (Class Osteichthyes) *(12 min.)*

The bony fishes, the fishes with which you are most familiar, inhabit virtually all the waters of the world and are the largest vertebrate group. They are economically important, commercially and as game species. The ancestors of cartilaginous fish gave rise to the first bony fishes about 415 million years ago. As their name suggests, the *skeleton* is at least *partly ossified (bony),* and the flat *scales* that cover at least some of the surface of most bony fishes are *bony* as well. Gill slits are housed in a common chamber covered by a bony movable flap, the **operculum.**

MATERIALS

Per student:

- scalpel
- compound microscope, lens paper, a bottle of lens-cleaning solution (optional), and a lint-free cloth (optional)
- clean microscope slide and coverslip
- dH_2O in dropping bottle
- forceps
- blunt probe or dissecting needle

Per student pair:

- dissection pan

Per student group:

- preserved specimens of yellow perch

Per lab room:

- collection of preserved bony fishes
- boxes of different sizes of vinyl or latex gloves
- box of safety goggles

PROCEDURE

1. Examine the assortment of bony fishes on display.
2. Examine a yellow perch as a representative advanced bony fish (Figure 22-5).

   (a) Identify the *spiny* **dorsal fin** located in front of the *soft* **dorsal fin.** Paired **pectoral** and **pelvic fins** are also present, and behind the **anus** is an unpaired **anal fin.** The **tail fin** consists of *dorsal* and *ventral lobes* of approximately equal size. The fins, as in the cartilaginous fishes, are used to brake, steer, and maintain an upright position in the water.

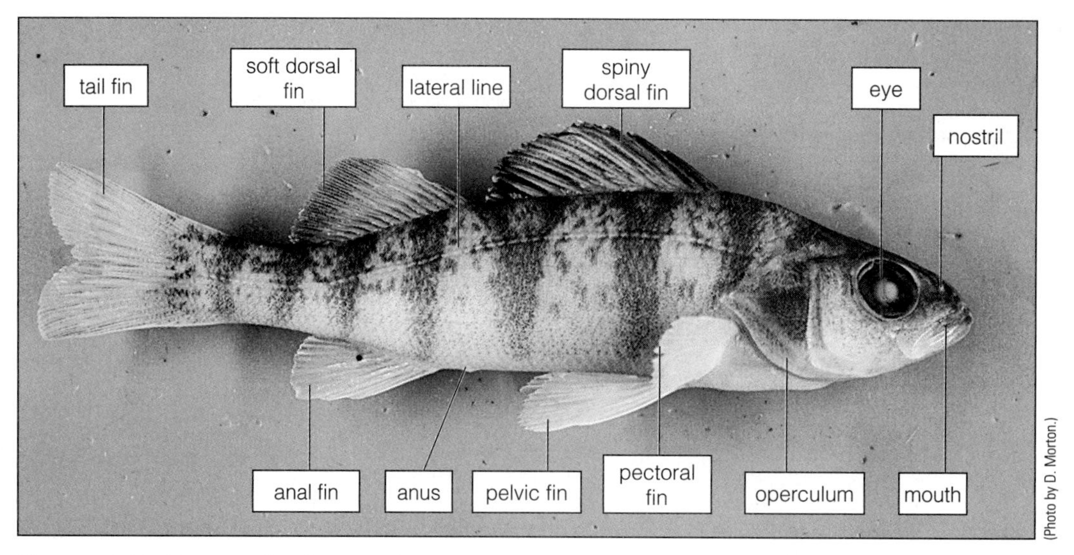

Figure 22-5 Preserved yellow perch.

(b) Note the double **nostrils,** leading into and out of *olfactory sacs.* The eyes resemble those in sharks, with no eyelids. With forceps, pry open the operculum to see the organs of respiration, the **gills.**

(c) Find the **lateral line.** This functions similarly to that of the shark.

(d) Bony scales cover the body. Using forceps, remove a scale, make a wet mount, and examine it with the compound microscope. Locate the *annual rings,* which indicate the age of the fish. How are the annual rings of the fish scale analogous to the annual ridges of the mussel or clam shell?

____

____

____

Draw the scale in Figure 22-6.

Figure 22-6 Drawing of yellow perch scale (_____×).

22.4 Amphibians (Class Amphibia) *(30 min.)*

The *amphibians* were the first vertebrates to assume a terrestrial existence, having evolved about 370 million years ago from a group of lobe-finned fishes. The paired appendages are modified as legs, which support the individual during movement on land. Respiration by one species or another or by the different developmental stages of one species is by lungs, gills, and the highly vascularized skin and lining of the mouth. There is a *three-chambered heart,* compared to the two-chambered heart of fishes, with two *circuits for the blood circulation,* compared to one circuit in fishes. Reproduction requires water, or at least moist conditions on land. Fertilized eggs hatch in water, and larvae generally live in water. The skeleton is bonier than that of the bony fishes, but a considerable proportion of it remains cartilaginous. The *skin* is usually *smooth and moist,* with mucous glands; scales are usually absent. This group of vertebrates includes the frogs, toads, salamanders, and tropical, limbless, burrowing, mostly blind caecilians.

MATERIALS

Per student:

- forceps
- dissecting scissors
- blunt probe or dissecting needle
- dissection pins

Per student pair:

- preserved leopard frog

Per lab room:

- collection of preserved bony fishes
- collection of preserved amphibians
- skeleton of frog
- skeleton of human
- boxes of different sizes of vinyl or latex gloves
- box of safety glasses

PROCEDURE

1. Examine preserved specimens of a variety of amphibians.
2. The leopard frog illustrates well the general features of the vertebrates and the specific characteristics of the amphibians. Obtain a preserved specimen and after rinsing it in tap water, examine its external anatomy (Figure 22-7).
   (a) Find the two *nostrils* at the tip of the head. These are used for inspiration and expiration of air. Just behind the eye locate a disklike structure, the *tympanum,* the outer wall of the middle ear. There is no external ear. The tympanum is larger in the male than in the female. Examine the frogs of other students in your lab.
   Is your frog a male or female?
   ____
   (b) At the back end of the body, locate the *cloacal opening.*
   (c) The forelimbs are divided into three main parts: the **upper arm, forearm,** and **hand.** The hand is divided into a **wrist, palm,** and **fingers** (digits). The three divisions of the hindlimbs are the **thigh, shank** (lower leg), and **foot.** The foot is further divided into three parts: the **ankle, sole,** and **toes** (digits).
3. Examine the internal anatomy of the frog.
   (a) With the scissors, cut through the joint at both angles of the jaw. Identify the structures labeled in Figure 22-8. The region containing the opening into the **esophagus** and the **glottis** is the **pharynx** (throat). The esophagus leads to the stomach and the rest of the digestive tract; the glottis leads into the blind respiratory tract.
   (b) Fasten the frog, ventral side up, with pins to the wax of a dissection pan. Lift the skin with forceps. Then make a superficial

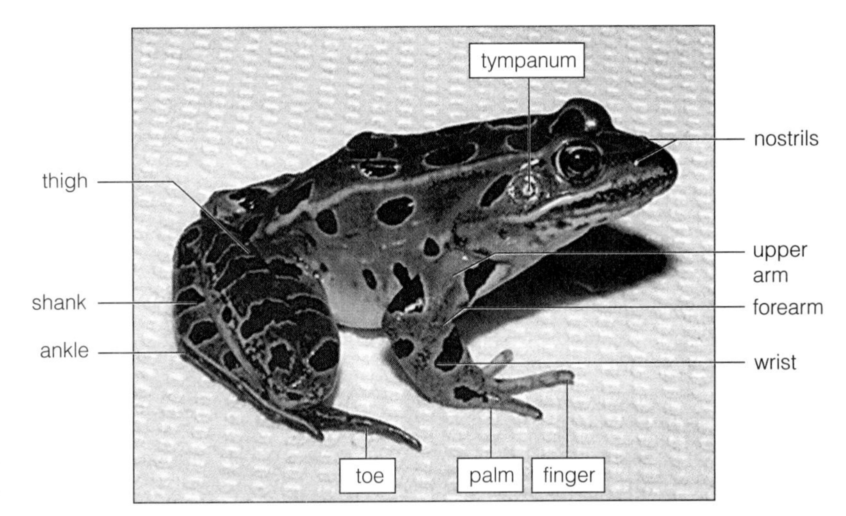

Figure 22-7 External anatomy of the frog. (Photo by D. Morton.)

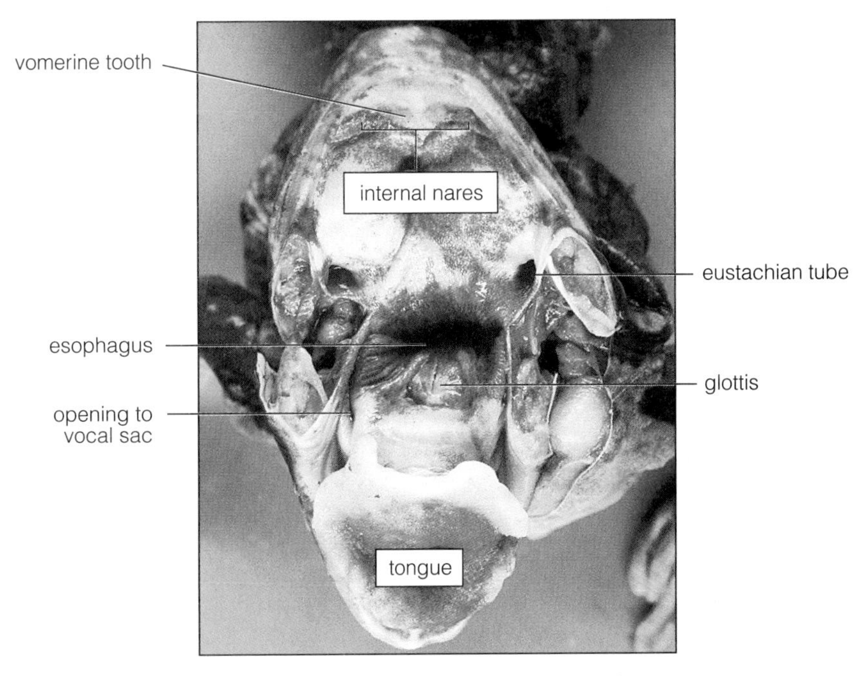

Figure 22-8 Oral cavity of the frog. (Photo by D. Morton.)

cut with your scissors from the lower abdomen forward, and just left or right of center, to the tip of the lower jaw. Pin back the skin on both sides to expose the large abdominal muscles. Lift these muscles with your forceps, and cut through the body wall with your scissors again from the lower abdomen to the tip of the lower jaw, cutting through the sternum (breastbone) but not damaging the internal organs. Pin back the body wall as you did the skin to expose the internal organs (Figure 22-9).

(c) Use Figure 22-9 to locate the *spleen* (Figure 22-9c) and the following organs of the digestive system: **stomach, small intestine, large intestine** (colon), **liver, gall bladder,** and **pancreas.** Trace the esophagus from its opening in the pharynx to the stomach. Parallel to its upper portion find the **bronchi** (the singular is *bronchus*), which lead toward the **lungs.** Are the bronchi dorsal or ventral to the esophagus? ______________________________

(d) Identify the two thin-walled **atria** and the thick-walled **ventricle** and the **conus arteriosus** of the heart along with and the right and left branches of the **truncus arteriosus** (Figure 22-9a). The latter structure carries blood to the paired **pulmocutaneous arches** and **systemic arches.** In frogs and other amphibians a *pulmocutaneous circuit* circulates blood to and from the lungs and the skin, and a *systemic circuit* services the rest of the body. Even though there is only one ventricle, because of the structure of the heart and its large arteries, most of the deoxygenated blood returning from the body is pumped into the pulmocutaneous circuit before the oxygenated blood is pumped into the systemic circuit. In comparison, a fish has a two-chambered heart with only one atrium and a one-circuit circulation. A two-circuit system is more efficient because blood pressure drops less after the blood is oxygenated. Partition of the ventricle of the heart and the further separation of the two circuits will continue in subsequent vertebrate classes.

(e) Refer to Figures 22-9 and 22-10 and find the major *arteries* listed in Table 22-1.

TABLE 22-1 Major Arteries of the Frog

Arteries	Location	Function Is to Convey Blood to
Pulmocutaneous arches	First pair of vessels branching off truncus arteriosus	Lungs and skin
Systemic arches	Second pair of vessels branching off truncus arteriosus	The rest of the body
Common carotids (carotid arches)	Branch from the systemic arches	Head
Subclavians	Branch from the systemic arches	Arms
Dorsal aorta	Formed by the fusion of the systemic arches along dorsal body wall in mid abdomen	Lower half of the body
Common iliacs	Division of dorsal aorta in lower abdomen	Legs

Which of these arteries carries deoxygenated blood? ________

(f) Locate the **anterior** and **posterior vena cavae,** which return deoxygenated blood from the systemic circuit to the right atrium of the heart (Figures 22-9c and 22-10). Other veins return blood from the head, arms, and legs to one of the vena cavae. Right and left **pulmonary veins** return oxygenated blood from the lungs to the left atrium.

(g) The excretory and reproductive organs together comprise the *urogenital system* (Figure 22-11). Locate the urine-producing excretory organs, the pair of **kidneys** located dorsally in the body cavity. A duct, the **ureter,** leads from the kidney to the **urinary bladder** (also Figure 22-9a), an organ that stores urine for resorption of water from the urine into the circulatory system. The bladder empties into the *cloaca.*

(h) Refer to Figure 22-11a and, in the male, locate the **testes** (also Figure 22-9c); these organs produce sperm that are carried to the kidneys through tiny tubules, the *vasa efferentia.* The ureters serve a dual function in male frogs, transporting both urine and sperm to the cloaca. Find the vestigial female oviducts (also Figure 22-9b), which are located lateral to the urogenital system.

(i) Refer to Figure 22-11b and, in the female, find the *ovaries* (also Figure 22-9d); these organs expel eggs into the *oviducts.* The oviducts lead into the *uterus.* As in the male, the reproductive tract ends in the cloaca.

(j) Locate the *fat bodies* (Figures 22-9c and d), many yellowish, branched structures just above the kidneys. They store food reserves for hibernation and reproduction.

(k) The nervous system of vertebrates is composed of (1) the *central nervous system,* the brain and spinal cord, and (2) the *peripheral nervous system,* nerves extending from the central nervous system. Turn your frog over and remove the skin from the dorsal surface of the head between the eyes and along the vertebral column. With your scalpel, shave thin sections of bone from the skull, noting the shape and size of the *cranium,* until you expose the *brain.* Pick away small pieces of bone with your forceps to expose the entire brain. Use the same procedure to expose the vertebrae and *spinal cord.* Note the *cranial nerves* coming from the brain and the *spinal nerves* coming from the spinal cord. Spinal nerves are also easy to see running along the dorsal wall of the abdomen.

(l) Dispose of your dissected specimen in the large plastic bag provided for this purpose.

4. Comparison of frog and human skeleton.

(a) The skeleton of vertebrates consists of the following: (1) the **axial skeleton,** consisting of the bones of the **skull, sternum, hyoid bone, vertebral column,** and rib cage (when present); and (2) the **appendicular skeleton,** with the bones of its **girdles** and their appendages. The **pectoral girdle** consists of the paired **clavicle** and **scapula** bones (**suprascapular** in the bullfrog). The latter articulates—forms a joint—with the **humerus.** The **pubis, ischium,** and **ilium** bones form the pelvic girdle, which articulates with the femur. Compare the skeleton of the bullfrog and human (Figure 22-12), and identify the axial and appendicular portions of the skeleton.

(b) There are differences in the two skeletons, but their basic plan is remarkably similar. To a large extent, the size and shape of the different parts of the skeleton of a vertebrate correlate with body specializations and behavior. List some reasons for the differences in the girdles, appendages, and cranium of the frog and human skeletons.

____

____

____

22.5 Reptiles (Class Reptilia) *(12 min.)*

Reptiles are the oldest group of vertebrates *adapted to living primarily on land,* although many do live in fresh water or seawater. They arose from the ancestors of amphibians about 300 million years ago. Their *skeleton* is bonier than that of amphibians. The *skin* is *dry* and covered by *epidermal scales.* There are virtually no skin glands. Reptiles lay *amniotic eggs* (Figure 22-13). Inside these eggs, embryos are suspended by fetal membranes in an internal aquatic environment surrounded by a shell to prevent their drying out.

There is no larval stage. The amniotic egg (characteristic of birds and some mammals as well as reptiles) and lack of larvae are adaptations to living an entire life cycle on land. The heart consists of two atria and a partially divided ventricle or two ventricles (in crocodiles). The nervous system, especially the brain, is more highly developed than that of amphibians.

Reptiles, like the fishes and amphibians, are largely **ectothermic,** without the capability to maintain a relatively high core-body temperature physiologically and with a greater dependence on gaining or losing heat from or to their external environment. Reptiles and other ectotherms have an adaptive advantage over endothermic mammals in warm and humid areas of the earth because they expend minimal energy on maintaining body temperature, which leaves more energy for reproduction. This advantage is reflected in the greater numbers of individuals and species of reptiles compared to mammals in the tropics. Mammals have the advantage in moderate to cold areas.

MATERIALS

Per student:
- scalpel

Per student pair:
- dissection pan

Per lab room:
- collection of preserved reptiles
- skeleton of snake
- turtle skeleton or shell
- boxes of different sizes of vinyl or latex gloves
- box of safety goggles

PROCEDURE

1. Examine an assortment of preserved reptiles and note their diversity as a group. This diverse group of vertebrates includes the turtles, lizards, snakes, crocodiles, and alligators.

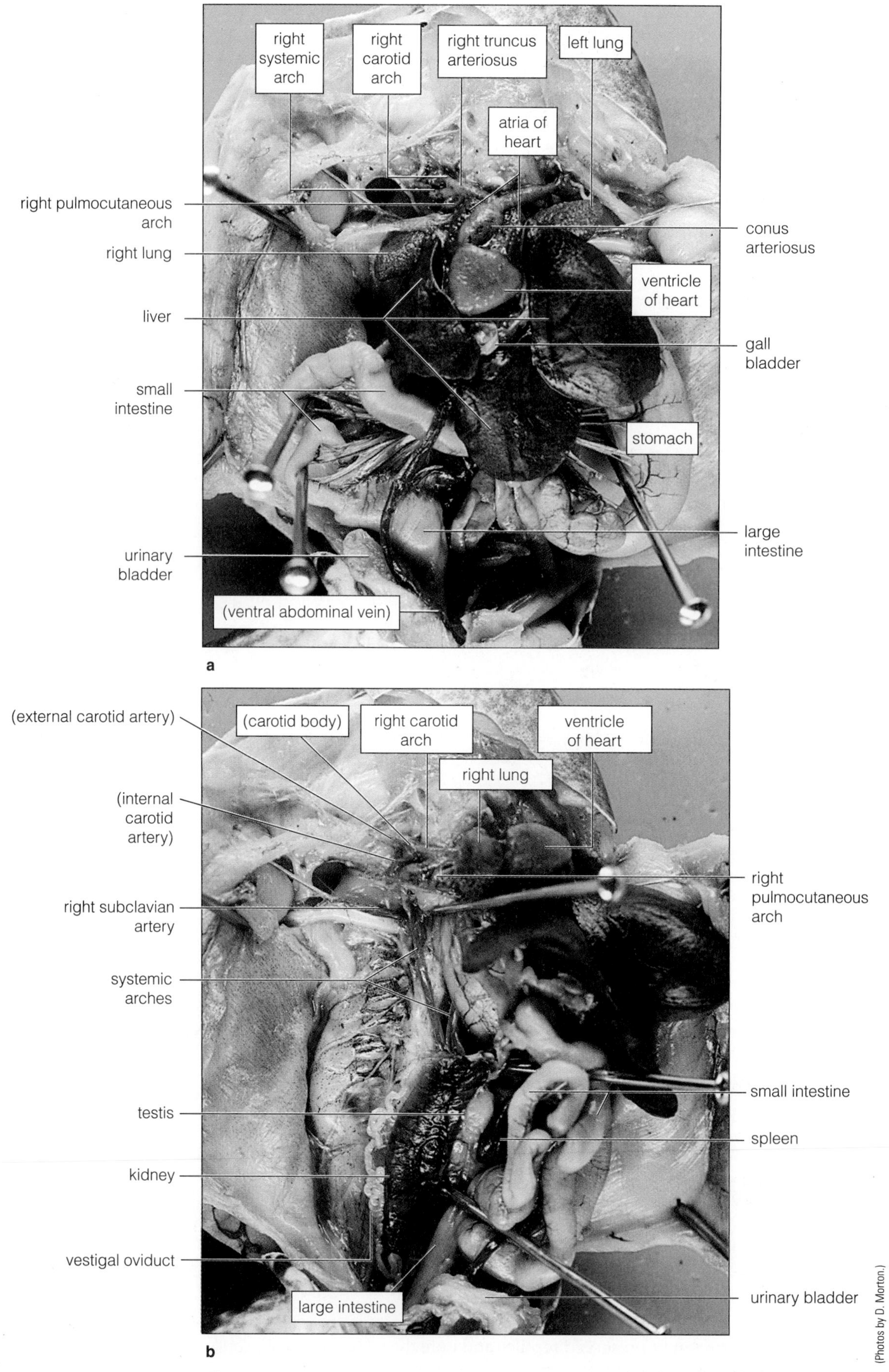

Figure 22-9 Abdominal views of dissected frogs. (**a**–**c**) Male. (**d**) Female. *Continues.*

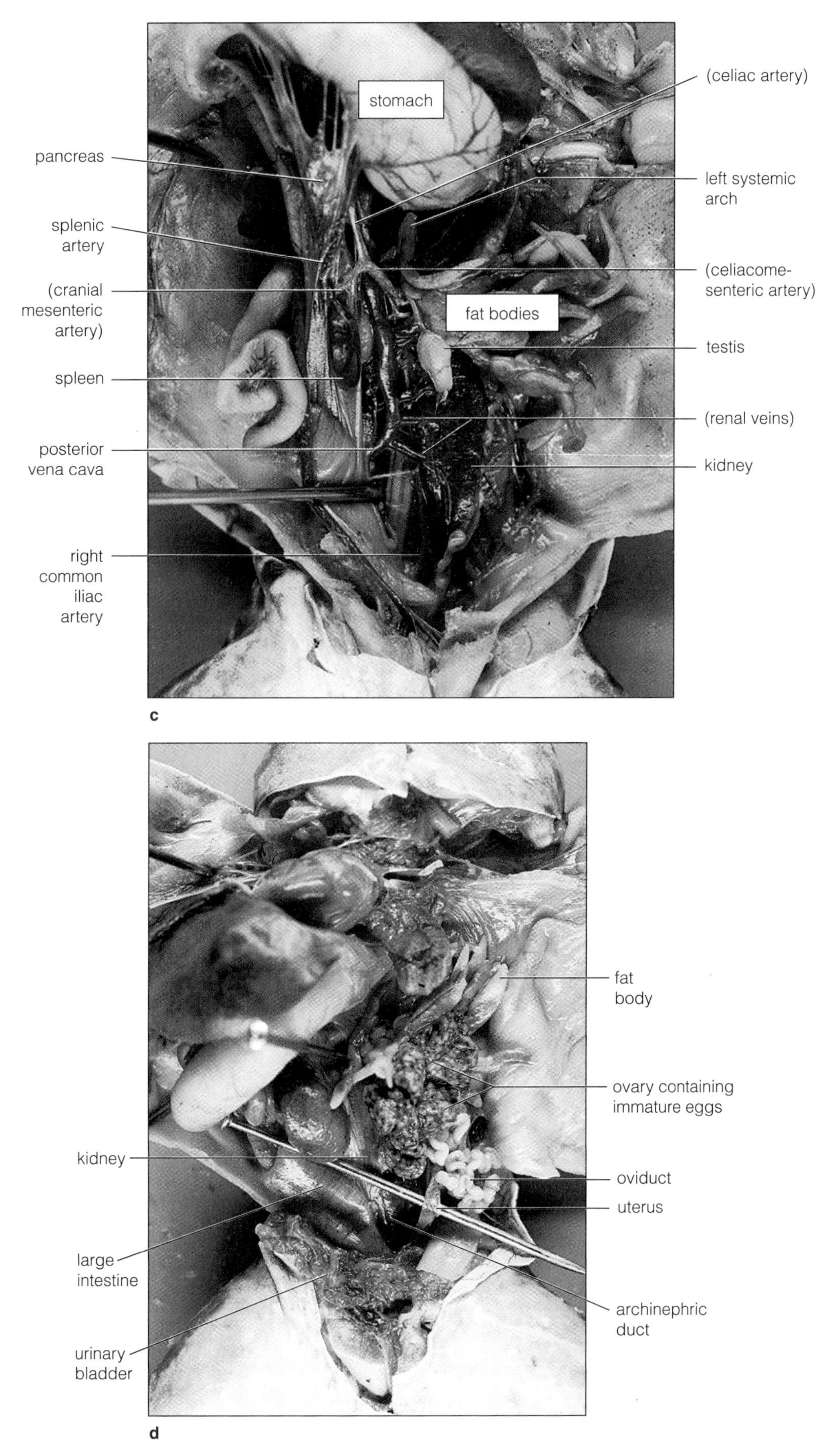

Figure 22-9 *Continued.*

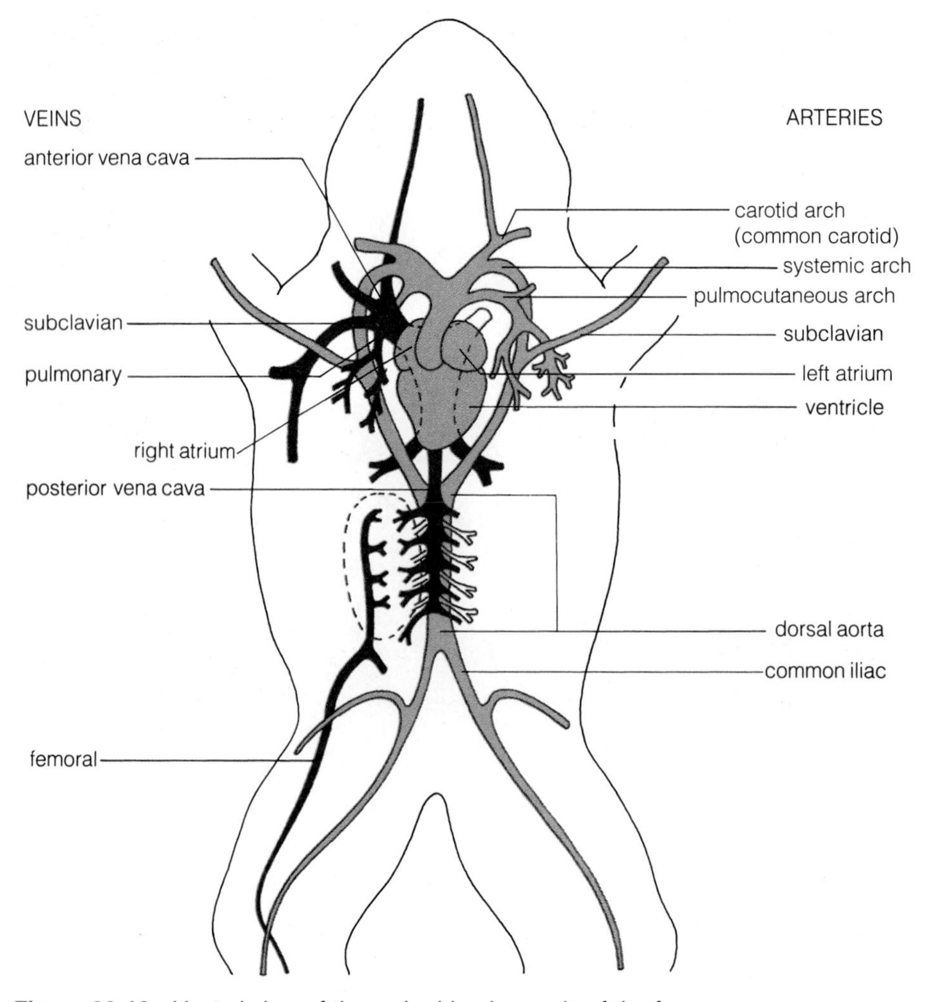

Figure 22-10 Ventral view of the major blood vessels of the frog.

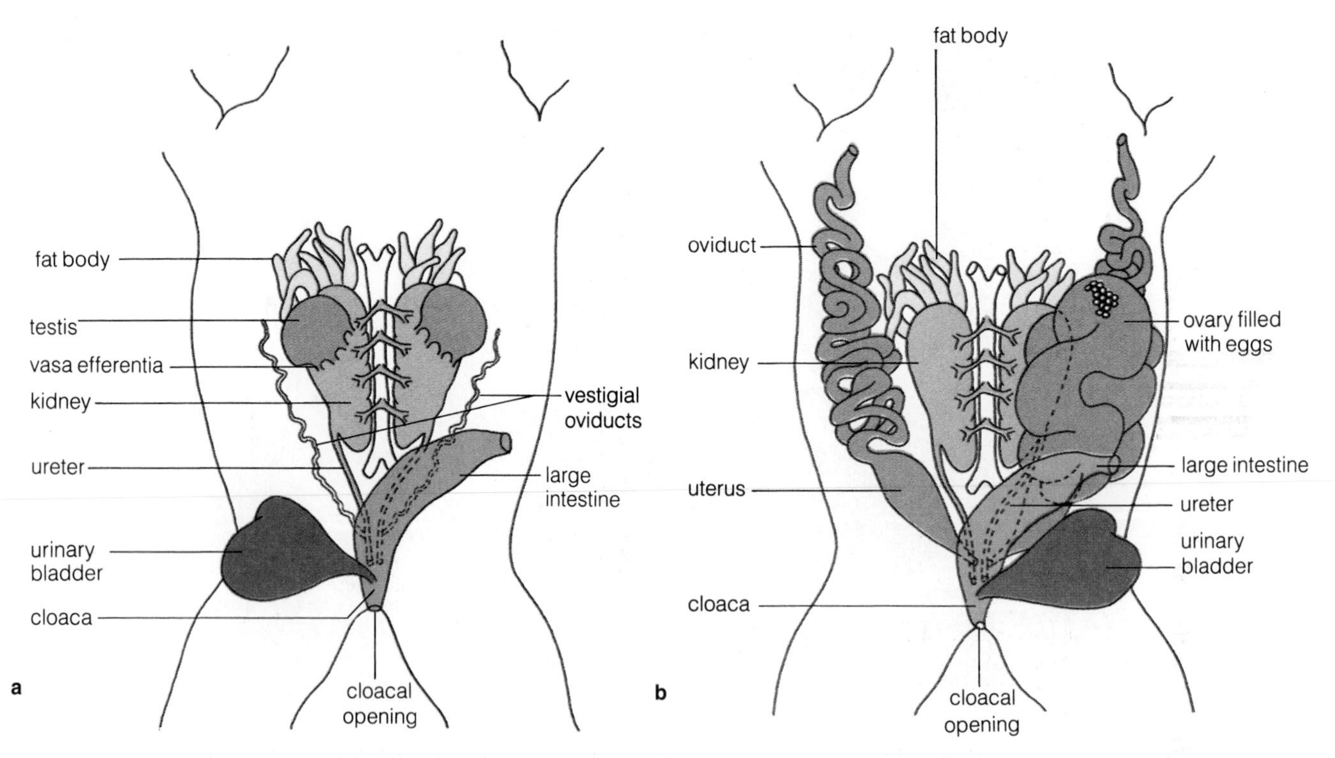

Figure 22-11 Ventral views of the urogenital systems of (a) male and (b) female frogs.

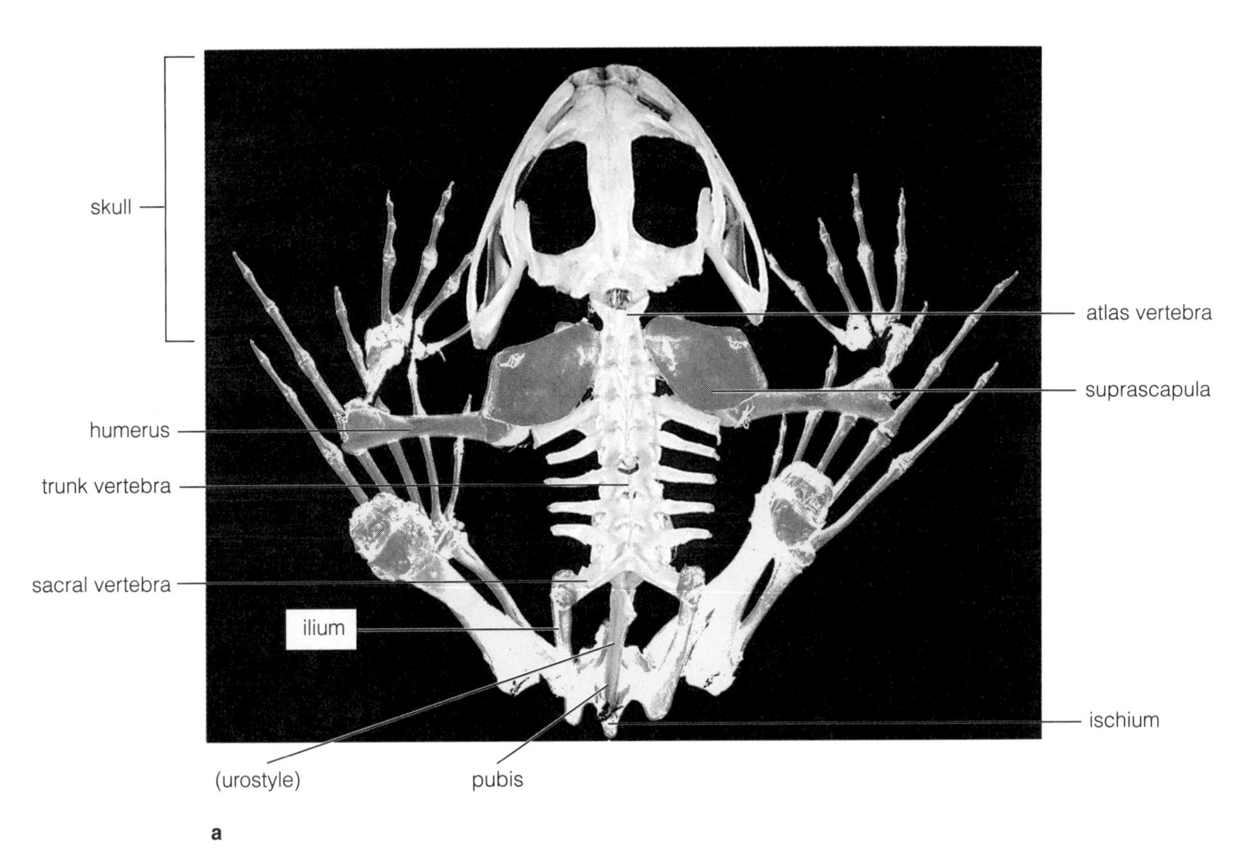

orbit
nasal cavity
skull
clavicle
hyoid
scapula
sternum
rib cage
humerus
(costal cartilages)
ilium
b

Figure 22-12 Skeletons of (**a**) a bullfrog and (**b,c**) a human. The appendicular skeleton is colored yellow and the axial skeleton is uncolored. (Photos by D. Morton.) *Continues.*

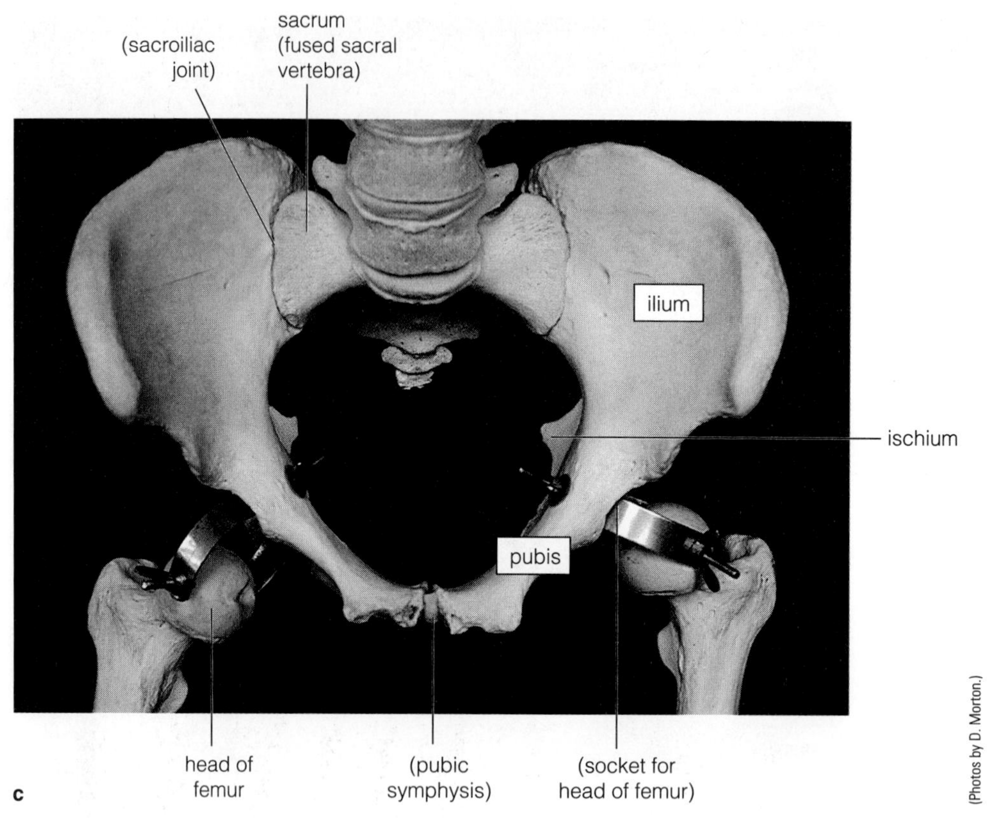

Figure 22-12 Skeletons of (**a**) a bullfrog and (**b,c**) a human. The appendicular skeleton is colored yellow and the axial skeleton is uncolored. *Continued.*

2. Compare the skeleton of a snake (Figure 22-14) with that of the frog. Describe any differences you see.

____
____
____
____
____

Identify the axial and appendicular portions of the skeleton.

3. Look at the inside of the *carapace* (the upper portion) of a turtle shell. What portions of the skeleton are incorporated into the shell?

____
____
____

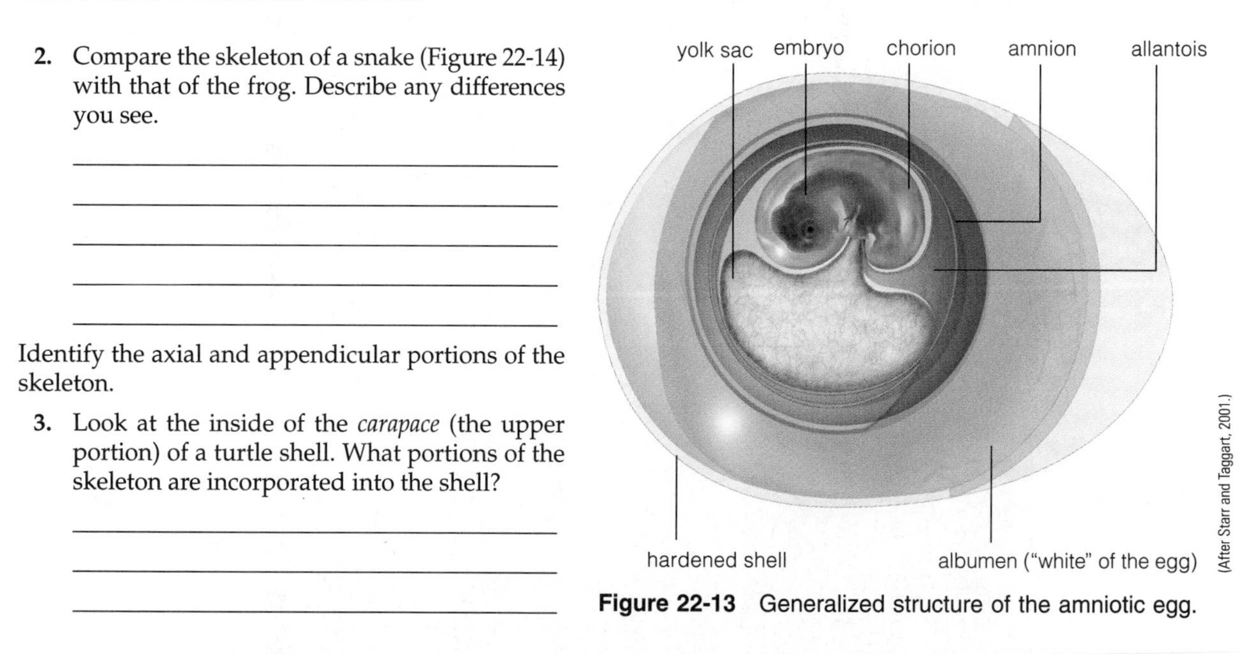

Figure 22-13 Generalized structure of the amniotic egg.

22.6 Birds (Class Aves) *(12 min.)*

Birds are likely descended from small bipedal (walked on two legs) reptiles about 160 million years ago and are thought to be closely related to dinosaurs. Their body is covered with *feathers;* and *scales,* reminiscent of their reptilian heritage, are present on the feet. The front limbs in most birds are modified as *wings* for flight. An additional internal adaptation for flight is the well-developed *sternum* (breastbone), with a *keel* for the attachment of powerful muscles for flight (Figure 22-15). Birds are **endothermic** vertebrates, capable of maintaining relatively high core-body temperatures physiologically.

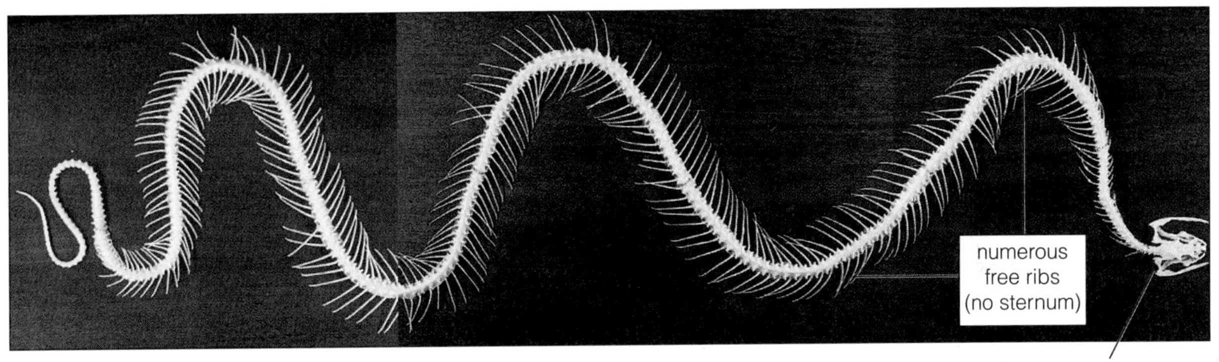

Figure 22-14 Snake skeleton.

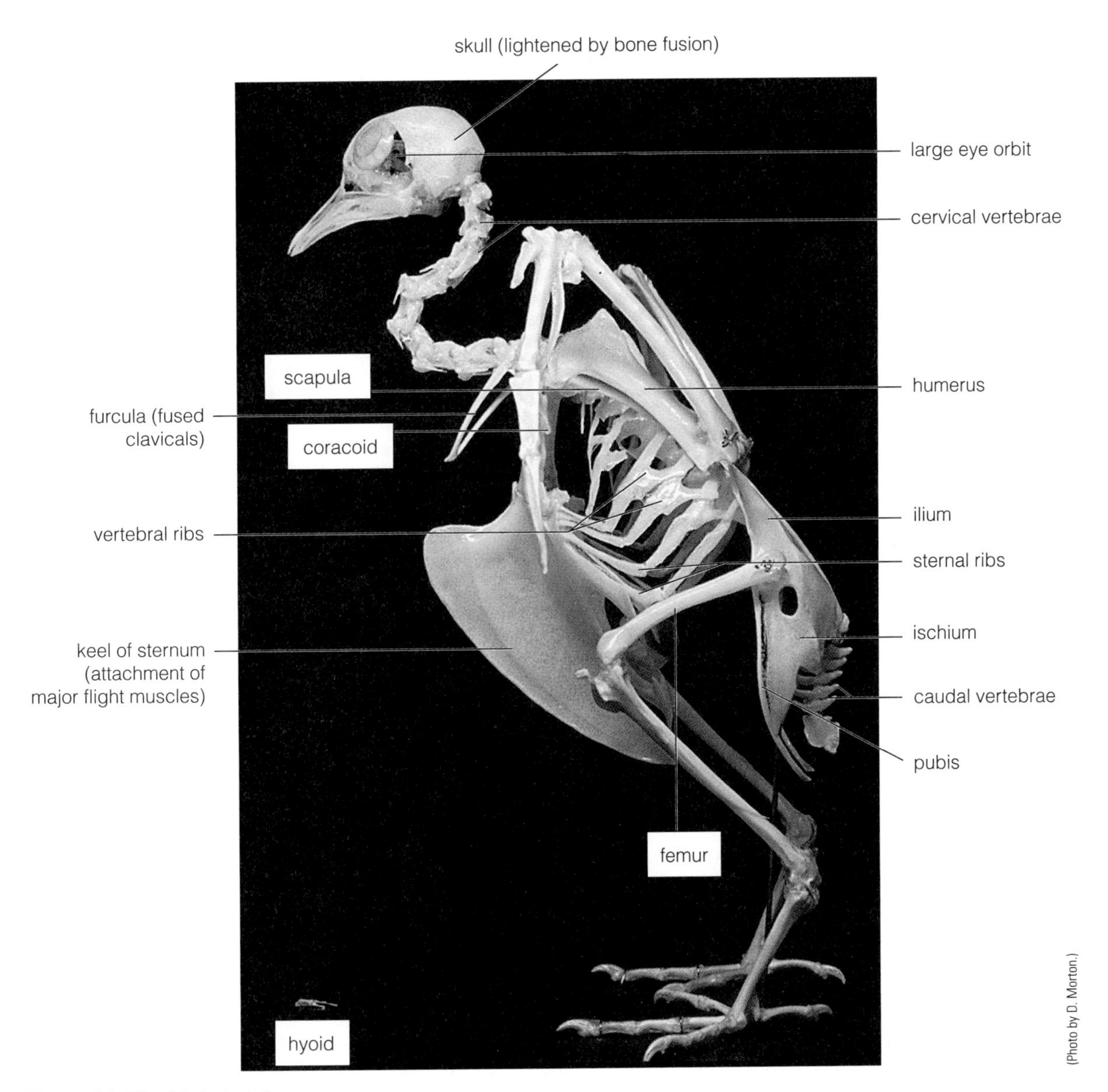

Figure 22-15 Bird skeleton.

The major *bones* of birds are *hollow* and contain *air sacs connected to the lungs.* Birds have a *four-chambered heart,* with two atria and two ventricles (Figure 22-16). This permits the complete separation of oxygenated and deoxygenated blood. Why is this circulatory arrangement an advantage for birds, as opposed to that found in amphibians and reptiles? (*Hint:* Recall that birds are endothermic and most can fly.)

___

___

___

The nervous system resembles that of reptiles. However, the brain is larger, permitting more sophisticated behavior and muscular coordination; the optic lobes are especially well developed in association with a keen sense of sight.

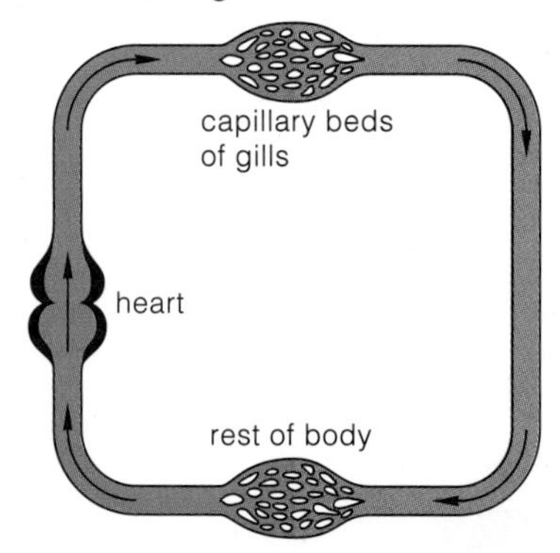

a In fishes, a two-chambered heart (atrium, ventricle) pumps blood in one circuit. Blood picks up oxygen in gills and delivers it to rest of body. Oxygen-poor blood flows back to heart.

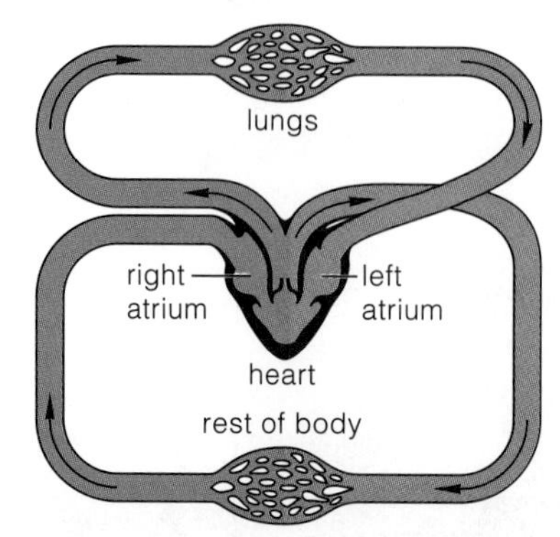

b In amphibians, a heart pumps blood through two partially separate circuits. Blood flows to lungs, picks up oxygen, and returns to heart. It mixes with oxygen-poor blood still in heart, flows to rest of body, and returns to heart.

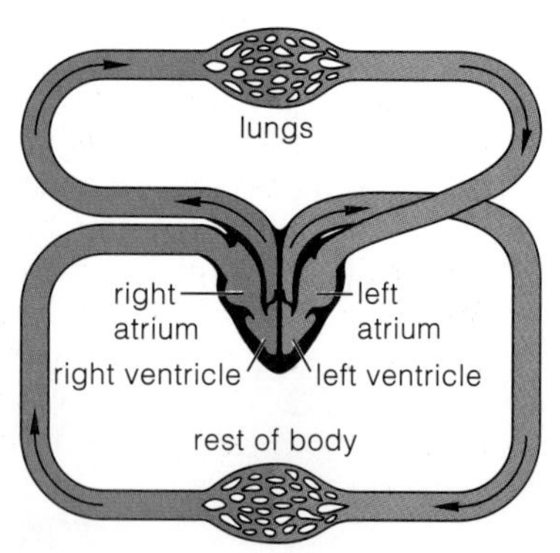

c In birds and mammals, the heart is fully partitioned into two halves. Blood circulates in two circuits: from the heart's right half to lungs and back, then from the heart's left half to oxygen-requiring tissues and back.

Figure 22-16 Circulatory systems of (**a**) fishes, (**b**) amphibians, and (**c**) birds and mammals. (After Starr and Taggart, 2001.)

MATERIALS

Per student:
- scalpel

Per student pair:
- dissection microscope

Per lab room:
- an assortment of feathers, stuffed birds, or bird illustrations
- several field guides to the birds

PROCEDURE

1. Examine a feather and identify the **rachis** (shaft) and **vane** (Figure 22-17). Examine the vane with the dissection microscope and note the **barbs** and **barbules.** What else do birds use their feathers for besides flight?

   ___

   ___

2. Identify the axial and appendicular portions of the skeleton (Figure 22-15). Note the various adaptations for flight, including the elongated wrists and ankles.
3. The upper and lower jaws are modified as variously shaped *beaks* or *bills,* with the shapes reflecting the feeding habits of the species. No teeth are present in adults. Examine the assortment of birds or bird illustrations and speculate on their food preferences by studying the configurations of their bills. Check your conclusions with a field guide or other suitable source that describes the food habits.

22.7 Mammals (Class Mammalia) *(12 min.)*

Mammals arose from an ancient branch of reptiles over 200 million years ago. The term *mammal* stems from the *mammary glands,* which all mammals possess. Female mammals feed their young milk produced by the mammary glands (Figure 22-18). The *brain* is relatively *larger than that of other vertebrates,* and its surface area is increased by grooves and folds (Figure 22-19). Behavior is more complex and flexible. *Hair* is present during some portion of the life span. Mammals are equipped with a variety of modifications and outgrowths of the skin in addition to hair, all made of the protein keratin. They include horns, spines, nails, claws, and hoofs. Mammalian dentition is intricate, with teeth specialized for cutting, grooming, wounding, tearing, slicing, and grinding. The teeth in the upper and lower jaws match up.

Like birds, mammals are *endothermic* and the *heart* is *four-chambered.* The ear frequently has a cartilaginous outer portion called the *pinna,* which functions to collect sound waves. In addition to mammary glands, there are three other types of skin glands: *sebaceous, sweat, and scent glands.*

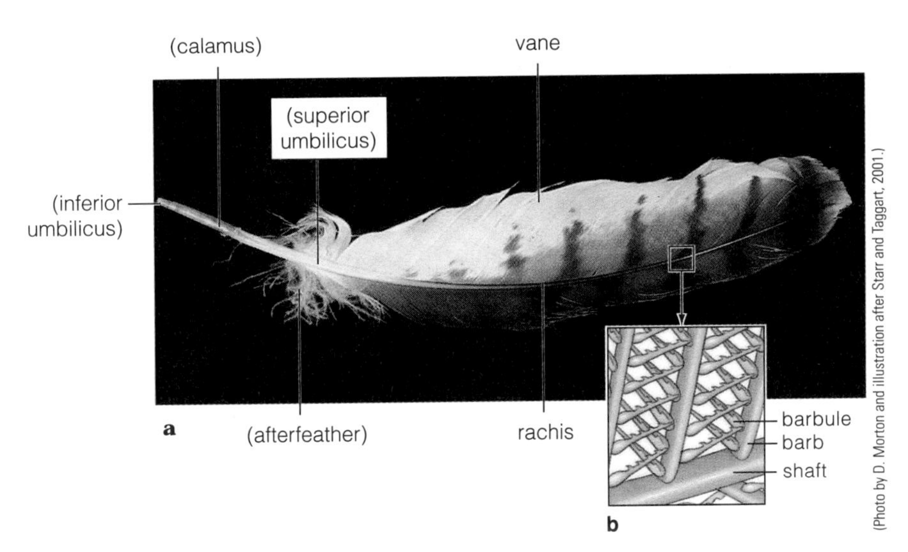

Figure 22-17 (**a**) Feather and (**b**) expanded view of the vane.

MATERIALS

Per student:

- scalpel

Per lab room:

- stuffed mammals or mammal illustrations
- mammalian placentas and embryos (in utero)

Figure 22-18 Female dog with nursing puppies.

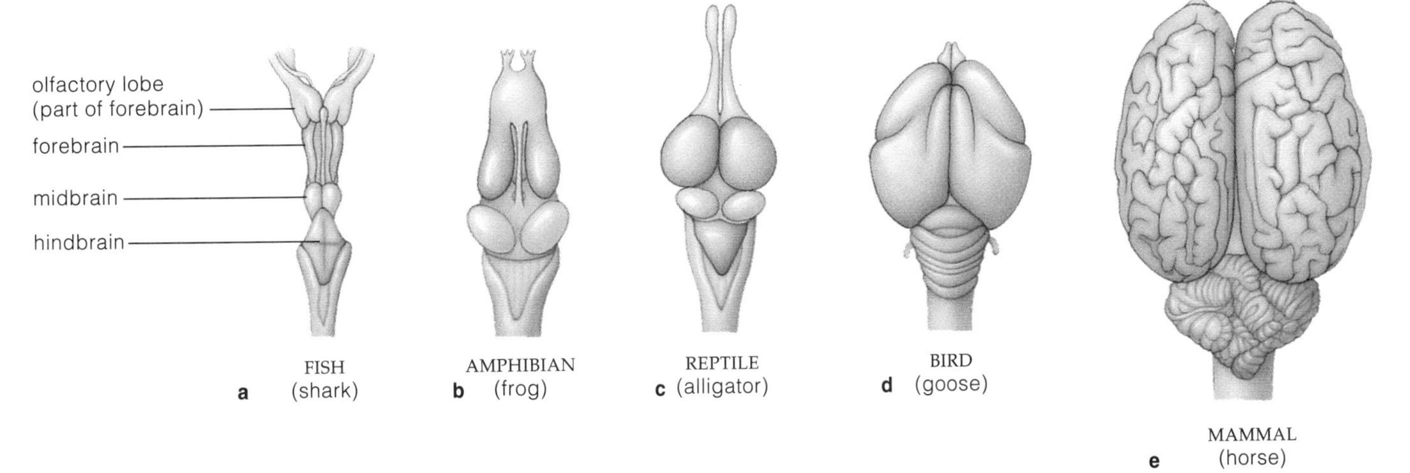

Figure 22-19 Evolutionary trend toward an expanded, more complex brain, as suggested by comparing the brain of existing vertebrates: (**a**) shark; (**b**) frog; (**c**) alligator; (**d**) goose; and (**e**) horse. These dorsal views are not to the same scale. (After Starr and Taggart, 2001.)

PROCEDURE

1. Study the assortment of mammals on display and list as many functions for hair as you can.

____

____

2. Except for two species of monotremes that lay eggs (platypus and spiny anteater), most mammals are **viviparous:** Females bear live young after supporting them in the uterus with a *placenta,* a connection between the mother and the embryo. The placenta exchanges gases with, delivers nutrients to, and removes wastes from the embryo, among other functions.

Some viviparous mammals (marsupials—kangaroos, opossums, and so on) give birth early, and along with spiny anteater hatchlings, continue development suckling milk in a skin pouch. Eutherian mammals give birth later, but the young of all mammals receive extended parental care. Examine the placentas and embryos in uteri of mammals on display in the lab. Can you suggest a relationship between the early development of mammals and the comparative sophistication of the nervous system and behavior of adult mammals?

____

____

3. Examine Figure 22-20 and identify the teeth present in the skull of the human skeleton. Fill in the numbers of each type in the upper and lower jaws and their functions in Table 22-2.

TABLE 22-2 Human Dentition

Type of Tooth	Number in Upper Jaw	Number in Lower Jaw	Function
Incisors			
Canines			
Premolars			
Molars			

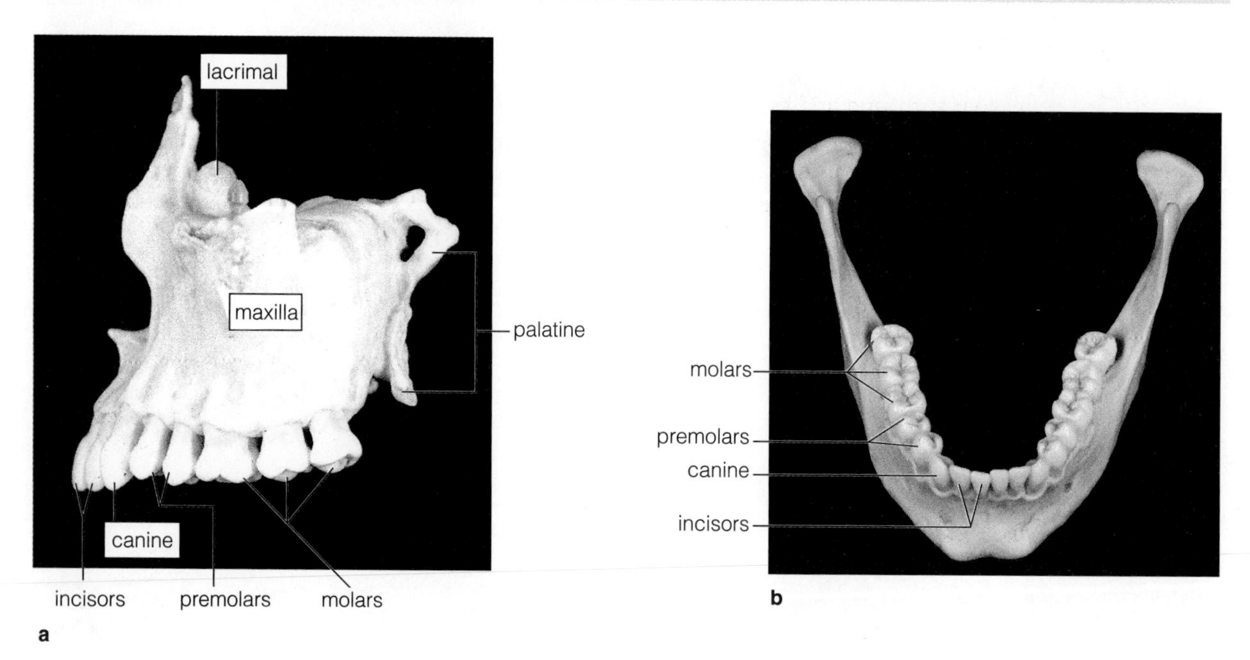

Figure 22-20 (**a**) Human maxilla in side view. (**b**) Top view of the mandible. (Photos by D. Morton.)

22.8 Construction of a Dichotomous Key to the Animals *(50 min.)*

Your instructor will provide you with instructions.

PRE-LAB QUESTIONS

_____ **1.** Animals that have the four basic characteristics of chordates plus a vertebral column are
(a) invertebrates
(b) hemichordates
(c) cephalochordates
(d) vertebrates

_____ **2.** The ammocoete is the larva of
(a) lampreys
(b) sharks
(c) bony fishes
(d) amphibians

_____ **3.** Placoid scales are characteristic of class
(a) Cephalaspidomorphi
(b) Chondrichthyes
(c) Osteichthyes
(d) Reptilia

_____ **4.** The bony movable flap that covers the gills and gill slits is called the
(a) operculum
(b) cranium
(c) tympanum
(d) colon

_____ **5.** Which group of animals is endothermic?
(a) fishes
(b) mammals
(c) reptiles
(d) amphibians

_____ **6.** The bones that surround and protect the brain are collectively called the
(a) spinal cord
(b) vertebral column
(c) pelvic girdle
(d) cranium

_____ **7.** The lampreys are unusual vertebrates in that they have no
(a) eyes
(b) jaws
(c) gill slits
(d) mouth

_____ **8.** The structure of the cartilaginous and bony fishes that detects vibrations in the water is the
(a) anal fin
(b) operculum
(c) nostrils
(d) lateral line

_____ **9.** Amphibians have
(a) a two-chambered heart
(b) a three-chambered heart
(c) a four-chambered heart
(d) none of the above

_____ **10.** The amniotic egg of reptiles, birds, and mammals is an adaptation to
(a) carnivorous predators
(b) a life on land
(c) compensate for the short period of development of the young
(d) protect the young from the nitrogenous wastes of the mother during the formation of the embryo

Name ______________________________ Section Number ______________________

EXERCISE 22

Vertebrates

POST-LAB QUESTIONS

Introduction

1. What characteristic of vertebrates is missing in all invertebrates?

2. List the basic characteristics of vertebrates.

22.1 Lampreys (Class Cephalaspidomorphi)

3. What is an ammocoete? What is its possible significance to the evolution of vertebrates?

4. How do lampreys differ from other vertebrates?

22.2 Cartilaginous Fishes (Class Chondrichthyes)

5. Explain how the adaptations of cartilaginous fishes make them better predators.

22.3 Bony Fishes (Class Osteichthyes)

6. Identify this vertebrate structure. What do the rings represent?

(Photo courtesy of Biodisc, Inc.)

(24×).

22.5 Reptiles (Class Reptilia)

7. Describe an amniotic egg. What is its evolutionary significance?

22.7 Mammals (Class Mammalia)

8. List the unique characteristics of mammals.

9. Describe the evolution of the heart in the vertebrates.

Food for Thought

10. About 65 million years ago, an asteroid impact may have caused a mass extinction, which led to the demise of dinosaurs and created new opportunities for surviving plants and animals, including mammals. Create a short evolutionary scenario for the next human-made or natural mass extinction event.

EXERCISE 23

Plant Organization: Vegetative Organs of Flowering Plants

OBJECTIVES

After completing this exercise, you will be able to

1. define *vegetative, morphology, dicotyledon (dicot), monocotyledon (monocot), cotyledon, taproot system, node, internode, adventitious root, herb (herbaceous plant), woody (woody plant), heartwood, sapwood, growth increment (annual ring)*;
2. identify and give the function of the external structures of flowering plants (those in **boldface**);
3. identify and give the functions of the tissues and cell types of roots, stems, and leaves (those in **boldface**);
4. determine the age of woody branches.

INTRODUCTION

It's difficult to overstate the importance of plant life. Are you sitting on a wooden chair? If so, you're perched on part of a tree. No, you say? Perhaps your chair is covered with fabric. If it's natural fabric other than wool, it's a plant product. If the chair is covered with plastic, that cover was made from petroleum products derived from plant material that lived millions of years ago. The energy used to create the chair was probably derived from burning petroleum products or coal, also derived from plant material.

Obviously, plants are an important part of our lives. This exercise introduces you to the external and internal structure of the **vegetative organs**—those not associated with sexual reproduction—of flowering plants. The organs are the roots, stems, and leaves. Flowers are the sexual reproductive organs, and were considered in detail in Exercise 21.

Each organ is usually distinguished by its shape and form, its **morphology.** But the cells and tissues comprising these three organs are remarkably similar. Each organ is covered by the protective dermal tissue; each possesses vascular tissue that transports water, minerals, and the products of photosynthesis; and each contains ground tissue, that which is covered by the dermal tissue and in which the vascular tissue is embedded. Examine Figure 23-1, which illustrates these relationships.

The organs of the plant body are more similar than they are dissimilar. In fact, for this reason the differences between a root and a stem or leaf are said to be *quantitative* rather than *qualitative.* That is, these differences are in the number and arrangement of cells and tissues, not the type. Consequently, the plant body is a continuous unit from one organ to the next.

In this exercise, we'll study the organs of two rather large assemblages of flowering plants, **dicotyledons** (dicots) and **monocotyledons** (monocots). **Cotyledons** are the leaves that are formed within the seed. Dicots have two seed leaves, monocots one. In addition to this fundamental difference, dicots and monocots have other dissimilarities, as shown in Figure 23-2. Refer to Figure 23-2 as you proceed through each section.

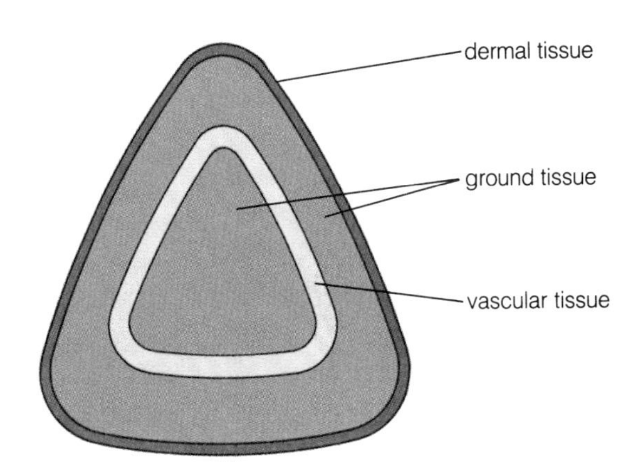

Figure 23-1 Model plant organ in cross section.

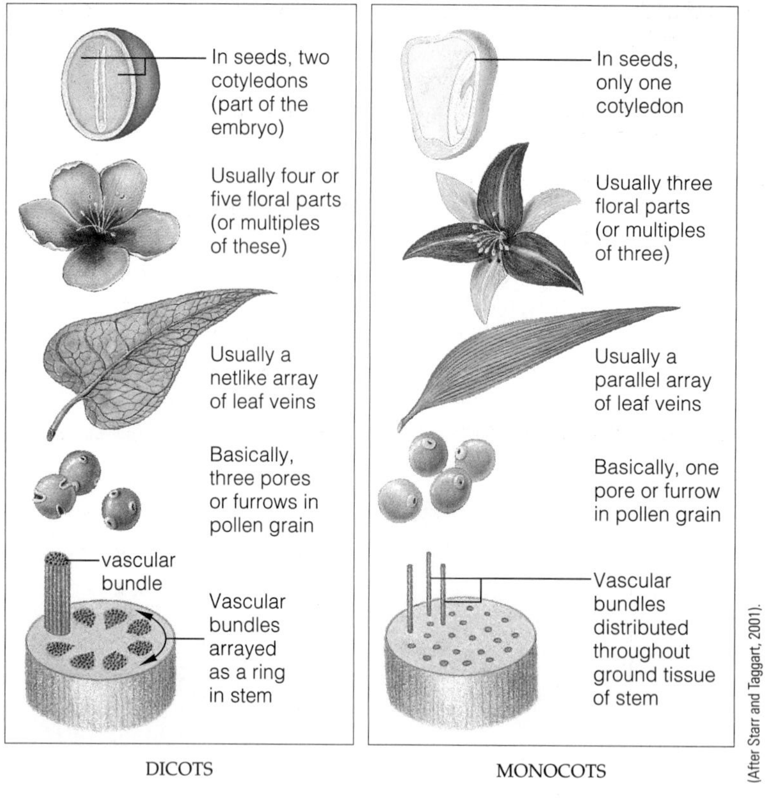

Figure 23-2 Comparison of dicot and monocot structure.

23.1 External Structure of the Flowering Plant *(About 30 min.)*

MATERIALS

Per student group (table):
- mature corn plant

Per lab room:
- living bean and corn plants in flats
- potted geranium and dumbcane plants
- dishpan half-filled with water

PROCEDURE

A. *Dicotyledons*

The common garden bean, *Phaseolus vulgaris,* is a representative dicot. Other familiar examples of dicotyledons (dicots) are sunflowers, roses, cucumbers, peas, maples, and oaks. Let's look at its external morphology. Label Figure 23-3 as you study the bean plant.

1. Obtain a bean plant by gently removing it from the medium in which it is growing. Wash the root system in the dishpan provided, not in the sink.
2. The plant consists of a **root system,** that portion usually below ground, and a typically aerial **shoot system.** Examine the root system first. The root system of the bean is an example of a **taproot system,** that is, one consisting of one large **primary root** (the **taproot**) from which **lateral roots** arise. Identify the taproot and lateral roots.
3. Now turn your attention to the shoot system, consisting of the stem and the leaves.
4. Identify the points of attachment of the leaves to the stem, called **nodes;** the regions between nodes are **internodes.**
5. Look in the upper angle created by the junction of the stem and leaf stalk to find the **axillary bud.** These buds give rise to branches and/or flowers.

6. Find the **terminal bud** at the very tip of the shoot system. The terminal bud contains an apical meristem that accounts for increases in length of the shoot system.

If lateral branches are produced from axillary buds, each lateral branch is terminated by a terminal bud and possesses nodes, internodes, and leaves, complete with axillary buds. As you see, the shoot system can be a highly branched structure.

7. Look several centimeters above the soil line for the lowermost node on the stem. If the plant is relatively young, you should find the **cotyledons** attached to this node. The cotyledons shrivel as food stored in them is used for the early growth of the seedling. Eventually the cotyledons fall off.

The cotyledons are sometimes called *seed leaves* because they are fully formed (although unexpanded) in the seed. By contrast, most of the leaves you're observing on the bean plant were immature or not present at all in the seed.

Now let's examine the other component of the shoot system, the foliage leaves (also called *true leaves* in contrast to the cotyledons). The first-formed foliage leaves (those nearest the cotyledons) are **simple leaves,** each leaf having one undivided blade.

8. Identify the **petiole** (leaf stalk) and **blade** on a simple leaf, and label it on Figure 23-3.

Figure 23-3 External structure of the bean plant.
Labels: root system, shoot system, primary root (taproot), lateral root, node, internode, axillary bud, terminal bud, cotyledon remnant, petiole, blade, simple leaf, compound leaf, leaflet

In bean plants, subsequently formed leaves are **compound leaves,** consisting of three **leaflets** per petiole. Each leaflet has its own short stalk and blade.

9. Identify a compound leaf, and label the petiole and leaflets in Figure 23-3.
10. Note the netted arrangement of veins in the blades. Veins contain vascular tissues—the xylem and phloem.
11. Identify the **midvein;** it is the largest vein and runs down the center of the blade, giving rise to numerous lateral veins.

Now that you have a general idea of the external structure of a typical dicot, let's look at another one, so that you can use your knowledge of a particular plant to make some generalizations.

12. Obtain a potted geranium plant and examine the external structure of the shoot. Identify the following parts, checking off each as you go along.

_____ node	_____ terminal bud
_____ internode	_____ leaf petiole
_____ axillary bud	_____ leaf blade

What color are the stems of the plants you've examined? ______________________________

What structure in the cytoplasm of the cells making up the stem is responsible for this color?

____

What is the function of this structure? ______________________________

What, then, is one function of the stem? ______________________________

13. Compare the leaves of the bean with those of the geranium.

Are the geranium leaves simple or compound? (You may wish to refer to some of the figures in Exercise 15, Taxonomy: Classifying and Naming Organisms, if you have difficulty deciding.) ______________________________

Is there a single midvein in the geranium, or are there many large veins? ______________________________

List all the features shared by the leaves of beans and geraniums.

____

List any differences you observe in the leaves.

____

B. Monocotyledons

Corn *(Zea mays)* is a **monocotyledon.** These plants have only one cotyledon (seed leaf). You're probably familiar with a number of monocots: lilies, onions, orchids, coconuts, bananas, and the grasses. (Did you realize that corn is actually a grass?)

Label Figure 23-4 as you study the corn plant.

1. Remove a single young plant from its growing medium and wash its root system in the dishpan. Note that the seed (technically this is a fruit called a "grain") is still attached to the plant.
2. Identify the **root system.** Note that there is no one particularly prominent root. In most monocots, the primary root is short lived and is replaced by numerous **adventitious roots.** Adventitious roots are roots that arise from places other than existing roots.

Figure 23-4 External structure of a corn plant. (**a**) Seedling (0.5×). (**b**) Lower portion of mature plant (0.1×).
Labels: root system, prop root, leaf sheath

3. Trace the adventitious roots back to the corn grain.

Where do they originate?

___

As the roots branch, they develop into a **fibrous root system,** one particularly well suited to prevent soil erosion.

4. Examine the mature corn plant. Identify the large **prop roots** at the base of the plant.

Where do prop roots arise from?

___

Would you classify these as adventitious roots? (yes or no) ___

5. The shoot system of a young corn plant appears somewhat less complex than that of the bean. There seem to be no nodes or internodes. Look at the mature corn plant again. You should see that nodes and internodes do indeed exist. In your young plant, elongation of the stem has not yet taken place to any appreciable extent.
6. Strip off the leaves of the young corn plant. Keep doing so until you find the shoot apex. (It's deeply embedded.)
7. Examine the leaves of the corn plant in more detail. Note the absence of a petiole and the presence of a **sheath** that extends down the stem. The leaf sheath adds strength to the stem. Look at the veins, which have a parallel arrangement. (Contrast this to the petioled, netted-vein arrangement of the bean leaves.)

You may wonder how representative corn is of all monocotyledons. As it turns out, it's quite representative of most grasses, but not particularly so of monocots as a whole. Let's look at another monocot, a common horticultural plant found in many homes, called dumbcane (*Diffenbachia*).*

8. Obtain a potted specimen of the dumbcane plant. Observe its external morphology, comparing it with the corn plant.

Does the dumbcane have sheathing leaves like corn, or does each leaf have a petiole?

___

Are the veins in the leaves parallel, or is netted venation present? ___
(*Hint:* Look on the lower surface of the leaves, where it is most obvious.)

Is there a midvein? (yes or no) ___

Is the terminal bud obvious or is it deeply embedded, as in the corn plant? ___

Are prop roots present on the dumbcane plant? ___

23.2 The Root System *(About 30 min.)*

MATERIALS

Per student:

- single-edged razor blade
- clean microscope slide
- coverslip
- prepared slide of buttercup (*Ranunculus*) root, c.s.
- prepared slide of corn (*Zea*) root, c.s.
- compound microscope

Per student pair:

- dH_2O in dropping bottles

Per lab room:

- germinating radish seeds in large petri dishes
- demonstration slide of Casparian strip in endodermal cell walls

PROCEDURE

A. Living Root Tip

1. Obtain a germinating radish seed. Identify the **primary root.** Its fuzzy appearance is due to the numerous tiny **root hairs** (Figure 23-5). Root hairs increase the absorptive surface of the root tremendously.
2. Using a razor blade, cut off the seed and discard it, then make a wet mount of the primary root. (Add enough water so that no air surrounds the root. *Do not* squash the root.)

___

* The common name *dumbcane* has its origin in a use by the ancient Greeks. When they tired of long orations by their senators, Greeks sometimes ground up parts of the shoot and added it to a drink. Because certain cells contain needle-shaped crystals, consumption caused a temporary paralysis of the larynx (voice box), ending the oration. Today, dumbcane is a hazard to young children. Ingestion causes throat swelling that can lead to suffocation.

3. Examine your preparation with the low-power objective of your compound microscope. (If you're having difficulty seeing the root clearly, increase its contrast by closing the microscope's diaphragm somewhat.) Locate the conical root tip.
4. At the very end of the root tip, identify the protective **root cap** that covers the tip. As a root grows through the soil, the tip is thrust between soil particles. Were it not for the root cap, the apical meristem containing the dividing cells would be damaged.
5. Find the root hairs.

Do they originate all the way down to the root cap? (yes or no) ______________________________

6. Examine the root hairs carefully.

What happens to their length as you observe them at increasing distance from the root tip?

____

____

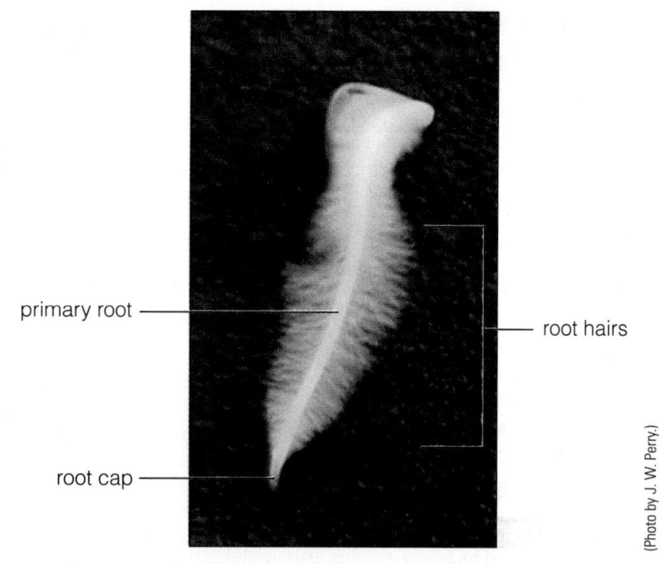

Figure 23-5 Living primary root (4×).

The youngest root hairs are the shortest. What does this imply regarding their point of origin and pattern of maturation?

____

____

B. Dicot Root Anatomy

Now let's see what the internal architecture of the root looks like.

1. Obtain a prepared slide containing a mature buttercup (*Ranunculus*) root in cross section. Refer to Figure 23-6 as you study this slide. Examine the slide first with the low-power objective of your compound microscope to gain an overall impression of the organization of the tissues present.
2. Starting at the edge of the root, identify the **epidermis.**
3. Moving inward, locate the **cortex** and the central **vascular column.** These regions represent the dermal, ground, and vascular tissue systems, respectively.

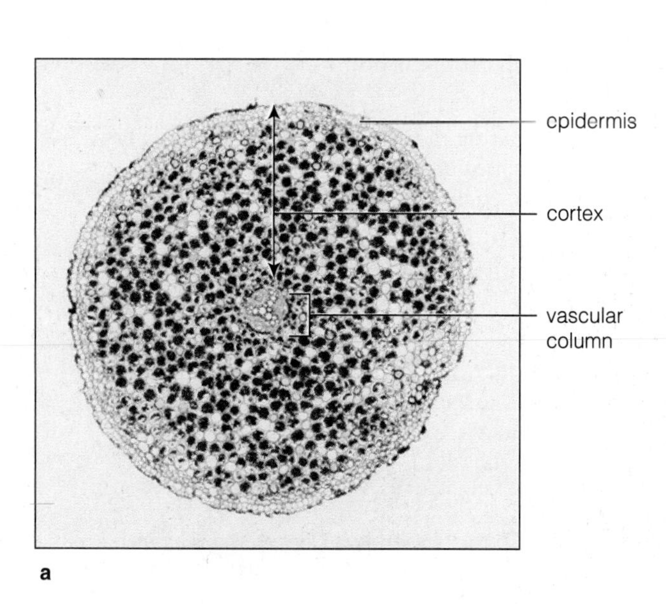

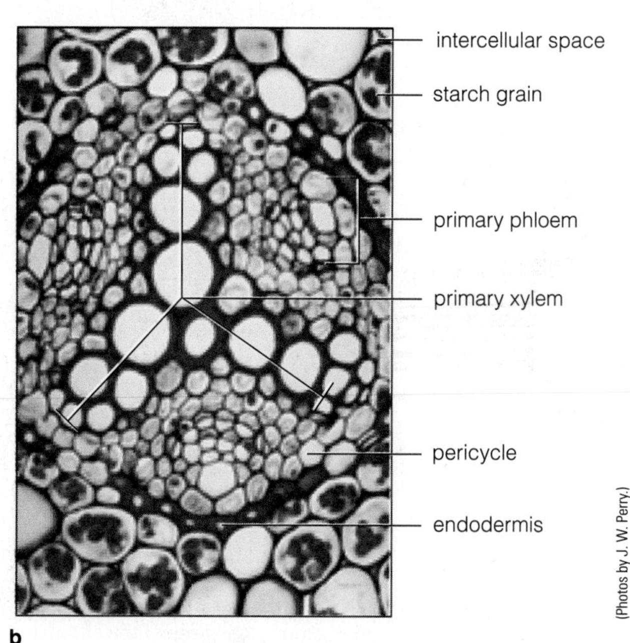

Figure 23-6 Cross section of buttercup (*Ranunculus*) root. (**a**) 45×. (**b**) Portion of cortex and vascular column (75×).

4. Switch to the medium-power objective for further study. Look at the outermost layer of cells, the **epidermis.**
5. Beneath the epidermis find the relatively wide **cortex**, consisting of parenchyma cells that contain numerous **starch grains.**

Based on the presence of starch grains, what would you suspect one function of this root might be?

____

____

6. Switch to the high-dry objective. Between the cells of the cortex, find numerous **intercellular spaces.**

The innermost layer of the cortex is given a special name. This cylinder, a single cell thick, is called the **endodermis.**

7. Locate the endodermis on your slide and in Figure 23-6b.

Unlike the rest of the cortical cells, endodermal cells *do not* have intercellular spaces between them. The endodermis regulates the movement of water and dissolved substances into the vascular column. Each endodermal cell possesses a **Casparian strip** within its radial and transverse walls. To visualize this arrangement, imagine a rectangular box (the endodermal cell) that has a rubber band (the Casparian strip) around its long dimension. Now imagine that the rubber band is actually part of the wall of the box. Figure 23-7 diagrams this arrangement.

The Casparian strip consists of waxy material. Water and substances dissolved in the water normally move through the cell *walls* as they flow from the edge of the root toward the vascular column. Because the Casparian strip consists of a waxy material, it effectively waterproofs the cell wall of the endodermal cells. Consequently, substances moving from the cortex toward the vascular cylinder (or vice versa) must flow through the cytoplasm of the endodermal cell.

As you may recall, the cytoplasm is bounded by the differentially permeable plasma membrane. Thus, dissolved substances are "filtered" through endodermal cells, which regulate what goes into or comes out of the vascular column.

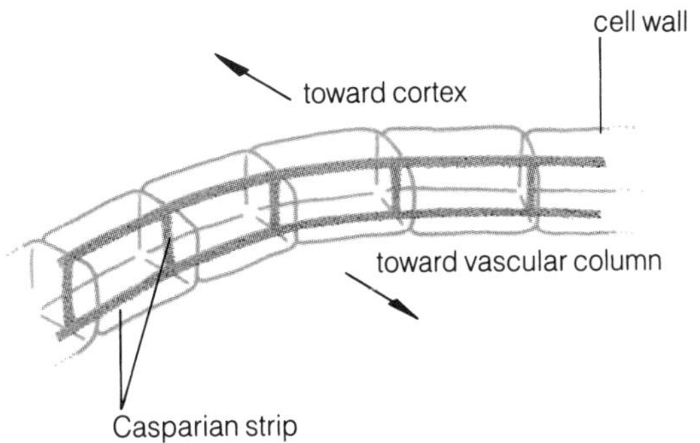

Figure 23-7 Diagrammatic representation of endodermal cells with Casparian strip.

8. In many cases, the Casparian strip is difficult to distinguish. With the highest magnification available, look for a red "dot" on the radial wall of the endodermal cells. If you cannot distinguish it on your own slide, examine the demonstration slide (Figure 23-8), which has been selected to show the Casparian strip. On this slide, you should be able to see its bandlike nature.
9. Return to your own slide, switch back to the medium-power objective, and focus your attention on the central vascular column. The cell layer immediately beneath the endodermis is the **pericycle.** Cells of the pericycle have the capacity to divide and produce lateral roots, like those you saw in the bean plant (Figure 23-3).
10. Finally, find the **primary xylem,** consisting of three or four ridges of thick-walled cells. (The stain used by most slide manufacturers stains the xylem cell walls red.) The xylem is the principal water-conducting tissue of the plant.
11. Between the "arms" of the xylem find the **primary phloem,** the tissue responsible for long-distance transport of carbohydrates produced by photosynthesis (known as photosynthates).

Primary xylem and phloem make up the *primary* vascular tissues, which are those produced by the apical meristems at the tips of the root and shoot. Later, we'll examine a *secondary* vascular tissue.

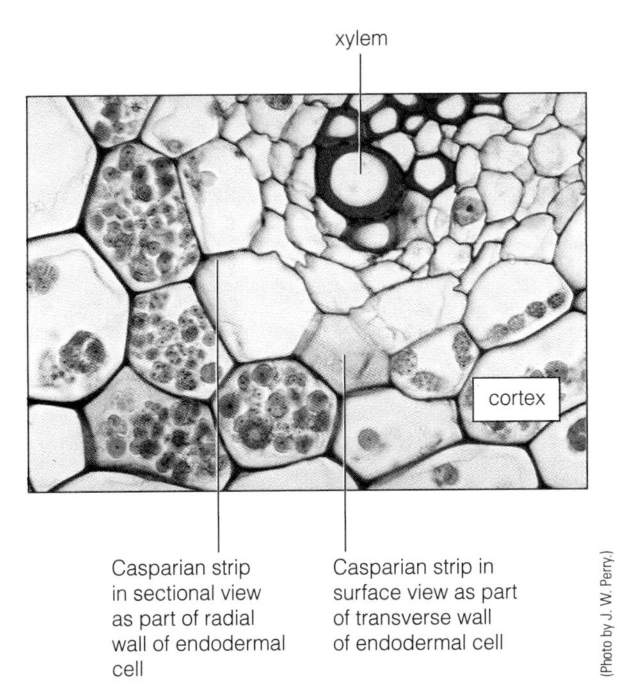

(Photo by J. W. Perry.)

Figure 23-8 Endodermal cells showing Casparian strip (296×).

C. Monocot Root Anatomy

Now that you have an understanding of the internal structure of a dicot root, let's compare and contrast it with that of a monocot. Draw and label the monocot root in Figure 23-9 as you proceed with your study.

1. Obtain a prepared slide of a cross section of a corn (*Zea mays*) root. Examine it with the low-power objective of your compound microscope. Draw the general shape of the corn root in Figure 23-9.
2. Identify and draw the **epidermis.**
3. Next locate the **cortex**. Count the number of cell layers and draw the cortex.

Do you find any starch grains in the cortex of the corn stem? ______________________________

4. Now locate the **endodermis.** If the root you have is a mature one, you will see that the endodermis has very thick walls on three sides.
5. The vascular column of a monocot is quite different from a dicot in several ways. First note that the center is occupied by thin-walled cells. These cells make up a region called the **pith.**
6. Draw the pith.
7. Next note that the **xylem** of this monocot root consists of scattered cells. Look at Figure 23-10, a high magnification photomicrograph of the xylem and phloem of this root.
8. Draw the xylem.

Figure 23-9 Drawing of a corn (monocot) root, c.s. (____×).
Labels: epidermis, cortex, endodermis, pith, xylem, phloem

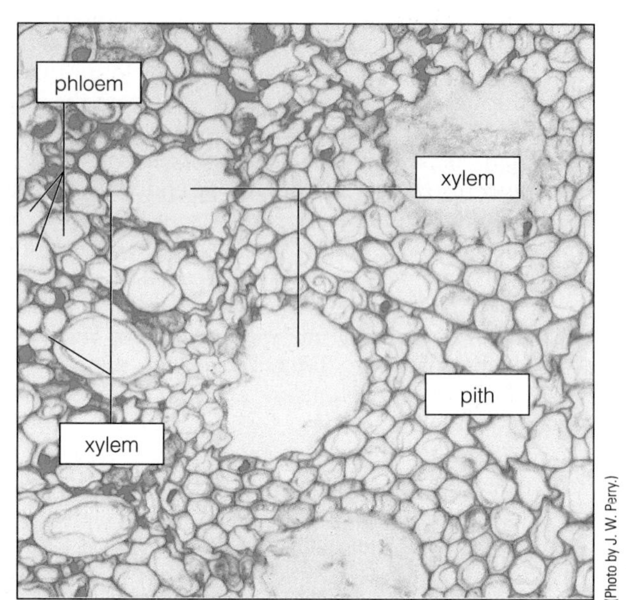

Figure 23-10 Portion of a cross section of a corn (monocot) root showing vascular tissue (140×).

23.3 The Shoot System: Stems *(About 50 min.)*

MATERIALS

Per student:

- prepared slides of
  - herbaceous dicot stem, c.s. (flax, *Linum;* or alfalfa, *Medicago*)
  - monocot stem, c.s. (corn, *Zea*)
  - woody stem, c.s. (basswood, *Tilia*)
- woody twig (hickory, *Carya;* or horse chestnut, *Aesculus*)
- metric ruler or meter stick
- compound microscope

Per student pair:

- cross section of woody branch (tree trunk)
- dissecting microscope

Per lab room:

- demonstration slide of lenticel

PROCEDURE

A. Dicot Stem: Primary Structure

Remember that the root and shoot are basically similar in structure; only the arrangement of tissues differs. Dicot stems have their vascular tissues arranged in a more or less complete ring of individual bundles of vascular tissue (called vascular bundles). Moreover, the ground tissue of dicots can be differentiated into two regions: **pith** and **cortex.**

You've probably heard the term *herb.* An *herb* is an **herbaceous plant,** one that develops no or very little wood. Herbaceous plants have only primary tissues. Woody plants develop secondary tissues—wood and bark.

Beans, flax, and alfalfa are examples of herbaceous dicots. Maples and oaks are woody dicots. Let's look at an herbaceous stem first.

1. Obtain a slide of an herbaceous dicot stem (flax or alfalfa). This slide may be labeled "herbaceous dicot stem." As you study this slide, refer to Figure 23-11, of a partial section of an herbaceous dicot stem.
2. Using the low- and medium-power objectives, identify the single-layered **epidermis** covering the stem, a multilayered **cortex** between the epidermis and **vascular bundles,** and the **pith** in the center of the stem.
3. Observe a vascular bundle with the high-dry objective (Figure 23-12). Adjacent to the cortex, find the **primary phloem.**
4. Just to the inside of the primary phloem, locate the thin-walled cells of the **vascular cambium.** (The vascular cambium is the lateral meristem that produces wood and secondary phloem, both secondary tissues. Despite your specimen being an herbaceous plant, a vascular cambium may be present. Generally, this meristem does not produce enough secondary tissue to result in the plant being considered woody.)
5. Locate the thick-walled cells of the **primary xylem.** The primary xylem is that vascular tissue closest to the pith. (The wall of the xylem cells is probably stained red.)

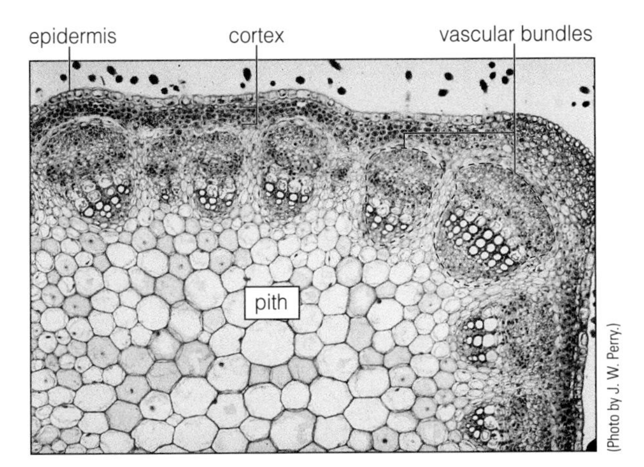

Figure 23-11 Portion of a cross section of an herbaceous dicot stem (75×).

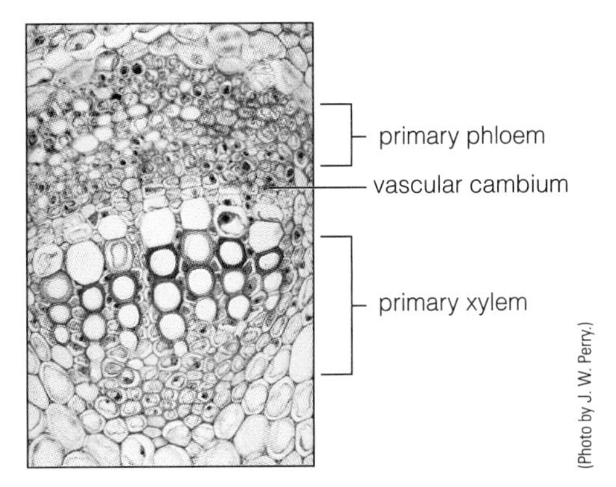

Figure 23-12 Vascular bundle of an herbaceous dicot stem (192×).

B. External Features of Woody Dicot Stems

Woody plants are those that undergo secondary growth, producing the secondary tissues. Both roots and shoots may have secondary growth. This growth occurs because of activity of the two meristems near the edge of the plant—the vascular cambium and the cork cambium. The vascular cambium produces secondary xylem (wood) and secondary phloem, while the cork cambium produces a portion of the bark called the **periderm.**

1. Examine a twig of hickory (*Carya*) or buckeye (*Aesculus*) that has lost its leaves. Label Figure 23-13 as you study the twig.
2. Find the large **terminal bud** at the tip of the twig. If your twig is branched, each branch has its own terminal bud.
3. Identify the shield-shaped **leaf scars** at each **node.** (Remember, a node is the region where a leaf attaches to the stem.) Leaf scars represent the point at which the leaf petiole was attached on the stem.
4. Within each leaf scar, note the numerous dots. These are the **vascular bundle scars.** Immediately above and adjacent to most leaf scars should be an **axillary bud.**

5. In the **internode** regions of the twig, locate the small raised bumps on the surface; these are **lenticels,** the regions of the periderm that allow for exchange of gases.
6. Return to the terminal bud. Note that the bud is surrounded by **bud scales.** When these scales fall off during spring growth, they leave **terminal bud scale scars.** Because a terminal bud is produced at the end of each growing season, you can use groups of terminal bud scale scars to determine the age of a twig.

   If the most recent growth took place during the last growing season (summer), when was the portion of the twig immediately adjacent to the cut end produced? ______________________

(Remember to date all regions between successive bud scale scars.)

Now that you've got some idea of the structure of a particular woody plant, let's see what amount of growth takes place in several different species of plants.

7. Measure the distance between a number of successive terminal bud scale scars in your specimen. Average the results to get an approximate idea of how much growth is produced annually. Record your results in Table 23-1.
8. Now obtain twigs of several other species, including the tree of heaven (*Ailanthus*), and do the same.

Would you say the growth rate is similar or quite variable among species that grow in your area? ______________
____

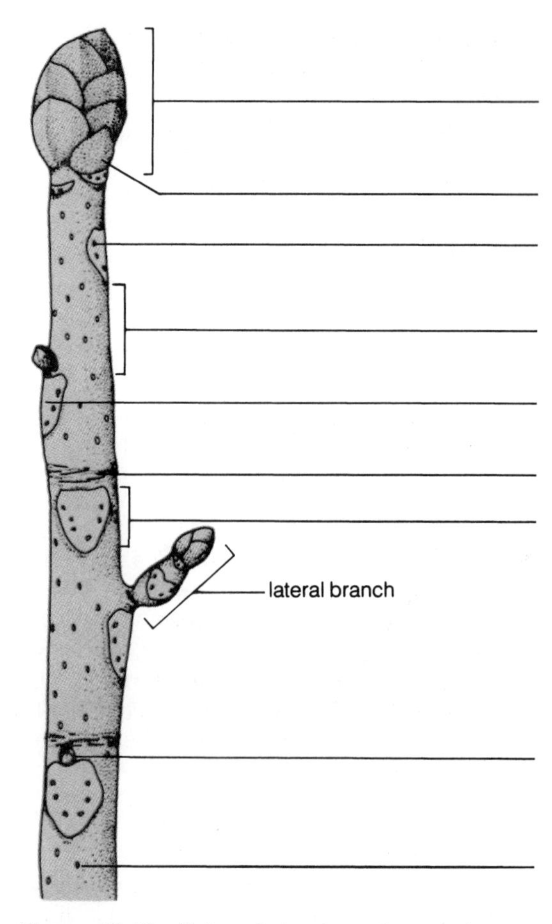

Figure 23-13 External structure of a woody stem. **Labels:** terminal bud, leaf scar, node, vascular bundle scar, axillary bud, internode, lenticel, bud scale, terminal bud scale scars

TABLE 23-1 Average Annual Terminal Growth in Woody Stems

Species	Average Distance Between Terminal Bud Scale Scars (cm)
Carya (or *Aesculus*)	
Ailanthus	

C. Secondary Growth: Gross Anatomy of a Woody Dicot Stem

1. Examine a cross section of a tree trunk (Figure 23-14; *trunk* is the nonscientific term for any large, woody stem). A tiny region in the center of the stem is the **pith.** Most of the trunk is made up of **wood** (also called *secondary xylem*). Locate the pith and the wood.
2. Now identify the **bark.**

As a woody stem grows larger, it requires additional structural support and more tissue for transport of water. These functions are accomplished by the wood. However, it is only the outer few years' growth of wood that is involved in water conduction. In some tree species, like that pictured in Figure 23-14, the distinction between conducting and nonconducting wood is obvious because of the color differences in each. The nonconducting wood, very dark in Figure 23-14, is called the **heartwood.**

3. Locate the heartwood in your stem.

The heartwood of many species, like black walnut or cherry, is highly prized for furniture making. Although virtually all trees develop heartwood, not all species have heartwood that is *visibly* distinct from the conducting wood.

4. The conducting wood, significantly lighter in color in Figure 23-14, is the **sapwood.** Examine the photo closely, because the color difference between the sapwood and the inner layers of the bark is subtle. Attempt to identify the sapwood on the stem you are examining.
5. Count the **growth increments (annual rings)** within the wood to estimate the age of the stem when the section was cut. (In most woods, a growth increment includes *both* a light and a dark layer of cells.)

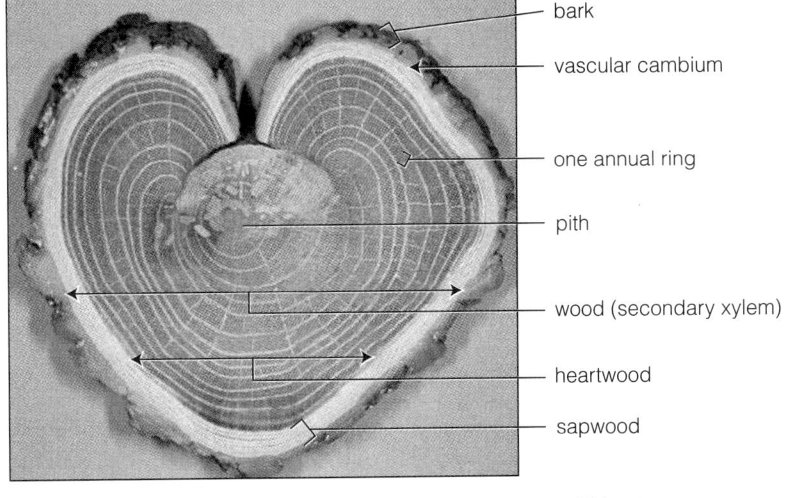

Figure 23-14 Cross section of a woody locust stem. This stem was injured several years into its growth, causing the unusual shape (0.5×).

How old would you estimate your section to be? _____________ years

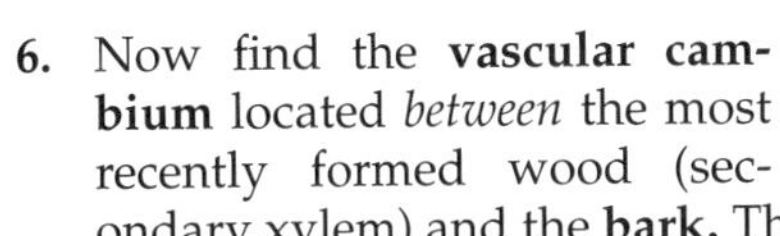

6. Now find the **vascular cambium** located *between* the most recently formed wood (secondary xylem) and the **bark.** The bark is everything *external* to the vascular cambium.
7. Identify the bark, consisting of **secondary phloem** and **periderm.** (As the vascular cambium produces secondary phloem to the outside, the primary phloem, cortex, and epidermis are sloughed off, much as dead skin on your body is shed.)

The periderm performs the same function as the epidermis before the epidermis ruptured as a result of the stem's increase in girth. What is the function of the periderm?

____

8. Within the wood, find the **rays,** which appear as lines running from the center toward the edge of the stem.
9. Examine the wood with a dissecting microscope and locate the numerous holes in the wood. These are the cut ends of the water-conducting cells, often called **pores.**

D. Secondary Growth: Microscopic Anatomy of a Woody Dicot Stem

Now that you've got an idea of the composition of a woody stem, let's examine one with the microscope.

1. Obtain a prepared slide of a cross section of a woody stem (basswood, *Tilia,* or another stem). Examine it with the various magnifications available on the dissecting microscope. Use Figure 23-15 as a reference.
2. Starting at the edge, identify the **periderm** (darkly stained cells). Depending on the age of the stem, **cortex** (thin-walled cells with few contents) may be present just beneath the periderm. Now find the broad band of **secondary phloem** (consisting

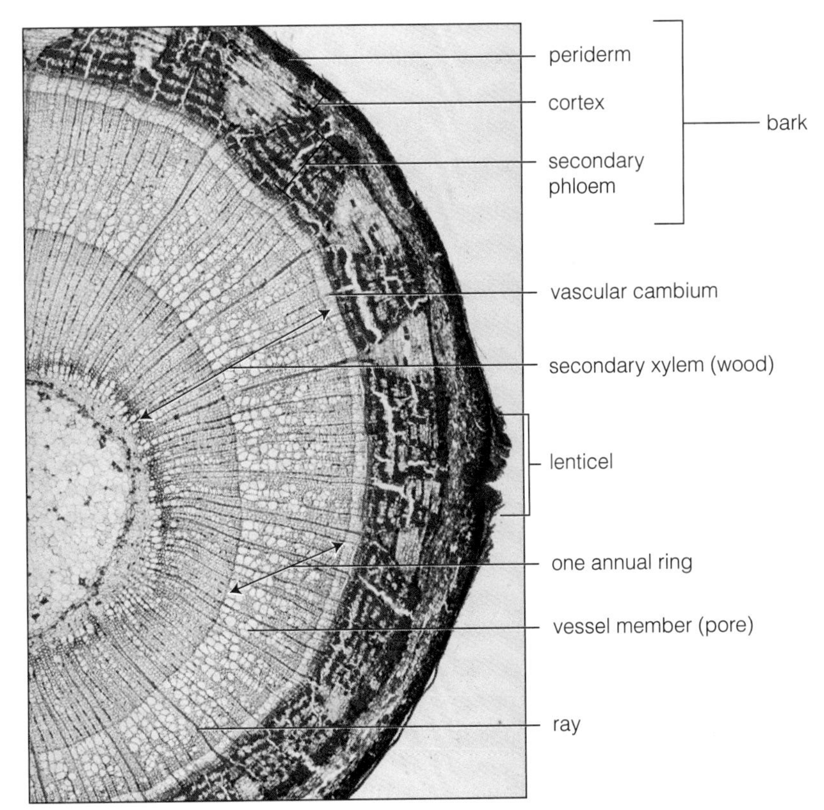

Figure 23-15 Cross section of a woody dicot stem (15×).

of cells in pie-shaped wedges). Identify the **vascular cambium** (a narrow band of cells separating the secondary phloem from secondary xylem), **secondary xylem (wood),** and pith (large, thin-walled cells in the center). Count the number of growth increments (annual rings). Note the largest, thick-walled cells in the wood. These are the pores.

How old is this section? _______________ years

3. Find the **rays** running through the wood. Rays are parenchyma cells that carry water and photosynthates *laterally* in the stem. (For the most part, the xylem and phloem carry substances *vertically* in the plant.)

Recall that the periderm replaces the epidermis as secondary growth takes place. The epidermis had stomata, which allowed the exchanging of gases between the plant and the environment. When the epidermis was shed, so were the stomata. But the need for exchange of gases still exists because the living cells require oxygen for respiration (Exercise 10). The plant has solved this problem by having special regions, **lenticels,** in the periderm. Lenticels are groups of cells with lots of intercellular space, in contrast to the tightly packed cells in the rest of the periderm.

4. Identify a lenticel on your slide. If none is found, examine the *demonstration* slide, specifically chosen to demonstrate this feature.

E. Structure of a Monocot Stem

Monocot and dicot stems differ in several ways. Monocot stems lack differentiation of the ground tissue into cortex and pith. Unlike the dicot stem whose vascular tissues are arranged in a ring, the vascular bundles of monocots are scattered throughout the ground tissue.

As you study the monocot stem, draw what you see in Figure 23-16.

1. Obtain a prepared slide of a monocot (corn, *Zea mays*) stem cross section. Observe the slide first using the low-power objective of your compound microscope, or perhaps even a dissecting microscope.
2. Draw its general appearance in Figure 23-16, labeling the **epidermis** on the outside, the thin-walled cells comprising the **ground tissue,** and the numerous **vascular bundles** scattered within the vascular tissue.
3. Switch to the medium-power objective of your compound microscope to study a single vascular bundle more closely. Draw the detail of one vascular bundle in Figure 23-17. (The vascular bundles of monocots look a lot like the face of a human or other primate.)
4. Observe the thick-walled cells surrounding the vascular bundle. These cells make up the bundle sheath. (The walls of these cells are often stained red.)
5. Note the very large cells that give the impression of the vascular bundle's "eyes, nose, and mouth." These cells make up the water-conducting xylem. (The cell walls are often stained red.) Draw and label these xylem.
6. Finally, find the phloem, located on the vascular bundle's "forehead." (Phloem cell walls are typically stained green or blue.)

Figure 23-16 Drawing of a corn (monocot) stem, c.s. (____×).
Labels: epidermis, ground tissue, vascular bundles

Figure 23-17 Drawing of the vascular bundle of a corn (monocot) stem, c.s. (____×).
Labels: bundle sheath, xylem, phloem

23.4 The Shoot System: Leaves *(About 30 min.)*

Leaves make up the second part of the shoot system. You examined the external morphology of the leaves in Section 23.1. Now let's look inside leaves, starting with a typical dicot leaf.

MATERIALS

Per student:

- prepared slide of dicot leaf, c.s. (lilac, *Syringa*)
- prepared slide of monocot leaf, c.s. (corn, *Zea*)
- compound microscope

PROCEDURE

A. Dicot Leaf Anatomy

1. Obtain a prepared slide of a cross section of a dicot leaf (lilac or other). Refer to Figure 23-18 as you examine the leaf.

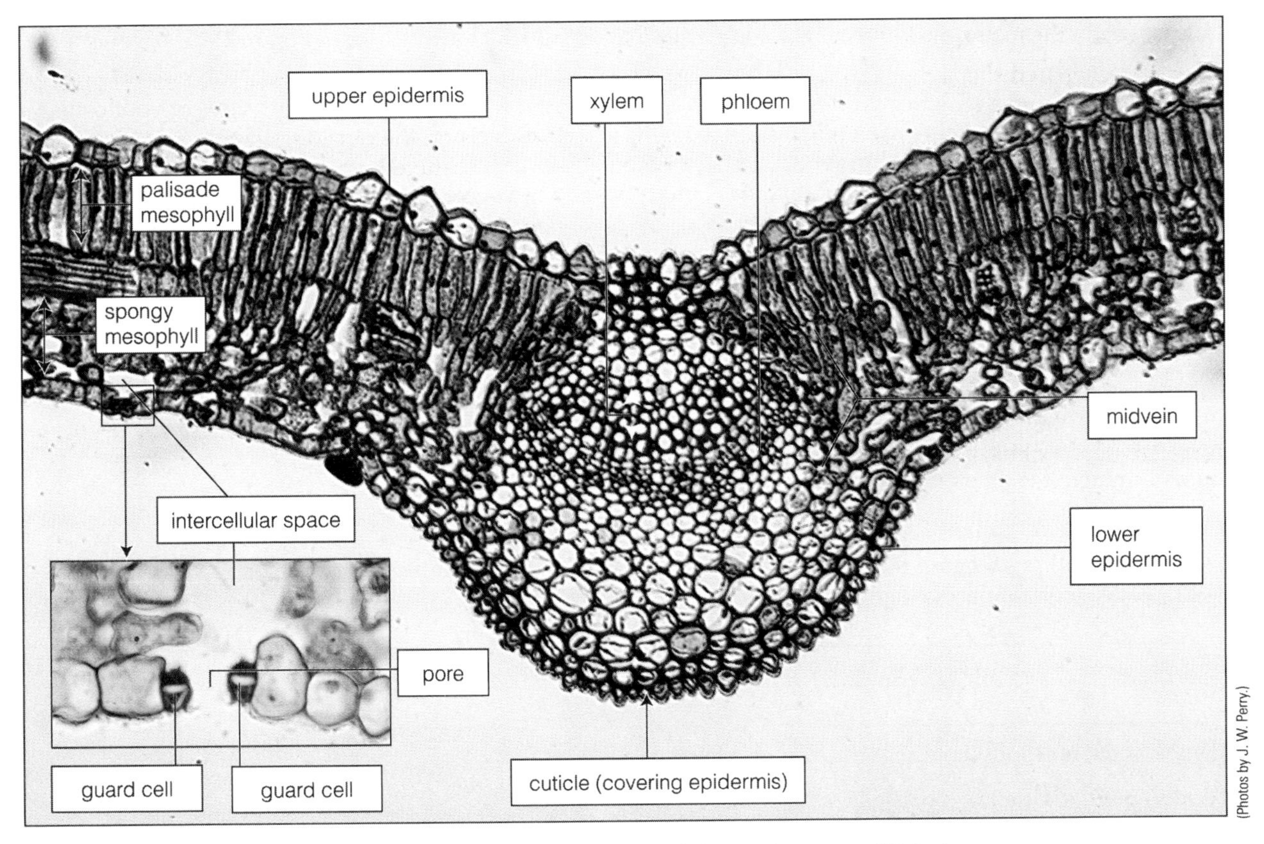

Figure 23-18 Cross section of a dicot leaf with midrib (105×); inset of a stoma (480×).

2. Examine the leaf first with the low-power objective to gain an overall impression of its morphology. Note the size and orientation of the **veins** within the leaf. The veins contain the xylem and phloem.
3. Find the centrally located **midvein** within the **midrib,** the midvein-supporting tissue. You might think of the midvein as the major pipeline of the leaf, carrying water and minerals to the leaf and materials produced during photosynthesis to sites where they will be used during respiration.
4. Use the high-dry objective to examine a portion of the blade to one side of the midvein. Starting at the top surface of the leaf, find the **cuticle,** a waxy, water-impervious substance covering the **upper epidermis.** The epidermis is a single layer of tightly appressed cells.

5. The ground tissue of the leaf is represented by the **mesophyll** (literally "middle leaf"). In dicot leaves, the mesophyll is usually divided into two distinct regions; immediately below the upper epidermis, find the two layers of **palisade mesophyll.** These columnar-shaped cells are rich in chloroplasts.
6. Below the palisade mesophyll, find the loosely arranged **spongy mesophyll.** Note the large volume of **intercellular air space** within the spongy mesophyll.

Does the spongy mesophyll contain any chloroplasts? (yes or no) ______________________________

What is one function that occurs within the spongy mesophyll? ______________________________

7. In the **lower epidermis,** find a **stoma** (plural: *stomata*) with its **guard cells** and the **pore** (inset, Figure 23-18). Large **epidermal hairs** shaped somewhat like mushrooms are usually found on the lower epidermis.

Is the lower epidermal layer covered by a cuticle? ______________________________

8. Compare the abundance of stomata within the lower epidermis with that in the upper epidermis.

Which epidermal surface has more stomata? ______________________________

9. Examine the midvein in greater detail. Find the thick-walled xylem cells (often stained red).
10. Below the xylem, locate the **phloem** (usually stained green).
11. Now identify and examine the smaller veins within the lamina (blade). Note that these, too, contain both xylem and phloem.

B. Monocot Leaf Anatomy

Let's compare the structure of a monocot grass leaf with that of a dicot. Draw the leaf in Figure 23-19 as you proceed with your study.

1. Obtain a prepared slide of a corn (*Zea mays*) leaf. Examine it first with the low-power objective of your compound microscope to gain an overall impression of its structure. Draw what you see in Figure 23-19.
2. Identify the **upper** and **lower epidermis.** Label them in Figure 23-19.
3. Search on both epidermal layers for **stomata.** Count them on each layer.

Are there more, fewer, or about the same number on the upper epidermis versus the lower?

____

4. Think about the orientation of the leaf blade on the plant. (If necessary, examine a living or dried corn plant to gain an impression of the orientation.)

Is the orientation of the corn leaves the same or different from that of a dicot, such as the bean plant?

____

If you determined it is different, in what way is it different?

____

5. Make a hypothesis relating the orientation of the leaf blade and the number of stomata on each surface of a leaf blade.

____

____

____

6. Look at the **mesophyll** of the leaf. Draw and label it in Figure 23-19.

Is the mesophyll divided into palisade and spongy layers? (yes or no) ______________________________

7. Note the **intercellular air spaces** within the mesophyll. Draw and label the intercellular air spaces.
8. Examine a single vascular bundle. Note the larger cells that surround the vascular bundle. This "wreath" of cells is the **bundle sheath,** a layer that is exceptionally conspicuous in grasses. Draw and label the bundle sheath.
9. Note that the bundle sheath cells are tightly packed, with no intercellular spaces separating them. The bundle sheath serves a function like that of the endodermis in the root.
10. Finally, identify the larger **xylem** cells and **phloem** cells within the vascular bundle. Draw and label these tissues in Figure 23-19.

Figure 23-19 Drawing of a corn (monocot) leaf, c.s. (____×).
Labels: cuticle, epidermis, stoma, guard cells, mesophyll, bundle sheath, xylem, phloem

PRE-LAB QUESTIONS

_____ **1.** The study of a plant's structure is
(a) physiology
(b) morphology
(c) taxonomy
(d) botany

_____ **2.** A plant with two seed leaves is
(a) a monocotyledon
(b) a dicotyledon
(c) exemplified by corn
(d) a dihybrid

_____ **3.** A taproot system lacks
(a) lateral roots
(b) a taproot
(c) both of the above
(d) none of the above

_____ **4.** Which structure is not part of the shoot system?
(a) stems
(b) leaves
(c) lateral roots
(d) axillary buds

_____ **5.** An axillary bud
(a) is found along internodes
(b) produces new roots
(c) is the structure from which branches and flowers arise
(d) is the same as a terminal bud

_____ **6.** The endodermis
(a) is the outer covering of the root
(b) is part of the vascular tissue
(c) contains the Casparian strip, which regulates the movement of substances
(d) is none of the above

_____ **7.** Meristems are
(a) located at the tips of stems
(b) located at the tips of roots
(c) regions of active growth
(d) all of the above

_____ **8.** To determine the age of a woody twig, one counts the
(a) nodes
(b) leaf scars
(c) lenticels
(d) regions between sets of terminal bud scale scars

_____ **9.** The midrib of a leaf
(a) contains the midvein
(b) contains only xylem
(c) is part of the spongy mesophyll
(d) contains only phloem

_____ **10.** The bundle sheath in a monocot leaf
(a) is filled with intercellular space
(b) is the location of stomata
(c) is covered with a cuticle to prevent water loss
(d) functions in a manner somewhat similar to the root endodermis

Name ______________________ Section Number ______________________

EXERCISE 23

Plant Organization: Vegetative Organs of Flowering Plants

POST-LAB QUESTIONS

23.1 External Structure of the Flowering Plant

1. What type of root system do you see on the dandelion at the right?

(0.25×). (Photo by J. W. Perry.)

23.2 The Root System

2. Describe the location, structure, and importance of the Casparian strip.

23.3 The Shoot System: Stems

3. On the figure below, identify structures **a**, **b**, and **c**.

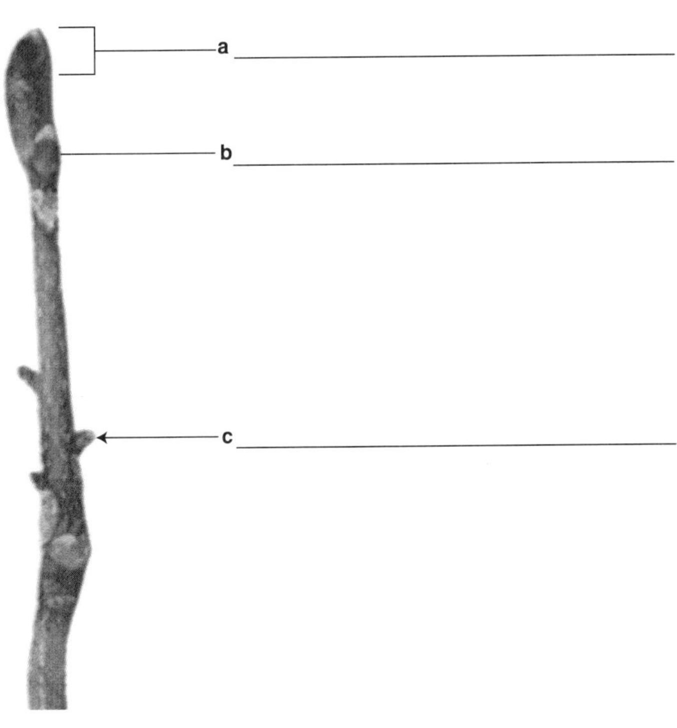

(0.5×). (Photo by J. W. Perry.)

4. Identify the structures labeled **a** and **b** on the figure below.

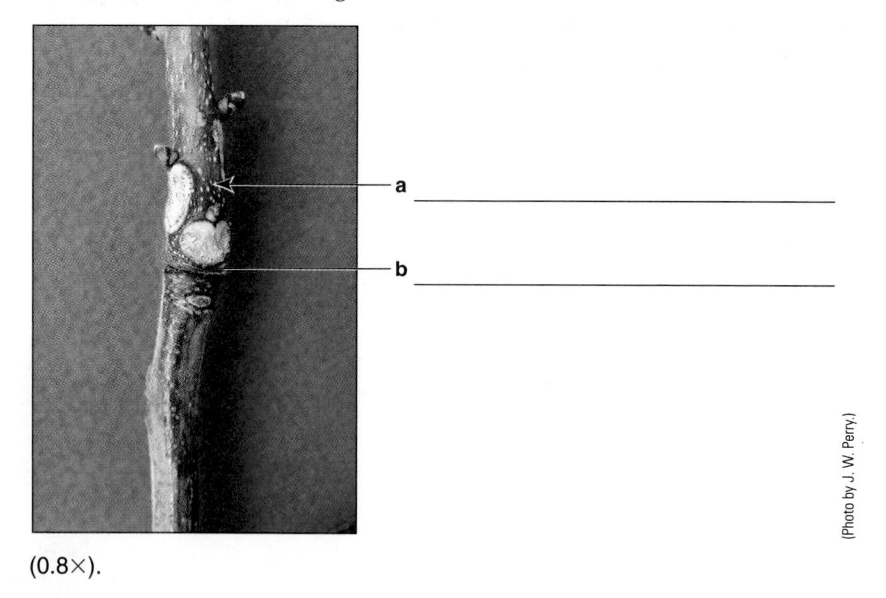

(0.8×).

5. What feature(s) would you use to determine the age of a woody twig?

6. Label the diagram using the following terms: bark; heartwood; sapwood; secondary phloem; vascular cambium.

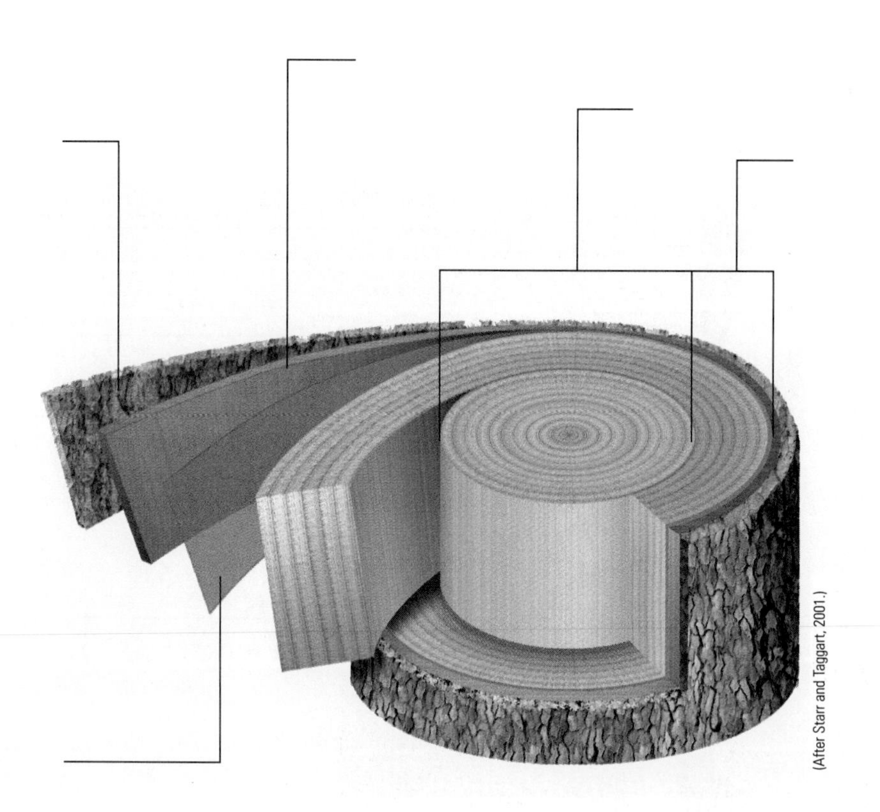

7. The photo shows the microscopic appearance of maple wood. Using your knowledge of the woody stem section, identify cell type **a** and the "line" of cells at **b.** (Note: The outside of the tree from which this section was taken is shown in the photo that accompanies question 8.)

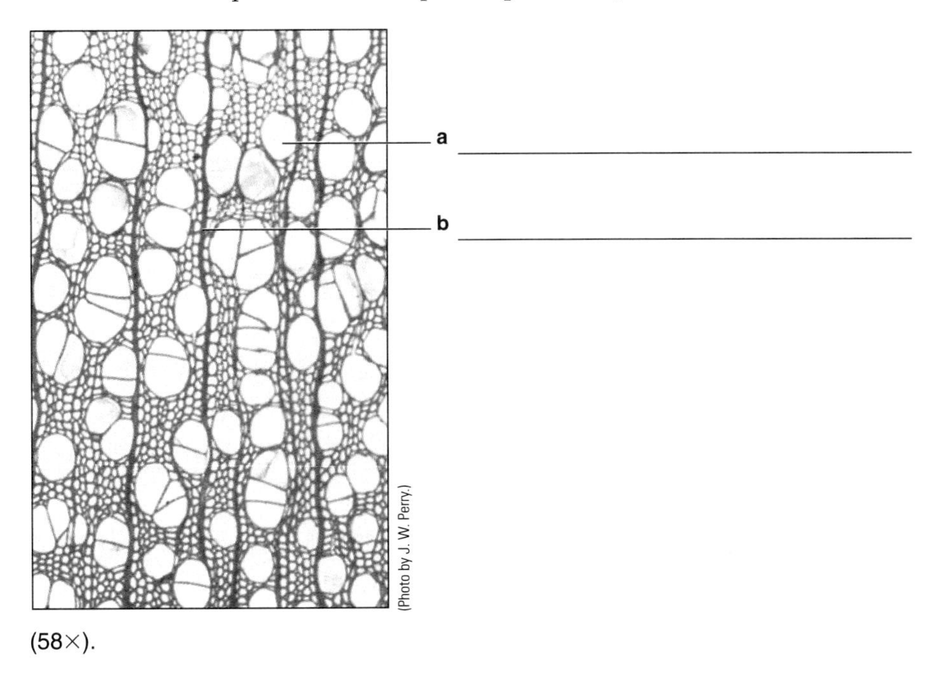

(58×).

8. A section cut from an ash branch is shown below.

   a. Identify region **a.**

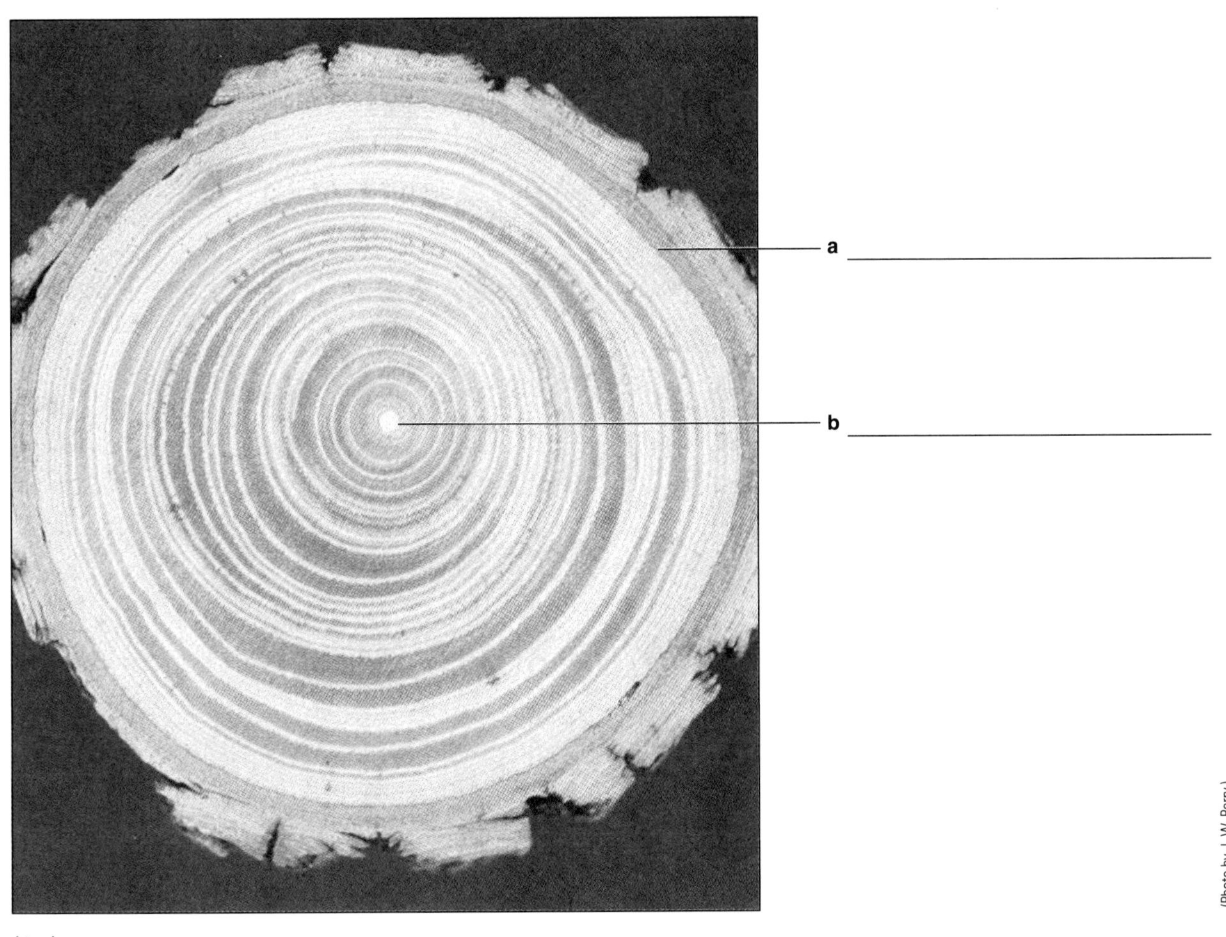

(1×).

(Photo by J. W. Perry.)

b. Which meristem is located at **b**?

c. Within ±3 years, how old was this branch when cut?

23.4 The Shoot System: Leaves

9. The following photo shows a section of a leaf from a dicot that is adapted to a dry environment. Its lower epidermis has depressions, and the stomata are located in the cavities. Even though it's different from the leaf you studied in lab, identify the regions labeled **a, b,** and **c.**

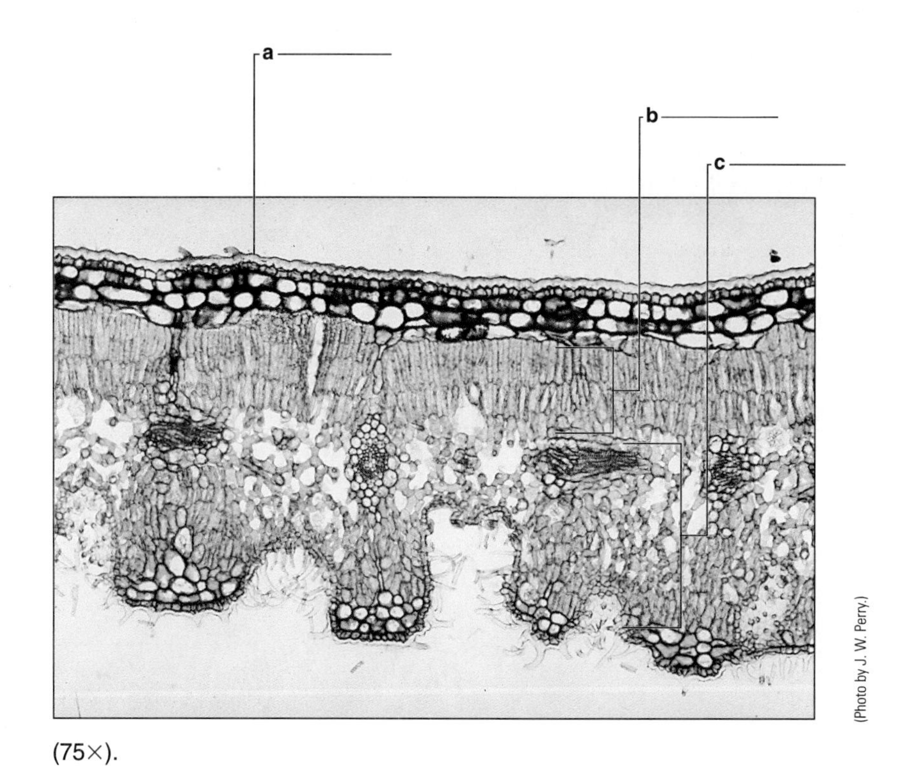

(75×).

Food for Thought

10. A major problem for land plants is water conservation. Most water is lost through stomata due to evaporation at the surface of the leaf. Many plants, including lilac (the leaf section you examined), orient their leaves perpendicular to the drying force of the sun's rays. What did you observe about the relative abundance of stomata in the lower epidermis versus the upper epidermis? Why do you think this distribution has evolved?

APPENDIX 1

Measurement Conversions

Metric to American Standard	***American Standard to Metric***
Length	*Length*
1 mm = 0.039 inch	1 inch = 2.54 cm
1 cm = 0.394 inch	1 foot = 0.305 m
1 m = 3.28 feet	1 yard = 0.914 m
1 m = 1.09 yards	1 mile = 1.61 km
1 km = 0.622 miles	
Volume	*Volume*
1 mL = 0.0338 fluid ounce	1 fluid ounce = 29.6 mL
1 L = 4.23 cups	1 cup = 237 mL
1 L = 2.11 pints	1 pint = 0.474 L
1 L = 1.06 quarts	1 quart = 0.947 L
1 L = 0.264 gallon	1 gallon = 3.79 L
Mass	*Mass*
1 mg = 0.0000353 ounce	1 ounce = 28.3 g
1 g = 0.0353 ounce	1 pound = 0.454 kg
1 kg = 2.21 pounds	

APPENDIX 2

Genetics Problems

You may find it helpful to draw your own Punnett squares on a separate sheet of paper for the following problems.

Monohybrid Problems with Complete Dominance

1. In mice, black fur (*B*) is dominant over brown fur (*b*). Breeding a brown mouse and a homozygous black mouse produces all black offspring.

   a. What is the genotype of the *gametes* produced by the brown-furred parent? ______

   b. What genotype is the brown-furred parent? ______

   c. What genotype is the black-furred parent? ______

   d. What genotype is the black-furred offspring? ______

   e. If two F_1 mice are bred with one another, what phenotype will the F_2 offspring be, and in what proportion?

   phenotype ______

   proportion ______

2. The presence of horns on Hereford cattle is controlled by a single gene. The hornless (*H*) condition is dominant over the horned (*h*) condition. A hornless cow was crossed repeatedly with the same horned bull. The following results were obtained in the F_1 offspring:

   8 hornless cattle

   7 horned cattle

   What are the parents' genotypes?

   cow ______

   bull ______

3. In fruit flies, red eyes (*R*) are dominant over purple eyes (*r*). Two red-eyed fruit flies were crossed, producing the following offspring:

   76 red-eyed flies

   24 purple-eyed flies

   a. What is the approximate ratio of red-eyed to purple-eyed flies? ______

   b. Based on your experience with previous problems, what two genotypes give rise to this ratio? ______

   c. What are the parents' genotypes? ______

   d. What is the genotypic ratio of the F_1 offspring? ______

   e. What is the phenotypic ratio of the F_1 offspring? ______

Monohybrid Problems with Incomplete Dominance

4. Petunia flower color is governed by two alleles, but neither allele is truly dominant over the other. Petunias with the genotype R^1R^1 are red-flowered, those that are heterozygous (R^1R^2) are pink, and those with the R^2R^2 genotype are white. This is an example of **incomplete dominance.** (Note that superscripts are used rather than upper- and lowercase letters to describe the alleles.)

   a. If a white-flowered plant is crossed with a red-flowered petunia, what is the genotypic ratio of the F_1 offspring? ______

   b. What is the phenotypic ratio of the F_1 offspring? ______

   c. If two of the F_1 offspring are crossed, what phenotypes will appear in the F_2 generation? ______

   d. What will be the genotypic ratio in the F_2 generation? ______

Monohybrid Problems Illustrating Codominance

5. Another type of monohybrid inheritance involves the expression of *both* phenotypes in the heterozygous situation. This is called **codominance.**

One well-known example of codominance occurs in the coat color of Shorthorn cattle. Those with reddish-gray roan coats are heterozygous (RR'), and result from a mating between a red (RR) Shorthorn and one that's white ($R'R'$). Roan cattle don't have roan-colored hairs, as would be expected with incomplete dominance, but rather appear roan as a result of having both red *and* white hairs. Thus, the roan coloration is not a consequence of pigments blending in each hair. Because the R and R' alleles are *both* fully expressed in the heterozygote, they are codominant.

a. If a roan Shorthorn cow is mated with a white bull, what will be the genotypic and phenotypic ratios in the F_1 generation?

genotypic ratio ______

phenotypic ratio ______

b. List the parental genotypes of crosses that could produce at least some

white offspring ______

roan offspring ______

Monohybrid, Sex-linked Problems

6. In humans, as well as in many other animals, sex is determined by special sex chromosomes. An individual containing two X chromosomes is a female, while an individual possessing an X and a Y chromosome is a male. (Rare exceptions of XY females and XX males have recently been discovered.)

I am a male/female (circle one).

a. What sex chromosomes do you have? ______

b. In terms of sex chromosomes, what type of gametes (ova) does a female produce? ______

c. What are the possible sex chromosomes in a male's sperm cells? ______

d. Which parent's gametes will determine the sex of the offspring? ______

7. The sex chromosomes bear alleles for traits, just like the other chromosomes in our bodies. Genes that occur on the sex chromosomes are said to be sex-linked. More specifically, the genes present on the X chromosome are said to be X-linked. Many more genes are present on the X chromosome than are found on the Y chromosome. Nonetheless, those genes found on the Y chromosome are said to be Y-linked.

The Y chromosome is smaller than its homologue, the X chromosome. Consequently, most of the loci present on the X chromosome are absent on the Y chromosome.

In humans, color vision is X-linked; the gene for color vision is located on the X chromosome but is absent from the Y chromosome.

Normal color vision (X^N) is dominant over color blindness (X^n). Suppose a color-blind man fathers the children of a woman with the genotype X^NX^N.

a. What genotype is the father? ______

b. What proportion of daughters will be color-blind? ______

c. What proportion of sons will be color-blind? ______

8. One daughter from the preceding problem marries a color-blind man.

a. What proportion of their sons will be color-blind? (Another way to think of this is to ask, What are the *chances* that their sons will be color-blind?) ______

b. Explain how a color-blind daughter might result from this couple.

______

______

Dihybrid Problems

Recall that pigmented eyes (P) are dominant to nonpigmented (p), and dimpled chins (D) are dominant to nondimpled chins (d).

9. A pigment-eyed, dimple-chinned man marries a blue-eyed woman without a dimpled chin. Their first-born child is blue-eyed and has a dimpled chin.

a. What are the possible genotypes of the father? ______

b. What genotype is the mother? ______

c. What alleles may have been carried by the father's sperm? ______

10. Suppose a dimple-chinned, blue-eyed man whose father lacked a dimple marries a woman who is homozygous recessive for both traits.

a. What is the expected genotypic ratio of children produced in this marriage? ______

b. What is the expected phenotypic ratio? ______

11. In his original work on the genetics of garden peas, Mendel found that yellow seed color (*YY*, *Yy*) is dominant over green seeds (*yy*) and that round seed shape (*RR*, *Rr*) is dominant over shrunken seeds (*rr*). Mendel crossed pure-breeding (homozygous) yellow, round-seeded plants with green, shrunken-seeded plants.

a. What will be the genotype and phenotype of the F_1 produced from such a cross?

genotype ______

phenotype ______

b. If the F_1 plants are crossed, what will be the expected phenotypic ratio of the F_2 generation?

______

Multiple Alleles

12. The major blood groups in humans are determined by **multiple alleles;** that is there are *more than* two possible alleles, any one of which can occupy a locus.

In this ABO blood group system, a single gene can exist in any of three allelic forms: I^A, I^B, or *i*. The alleles *A* and *B* code for production of antigen A and antigen B (two proteins) on the surface of red blood cells. Alleles *A* and *B* are codominant, while allele *i* is recessive.

Four blood groups (phenotypes) are possible from combinations of these alleles (Table 1).

TABLE 1 The ABO Blood Groups

Blood Type	Anitgens Present	Antibody Present	Genotype
O	Neither A nor B	A and B	*ii*
A	A	B	I^AI^A or I^Ai
B	B	A	I^BI^B or I^Bi
AB	AB	Neither A nor B	I^AI^B

a. Is it possible for a child with blood type O to be produced by two AB parents? ______ (yes or no) Explain

______

______

______

b. In a case of disputed paternity, the child is type O, the mother type A. Could an individual of the following blood types be the father? ______ Explain each possibility.

O ______

A ______

B ______

AB ______

Chi-Square Analysis

13. In fruit flies, red eyes (*R*) are dominant over white eyes (*r*). A student performs a cross between a heterozygous red-eyed fly and a white-eyed fly. The student counts the offspring and finds 65 red-eyed flies and 49 white-eyed flies.

    a. What is the expected phenotypic ratio of this cross? ________

    b. Using a χ^2 test, determine whether the deviation between the observed and the expected is the result of chance.

    χ^2 = ______

    c. Conclusion

    ____

    ____

14. In fruit flies, gray body (*G*) is dominant over ebony body (*g*).

    a. A red-eyed, gray-bodied fly known to be heterozygous for both traits is mated with a white-eyed fly that is heterozygous for body color.

    What is the expected phenotypic ratio for this mating? ______

    b. The observed offspring consist of 15 white-eyed, ebony-bodied flies; 31 white-eyed, gray-bodied flies; 12 red-eyed, ebony-bodied flies; and 38 red-eyed, gray-bodied flies.

    What is the χ^2 value for this cross? ______

    c. Is it likely that the observed results "fit" the expected values? ______

APPENDIX 3

Terms of Orientation in and Around the Animal Body

Body Shapes

Symmetry. The body can be divided into almost identical halves.

Asymmetry. The body cannot be divided into almost identical halves (for example, many sponges).

Radial symmetry. The body is shaped like a cylinder (for example, sea anemone) or wheel (for example, sea star).

Bilateral symmetry. The body is shaped like ours in that it can be divided into halves by only one symmetrical plane (midsagittal).

Directions in the Body

Dorsal. At or toward the back surface of the body.

Ventral. At or toward the belly surface of the body.

Anterior. At or toward the head of the body—ventral surface of humans.

Posterior. At or toward the tail or rear end of the body—dorsal surface of humans.

Medial. At or near the midline of a body. The prefix *mid-* is often used in combination with other terms (for example, midventral).

Lateral. Away from the midline of a body.

Superior. Over or placed above some point of reference—toward the head of humans.

Inferior. Under or placed below some point of reference—away from the head of humans.

Proximal. Close to some point of reference or close to a point of attachment of an appendage to the trunk of the body.

Distal. Away from some point of reference or away from a point of attachment of an appendage to the trunk of the body.

Longitudinal. Parallel to the midline of a body.

Axis. An imaginary line around which a body or structure can rotate. The midline or *longitudinal axis* is the central axis of a symmetrical body or structure.

Axial. Placed at or along an axis.

Radial. Arranged symmetrically around an axis like the spokes of a wheel.

Planes of the Body

Sagittal. Passes vertically to the ground and divides the body into right and left sides. The *midsagittal* or *median plane* passes through the longitudinal axis and divides the body into right and left halves.

Frontal. Passes at right angles to the sagittal plane and divides the body into dorsal and ventral parts.

Transverse. Passes from side to side at right angles to both the sagittal and frontal planes and divides the body into anterior and posterior parts—superior and inferior parts of humans. This plane of section is often referred to as a cross section.

Illustration References

Abramoff, P., and R. G. Thomson. 1982. *Laboratory Outlines in Biology III.* New York: W. H. Freeman.

Boolootian, R. A., and K. A. Stiles Trust. 1981. *College Zoology.* Tenth Edition. New York: Macmillan.

Case, C. L., and T. R. Johnson. 1984. *Experiments in Microbiology.* Menlo Park, California: Benjamin/Cummings.

Fowler, I. 1984. *Human Anatomy.* Belmont, California: Wadsworth.

Gilbert, S. G. 1966. *Pictorial Anatomy of the Fetal Pig.* Second Edition. Seattle, Washington: University of Washington Press.

Glase, J. C., et al. 1975. *Investigative Biology.* Ithaca, New York.

Hickman, C. P. 1961. *Integrated Principles of Zoology.* Second Edition. St. Louis, Missouri: C. V. Mosby.

Hickman, C. P., et al. 1978. *Biology of Animals.* Second Edition. St. Louis, Missouri: C. V. Mosby.

Jensen, W. A., et al. 1979. *Biology.* Belmont, California: Wadsworth.

Kessel, R. G., and R. H. Kardon. 1979. *Tissues and Organs.* New York: W. H. Freeman.

Kessel, R. G., and C. Y. Shih. 1974. *Scanning Electron Microscopy in Biology.* New York: Springer-Verlag.

Lytle, C. F., and J. E. Wodsedalek, 1984. *General Zoology Laboratory Guide.* Complete Version. Ninth Edition. Dubuque, Iowa: Wm. C. Brown.

Patten, B. M. 1951. *American Scientist* 39: 225–243.

Scagel, R. F., et al. 1982. *Nanvascular Plants.* Belmont, California: Wadsworth.

Sheetz, M., et al. 1976. *The Journal of Cell Biology* 70:193.

Shih, C. Y., and R. G. Kessel. 1982. *Living Images.* Boston: Science Books International/Jones and Bartlett Publishers.

Stanier, R., et al. 1986. *The Microbial World.* Fifth Edition. Englewood Cliffs, New Jersey: Prentice-Hall.

Starr, C., and R. Taggart. 1984. *Biology.* Third Edition. Belmont, California: Wadsworth.

Starr, C., and R. Taggart. 1987. *Biology.* Fourth Edition. Belmont, California: Wadsworth.

Starr, C., and R. Taggart. 1989. *Biology.* Fifth Edition. Belmont, California: Wadsworth.

Starr, C., and R. Taggart. 2001. *Biology: The Unity and Diversity of Life.* Ninth Edition. Belmont, California: Wadsworth.

Starr, C. 1991. *Biology: Concepts & Applications.* Belmont, California: Wadsworth.

Starr, C. 2000. *Biology: Concepts and Applications.* Fourth Edition. Belmont, California: Wadsworth.

Steucek, G. L., et al. 1985. *American Biology Teacher* 471: 96–99.

Storer, T., et al. 1979. *General Zoology.* New York: McGraw-Hill.

Villee, C. A., et al. 1973. *General Zoology.* Fourth Edition. Philadelphia: W. B. Saunders.

Weller, H., and R. Wiley. 1985. *Basic Human Physiology.* Boston: PWS Publishers.

Wischnitzer, S. 19979. *Atlas and Dissection Guide for Comparative Anatomy.* Third Edition. New York: W. H. Freeman.

Wolfe, S. L. 1985. *Cell Ultrastructure.* Belmont, California: Wadsworth.